Ecosystem Crises Interactions

AF794914

Ecosystem Crises Interactions

Human Health and the Changing Environment

Merrill Singer, PhD

Department of Anthropology
Emeritus Professor
University of Connecticut
Storrs, CT, USA

WILEY

This edition first published 2021
© 2021 John Wiley & Sons, Inc.

All rights reserved. No part of this publication may be reproduced, stored in a retrieval system, or transmitted, in any form or by any means, electronic, mechanical, photocopying, recording or otherwise, except as permitted by law. Advice on how to obtain permission to reuse material from this title is available at http://www.wiley.com/go/permissions.

The right of Merrill Singer to be identified as the author of this work has been asserted in accordance with law.

Registered Office
John Wiley & Sons, Inc., 111 River Street, Hoboken, NJ 07030, USA

Editorial Office
111 River Street, Hoboken, NJ 07030, USA

For details of our global editorial offices, customer services, and more information about Wiley products visit us at www.wiley.com.

Wiley also publishes its books in a variety of electronic formats and by print-on-demand. Some content that appears in standard print versions of this book may not be available in other formats.

Limit of Liability/Disclaimer of Warranty
While the publisher and authors have used their best efforts in preparing this work, they make no representations or warranties with respect to the accuracy or completeness of the contents of this work and specifically disclaim all warranties, including without limitation any implied warranties of merchantability or fitness for a particular purpose. No warranty may be created or extended by sales representatives, written sales materials or promotional statements for this work. The fact that an organization, website, or product is referred to in this work as a citation and/or potential source of further information does not mean that the publisher and authors endorse the information or services the organization, website, or product may provide or recommendations it may make. This work is sold with the understanding that the publisher is not engaged in rendering professional services. The advice and strategies contained herein may not be suitable for your situation. You should consult with a specialist where appropriate. Further, readers should be aware that websites listed in this work may have changed or disappeared between when this work was written and when it is read. Neither the publisher nor authors shall be liable for any loss of profit or any other commercial damages, including but not limited to special, incidental, consequential, or other damages.

Library of Congress Cataloging-in-Publication Data

Names: Singer, Merrill, author.
Title: Ecosystem crises interactions : human health and the changing environment / Merrill Singer, PhD, Department of Anthropology, Emeritus Professor, University of Connecticut Storrs, CT, USA.
Description: First edition. | Hoboken : Wiley, 2021. | Includes index.
Identifiers: LCCN 2020040393 (print) | LCCN 2020040394 (ebook) | ISBN 9781119569541 (paperback) | ISBN 9781119570028 (adobe pdf) | ISBN 9781119570011 (epub)
Subjects: LCSH: Environmental health. | Human beings–Effects of environment on. | Species–Effects of environment on.
Classification: LCC RA565 .S532 2021 (print) | LCC RA565 (ebook) | DDC 613/.1–dc23
LC record available at https://lccn.loc.gov/2020040393
LC ebook record available at https://lccn.loc.gov/2020040394

Cover Design: Wiley
Cover Image: © Jackal Yu/Shutterstock

Set in 9.5/12.5pt STIXTwoText by SPi Global, Pondicherry, India
Printed and bound by CPI Group (UK) Ltd, Croydon, CR0 4YY

C095794_050321

This book is dedicated to Ann Atwater and C.P. Ellis, unlikely allies in the struggle for justice, and to the global youth fighting to rescue our fragile environment.

Contents

Preface

This book appears at a moment in which leading climate scientists from around the world are warning that, if nothing else changes, there are only about a dozen years for global temperature to be kept at a level beyond which, even if it rises by only half a degree, there will be a significant worsening of the health and social impacts of drought, floods, extreme heat, ecosystem destruction, and poverty for hundreds of millions of people. Earth will not end, but for people around the planet, life as we know it today will be history. This dire warning that urgent and unprecedented changes are needed to avoid catastrophe was issued in a landmark report by the UN Intergovernmental Panel on Climate Change in 2018. This report does not stand alone; rather, the pace of critical findings on the effect of human-caused climatic and other environmental changes is accelerating. In my online news feed, I regularly find new articles affirming that we are at a critical juncture, a consequential threshold for humans and other earthlings.

Of special concern in this volume is the issue of ecocrises interaction: the fact that not only do we face a litany of negative environment changes, but these changes are made worse through interface. Put simply, 1 ecocrisis + 1 ecocrisis results in more adverse outcomes than the additive effects of 1+1 because of interaction. For example, the loss of coral reefs, the most biodiverse marine ecosystems in the world, upon which millions of people depend for their food and livelihoods, is being driven by the enhanced interacting effects of warming oceans, absorption of carbon dioxide into those oceans, overfishing, and water pollution. Interaction of these stressors severely compromises the ability of corals to grow, reproduce, and thrive. As a result, as much as one-third of all reef-building corals are currently at grave risk of extinction. Nor are corals alone in this; numerous ecosystems that are critical to human populations are confronted by similar risk from interacting stressors of human origin.

Thus, interacting ecocrises constitute one of the most serious threats to health around the globe. These pressures come in the form of flooding, heat stroke, loss of food sources and production, spread of infectious diseases, and enhanced armed conflict. Efforts to address environmental risk have been stymied by a business-as-usual stance among elite polluters that benefit financially from their claimed 'right to pollute' and the investment of some of their profits in the organized denial of climate change and environmental disruption. However, this corporate strategy, which has been supported through lobbying for government inaction, is meeting a growing popular resistance around the world, especially among youth. This book, which provides a detailed examination of all of these issues, is intended to arm readers with the knowledge required to make vital decisions about the kind of world Earth will be over the coming decades.

1

Introduction: public health, EcoHealth, planetary health, and you

For decades, many of us have turned a blind eye to what is happening to the planet. But now, given that Earth [as we know it] may well be dying, we may be ready to stand up to protect what we love. An extraordinary alchemy can take place when people follow their inner directives to stand up and face squarely the dire odds of biosphere survival. These actions involve extraordinary outer and inner courage, which can nurture a profound activism.

Dahr Jamail (2019)

1.1 Connections

This book is about connections, often unexpected links in a complex world. It brings together detailed information about seemingly disparate subjects. In the process, the growing air traffic of Miami International Airport is shown to be connected to the city's low-lying status in the ocean, to the use of sediment rock in the area's hydraulic and drinking water filtration systems, and to king and red tides. Elsewhere, the connections are examined between the climate of Earth and human evolution, the peopling of the world, and the historic development of modern society. Also of concern are the multiple ways national and international development under capitalism are entwined with deforestation, coral die-off, emergent infectious disease, and mounting food insecurity around the world. This list could go on, but the basic point is that a holistic perspective on diverse issues is needed in the investigation of the interaction of contemporary ecological dangers, human behavior, and health.

In visiting ecocrises around the globe, from U.S. nuclear waste buried below the melting glaciers of Iceland to the deteriorating asbestos cement used to construct homes in Australia, and many other catastrophes in many places in between, a considerable amount of detail is offered. Why such a thoroughgoing approach to cases and examples? Because, while there are redundancies reflective of the unfortunate fact that in countless ways adverse elements of environmental history repeat themselves, it is by knowing a multitude of events that the patterns and their determinants are brought into sight. In this way, we begin to see the forest through the trees. It is a critical moment for this task, as the logger's saw, intensifying wildfires, climate change-induced pest invasions, drought, global warming, and over-extraction conspire to denude the planet of oxygen-generating forests, depopulate the oceans, melt the cryosphere, and put human well-being at risk.

1.2 Is this a dangerous book?

In an essay entitled "The Danger of Environmentalism," Michael Berliner (2020), senior advisor to the Ayn Rand Archives, writes: "Earth Day approaches, and with it a

Ecosystem Crises Interactions: Human Health and the Changing Environment, First Edition. Merrill Singer.
© 2021 John Wiley & Sons, Inc. Published 2021 by John Wiley & Sons, Inc.

grave danger faces mankind. The danger is not from acid rain, global warming, smog, or the logging of rain forests, as environmentalists would have us believe. The danger to mankind is from environmentalism." The perception, by some, of the danger inherent in being concerned about the industrial destruction of the environment is what has led covering the environment to be one of the most risky assignments in journalism. It is estimated that 40 reporters around the world died between 2005 and 2016 because of their environmental reporting. This is more than were killed covering the U.S. war in Afghanistan. Notes journalist Saul Elbein (quoted in Warren 2016): "In Cambodia and in remote forests elsewhere, a rising boom in the illegal sale of wood, land and minerals has turned the environmental beat into a new sort of conflict journalism. The dead have overwhelmingly been local reporters, covering illegal mining or logging. They are largely independent, poorly educated, untrained and despised by their nations' establishment media. Reporting on a violent, corrupt frontier, they are never sure when they'll cross a line and end up dead. Their lives in their hands, they head into the woods." Newsworthy environmental controversies often involve powerful business and economic interests, explosive political clashes, criminal gangs, anti-government rebels, or outright corruption. Often, they focus on struggles between elites and local people over indigenous rights to land and natural resources. In wealthy and developing countries alike, journalists covering environment issues find themselves the targets of attack by military, police, and paramilitary forces. In 2013, for example, independent journalist Miles Howe was assigned to cover protests by the Elsipotog First Nation in New Brunswick, Canada against hydraulic fracturing for natural gas. In his reporting, Howe sought to highlight unreported and underreported incidents. He recalls (quoted in Freedman 2018): "Many times I was the only accredited journalist witnessing rather violent arrests, third-trimester pregnant women being locked up, guys tackled to the ground." On one occasion, a member of the Royal Canadian Mounted Police pointed at him and shouted, "He's with them!" His equipment was taken and the police searched his home. In addition, they offered to pay him to spy on the protesters—an offer that he rejected. In 2014, the body of Taing Try, a journalist covering the illegal logging of the forests of Cambodia, was found by local farmers in Kratie Province. He was lying face down in the mud on a logging road, a bullet in the back of his head. Try's fate was not his alone. In 2008, Mikhail Beketov, a Russian journalist who wrote about the destruction of the Khimki Forest, part of the green belt surrounding Moscow, to make way for the Moscow–Saint Petersburg motorway, died of injuries suffered several years earlier after unknown attackers crushed his skull, broke his legs, and left him for dead in his own front yard. Similarly, in 2012, Chandrika Rai, an Indian reporter for the Hindi-language newspaper *Navbharat* and the English-language *The Hitavada*, who covered illegal coal mining, was bludgeoned to death along with his wife and their two teenage children in their home in Umaria, a small city in Madhya Pradesh state in central India. In 2017, Colombian radio journalist Efigenia Vásquez Astudillo was shot to death while covering the emergence of an indigenous movement to reclaim ancestral land that had been stolen and converted to large farms, resorts, and sugar plantations (Elbein 2016; Freedman 2018). The list goes on. Environmental reporters are often put at risk because they stand between the public and its need to know about environmental destruction and business interests or corrupt and criminal organizations intent on profit or personal gain. Degraders of the environment have a great deal invested in preventing the public from knowing the truth, or from knowing that is unencumbered by corporate spin (Conway 2018).

While environmental journals have been subjected to the most brutal forms of violence, environmental scientists—especially climate scientists—have been the targets of violent intimidation. The Climate Science Legal Defense Fund (2019), which has assisted numerous researchers who have been harassed and attacked through email, social media, and other online platforms or subjected to more explicit politically motivated personal threats, notes that:

> Vicious hate mail and death threats are common occurrences for climate scientists: One was the victim of an anthrax scare. Another had a dead animal dropped on their doorstep—while their child was at home. There are websites devoted to publishing scientists' contact information, a practice known as doxxing, so that the followers of these sites can flood scientists' inboxes with threatening messages. Other internet harassment tactics involve posting private details about researchers, such as the names of their family members or home addresses, for harassers to use against them.

These forms of intimidation have also been directed at other kinds of scholars who have written about climate change. A founding member and current board member of the Climate Science Legal Defense Fund, Naomi Oreskes, a science historian at Harvard University, became a target of the anti-climate science movement in 2004 when she published an article in *Science* documenting the broad scientific agreement on climate change. Based on an analysis of 928 scientific abstracts with the keywords "global climate change," she concluded that "there is a scientific consensus on the reality of anthropogenic climate change. Climate scientists have repeatedly tried to make this clear. It is time for the rest of us to listen" (Oreskes 2004, p. 1686). In response to her publication, Oreskes began receiving hate mail, and extremely hostile letters were sent to *Science*. Many of the letters to the journal were penned by economists and others affiliated with pro-free market think tanks, such as the Competitive Enterprise Institute. James Inhofe (R-Oklahoma), a recipient of significant campaign funding from the fuel industry and a science denier who has called global warming a "hoax," attacked Oreskes by name on the floor of the U.S. Senate. He cited Benny Peiser, a British social scientist, who claimed that his own review showed that fewer than 2 percent of climate studies endorsed the so-called "consensus view" that human activity is driving global warming, and some actually opposed it. Inhofe was unaware of or failed to acknowledge the fact that Peiser later admitted he could only find one "peer-reviewed" article that challenged the anthropogenic climate change consensus; this turned out to be published in a journal that is not peer-reviewed and is owned by a fuel industry organization, the American Association of Petroleum Geologists (Littlemore 2006). Moreover, not a single climate scientist subsequently disputed Oreskes' finding. Oreskes concluded: "We weren't being attacked because we'd done something wrong . . . We were being attacked because we'd done something right. Because we'd explained something significant, we'd laid facts on the table, those facts had implications, and some people were threatened by those implications" (quoted in Cho 2018).

Tyrone Hayes, a biologist at the University of California, Berkeley, notes that the history of such abuse predates the Internet: "Especially in the area of environmental science, it's mostly been women that have been the target, going all the way back to Rachael Carson" (quoted in Begos 2014). When her book *Silent Spring*—a groundbreaking science-based exposé of the environmental harms of pesticide use—was published in 1962, Carson was subjected to vicious personal assaults. The late novelist Michael Crichton—himself an outspoken climate change denier—even branded her "a mass murderess," directly responsible for the deaths of millions of African children from malaria because her efforts led to a ban on the pesticide DDT in the United States (a chemical to which mosquitoes in Africa already were evolving resistance) (Graham 2012).

Hayes came to personally experience the vitriol of the agricultural chemical industry when he began publishing his research showing that the widely used herbicide atrazine (2-chloro-4-(ethylamino)-6-(isopropylamino)-S-triazine) is an endocrine disruptor that demasculinizes and feminizes male frogs. He even wrote a poem, "The Atrazine Rap" (Hayes 2008), about ecocrises interaction, which includes the lyrics:

> Atrazine ain't a good thing
> it causes male frogs to grow eggs
> contributes to extra legs
> and exposed males don't want to sing
> If that ain't enough
> when you combine the stuff
> with a few other pesticides
> it causes greater than additive effects
> unpredictable defects
> exposed larvae don't grow
> and they develop slow
> and they contract diseases that otherwise
> could be beaten

Appreciating neither his well-documented science nor his lyricism, Syngenta Crop Protection, the Basel, Switzerland-based manufacturer of atrazine and initial funder of Hayes' research, unleashed a barrage of attacks against him through press releases, letters to the editor, and a formal ethics complaint filed with the University of California, Berkeley (Aviv 2014). Internal memos, notes, and e-mails from Syngenta released during a subsequent class-action lawsuit revealed how it had conspired to discredit Hayes, including attempting to get scientific journals to retract his publications, investigating his funding, and examining his private life. The documents also showed that Syngenta's P.R. team had proposed that the company purchase "Tyrone Hayes" as a search term on the Internet, so that any time someone searched for Tyrone's research findings, the first thing they saw was an advertisement saying, "Tyrone Hayes Not Credible." Finally, Syngenta launched a multimillion-dollar no-holds-barred campaign to protect its considerable atrazine profits, hiring a detective agency to investigate scientists on a federal advisory panel, digging into the personal life of a judge, and commissioning a psychological profile of Hayes. In 2003, Hayes received a job offer from Duke University. Duke is close to Syngenta's headquarters in Greensboro, North Carolina and the company's research center in Research Triangle Park. When Syngenta officials got wind of the offer, they contacted university administrators and had it withdrawn (Howard 2013; Rohr 2018). Subsequently, Steven Milloy, a freelance science columnist and lobbyist with ties to the tobacco and fossil fuel industries and the head of a nonprofit organization, CSRWatch.com, to which Syngenta has given tens of thousands of dollars, wrote an article for Fox News titled "Freaky-Frog Fraud" in which he attacked a Hayes paper published in *Nature*, characterized Hayes as a "junk scientist," ridiculed his "shoddy write-up [and] shaky research," and dismissed his "lame" conclusions as "just another of Hayes' tricks" (Milloy 2015). Hayes weathered the storm and continued his research and publications. By 2001, atrazine was the most commonly detected pesticide contaminating U.S. drinking water, and its safety remains controversial. Ultimately, Syngenta was forced to pay $105 million to more than a thousand Midwestern water utilities in reimbursement for the cost of filtering atrazine out of public drinking water (Duhigg 2009).

The eminent novelist and journalist George Orwell (1953) explained that there were four reasons why he wrote: sheer egoism, aesthetic enthusiasm in words and their right arrangement, a historical impulse to find out true facts and store them up for the use of posterity, and political purpose: to push the world in a certain direction. This book was motivated by similar impulses. The degree to which it is deemed to be dangerous is the degree to which it succeeds in effectively communicating to readers the true and pressing dangers of anthropogenic environmental destruction and motivating them to action. Most desirably, it will be a book of the sort Mary Harris "Mother" Jones (1837–1930), a beloved activist and union organizer who was committed to pushing the world in a certain direction, had in mind when she advised, "Sit down and read. Educate yourself for the coming conflicts" (quoted in Fetherling 2010).

1.3 Three alternative approaches to health and the environment

There is a wealth of evidence connecting ecosystem change with human and nonhuman health impacts. The following discussion of this relationship is organized around the examination of three holistic and interdisciplinary approaches to health: EcoHealth, One Health, and planetary health. All of these, rooted philosophically in wildlife ecologist Aldo Leopold's (1949) concept of "land health"—the maintenance capacity of organisms and natural systems for internal self-renewal—reflect an ecological turn that occurred in public health beginning in the 1980s. Deeper roots extend to early thinkers such as Calvin Schwabe (who worked with pastoralists in East Africa and coined the concept of "one medicine," linking biomedicine and veterinary medicine) and Rudolf Virchow (who pointed to the role of the socioeconomic environment in the origins of infectious and other disease outbreaks), both of whom recognized the close relationship between the health of humans and work done in veterinary medicine (Zinsstag *et al.* 2010).

Expressing a holistic vision, these approaches seek to break down the historic compartmentalization that separated specialized understandings of the environment and specialized understandings in health. While their points of origin differ, "the relative synchrony of the success of [these] paradigms . . . did not come about purely by chance: they respond to a growing common perception of the complexity of the linkages between animal and human health and their proximate and distant socio-ecological environment" (Roger *et al.* 2016). All three were influenced by the Ottawa Charter of 1986, which highlighted the centrality of ecosystems and sustainable resources for human health. They share: 1) a focus on environmental factors in health; 2) a recognition of the increasing interconnectedness of ecosystems in a globalizing world; and 3) the underlying assumption that humans and other animals face many of the same environmental challenges (e.g., infectious disease, food and nutrition, contention with weather). Roger *et al.* (2016) note that:

> Despite their different origins, One Health and EcoHealth are convergent in their vision and goals to reposition animal and public health within their broader context. Both are motivated by the conviction that health concerns must be addressed at the human–animal interface within their broader natural and social environments (i.e., socio-ecosystem approach). Both try to integrate scientific disciplines combining multi- and cross-disciplinary approaches. Both aim to mitigate the risks threatening ecosystems and public health, including veterinary public health. Both deal with the complexity of diseases and health . . . Finally, both struggle to properly define the boundaries of their paradigms despite their apparent similarities regarding principles and objectives.

The Ottawa Charter was a product of the first International Conference on Health Promotion, which took place in Ottawa, Canada in November 1986. Its objective was to expedite the achievement of the U.N. goal of Health For All. A response to the growing expectations for a new public health movement around the world, signatories to the Charter recognized that a key requirement of health is having a supportive environment.

While the primary focus was on the built or human-created environment (e.g., cities and industrial workplaces), there was also a recognition of the need for a stable ecosystem and sustainable natural resources. Subsequent key events in the study of ecology and health included the Rio Declaration of 1992, which was a product of the U.N. Conference on Environment and Development, the launching of the journal *EcoHealth* in 2004, and the founding of the International Association for Ecology & Health (IAEH) in 2009. These developments have been described as representing a "transdisciplinary imperative for a sustainable future" (Wilcox *et al.* 2004).

Because of the similarities among them, some have urged a merging of these three alternative approaches to health promotion (Parkes 2012; Zinsstag *et al.* 2012; Roger *et al.* 2016). Despite their shared features, however, they are not synonymous. One critical factor that differentiates them is their handling of the human–animal–environment interface (see later). The effort to discuss and compare these approaches is complicated by the various definitions that exist for each of them. Following Lerner & Berg (2017), the focus here is on their respective core values.

1.3.1 EcoHealth

The multidisciplinary EcoHealth approach affirms that humans, animals, and their environment form complex socioecological systems and that destabilization can lead to disease. EcoHealth "examines the complex relationships among humans, animals, and the environment, and how these relationships affect the health of each of these domains" (Listza & Wolbring 2018). It provides a platform to bring together physicians, veterinarians, ecologists, economists, social scientists, health planners, and others to study, understand, and respond to the effects of ecosystem changes on human health. Moreover, it facilitates the integration of different epistemologies, including local, traditional, and indigenous knowledges (Rapport *et al.* 1998). EcoHealth seeks to build participatory frameworks that include more than conventional expert-led research toward a richer participatory process that includes relevant community actors in everything from initial problem definition to project design to the interpretation and sharing of findings (Mertens *et al.* 2005). Underlying all this is the insight that interventions for improving health are more effective if they respond to local realities. In Waltner-Toews' (2009, p. 519) assessment, the goal of EcoHealth is "sustainable human and animal health and well-being, [achieved] through healthier ecosystems."

The leading journal of the approach, *EcoHealth* (n.d.), defines it as being:

> committed to fostering the health of humans, animals, and ecosystems and to conducting research which recognizes the inextricable linkages between the health of all species and their environments. A basic tenet held is that health and well-being cannot be sustained in a resource depleted, polluted, and socially unstable planet.

Alternately, Waltner-Toews (2009, p. 520) states that EcoHealth:

> can be defined as systemic, participatory approaches to understanding and promoting health and wellbeing in the context of social and ecological interactions. They reflect not only an understanding of social systems and ecosystems, and how they interact, but also a convergence of applied, organizational ideas from business management, environmental planning, community operations research, participatory action research, critical systems theories and a variety of other fields

Within veterinary medicine, EcoHealth is viewed as an expansion of the scope of the field from a focus on the health of individual animals to a concern about animal populations and herd health. Promoted especially by the International Development Research Centre, Canada (IDRC), the approach appraises "the effective multi-sectoral collaboration and the engagement of multiple types of stakeholder" (IDRC 2008) through its six integrated principles—namely, system thinking, transdisciplinary research, participation, gender and social equity, knowledge to action, and sustainability—in order to provide a consideration of the human–animal disease interface. A medical perspective on EcoHealth is offered by the International Society of Doctors for the Environment (ISDE), a not-for-profit organization based in Switzerland with member groups in 38 countries. The Australian chapter, for example, Doctors for the Environment Australia (DEA), defines its mission as being to publicize "the relationship between the condition of the environment and human health, promote environmentally friendly behaviour amongst physicians, patients, and the public, and [cooperate] at all political levels in the reduction of harmful environmental influences on health" (DEA 2018). In the perspective of the DEA, doctors cannot narrowly focus on the health of individual patients in a world in which a descent into environmental chaos will seriously imperil their ability to use the findings of medical research and advances in patient care to relieve human suffering.

The IDRC supports research in developing countries by offering financial resources, advice, and training to address local problems. Since its foundation in 1970, it has provided funding for multiple projects. In 2002, for example, it funded a study in Molango, Mexico that traced the transport of locally mined manganese through various components of the ecosystem and examined its impact on human health. Researchers were able to track manganese from mine smokestacks, through wind and dust and along trucking corridors, into people's homes. They showed that exposure to contaminated air led to motor skill damage in adults, especially in women, and they identified nervous system risks in children. These findings prompted changes in local policies (IDRC 2008). Similarly, in Lebanon, IDRC-supported researchers implemented EcoHealth approaches to address water sanitation and waterborne diseases. In the town of Bebnine, they found that women played the greatest role in managing water at the household level. Consequently, they recruited more than 25 Bebnine women to shared their experiences working in the community. According to Iman Nuwayhid, who led the research, "In this way, researchers learned that some women preferred to use untreated well water rather than to pay for access to a new water network, even when they could afford to do so. They rejected the new network because they believed it was affiliated with certain political views or families, and refused to contribute to its shared costs" (IDRC 2010). Further, monthly water quality monitoring in 423 households revealed high levels of fecal contamination. These findings led the local municipality to install a water disinfection unit and to develop a health promotion communication strategy (IDRC 2010). In 2008, the IDRC co-convened the International Ecohealth Forum in Mérida, Mexico with participants from around the world in order to consolidate an international community of practitioners around the EcoHealth perspective. It has provided funding to the Communities of Practice in Ecosystem Approaches to Health in Canada, Latin America, the Caribbean, Africa, and the Middle East, as well as to a Field Building Leadership Initiative on EcoHealth in Asia. Its work of this sort has continued through to the present.

Another critical step in the development of the EcoHealth approach occurred at the "Healthy Ecosystems, Healthy People" conference in Washington, D.C. in 2002, which initiated a process that led both to the formation of the professional journal *EcoHealth* and to the eventual establishment of the IAEH. The IAEH seeks to support transdisciplinary initiatives that combine the strengths of various scholarly disciplines in order to achieve hybridized innovative approaches to problem-solving. In the organization's perspective, "Today's complex environmental public health challenges demand this level of strategic team-based approaches to the tightly coupled human-natural system interactions underlying many of our most pressing threats to our sustained health and environment" (IAEH 2019). The IAEH has sponsored a series of annual conferences at sites around the world. The theme of its 2018 conference, for example, hosted by the University of Valle, Cali, Colombia, was "Environmental and Health Equity: Connecting Local Alternatives in a Global World." This theme emphasized the need to connect local initiatives in a world with global drivers that threaten healthy ecosystems and populations in pursuit of both health and ecological justice.

The first issue of the journal *EcoHealth*, published by the International Society for Ecosystem Health (ISEH), appeared in March 2004. It provides similar-minded researchers across disciplines a venue in which to communicate and participate in sometimes lively debate. As Wilcox *et al.* (2012, p. 27) observe:

> [t]he rationale for ecohealth articulated in the first issue of the journal *EcoHealth* centered on the increasing need for applying ecological concepts to understand and nurture sustainable human health and, conversely, applying health concepts to understand ecosystems; collectively, this was referred to as "integrative studies in ecology, health, and sustainability." Issues identified as within the scope of the journal included: 1) global loss of biological diversity; 2) the drivers of human-induced changes in climate and the degradation and pollution of natural habitats; 3) health of human populations influenced by large-scale environmental changes, including the increasing gap between the rich and poor; and 4) interactions between environment, development, and health.

In sum, the core value of EcoHealth is addressing and improving human health through participation in collaborative efforts to protect biodiversity in the context of development, as expressed in the preceding review, and as derived from its grassroots practical focus on the relationship among health, ecosystems, and sustainable development.

1.3.2 One Health

Development of the One Health approach was motivated at official levels (e.g., the World Health Organization, WHO; the U.N. Food and Agriculture Organization, FAO; the World Organisation for Animal Health, OIE) by recognition that many of the same microbes infect both animals and humans. Diseases like COVID-19 that have passed from nonhuman animal to human populations are known as zoonotic diseases. Almost two-thirds of human pathogens have a zoonotic origin, and almost three-quarters of emerging and re-emerging infectious diseases of humans are zoonoses (Woolhouse & Gowtage-Sequeria 2005). These include both newer human threats, such as COVID-19, severe acute respiratory syndrome (SARS), West Nile virus, Nipah virus, and bovine spongiform encephalopathy (BSE), and ancient zoonotic maladies, such as rabies, anthrax, brucellosis, bovine tuberculosis, and zoonotic trypanosomiasis. The incidence of both types has risen significantly over time, contributing to a significant contemporary burden on global public health (Morens *et al.* 2004). Moreover, their occurrence is significantly correlated with identifiable socioeconomic, environmental, and ecological factors, especially social inequality, environmental and habitat degradation, and ecosystem disruption. In the case of the emergence of COVID-19 as a human disease, for example, there is growing evidence indicating the role of deforestation and land use for intensive farming in the pandemic (Arora & Misra 2020). Many of these multihost dangers require multisectoral responses, as efforts by just one sector

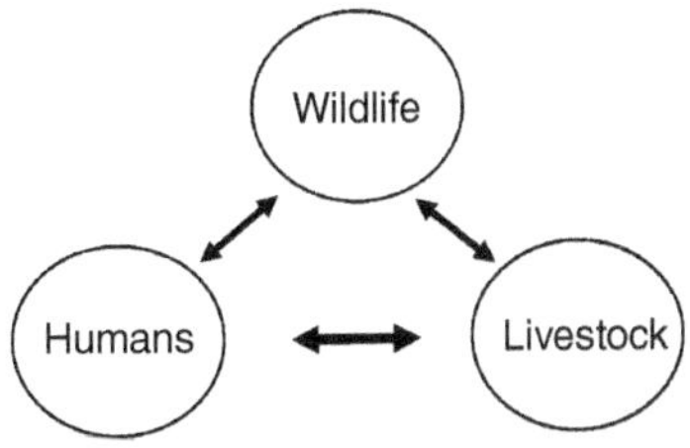

Fig. 1.1 Zoonotic disease triad.

(e.g., biomedicine or veterinary medicine) cannot prevent or eliminate the problem. In the case of rabies in humans, for example, targeting of the animal source of the virus (e.g., vaccinating dogs) is required to control incidence (Lloyd-Smith *et al.* 2009). At any point in time, particular zoonotic threats may be wide-ranging (e.g., the influenzas globally), regional (e.g., Middle East respiratory syndrome, MERS), or local (e.g., Kyasanur forest disease in parts of India). Different forms of monitoring at the human–animal interface are required depending on the conditions and route of transmission. Moreover, local and regional zoonotics may spread widely beyond their point of origin. While found primarily in Middle Eastern countries, for example, confirmed cases of MERS have been identified in Austria, China, France, Germany, Greece, Italy, Malaysia, the Netherlands, the Philippines, the Republic of Korea, Thailand, the United Kingdom, and the United States (CDC 2017). In the case of COVID-19, spread has included all nations. All disease vectors (e.g., mice, deer) and pathogenic diseases have their own features, and thus there is no single surveillance, prevention, or intervention strategy to tackle all zoonotic threats to health. One Health stresses expanding awareness of the conscious and less overt ways in which humans interact with animals in their daily lives (e.g., livestock for food, pets, hunting) (see Fig. 1.1). While animals represent part of the human physical environment, they also are deeply embedded in the human social environment. Increasing movements of people and an increased trade in animals and animal products are contributing to emerging zoonoses.

The Washington, D.C.-based One Health Commission defines One Health as "the collaborative effort of multiple health science professions, together with their related disciplines, and institutions—working locally, nationally, and globally—to attain optimal health for people, domestic animals, wildlife, plants, and our environment" (Roger *et al.* 2016). An outgrowth of collaboration between the American Veterinary Medical Association (AVMA) and the American Medical Association (AMA), and ultimately the American Public Health Association (APHA) as well, the One Health Commission seeks to break down disciplinary silos and establish closer professional interactions, collaborations, and educational and research opportunities across the health science professions. Alternatively, the One Health approach has been defined by the One Health Global Network—a worldwide computer-linked assemblage of organizations (e.g., One Health Alliance of South Asia, One Health Central and Eastern African Network, International Federation of Red Cross and Red Crescent Societies, Centers for Disease Control and Prevention One Health)—as an effort to "improve health and well-being through the prevention of risks and the mitigation of effects of crises that originate at the interface between humans, animals, and their various environments" (Whitmee *et al.* 2015). In both of these definitions, focus is placed on a "whole of society" orientation in which all of the health sciences work collaboratively across traditionally well-guarded disciplinary borders not only to improve the health of humans and animals but also to address biodiversity, ecology disruption, climate change, urbanization, and the changing impacts of expanding agricultural systems.

In Lerner & Berg's (2017) assessment, the core values of One Health involve safeguarding the health—and especially the individual health—of vertebrate species. There have been calls to expand the One Health concept. Keune *et al.* (2017) identify ecology, agriculture, and especially the social sciences as missing links for a coherent One Health approach. From their perspective, "[s]ocial and cultural drivers of health are often neglected and considered only in case of programme failure" (Kene *et al.* 2017, pp. 7–8).

1.3.3 Planetary health

> Global health is not merely a constellation of diseases, a collection of national health systems, or even a set of values. It is a way of looking at our world. It seeks to observe, document, monitor, interpret, and eliminate the harms that accrue from national and transnational forces inimical to health—political, commercial, military, financial, diplomatic, legal, intersectional, and cultural. Global health is about power and poverty, violence and exploitation, oppression and silence, and collusion and exclusion. (Horton 2014)

Planetary health, the newest of the three concepts under consideration, begins with the understanding that human health and the health of the planet are intimately connected. One might say, this is the "one health" of this approach. The origins of this line of medical thinking lie in the conclusion drawn by the Commission on Managing the Health Effects of Climate Change, organized by *The Lancet*, a British medical journal, and academics from many disciplines across University College London that climate change is the biggest threat to human health in the 21st century (UCL News 2009). The Commission, made up of faculty from the fields of health, anthropology, geography, engineering, economics, law, and philosophy, focused on six key areas: patterns of disease and mortality, food security, water and sanitation, shelter and human settlements, extreme events, and population migration. According to Anthony Costello, a world leader on issues affecting the health of children and the lead author of the Commission's report: "The big message of this report is that climate change is a health issue affecting billions of people, not just an environmental issue about polar bears and deforestation . . . The impacts will be felt not just in the U.K., but all around the world—and not just in some distant future but in our lifetimes and those of our children" (UCL News 2009). The Commission examined the global health implications of the Intergovernmental Panel on Climate Change (IPCC) projections and considered a myriad of pathways through which climate change exerts its effects on health.

Health clearly has entered a new epoch when environmental factors, under ever mounting adverse human influence, have outpaced other threats to human well-being. This recognition sparked the planetary health initiative, spearheaded by *The Lancet* and the Rockefeller Foundation, which is guided by acceptance of the fundamental need for collective achievement of a world "that nourishes and sustains the diversity of life with which we coexist and on which we depend" (Horton 2014). From the perspective of planetary health:

> By unsustainably exploiting nature's resources, human civilisation has flourished but now risks substantial health effects from the degradation of nature's life support systems in the future. Health effects from changes to the environment including climatic change, ocean acidification, land degradation, water scarcity, overexploitation of fisheries, and biodiversity loss pose serious challenges to the global health gains of the past several decades and are likely to become increasingly dominant during the second half of this century and beyond. These striking trends are driven by highly inequitable, inefficient, and unsustainable patterns of resource consumption and technological development, together with population growth. (Rockefeller Foundation 2020)

The emergence of planetary health has been described as constituting a fifth stage in the historic evolution of the modern population health paradigm (Singer 2014). This mode of thinking about health as more than an individual-level condition—which historically was the prevailing orientation of biomedicine—has its roots in local public health efforts, which began especially in Europe during the 19th century, transitioned to tropical medicine as a consequence of the European colonial encounter with the infectious diseases of tropical colonial environments, blossomed during the post-World War II period into international health with the expansion of multinational health initiatives like the World Health Organization (WHO), and emerged during the 1990s as global health, a reflection of the consolidation of a global neoliberal economy and global communication systems, combined with a recognition that risks to health transcend borders and require multilateral responses. The shift to a planetary health understanding as a further advance in population health thinking was motivated by the growing awareness that not only are human communities worldwide now multiply linked together by flows of commodities, ideas, people, and health-related influences from vectors to medicines, but that the health and well-being of human communities are increasing linked to the environment and to other species. As stated in the planetary health manifesto: "Our patterns of overconsumption are unsustainable and will ultimately cause the collapse of our civilisation. The harms we continue to inflict on our planetary systems are a threat to our very existence as a species" (Horton *et al.* 2014, p. 847). A planetary health perspective, in other words, reveals the fundamental ways in which human beings are not just agents of environmental change, but also vulnerable objects of that change.

Pivotal to the planetary health movement is a vocal commitment to equity in a world of unjust societies and unequal relations among societies. As articulated in the planetary health manifesto: "The discipline of public health is critical to this vision because of its values of social justice and fairness for all, and its focus on the collective actions of interdependent and empowered peoples and their communities" (Horton *et al.* 2014, p. 847). The evidence on anthropogenic climate change demonstrates the consequences of inequity. Translated into stark moral and palpably human terms, anthropogenic climate change can be assessed as a form of "social murder," a term introduced by Frederick Engels in 1845 to characterize the impact of the corporate-owning class on the health and survival of poor and working people. Chernomas & Hudson (2009) redeploy the term to label contemporary corporate policies designed to maximize the accumulation of private profit while socializing associated risks and costs.

To advance the perspective, *The Lancet* initiated a new journal called *The Lancet Planetary Health* in 2017. Exemplifying its content, in its second year it published a study of the adverse health effects of air pollution on diabetes in China (Yang *et al.* 2018). This study's cohort included over 15 000 people aged 18–74 years from three cities in Liaoning province, in the northeastern sector of the country. Diabetes was defined according to American Diabetes Association (ADA) recommendations. Exposure to air pollutants (sulfur dioxide, nitrogen dioxide, and ozone) was measured by monitoring stations. The researchers found that all of the pollutants studied were significantly associated with increased diabetes prevalence. They were also associated with higher concentrations of fasting glucose. Stratified analyses indicated greater effects on those individuals who were younger (<50 years) or overweight or obese. The researchers concluded that long-term exposure to air pollution is significantly associated with increased risk of diabetes, especially in some subgroups. Other articles in the journal have documented the health benefits of other kinds of environmental exposure. Crouse *et al.* (2017), for example, reported the findings of a study in

30 cities in Canada that showed that increased amounts of residential greenness (based on a vegetation index) were associated with reduced risks of dying from several common causes of death (e.g., cardiovascular disease, diabetes, ischemic heart disease) among Canadian urbanites. While affirming the effects of social inequalities and personal socioeconomic status on health, in terms of both exposures to greenness and mortality risks, the study found that greener cities were healthier cities.

The three somewhat different approaches to health and the environment described in this section are all products of the increasing risk of anthropogenic environmental and climatic changes on species health on the one hand, and the growing awareness of this fact and of living in the Anthropocene on the other. The objective in introducing these concepts, their histories, and their core values was not to choose the best or most useful among them but to illuminate the ways our comprehension of our role in the environment and of the fundamental role of the environment in health have been changing in the contemporary world. Collectively, the emergence of these approaches affirms the importance of accessing not simply the health–environment nexus but specifically within this initiative the role of the growing number of interacting ecocrises in health.

1.4 Global warming or climate change?

Anthony Leiserowitz is a Senior Research Scientist and Director of the Yale Program on Climate Change Communication. His research investigates the psychological, cultural, and political factors that influence environmental beliefs, attitudes, support of policy, and human behavior. His scholarly work is widely published and often cited by his peers. He is invited regularly to speak at scholarly conferences and concerned citizen meetings, and with CEOs and politicians on climate issues. After he gives an address, one of the most common questions he encounters is: "Is it global warming or climate?" (Christensen 2019). Both of these terms have become familiar to people around the world in the last 25 years, but is one of them more accurate or preferred?

The climate is changing and Earth is warming. In fact, however, neither term fully captures the entangled set of significant atmospheric and environmental changes that stem from the release of greenhouse gases (e.g., ocean acidification) or other changes such as the effects of black carbon pollution on solar reflection. Warming of the planet certainly is the force driving a complex array of transformations taking place around us, but the everyday understanding of warming does not adequately describe the erratic nature of what is occurring, nor does it quite encompass the sudden and ever more frequent occurrence of extreme weather events, including extreme cold spells. The world is not just increasing in temperature—weather is becoming less stable, less predictable, weirder, and more deadly in various ways (Climate Central 2012). And while it is accurate to say that the climate is changing, this has been the case for the whole history of Earth (Brooke 2014). What is critical is the clear pattern of that change (an uptrend in average annual temperature) over the last 150 years since the beginning of the Industrial Revolution, and especially the last 50, and our own role in pushing that change. Moreover, climate-related changes are not occurring in isolation from other dramatic planetary changes resulting from human activity. In short, it is not all about warming or climate. Both terms only cover part of the contemporary problem of environmental/climate turmoil (Baer & Singer 2018).

The first recognition of the relationship between carbon dioxide in the atmosphere and atmospheric temperature dates to the work of Swedish scientist Svante Arrhenius (winner of the Nobel Prize in Chemistry), who in 1896 linked what would come to be called the greenhouse effect to the use of fossil fuels. Of historical note, early on Arrhenius "strongly advocated remedies to stop the indiscriminate waste of fossil energy and chemical resources, oil, gas and coal" (Arrhenius *et al.* 2008). By the 1970s, the connections among fossil fuel burning, CO_2 emission, and climate were becoming clear to scientists, and the terms "global warming" and "climate change" were becoming increasingly frequent in the scientific literature. Both terms began to filter into the mass media during the 1990s, and then into everyday discourse. In the contemporary scientific literature, "climate change" is generally used unless the focus is on the pattern of increase in Earth's surface temperature. Those who deny or downplay the changes that are occurring on Earth tend to prefer "global warming," however, as it allows them to argue during extreme cold events that surely the planet cannot be warming, just step outside your front door—or, in the case of Senator James Inhofe of Oklahoma, author of *The Greatest Hoax* (2012), bring a snowball on to the floor of the U.S. Senate (Bump 2015). In Inhofe's view, it is a "scientific scandal . . . to make it sound like man is responsible for all these things [i.e., climate change]" (MSNBC 2015).

In this book, both terms are used, but it is with the understanding that there is a need for a broad picture of the patterns of change, the components of change, the causes of change, the consequences of change, and, most especially, the interactions at the heart of the vast changes occurring around and because of us.

1.5 Depth of the human footprint

As Lewis (2006, p. 195) states:

> Humans moved more rock, sediment and soil than all natural processes combined, by an order of magnitude . . . Between a third and half of all land was appropriated for human use . . . By the [21st] century's end three to six times as much water was held in reservoirs as in natural rivers . . . Two predictions stand out: changes in land-use will cause the sixth mass extinction in evolutionary history . . . while atmospheric CO_2 concentrations will reach their highest levels for 60 million years.

The human footprint on Earth is already very deep, and each year that passes without significant reversal of course will make our presence an ever-greater disrupter of Earth ecological, climatic, atmospheric, hydraulic, and other systems. As Vitousek *et al.* (1997, p. 494) observe, "[u]ntil recently, the term 'human-dominated ecosystems' would have elicited images of agricultural fields, pastures, or urban landscapes; now it applies with greater or lesser force to all of Earth." Quite literally no location on the planet is totally untouched by our species. One place you would not expect to find a human footprint is at the bottom of the ocean. While our knowledge about the complexities and lifeforms of the deep ocean (below 10 000 feet) remains limited, of one thing we are certain: human activities are increasingly affecting deep-sea habitats. This is the conclusion of an international study conducted by over 20 deep-sea scientific experts who participated in the Census of Marine Life project SYNDEEP (Towards a First Global Synthesis of Biodiversity, Biogeography, and Ecosystem Function in the Deep Sea). They grouped the impacts they detected into three types: waste and litter dumping, resource exploitation, and evidence of the effects of climate change (Ramirez-Llodra *et al.* 2011).

The routine dumping of many types of waste from ocean-going ships was legally banned by the London Convention of 1972. An even stricter convention went into force in 2006. Before these bans, a wide array of human litter regularly was dumped from ships, large and small, in ocean transit. New litter continues to accumulate as a result of illegal disposal, lost or discarded fishing gear, and being washed from the coast and through river discharges. It is estimated that 6.4 million tons of litter is deposited into the oceans each year, part of which sinks to the deep ocean bottom (Jeftic *et al.* 2009). In places like the Mediterranean and North East Atlantic, the most common litter types found on the deep ocean floor are soft plastic (e.g., bags), hard plastic (e.g., bottles, containers), and glass and metal (e.g., cans) (Galil *et al.* 1995; Galgani *et al.* 2000).

Expanding technological capabilities have facilitated the exploitation of biological, mineral, and petrochemical resources from the oceans. Trawl fishing can reduce the biodiversity and biomass of deep-water invertebrates, such as cold-water corals, with long-term effects persisting even after fishing has ended. Deep ocean mining is also of increasing concern. Three forms of mineral resources have been identified for commercial exploitation: manganese nodule mining on deep ocean plains, cobalt-rich crusts on seamounts, and polymetallic sulfide deposits at sites of hydrothermal vents. An additional threat is ultra-deep-water oil and gas drilling, the potential adverse consequences of which were exposed by the 2010 *Deepwater Horizon* catastrophe in the Gulf of Mexico (Michel *et al.* 2013).

Rising levels of anthropogenic (human-generated) greenhouse gases have made parts of the North Atlantic and hot spots in the Northern Atlantic, Indian, and Pacific Oceans so acidic that the chalky white form of calcium carbonate known as calcite that makes up the seafloor is dissolving. This dissolution is "altering the geological record of the deep sea" (Sulpis *et al.* 2018, p. 11 700). Moreover, in the European sector of the North East Atlantic, the extent of the effects of bottom trawling by fishing boats is significant—an order of magnitude greater than the total impact of all other activities (e.g., waste disposal, telecommunication cables) (Benn *et al.* 2010).

As these examples suggest, and as illustrated in Fig. 1.2, not only are human impacts varied and widespread, but they reveal "how, even on the grandest scale, most aspects of the structure and functioning of Earth's ecosystems cannot be understood without accounting for the strong, often dominant influence of humanity" (Vitousek *et al.* 1997, p. 494).

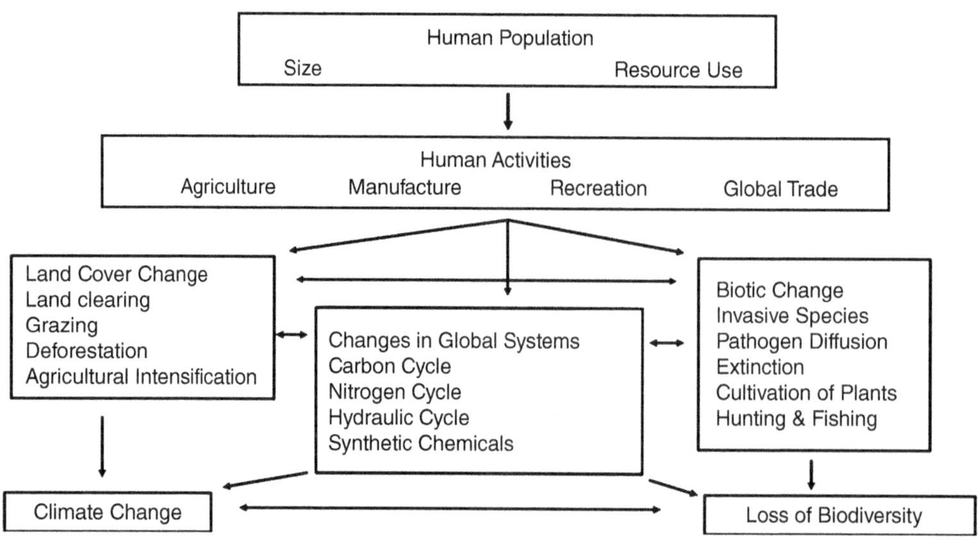

Fig. 1.2 Model of direct and indirect human effects on earth systems. *Source:* Modified from Vitousek, P., Mooney, H., Lubchenco, J., & Melillo, J. (1997). Human domination of the Earth's ecosystems. *Science* 277: 494–499.

1.6 Introducing ecocrises interactions and health

Environmentally speaking, what is a crisis? Further, how do we know we are in a crisis, and when is it time to sound the alarm? What kinds of environmental thresholds must be crossed to recognize we have passed from a noncrisis period into a time of such turbulence and threat that it is aptly called a crisis? Historically, in medicine, dating to the 15th century, "crisis" has referred to a "decisive point in the progress of a disease," the point at which change must come, for better or worse—toward a return to wellness or toward death (though a new up-and-down or slow-progression "steady state" of chronic illness also occurs). Beyond medicine, the word "crisis" has come to mean an unstable or crucial time or state of affairs in which the possibility of a highly undesirable outcome is possible. With regard to the environment, as Moore (2016) comments, "[t]he conditions of life on planet Earth are changing, rapidly and fundamentally. Awareness of this difficult situation has been building for some time. But the reality of a crisis—understood as a fundamental turning point in the life of a system, any system—is often difficult to understand, interpret, and act upon."

Part of this difficulty is a result of the complexity of environmental systems and the fact that the environment is constantly changing. Cycles of disruption and recovery are the usual state of environmental affairs (Paine *et al.* 1998). Another part occurs because we never know everything at once—new and perhaps initially controversial scientific findings are constantly coming to light. Additionally, humans are frequently unready to recognize both the full power of nature and our own capacity to adversely influence it. Finally, there is the intentional denial or even falsification of environmental crises because of the short-term benefits and profits of maintaining business-as-usual practices. As a result, "reasonable treatment of environmental concerns often falls prey to the political agendas of those who have a vested interest in an unsustainable, resource-extractive approach to economic development" (Hudson 2001).

While cycles of change are the "normal" state in nature, "rapidly compounded perturbations have more serious implications for long-term alterations of community state, occasionally or even often generating a different assemblage of species" (Paine *et al.* 1998, p. 536) because the "effect of compounded perturbations is multiplicative, not additive" (Paine *et al.* 1998, p. 537). In this book, I refer to compounded environmental disturbances as "ecocrises interactions." Ecocrises interaction—the synergistic interface of two or more environmental events or pollutants that multiply resultant harmful health effects beyond their additive impact—puts both humans and other species at catastrophic risk. This book is designed to increase understanding of the link between health and the environment in times of the ever-more-pervasive anthropogenic impacts on natural assemblages. Its specific concern is that the diverse ecological calamities we face are encountered not as standalone threats, which has been the conventional perspective to date, but as adversely interacting events and processes with magnified hazardous effects. The sixth Global Environment Outlook issued by the U.N. Environment Program (2019), a continent-by-continent 740-page assessment written by 250 scientists and experts from more than 70 countries, concludes that because of the perilous combination of climate change, pollution, and the growing human population, the damage already done to the planet is so severe that people's health will be increasingly threatened unless rapid action is taken. In the view of the report's authors, without drastically scaled-up implementation of environmental protections, cities and regions in Asia, the Middle East, and Africa may see millions of premature deaths by mid-21st century. This discussion raises essential questions about the extinction of populations, species, and ways of life.

Ecocrises interactions occur when the various pollutants and environmental disruptions generated especially today by fossil capitalism are not only co-present in the same environmental context but combine to uni- or multidirectionally magnify their impact on human health and society, and on the health and survival of other species as well. Critical to explicating ecocrises interaction is the development of an understanding of the pathways and mechanisms through which two or more ecocrises deleteriously intermingle in the air, on land, and in bodies of fresh or salt water. For example, respiratory health around the world has been growing worse in part because of the combined effects of heavier pollen loads caused by global warming and increased air pollution linked to prevailing fossil fuel use at a time when more intense wildfires are being triggered by climate change (Fig. 1.3). These kinds of complex sets of interactions between global warming and anthropogenic (human-caused) impairments to clean air can be seen in various planetary ecozones and involve a diverse set of environmental corridors of interaction. Most importantly, they increase human vulnerability to a range of debilitating and deadly diseases.

Similarly, there is a perilous interaction between meat production, nutrient pollution, and climate change. In the Gulf of Mexico off the coast of Louisiana, there is an 8000-square-mile area in which marine life is suffocating due to nutrient pollution. The nutrients, which feed a dead zone of algae bloom—a process known as eutrophication—flow down the Mississippi River in the form of excess nitrogen and phosphorus in fertilizers used on industrial corn farms. As meat production increases, the demand for corn rises, leading to mounting nutrient pollution. In 2017, the largest dead zone on record occurred in the Gulf region. More meat production also increases the release of methane, a powerful greenhouse gas (Weiss 2013).

Fig. 1.3 Smokestack pollution from a power plant. *Source:* bhumann34/Pixabay.

In this book, the familiar term "pollution" is used specifically to refer to "unwanted, often dangerous, material that is introduced into Earth's environments as the result of human activity, that threatens human health, and that harms ecosystems" (Landrigan *et al.* 2018). Diseases caused by pollution resulted in an estimated 9 million premature deaths in 2015, which is 6 percent of the total number of deaths that occurred worldwide. The significance of this rate of mortality is seen in the fact that it is three times higher than that of AIDS, tuberculosis, and malaria combined, and 15 times higher than that of all wars and other forms of physical violence. In some countries, primarily in South East Asia and the western Pacific, pollution-related disease is responsible for 25 percent of all deaths (Landrigan *et al.* 2018). Further, pollution is responsible for more deaths globally per year than a high-sodium diet, obesity, alcohol consumption, vehicular accidents, or child and maternal malnutrition.

Of special note are the populations globally most adversely impacted by pollution. Over 90 percent of pollution-related deaths are in low- and middle-income countries. In all nations, pollution-related disease is most prevalent among ethnic minorities, the poor, and the socially marginalized. In particular, children are at highest risk of diseases linked to pollution. In utero and in early infancy—periods of high vulnerability—exposures to even very low doses of pollutants can lead to disease, disability, and death, with impacts that occur throughout an individual's lifespan. What's more, "ambient air pollution, chemical pollution, and soil pollution—the forms of pollution produced by industry, mining, electricity generation, mechanised agriculture, and petroleum-powered vehicles—are all on the rise, with the most marked increases in rapidly developing and industrialising low-income and middle-income countries" (Landrigan *et al.* 2018). Cities, especially the most rapidly growing cities in developing nations, are severely affected by the production and health risks of pollution. This constitutes one of the great global existential challenges of the Anthropocene.

These facts notwithstanding, pollution remains an understudied and often ignored cause of human morbidity and mortality. Even though 70 percent of pollution-caused diseases fall into the category of noncommunicable diseases—those such as cardiovascular diseases, cancers, chronic respiratory diseases, and diabetes not transmitted from human to human or animal to human—there is rare mention of interventions to address them in the WHO's Global Action Plan for the Prevention and Control of Non-Communicable Diseases 2013–2020, and pollution is often downplayed in both global health assessments (WHO 2013) and evaluations of the full social costs of fossil fuel production (NAS 2010).

To the degree that environmental pollution has been considered as a driving force in global health, it has been treated as an inevitability, an inescapable blemish of socioeconomic development. This element of conventional wisdom is referred to as the "environmental Kuznets hypothesis" (Van Alstine & Neumayer 2010), a label stemming from the graphic known as the Kuznets curve, developed by Nobel Memorial Prize-winning economist and statistician Simon Kuznets (1955). Specifically, the curve, in the form of an inverted U-shape, reflects Kuznets' hypothesis that, as a society develops from a primarily agrarian to an industrialized economy, market forces first increase and then later decrease its overall degree of economic inequality. This hypothesis has been applied to environmental economics, where it is postulated that pollution and environmental degradation will unavoidably increase during the early stages of industrial economic development, and that pollution will continue to rise until the threshold of per capita income is achieved, after which the level of pollution will decrease despite continued economic growth. If the environmental Kuznets hypothesis is accurate, rather than being a growing threat to the environment, economic growth should be seen as an anthropogenic means of eventual environmental improvement. As Stern (2004, p. 1419) points out, "[t]he possibility of achieving sustainability without a significant deviation from business as usual was an obviously enticing prospect for many—letting humankind 'have our cake and eat it'."

Despite its comfortable appeal, the environmental Kuznets hypothesis is not confirmed by environmental health research, and the data used to support it have been criticized as being econometrically weak (Perman & Stern 2003). Part of the problem is that a reduction in pollution levels in wealthy countries often results from the outsourcing of polluting production to cheaper-labor developing ones (Carmin & Agyeman 2011). Indeed, enhanced environmental regulation designed to limit pollution in wealthier countries may contribute to corporate outsourcing to developing countries as a cost-cutting/profit-enhancing strategy (Lucas *et al.* 1992; Cole & Elliott 2003). The Lancet Commission on Pollution and Health concluded that "pollution is not the unavoidable consequence of economic development, and that it is much more important to formulate sound laws, policies, and regulations to control pollution than to wait for an economy to reach a magical tipping point that will solve the problems of environmental degradation and pollution-related disease" (Landrigan *et al.* 2018, p. 467).

1.7 Thresholds in the environment

Like pollution, other forms of anthropogenic environment disruption are not inevitable, unavoidable, or too costly to avoid. The full costs of pollution and disruption are manifest in ecocrises interactions. Critical to the development of many ecocrises interactions is the crossing of the thresholds of planetary boundaries. Environmental conditions are defined as intrinsic features at local, regional, or global scales that constitute positions along one or more control variables, such as temperature and the albedo (solar reflection) feedback pathway involving floating sea ice. Icebergs in the ocean, being lighter in color than the surrounding water, reflect more solar radiation. Consequently, the significant rate at which sea ice is melting—a phenomenon driven by planetary warming—represents a negative feedback loop that feeds further warming. Once the volume of sea ice threshold is passed, environmental degradation will continue even without further human input, reducing the possibility for mitigation. Staying within such thresholds, as a result, is critical for human well-being and the well-being of the other lifeforms we depend upon and with which we share this planet.

On July 28, 2000, the U.N. Economic and Social Council established the Permanent Forum on Indigenous Issues to discuss and advise on issues facing indigenous peoples of the world related to social development, culture, environment, education, health, and human rights. Nine years later, on May 27, 2009, during the eighth meeting of the Permanent Forum in New York, one of the speakers was Nicolas Lucas Ticum, a priest and researcher from Guatemala of Maya ancestry. In seeking to reframe the human/environment relationship, Ticum stated: "All of humanity must work together to re-establish harmony and unity with the natural environment . . . The Earth does not belong to human beings. Human beings belong to the Earth" (Economic and Social Council 2009). The alternative is ever more devastating anthropogenic ecocrises.

1.8 Sustainability of human life on Earth

Ultimately, ecocrises threaten the sustainability of human life on Earth. Already, research suggests that Earth may be in the midst of its sixth mass extinction of species, in no small part because of the human reshaping of the environment. At this point in our history, the idea of pristine environments untouched by our presence is an illusion. We have left an indelible impression everywhere on the planet; there are no hidden natural sanctuaries that are unsullied by the effects of human society and activity. Archeological, paleoecological, and genetic research suggests that:

> As our planet experiences its sixth "mass extinction event" . . . the effect of anthropogenic landscape modification, habitat fragmentation, overexploitation, and species invasions could not be more apparent . . . These transformations are linked largely to the industrial economies, burgeoning populations, and dense transport networks of contemporary human societies. (Boivan *et al.* 2016)

In fact, Boivan *et al.* (2016) stress, the human-mediated alteration of Earth long predates the rise of industrial economies. Relative to our impact on other species, they identify four pivotal phases in the loss of pristinity: 1) the Late Pleistocene near-global dispersal of our ancestors; 2) the emergence and spread of agriculture and livestock domestication beginning in the Early Holocene; 3) the human colonization of the world's islands; and 4) the premodern expansion of urbanization and trade beginning 5000 years ago during the Bronze Age. While socioeconomic developments since the Bronze Age have dramatically accelerated the magnitude and range of the human impact, all of these prior moments in the making of a human-dominated Earth are significant.

A significant biological effect of the Late Pleistocene expansion of our species, for example, was felt by larger animals or megafauna, those weighing over 100 pounds. While there is debate concerning the extent of the human role in the extinction of about 90 megafauna genera in the period between 50 000 and 10 000 years ago (Koch & Barnosly 2006), new analytic approaches affirm at least some direct human role through hunting in the disappearance of animals like the giant ground sloth, woolly mammoth, saber-tooth cat, a giant 6-foot-long beaver, and the 3-ton Diprotodon, the largest marsupial that ever lived. This research suggests an inverse relationship between the severity of the extinction and the duration of hominin–megafauna contact.

If hunting only partly accounts for the megafauna extinctions, what else was involved? According to Koch & Barnosky (2006, p. 239):

> Taken as a whole, recent studies suggest that humans precipitated the extinction in many parts of the globe through combined direct (hunting) and perhaps indirect (competition [for food or habitat], habitat alteration and fragmentation) impacts, but that . . . environmental change influenced the timing, geography, and perhaps magnitude of extinction. Put another way, absent the various impacts of Homo sapiens . . . it is highly unlikely global ecosystems would have experienced a mass extinction of large, slow-breeding animals in the late Quaternary. But, absent concurrent rapid climatic change evident in many parts of the globe, some species may have persisted longer.

In other words, it is likely that it was the interaction of human hunting and climate change that eliminated or fragmented the environmental conditions necessary to maintain the megafauna, leading to their extinction.

Without question, the vast majority (99 percent) of species that have ever lived are now extinct (Novacek 2001). Human presence on the planet is not guaranteed; indeed, like all species, and despite our comparatively big brains and powerful technologies, we are dependent upon our fragile relationship with the environment. Failure to fully appreciate this simple truth and move with determination to more sustainable socioeconomic systems will place our near future at grave risk. This idea touches on a broader question about lifeforms elsewhere in our galaxy: namely, why, with all of our searching (e.g., NASA probes, monitoring of electromagnetic radiation, the work of the SETI Institute), have we not encountered signs of alien civilizations? One explanation, known as the Great Filter theory, is that extraterrestrial civilizations may emerge, develop, and destroy their environments, and hence themselves, too quickly for us to find them, or vice versa. In other words, the Milky Way may be sprinkled with extinct alien civilizations that were wiped out by self-induced environmental destruction.

1.9 How did things get this bad?

Promoting the vital need to develop a deeper awareness of our relationship with the environment and assisting students to make informed decisions for action are the ultimate objectives of this book. While there is mounting concern over the ways anthropogenic environmental change is putting human health at risk, the specific rationale for this volume is the inadequate consideration given to the special risk of ecocrises interaction in the existing environmental literature.

The following chapters trace the history of that risk, beginning with the sweeping environmental and social changes launched by the Industrial Revolution. These changes were so immense that many environmental scientists see them as constituting a new geological epoch, one that highlights the dominant role now played by human activity in shaping the very geology and ecology of Earth. Referred to as the Anthropocene, it was triggered in the late eighteenth century with the invention of the steam engine, followed in relatively short order by the production and use of fossil fuels, which contributed to an intense rise in the amount of carbon dioxide and other pollutants in the atmosphere and the oceans. Industrial production, in turn, contributed to an array of other environmental crises, including fuel spills, waste dumping and leakages (Fig. 1.4), deforestation, the spread of infectious diseases, loss of biodiversity, and the devastation of marine life.

But how did all of this come about? What steps along the way pushed our arrival at this troublesome destination? Having come to this point, how can we avoid catastrophe? To adequately address these urgent questions, it is necessary to tease apart and rigorously examine a linked set of environmental, biological, social, political, and economic issues in historic context. This is the approach taken in explicating the health perils of ecocrises interaction in this volume.

Fig. 1.4 Toxic waste barrel. *Source:* EliasSch/Pixabay.

1.10 Age of the Anthropocene

The term "Anthropocene" unites the word *anthropo*, ancient Greek for "humankind", with the root *-cene*, the standard suffix to denote an epoch in geologic time. Another Greek term of relevance is *kairos*, which signifies a "moment of transition." As Moore (2016) remarks, "the Anthropocene concept [is] the most influential concept in environmental studies over the past decade." But the Anthropocene is more than a concept: it also increasingly constitutes a major influence on the lived experience of Earthlings of all species (Kelly & McDonald 2018), including playing a role in the ending of such experience through species extinction.

While "global-scale human influence on the environment has been recognized since the 1800s, the term Anthropocene . . . has only recently [during the 21st century] become widely, but informally, used in the global change research community" (Steffen *et al.* 2011). It was first introduced by the Nobel Prize-winning atmospheric chemist Paul Crutzen and the biologist Eugene Stoermer (Crutzen & Stoermer 2000; Crutzen 2002). A primary driver of its acceptance as a meaningful and needed concept is the multiple Earth-changing effects of climate change. But "climate change is only the tip of the iceberg" (Steffen *et al.* 2011). Human actions have also changed the life-sustaining cycles of key elements such as nitrogen, phosphorus, and sulfur. The nitrogen cycle has been altered through the production and use of fertilizer, deforestation, and the burning of fossil fuels. The massive amounts of nitrogen that humans produce—they at their highest level today for 2.5 billion years—are reshaping the world's ecosystems (Lewis & Maslin 2015). The phosphorous cycle has been changed by fertilizer use and the rearing of livestock (especially hogs), as well as by the use of detergents containing sodium tripolyphoshate. The sulfur cycle has been changed by the burning of fossil fuels, which increases the amount of sulfur in both the atmosphere and the oceans. Moreover, human activities have altered the terrestrial water cycle via dam building, river course modification, and the elimination of riverside wetlands, as well as through changes in land cover that modify the flow of water vapor from the land to the atmosphere. Another human impact is an acceleration of species extinctions to such a degree that it has raised questions about whether we are on the threshold of a mass extermination on par with the one that nearly eliminated the dinosaurs (nearly, and not totally, because the dinosaur ancestors of modern birds survived). In short, human activities are so widespread and so profound in the ways they are changing Earth that they have begun to threaten the very life-support systems upon which we and all other species depend (Steffen *et al.* 2005; Waters *et al.* 2016). The concept of the Anthropocene was developed to call attention to this human-driven quantitative shift in the geochronology of the planet. As Meyer (2018) comments, "[i]t is a stirring idea: that humans are not a momentary blip in the long procession of Earth's history, but a new and fundamental driver of planetary change, equal in stature to volcanoes and tectonic plates." In terms of the actual effects we are having on the planet, it also is a disturbing one.

How do earth scientists determine geochronology? The process involves ongoing data collection, international expert discussion, and committee consensus. Critical components occur within the International Union of Geological Sciences (IUGS), an international nongovernmental organization concerned with promoting international cooperation in the scientific field of geology. The Union manages six international commissions, including the International Commission on Stratigraphy (ICS), which is responsible for setting the stages and boundary markers of the geochronology of Earth history and for naming geological eras. In the language of the ICS, a historical epoch is defined as a subdivision of the geologic timescale that is longer than an age but shorter than a period. By way of analogy, in measuring the dimension of length, a foot is longer than an inch but shorter than a yard.

The establishment of an epochal change the geologic timescale generally is based on the identification of significant changes in the rock layers of Earth's stratigraphy. The conceptualization of the Anthropocene hinges on the idea that the geological epoch known as the Holocene (Greek for "entirely recent"), which began approximately 11 650 calendar years before the present, has ended, because "humankind has become a global geological force in its own right" (Steffen *et al.* 2005). The ICS has not as yet accepted the Anthropocene as a formally defined geological unit within the Geological Time Scale, though its Anthropocene working group has concluded that it is a plausible new layer in Earth geochronology. Establishing the Anthropocene as an accepted geologic unit of time requires that the ICS either: 1) identify a specific location in rock, sediment, or glacier ice that marks the transition to a new epoch (e.g., the Precambrian–Cambrian boundary marker has been placed at Fortune Head, Newfoundland, where there is clear fossil evidence of a life-form transition); or 2) agree on a date for this transition, based on a survey of the available stratigraphic evidence. Various dates, such as the beginning of the Industrial Revolution, with its ever-accelerating level of fossil fuel use and abundant greenhouse gas emissions (Crutzen 2002), or the period between 1945 and 1988, when nuclear weapons were tested with such frequency that their worldwide fallout left an identifiable mark in the chemostratigraphic record (Zalasiewicz *et al.* 2015), have been proposed.

One of the most curious features in the evidence for the Anthropocene involves the common domestic broiler chicken, the source of everything from buffalo wings to pre-sliced deli meat chicken sandwiches. While thus far no individual taxa have been suggested as the distinct marker of the Anthropocene, Bennett *et al.* (2018) propose *Gallus gallus domesticus* as the best choice. Driving the rise and global spread of this heavily engineered bird are human population growth, commercially impacted human consumption trends, and drops in the populations of wild animals. It is now by far the most numerous bird species around, with an estimated population of almost 23 billion individuals. This is an order of magnitude larger than stocks of the most abundant existing wild bird species, such as house sparrows, common pigeons, and red-billed quelea, as well as the extinct passenger pigeon, and is likely the largest population size ever achieved by a single species in the 160 million-year history of birds. Because of selective industrial breeding, the broiler chicken is distinctive in overall size, weight, bone size, genetic make-up, and appearance from its wild progenitor, the red jungle fowl. In 2016, over 65 billion broilers were consumed globally. Production and consumption data mean "that the potential rate of carcass accumulation of chickens is unprecedented in the natural world" (Bennett *et al.* 2018). Much of this global accumulation finds its way into landfill sites by way of domestic garbage removal systems. Organic materials like chicken bones are often well preserved in landfills because their anaerobic conditions tend to mummify organic deposits (Rathje & Murphy 2001). Consequently, the broiler chicken produces a particularly widely distributed and distinctive biostratigraphic signal in the sedimentary record, allowing it to serve as a key fossil index taxon of the Anthropocene.

While the ICS has not recognized the Anthropocene (Finney & Edwards 2016; Gibbard 2018), many scientists have. The concept has gained de facto authority even if it is not officially sanctioned, as reflected in the launching of multiple new academic journals, such as *The Anthropocene*, *The Anthropocene Review*, *Elementa: Science of the Anthropocene*, and *Anthropocene Coasts,* the organization of professional conferences sponsored by universities and scholarly groups, and the publication of an ever-growing international and multidisciplinary scientific literature on the topic (e.g.,

Zalasiewicz *et al.* 2017; Hughes *et al.* 2018; Steffen *et al.* 2018; Tucker *et al.* 2018).

Ultimately, whether or not the ICS decides that the Anthropocene meets the criteria for the establishment of an epoch in Earth history, it does not change the reality of the "perfect storm" of interacting human impacts on the planet, its global processes and ecosystems, and its lifeforms. For those at gravest risk from ecocrises interaction, whether their loss of access to adequate food, breathable air, survivable temperatures, or dry land occurs during a particular scientifically defined segment of geological time on ICS' International Chronostratigraphic Chart is of negligible importance. What does matter is the utility of Anthropocene recognition in spurring useful new research and practical political and social action to avert a slide into hothouse Earth conditions.

1.11 The hottest year on record

Most significant among the ecological crises Earthlings now face is global warming. Earth's temperature has risen by 1.4 °F over the last 100 years, and is predicted to rise to 11.5 °F in the next few decades. Despite various disagreements about regional and temporal variations in temperature, the vast majority of climate scientists recognize both that the planet is getting hotter and that fossil fuel technology is playing a fundamental role in this risky process. Today, the carbon dioxide level in the atmosphere is the greatest it has been in 400 000 years, having jumped from about 280 parts per million (ppm) before the Industrial Revolution to about 410 ppm today. In the final report of The Lancet Commission on Managing the Health Effects of Climate Change (Costello *et al.* 2009), the warming of Earth was labeled "the biggest global health threat of the 21st century," a conclusion affirmed by the Commission in 2015 (Watts *et al.* 2015). The WHO (2018) estimates there will be approximately 250 000 additional deaths due to climate change per year between 2030 and 2050, due to malnutrition, malaria, diarrhea, and heat stress. People living in coastal regions, megacities, island communities, and mountainous and polar regions are particularly vulnerable, especially children and elderly people. Notably, disproportionate morbidity and mortality rises due to climate change will affect the poorest and least powerful global citizens (Singer 2018).

The hottest year on record was 2016, with 2014, 2015, 2017, and 2018 representing the next four hottest since the measurement of annual records began 139 years ago. As Climate Central (2018) points out "[w]ith the five warmest years occurring during the past five years—and the 20 warmest occurring over the past 22—a consistent warming trend couldn't be clearer." The trend continues. June of 2019 was the warmest June ever recorded. For climate scientists like Sarah Green, an environmental chemist at Michigan Technological University, "[a]t this point, the inexorable increase in global temperatures is entirely predictable" (quoted in Kaufman 2019).

1.12 Organization of this book

The following chapters lay out an environmental health approach to ecocrises interaction that is informed by anthropology, ecology, climate science, history, and political economy. This begins in Chapter 2 with a dive into the intricacies of planetary ecosystems and the rise of the field of ecological study. Under consideration is the nature of nature, specifically the interrelationships among species in an environmental context, as well as biodiversity and environmental regional and planet-wide structures (e.g., air currents, hydraulics). Chapter 3 examines the social, economic, and technological bases of the contemporary environmental crisis, including the role played by the ongoing development of capitalism and the growth of a class of wealthy polluting elites, who represent the primary drivers of catastrophic environmental pollution. Chapter 4 provides an overview of the environment catastrophe unfolding during the Anthropocene, including the patterns of sweeping change reshaping life on the planet: pollution of the water, the air, and the land; habitat restructuring and ecological simplification; species movement; and species die-offs and extinctions. Chapter 5 offers a specific investigation of the major environmental crises endangering Earth as our planetary home, including the manifold expressions of contemporary environmental disruption. Building on this background, Chapter 6 explores the nature and threat of ecocrises interaction and presents several case studies of the physical, chemical, and biological factors involved in these perilous processes. Chapter 7 addresses the mechanisms and pathways of adverse interaction. Chapter 8 provides a detailed examination of climate change as the primary driver of ecocrises interaction. To explicate our response to climate change and why it progresses despite growing awareness of the risks it is ushering in, Chapter 9 scrutinizes the growth of climate change denial and the offensive against environmental science generally. Chapter 10 looks at how we continue to cross vital thresholds and approach identified planetary boundaries for human life as we know it. Finally, Chapter 11 offers a consideration of how you can participate in seeking to avoid the worst perils we face based on a review of human movements for change and alternative models for a sustainable future.

References

Arora, N. & Misra, J. (2020). COVID-19 and importance of environmental sustainability. *Environmental Sustainability* 3(2): 117–119.

Arrhenius, G., Caldwell, K. & Wold, S. (2008). *A Tribute to the Memory of Svante Arrhenius.* Stockholm: Royal Swedish Academy of Engineering Sciences.

Aviv, R. (2014). A valuable reputation: after Tyrone Hayes said that a chemical was harmful, its maker pursued him. *New Yorker.* https://www.newyorker.com/magazine/2014/02/10/a-valuable-reputation.

Baer, H. & Singer, M. (2018). *The Anthropology of Climate Change: An Integrated Critical Perspective.* Abingdon, Oxford: Routledge, Earthscan.

Begos, K. (2014). Adjusting to Threats and Harassment. National Association of Science Writers. https://www.nasw.org/article/sciencewriters-adjusting-threats-and-harassment.

Benn, A., Weaver, P., Billet, D., van den Hove, S., Murdock, A. & Gemma B. (2010). Human activities on the deep seafloor in the north East Atlantic: an assessment of spatial extent. *PLoS ONE* 5(9): e12730.

Bennett, C., Thomas, R., Williams, M., Zalasiewicz, J., Edgeworth, M., Miller, H. *et al.* (2018). The broiler chicken as a signal of a human reconfigured biosphere. *Royal Society Open Science* 5(12).

Berliner, M. (2009). The danger of radical environmentalism. *Capitalism Magazine.* https://www.capitalismmagazine.com/2020/04/the-danger-of-environmentalism/.

Boivan, N., Zeder, M., Fuller, D., Crowther, A., Larson, G., Erlandson, J. *et al.* (2016). Ecological consequences of human niche construction: examining long-term anthropogenic shaping of global species distributions. *Proceedings of the National Academy of Sciences U.S.A.* 113(23): 6388–6396.

Brooke, J. (2014). *Climate Change and the Course of Global History.* Cambridge: Cambridge University Press.

Bump, P. (2015). Jim Inhofe's snowball has disproven climate change once and for all. *Washington Post.* https://www.washingtonpost.com/news/the-fix/wp/2015/02/26/jim-inhofes-snowball-has-disproven-climate

-change-once-and-for-all/?utm_term=.3f25 4dbcb836.

Carmin, J. & Agyeman, J. (2011). *Environmental Inequalities Beyond Borders: Local Perspectives on Global Injustices*. Cambridge, MA: MIT Press.

CDC. (2017). Middle East Respiratory Syndrome (MERS). Centers for Disease Control and Prevention. https://www.cdc.gov/coronavirus/mers/index.html.

Chernomas, R. & Hudson, I. (2009). Social murder: the long-term effects of conservative economic policy. *International Journal of Health Services* 39(1): 107–121.

Cho, R. (2018). Perspectives of scientists who become targets: Naomi Oreskes. Climate Science Legal Defense Fund Blog. https://www.csldf.org/2018/01/10/perspectives-scientists-become-targets-naomi-oreskes/.

Christensen, J. (2019). Is it climate change or global warming? How science and a secret memo shaped the answer. *CNN*. https://www.cnn.com/2019/03/02/world/global-warming-climate-change-language-scn/index.html.

Climate Central. (2012). *Global Weirdness*. Grand Haven, MI: Brilliance Audio.

Climate Central. (2018). The 10 hottest global years on record. https://www.climatecentral.org/gallery/graphics/the-10-hottest-global-years-on-record.

Climate Science Legal Defense Fund. (2019). Doxxing and death threats. No longer available online.

Cole, M. & Elliott, R. (2003). Determining the trade–environment composition effect: the role of capital, labor and environmental regulations. *Journal of Environmental Economics and Management* 46: 363–383.

Conway, M, (2018). Why environmental reporting is so intensely dangerous. *Nonprofit Quarterly*. https://nonprofitquarterly.org/why-environmental-reporting-is-so-intensely-dangerous/.

Costello, A., Abbas, M., Allen, A., Ball, S., Bell, S., Bellamy, R. *et al.* (2009). Managing the health effects of climate change. *Lancet Commissions* 373(9676): 1693–1733.

Crouse, D., Pinault, L., Balram, A., Hystad, P., Peters, P., Chen, H. *et al.* (2017). Urban greenness and mortality in Canada's largest cities: a national cohort study. *Lancet Planetary Health* 1(7): e289–e297.

Crutzen, P. (2002). Geology of mankind—the Anthropocene. *Nature* 415(6867): 23.

Crutzen, P. & Stoermer, E. (2000). The Anthropocene. *Global Change Newsletter* 41: 17–18.

DEA. (2018). Origin and history of Doctors for the Environment Australia (DEA). https://www.dea.org.au/history/.

Duhigg, C. (2009). Debating how much weed killer is safe in your water glass. *New York Times*. https://www.nytimes.com/2009/08/23/us/23water.html.

EcoHealth. (n.d.). About the EcoHealth Journal. https://ecohealth.net/en/.

Economic and Social Council. (2009). "The Earth does not belong to human beings; human beings belong to the Earth," Permanent Forum hears as it takes up issues of climate change, land tenure. Press release. https://www.un.org/press/en/2009/hr4988.doc.htm.

Elbein, S. (2016). Cambodian journalists are dying trying to save the country's forests. *Vice*. https://www.vice.com/en_us/article/4w54nb/cambodian-journalists-are-dying-trying-to-save-the-countrys-forests-v23n6.

Fetherling, D. (2010). *Mother Jones, the Miners' Angel*. Carbondale, IL: Southern Illinois University Press.

Finney, S. & Edwards, L. (2016). The "Anthropocene" epoch: scientific decision or political statement? *GSA Today* 26(2–3): 4–10.

Freedman, E. (2018). Why covering the environment is one of the most dangerous beats in journalism. *The Conversation*. https://theconversation.com/why-covering-the-environment-is-one-of-the-most-dangerous-beats-in-journalism-105477.

Galgani, F., Leaute, J., Moguedet, P., Souplet, A., Verin, Y., Carpentier, A. *et al.* (2000). Litter on the sea floor along European coasts. *Marine Pollution Bulletin* 40(6): 516–527.

Galil, B., Golik, A. & Turkay M (1995). Litter at the bottom of the sea: a sea bed survey in the Eastern Mediterranean. *Marine Pollution Bulletin* 30(1): 22–24.

Gibbard, P. (2018). Formal Subdivision of the Holocene Series/Epoch. International Commission on Stratigraphy.

Graham, F. (2012). Fifty years after silent spring, attacks on science continue. YaleEnvironment[360]. https://e360.yale.edu/features/fifty_years_after_rachel_carsons_silent_spring_assacult_on_science_continues.

Hayes, T. (2008). The atrazine rap. https://www.youtube.com/watch?v=3MxrH4lN0-A.

Horton, R. (2014). Reimagining the meaning of health. *Lancet* 384: 218.

Horton, R., Bealehole, R., Bonita, R., Raeburn, J., McKee, M. & Wall, S. (2014). From public to planetary health: a manifesto. *Lancet* 383: 847.

Howard, C. (2013). Pest control: Syngenta's secret campaign to discredit Atrazine's critics. 100 Reporters. https://100r.org/2013/06/pest-control-syngentas-secret-campaign-to-discredit-atrazines-critics/.

Hudson, S. (2001). Challenges for environmental education: issues and ideas for the 21st century: environmental education, a vital component of efforts to solve environmental problems, must stay relevant to the needs and interests of the community and yet constantly adapt to the rapidly changing social and technological landscape. *BioScience* 51(4): 283–288.

Hughes, T., Anderson, K. Connolly, S., Heron, S. & Kerry, J. (2018). Spatial and temporal patterns of mass bleaching of corals in the Anthropocene. *Science* 359(637): 80–83.

IAEH. (2019). The International Association for Ecology and Health (EcoHealth). https://www.springer.com/environment?SGWID=0-198-2-454109-0.

IDRC. (2008). Ecohealth works: environmental pollution. International Development Research Centre, Canada. https://www.idrc.ca/sites/default/files/sp/Documents%20EN/ecohealth-works-environmental-pollution.pdf.

IDRC. (2010). Health in urban environments. International Development Research Centre, Canada. https://www.idrc.ca/en/research-in-action/health-urban-environments.

Inhofe, J. (2012). *The Greatest Hoax: How the Global Warming Conspiracy Threatens Your Future*. Medford, OR: WND Books.

Jamail, D. (2019). *The End of Ice: Bearing Witness and Finding Meaning in the Path of Climate Disruption*. New York: The New Press.

Jeftic, L., Sheavly, S. & Adler, E. (2009). *Marine Litter: A Global Challenge*. Nairobi: United National Environmental Program.

Kaufman, M. (2019). June was the warmest June ever recorded, but there's a bigger problem. *Mashable*. https://mashable.com/article/warmest-june-climate-change-trend-2019/.

Kelly, J. & McDonald, F. (2018). A multimodal approach to the Anthropocene. *American Anthropologist* 120(3): 583–595.

Keune, H., Flandroy, L., Thys, S., De Regge, N., Mori, M., van den Berg, T, *et al.* (2017) European OneHealth/EcoHealth Workshop Report. Brussels: Belgian Community of Practice Biodiversity and Health, Belgian Biodiversity Platform.

Koch, P. & Barnosly, A. (2006). Late quaternary extinctions: state of the debate. *Annual Review of Ecology, Evolution, and Systematics* 37: 215–250.

Kuznets S. (1955). Economic growth and income inequality. *American Economic Review* 45(1): 1–28.

Landrigan, P., Fuller, R., Acosta, N., Adeyi, O., Arnold, R., Basu, N. *et al.* (2018). The Lancet Commission on Pollution and Health. *Lancet* 391(10119): 462–512.

Leopold, A. (1949). *A Sand County Almanac*. New York: Oxford University Press.

Lerner, H. & Berg, C. (2017). A comparison of three holistic approaches to health: One Health, EcoHealth, and Planetary Health. *Frontiers in Veterinary Science* 4.

Lewis, S. (2006). Tropical forests and the changing earth system. *Philosophical Transactions B* 361(1465): 195–210.

Lewis, S. & Maslin, M. (2015). Defining the Anthropocene. *Nature* 519: 171–180.

Listza, A. & Wolbring, G. (2018). EcoHealth and the determinants of health: perspectives of a small subset of Canadian academics in the

EcoHealth community. *International Journal of Environmental Research in Public Health* 15(8): 1688.

Littlemore, R. (2006). A letter to Oklahoma Senator James Inhofe. Desmog. https://www.desmogblog.com/a-letter-to-oklahoma-senator-james-inhofe.

Lloyd-Smith, J., George, D., Pepin, K., Pitzer, V., Pulliam, J., Dobson *et al.* (2009). Epidemic dynamics at the human-animal interface. *Science* 326(5958): 1362–1367.

Lucas, R., Wheeler, D., & Hettige, H. (1992). Economic development, environmental regulation and the international migration of toxic industrial pollution: 1960–1988. In: P. Low (Ed.), International Trade and the Environment. Discussion Paper No. 159. Washington, D.C.: World Bank.

Mertens, F., Mergler, S.-C., Passos, C. & Lucotte, M. (2005). Network approach for analyzing and promoting equity in participatory ecohealth research. *EcoHealth* 2(2): 113–126.

Meyer, R. (2018). Geology's timekeepers are feuding. *Atlantic.* https://www.theatlantic.com/science/archive/2018/07/anthropocene-holocene-geology-drama/565628/.

Michel, J., Owens, E., Zengel, S., Graham, A., Nixon, Z., Allard, T. *et al.* (2013). Extent and degree of shoreline oiling: Deepwater Horizon oil spill, Gulf of Mexico, U.S.A. *PLoS ONE* 8(6): e65087.

Milloy, S. (2015). Freaky-frog fraud. *Fox News.* https://www.foxnews.com/story/freaky-frog-fraud.

Moore, J. (2016). *Anthropocene or Capitalocene? Nature, History, and the Crisis of Capitalism.* Oakland, CA: PM Press.

Morens, D., Folkers, K. & Fauci, A. (2004). The challenge of emerging and re-emerging infectious diseases. *Nature* 430(6996): 242–249.

MSNBC. (2015). Sen. James Inhofe's "snowball" speech, annotated. No longer available online.

NAS. (2010). *Hidden Costs of Energy: Unpriced Consequences of Energy Production.* Washington, D.C.: National Academies Press.

Novacek, M. (2001). *The Biodiversity Crisis: Losing What Counts.* New York: The New Press.

Oreskes, N. (2004). The scientific consensus on climate change. *Science* 304(5702): 1686.

Orwell, G. (1953). *Such, Such Were the Joys.* New York: Harcourt Brace Jovanovich.

Paine, R.T., Tegner, M.J., & Johnson, E.A. (1998). Compounded perturbations yield ecological surprises. *Ecosystems* 1(6): 535–545.

Parkes, M. (2012). Diversity, emergence, resilience: guides for a new generation of ecohealth research and practice. *EcoHealth* 8(2): 137–139.

Perman, R. & Stern, D. (2003). Evidence from panel unit root and cointegration tests that the environmental Kuznets curve does not exist. *Australian Journal of Agricultural and Resource Economics* 47: 325–347.

Ramirez-Llodra, E., Tyler, P., Baker, M., Bergstad, O., Clark, M. & Escobar, E. (2011). Man and the last great wilderness: human impact on the deep sea. *PLoS ONE* 6(8): e22588.

Rapport, D., Costanza, R. & McMichael, A. (1998). Assessing ecosystem health. *Trends in Ecology and Evolution* 13(10): 397–402.

Rathje, W. & Murphy, C. (2001). *Rubbish! The Archaeology of Garbage.* Tucson, AZ: University of Arizona Press.

Rockefeller Foundation. (2020). Safeguarding Human Health in the Anthropocene Epoch. https://www.rockefellerfoundation.org/report/safeguarding-human-health-anthropocene-epoch/.

Roger, F., Caron, A., Morand, S., Pedrono, M., de Garine-Wichatitsky, M., Chevalier, V. *et al.* (2016). One Health and EcoHealth: the same wine in different bottles? *Infection Ecology & Epidemiology* 6: 30978.

Rohr, J. (2018). Atrazine and amphibians: a story of profits, controversy, and animus. In: D. DellaSala & M. Goldstein (Eds.), *The Encyclopedia of the Anthropocene*, vol. 5, pp. 141–148. Oxford: Elsevier.

Singer, M. (2014). Climate change and planetary health. *Somatosphere.* http://somatosphere.net/2014/climate-change-and-planetary-health.html/.

Singer, M. (2018). *Climate Change and Social Inequality: The Health and Social Costs of Global Warming*. Abingdon: Routledge.

Steffen, W., Sanderson, A., Tyson, P, Jäger, J., Matson, P. & Moore, B. (2005). *Global Change and the Earth System: A Planet Under Pressure*. Berlin: Springer-Verlag.

Steffen, W., Grinevald, J., Crutzen, P. & McNeill, J. (2011). The Anthropocene: conceptual and historical perspectives. *Philosophical Transactions of the Royal Society A* 369(1938).

Steffen, W., Rockström, J., Richardson, K., Lenton, T., Folke, C., Liverman, D. *et al.* (2018). Trajectories of the earth system in the Anthropocene. *Proceedings of the National Academy of Sciences U.S.A.* 115(33): 8252–8259.

Stern, D. (2004). The rise and fall of the environmental Kuznets curve. *World Development* 32(8): 1419–1439.

Sulpis, O., Boudreau, B., Mucci, A., Jenkins, C., Trossman, D., Arbic, B. *et al.* (2018). Current CaCO3 dissolution at the seafloor caused by anthropogenic CO2. *Proceedings of the National Academy of Sciences U.S.A.* 115(46): 11 700–11 705.

Tucker, M., Böhning-Gaese, K., Fagan, W., Fryxell, J., Van Moorter, B, Alberts, S. *et al.* (2018). Moving in the Anthropocene: global reductions in terrestrial mammalian movements. *Science* 359(6374): 466–469.

UCL News. (2009). Climate change is the biggest health threat of the 21st century. https://www.ucl.ac.uk/news/2009/may/climate-change-biggest-health-threat-21st-century.

U.N. Environment Program. (2019). Our Planet: Healthy Planet, Healthy People. https://www.unenvironment.org/resources/report/our-planet-healthy-planet-healthy-people.

Van Alstine, J. & Neumayer, E. (2010). The environmental Kuznets curve. In: Gallagher, K. (Ed.), *Handbook on Trade and the Environment*. Cheltenham: Edward Elgar.

Vitousek, P., Mooney, H., Lubchenco, J., & Melillo, J. (1997). Human domination of the Earth's ecosystems. *Science* 277: 494–499.

Waltner-Toews, D. (2009). Eco-Health: a primer for veterinarians. *Canadian Veterinary Journal* 50: 519–521.

Warren, J. (2016). One of the most dangerous beats on journalism. *Vanity Fair*. https://www.vanityfair.com/news/2016/09/one-of-the-most-dangerous-beats-in-journalism-revealed.

Waters, C., Zalasiewicz, J., Summerhayes, C., Barnosky, A., Poirier, C., Gałuszka, A. *et al.* (2016). The Anthropocene is functionally and stratigraphically distinct from the Holocene. *Science* 351(6269): aad2622.

Watts, N., Adger, N., Agnolucci, P., Blackstock, J., Byass, P., Cai, W. *et al.* (2015). Health and Climate Change: Policy Responses to Protect Public Health. http://discovery.ucl.ac.uk/1479102/1/Colbourn_final%20commission%20-%20word.pdf.

Weiss, T. (2013). *The Ecological Hoofprint: The Global Burden of Industrial Livestock*. London: Zed Books.

Whitmee, S., Haines, A., Beyrer, C., Boltz. F., Capon, A., de Souza Dias, B. *et al.* (2015). Safeguarding human health in the Anthropocene epoch: report of The Rockefeller Foundation–Lancet Commission on planetary health. https://www.rockefellerfoundation.org/wp-content/uploads/PIIS0140673615609011.pdf.

WHO. (2013). Global Action Plan for the Prevention and Control of Non-communicable Diseases 2013–2020. Geneva: World Health Organization.

WHO. (2018). Climate Change and Health. World Health Organization. http://www.who.int/news-room/fact-sheets/detail/climate-change-and-health.

Wilcox, B., Aguirre, A., Daszak, P. Horwitz, P., Martens, M., Parkes, J. *et al.* (2004). EcoHealth: a transdisciplinary imperative for a sustainable future. *EcoHealth* 1(1): 3–5.

Wilcox, B., Aguirre, A. & Horowtz, P. (2012). ECOHEALTH: connecting ecology, health, and sustainability. In: A. Aguirre, R. Ostfeld & P. Dsazak (Eds.), *Conservation Medicine:*

Applied Cases of Ecological Health, pp. 17–32. Oxford: Oxford University Press.

Woolhouse, M. & Gowtage-Sequeria, S. (2005). Host range and emerging and reemerging pathogens. *Emerging Infectious Diseases* 11: 1842–1847.

Yang, B., Qian, Z., Li, S., Chen, G., Bloom, M., Elliott, M., *et al.* (2018). Ambient air pollution in relation to diabetes and glucose-homoeostasis markers in China: a cross-sectional study with findings from the 33 Communities Chinese Health Study. *Lancet Planetary Health* 2(2): e64–e73.

Zalasiewicz. J., Waters, C., Williams, M., Barnosky, A. Cearreta. A., Crutzen, P. *et al.* (2015). When did the Anthropocene begin? A mid-twentieth century boundary level is stratigraphically optimal. *Quaternary International* 383: 186–203.

Zalasiewicz, J., Waters, Colin, N., Wolfe, A., Barnosky, A., Cearreta, A. *et al.* (2017). Making the case for a formal Anthropocene Epoch: an analysis of ongoing critiques. *Ingenta* 50(2): 205–226.

Zinsstag, J., Schelling, E., Waltner-Toews, D. & Tanner, M. (2010). From "one medicine" to "one health" and systemic approaches to health and well-being. *Preventive Veterinary Medicine* 101(3–4): 148–156.

Zinsstag, J., Mackenzie, J., Jeggo, M., Hevmann, D., Patz, J. & Daszak, P. (2012). Mainstreaming one health. *EcoHealth* 9(2): 107–110.

Part 1

Impact on ecosystems

2

Intricacies of ecosystems

[E]cology, despite its fragmentary progress beginning with the environmental relations of plant life, is a study of the entire ecosystem. Of this this system, man is not just an observer and irresponsible exploiter but an integral part, now the world's dominant organism. He has come into the system and survived thus far by the bounty of that system plus his own marvelous power of adjustment. Even so, the historical record is replete with his failure.

Paul Sears (1964, p. 4)

2.1 The nature of nature and the pathway to understanding

This chapter examines the nature of nature, specifically the interrelationships among species in an environmental context, as well as biodiversity and environmental regional and planet-wide structures (e.g., air currents, hydraulics). The discussion begins with an introduction to the field of ecology and its conception of ecosystems in the world. The objective is to establish the deep connectivity in biotic/abiotic systems and identify threats to ecosystem sustainability, including the impacts these can have on species well-being and survival. This exploration begins with a historic examination of the near and distant roots that fed the development of the science of ecology.

2.2 Developing a historic understanding of ecology and ecosystems

The term "ecology" is derived from the ancient Greek *oikos*, meaning "house" or "environment." The field of ecology, as a specified domain or discipline of academic research, was defined in 1906 by Ernest Hackel, a German biologist, naturalist, and artist. He used the term "ecology" to label the rigorous science that seeks to comprehend the relations of organism to their surrounding environment, which includes "all conditions for existence" (quoted in Schulze *et al.* 2002, p. 1). As this meaning indicates, living organisms—even semi-living entities like viruses, and certainly all plants and animals, including humans—cannot live without resources provided by the physical world immediately around them. The deeper roots of ecological study trace back to the natural history observations of revered ancient Greek philosophers like Herodotus, Aristotle, and Hippocrates, and even further back to the need of all indigenous people to know their environment, as life itself depended upon it.

2.2.1 Ancient Greece

Some of the earliest Western conceptions of the environment, such as ideas about nature being characterized by balance and internal

Ecosystem Crises Interactions: Human Health and the Changing Environment, First Edition. Merrill Singer.
© 2021 John Wiley & Sons, Inc. Published 2021 by John Wiley & Sons, Inc.

regulation, have been attributed to Herodotus, though the word "ecology" does not appear in ancient Greek philosophical literature. Nonetheless, Herodotus provided one of the earliest written accounts of the relationship of mutualism, based on his observation of the way basking Nile crocodiles would open their mouths to give sandpipers safe access to remove leeches, which provided food for the birds and oral hygiene for the crocodiles (noted in Egerton 2001). Aristotle was an early influence on the philosophical development of ecology. He had a lifelong fascination with nature studies, and with his student Theophrastus, he carried out extensive observations on plant and animal migration, biogeography, physiology, and behavior (Hughes 1975, 1985).

Theophrastus had a long and productive life as a botanist, living to the age of 85. Following Aristotle's death, he succeeded him as a director of the Lyceum in Athens, a position he held for 35 years. Founded by Aristotle, the Lyceum was called the Peripatetic School (from the Greek *peripatos*, "stroll"), apparently after Aristotle's habit of meandering through its tree-filled groves discussing philosophy, mathematics, and rhetoric with his eager students. The Lyceum embraced an organized and rigorous approach to the study of nature and is considered by some to have pointed the direction for the modern scientific method. Theophrastus arrived there at the young age of 17 and became a dedicated student and disciple of both Aristotle and his approach. Considered "the father of botany," he conducted experiments and made and recorded observations on plant propagation, classification, physiology, and behavior. Unlike his teacher, however, he believed that animals could reason and therefore concluded that it was unethical to eat meat. Thus, he became a vegetarian and devoted his life to the study of vegetation.

As Hughes (1975) indicates, an underlying view of nature at the Lyceum, which became particularly pronounced in Theophrastus' writings on plants, was an understanding of the physical world "as the theatre of reason," a view "which made the beginnings of ecological thought possible." In his various writings, especially *De Historia Plantarum* (The History of Plants) and the first five books of *De Causis Plantarum* (The Causes of Plants), Theophrastus (Long *et al.* 1985) describes plants as having three aspects. The first is their inner nature, which modern scientists trained in genetics would call their genotype, but he called *telos*, which means purpose or goal. The second aspect is their environment, including soil, moisture, temperature, exposure to the elements, wind, and elevation. Theophrastus believed the environment might support or act against a plant's inner nature. Unlike later writers on the nature of nature, Theophrastus did not prescribe environmental determinism, but instead held an interactionist understanding of the distinctive features of plants and the distinctive features of their environments. Plants, he recognized, were adapted to local conditions, or what he called "mutation according to place." Consequently, he believed plants must be studied in their normal setting, in what modern ecologists would call their natural ecosystem. Part of the environment of a plant, Theophrastus realized, was other plants. He noted that olive and myrtle trees, like pine and bay trees, tended to grow together, and he observed that close-dwelling species might either compete with or benefit each other. Legumes, he understood, enrich the soil with nutrients absorbed by other plants. Ivy, however, kills the trees it grows on (Fig. 2.1). Moreover, aware of Aristotle's discussion of interspecies cooperation, he noted examples of birds contributing to the welfare of plants. Jays, for example, bury acorns, some of which sprout to grow new oak trees, which support jay populations. Thus, he was aware of the features of ecology now known as symbiosis and parasitism.

Theophrastus' final aspect of plants, one critical to the issues addressed in this book, is human agency, "which may have a goal quite at variance with both the plant and the environment" (Hughes 1985, p. 296). Unlike Aristotle, Theophrastus did not think plants

Fig. 2.1 Ivy growing on a tree. *Source:* Free-Photos/Pixabay.

existed to serve humans. In his *Metaphysics*, he asserted that:

> We must try to find a certain limit . . . both to final causation and to the impulse to the better. For this is the beginning of the inquiry about the universe, that is, of the effort to determine the conditions on which real things depend and the relations in which they stand to one another. (Ross & Forbes 1967, p. 92)

Moreover, he argued that the purpose of a plant has nothing to do with humans, but rather is just what plants will do without human agency: grow and reproduce.

Interestingly, Theophrastus realized that humans could have an impact, through acts like drainage, on regional climate. He stated that:

> The country around Larisa in Thessaly [in ancient and modern Greece], where formally, when there was much standing water and the plain was a lake, the air was thicker and the country warmer; but now that the water has been drained away and prevented from collecting, the country has become colder and freeing more common. In proof the fact is cited that formally there were fine tall olive trees in the city itself and elsewhere in the country, whereas now they are found nowhere, and the vine were never frozen before but freeze now often. (quoted in Hughes 1985, p. 303)

Further, he documented that deforestation could influence climate, noting that on the island of Crete, once heavily forested, extensive commercial logging and timber export had enhanced the severity of the wind, to the point that agriculture became unsustainable for many years. In Philippi, a city in eastern Macedonia, he asserted that the combined effects of drainage and deforestation—a case of ecocrises interaction—moderated the climate by allowing wind to blow away freezing air. These observations indicate that human awareness of anthropogenic climate change is over 5000 years old. In Hughes' (1985, pp. 304–305) view:

> If [Theophrastus] failed as a consistent ecologist, it was in concentrating his attention on individual species. Unlike modern ecologists, he did not emphasize the complex net of relationships in biotic communities . . . But it is hard to criticize him too severely, because he was the first to set out on the journey of inquiry that we call ecology.

In the millennia since Theophrastus lived, the Mediterranean environment he knew and studied has changed greatly. As Anne Ehrlich & Paul Ehrlich (1980) observe:

> The Greeks inherited a land covered by rich stands of oaks, pines, and other

> trees with thick, drought-resistant leaves called a "sclerophyllous" forest, in the jargon of [modern] plant ecologists. But, as the Greek population expanded, it progressively destroyed the forests for firewood, charcoal (needed in firing pottery and other industrial processes), and lumber. The great trees were often burned by accident, too, or as part of a military operation, or simply to create more open pastureland. Soil erosion on the slopes of the rugged Greek hills helped prevent reforestation as did grazing and browsing animals, which killed the seedlings before they could establish themselves. Especially prominent in the latter role were goats . . . Our overall impression was that the once rich area is now a badly deteriorated land inhabited by relatively impoverished peoples who, today, are partly dependent for their survival on the influx of tourists coming to see the physical monuments of past civilizations.

The Lyceum, for example, was excavated and opened for public viewing in 2014. Tourist websites encourage a visit to the free site, to stroll peaceful surroundings filled with fragrant plants (many species of which were planted there in Aristotle's time) and view the remains unveiled by archeologists. Meanwhile, in places like Crete, less than an hour's flight from Athens, sea level rise—due to the combined ecocrisis effects of climate change and extensive coastal construction—is beginning to challenge the sustainability of the beach-based tourist industry (Centre for Climate Adaptation 2018).

2.2.2 Indigenous environmental knowledge

All societies through time and location must contend with the opportunities and challenges presented by their environments. This requires the development of environmental knowledge and sustaining social practices. Failure to overcome challenges can lead to social collapse, leading in turn to reversion to less complex social organizations that put fewer demands on environmental resources. Failure can also lead to obliteration and complete depopulation. A case of social collapse and simplification is seen in Classic Mayan society. The indicators of this collapse are found in the archeological record, most notably in the abandonment of Maya cities like Palenque, Copán, Tikal, and Calakmulin, located in the southern Maya lowlands of Guatemala, Honduras, and Mexico, which were left to be swallowed up by the re-expansion of surrounding forests between the 8th and 9th centuries. After the collapse, there was a significant decline in large-scale architectural construction in the area, though Mayan communities continued to persist in the southern lowlands and Mayan people still inhabit the region today. There has been considerable debate among archeologists about the causes of the collapse of Classic Mayan society. Suggested reasons include a growth of population that could not be supported by the local environment given Mayan farming technology, epidemics that decimated the population, military conflict between competing city-states, deforestation, loss of plant and animal biodiversity, and catastrophic environmental events, such as climate change and intense drought (Hodell *et al.* 1995; Webster *et al.* 2007). Some of these factors may have been connected. War, for example, can be driven by the demand for environmental resources, while deforestation can lead to a loss of biodiversity (Durham 1976; Giam 2017). It is likely that the knowledge and practices needed to meet the environmental challenges faced by the heavily populated city-states of Classical Mayan society were inadequate, and that this played an important role in the collapse. This example underlines the importance of investigating a society's environmental knowledge/beliefs and their practical utility and contribution to sustainability in the encounter with environment opportunities and threats and changing conditions.

Robert Paine (2005, p. 280), coiner of the ecosystem concept "keynote species," proposed that:

> ecology with its roots in "natural history," whether that was stalking prey, keeping special mushroom (e.g., morel or truffle) sites secret, or the best spots to fish, was initially an individualistic endeavor. Deep knowledge of parts of the natural world was a requisite for survival.

However, precisely because it is a requisite for survival, deep knowledge of the natural world has always been social, not individual, as humans have survived as a species from our origin into the present because we live in interactive and mutually supportive groups.

Eriksen & Adams (2010, p. 1) note that various "terms have been used to describe the collective environmental knowledge and the shared cultural resource practices of indigenous peoples, including *indigenous environmental knowledge*, *traditional ecological knowledge*, and *ethnoecology*." Hardesty (1977, p. 291) defines ethnoecology, for example, as "the systems of knowledge developed by a given culture to classify the objects, activities, and events of its universe." Brosius *et al.* (1986, p. 187) bifurcate indigenous knowledge into two categories:

> On the one hand, there is knowledge that is inherited from generation to generation and is generally shared by most of the members of the society. This cultural information is often passed on as folk wisdom in the context of folktales or is preserved and transmitted in the context of rituals and various religious ceremonies. The second category includes information gained through individual experience-empirical observations made by individual[s] . . . during the course of [life] activities. The two categories of indigenous knowledge are not mutually exclusive. Information collected by an individual may become knowledge that is shared by members of the group through time.

There is then in cultural environmental knowledge an ongoing flow of information between individuals in a social group, some of which over time is inscribed into the cultural repertoire of the group and is transmitted to new generations as agreed upon reality. It should be stressed that not all knowledge in a group is equally or fully shared, as there may be some that is mostly retained by subgroups (e.g., traditional healers, mothers) within the wider society (e.g., knowledge about plants used to make traditional medicines, how to prepare them, and where they can be found). Moreover, traditional knowledge about the environment may contain entities not recognized by outsiders, such as spiritual beings and sacred places.

Human knowledge of the environment is conditioned by language and the ability to have complex thoughts entwined with memories, values, and cultural norms. Consequently, environmental perceptions are filtered through cultural perceptions. This point is emphasized by Vayda & Rappaport (1968) in the distinction they draw between the "operational environment," which they define as the sum total of all environmental features, whether or not they are locally comprehended, and the "cognized environment," which is the environment as it is perceived and understood within a social group. Notably, some features of an operational environment may affect a group even if its members are unaware of their existence (e.g., pathogens, disease vectors, climate change). Because of cultural filtering, people of different cultural backgrounds can view the same operational environment differently.

The study of indigenous environmental knowledge has a long history. One of the earliest efforts was conducted by Carl Linnaeus, the renowned 18th-century Swedish botanist, zoologist, and physician, and the creator of the binomial system we still use to give species their Latin names (e.g., *Homo sapiens*). While in his mid-20s, Linnaeus undertook a field study of the indigenous Sami of northern Sweden. The primary motivations for his six-month foray by foot, boat, and horseback to the northernmost Swedish province of Lapland were to discover new plants and animals and, if possible, find valuable minerals like silver, copper, and iron. At the same time, he was curious about the lifestyle and customs of the native Sami people, a reindeer-herding society of semi-nomads who traditionally followed their livestock across the tundra-covered areas of Sweden, Norway, Finland, and Russia in search of available grazing foods like mosses and lichen. Based on his journey, Linnaeus described about 100 formerly unidentified plants and wrote about them in his book, *Flora Lapponica* (Frängsmyr *et al.* 1983). In his grant application to the Royal Scientific Society, he described his interest in studying the plants of Lapland but additionally, if vaguely, noted that he also hoped to learn about the plant-based home remedies used by the Sami and what kinds of food they resorted to in times of scarcity. In fact, Linnaeus had numerous questions he hoped to answer about the Sami, including ones concerning their common foods and means of food acquisition, material culture (including housing, clothing, and handicraft), and belief system.

In the early days of his expedition, Linnaeus met a schoolteacher, Per Fjellström, who had learned the Sami language and gave him a long list of Sami names for plants, mammals, birds, and fish, a clear indicator that the Sami were quite acquainted with the other species with which they shared their environment, garnered from their past and ongoing empirical observations. Based on his time among them, Linnaeus was able to describe elements of the technologies the Sami constructed from their environment and used to survive in it. His writings provided systematic information about their sewn boats (used in fishing by forest-dwelling communities), their clothing, some of the traps they used to catch squirrels and forest birds, the harnesses they used with their reindeer when traveling with sledges, and their huts, tents, and storage structures. Moreover,

> he had ample opportunity to study the reindeer. He saw whole herds; saw how scared the animals were of the gadfly and what these insects did to their skins. He saw the milking of reindeer cows and the castration of bulls and watched one or two slaughterings . . . He also noted down different Sami words to denote the reindeer according to their age. He [also] got information about the tanning of leather [using tree bark]. (Zorgdrager 2008, p. 62)

While not a rigorous ethnography by today's standards, Linnaeus' account of the ecology of the Sami suggests both Sami ken awareness of their surroundings and their diverse interactions with the biotic and abiotic components therein. Lacking, however, is a textured holistic picture of Sami understanding of their world and its biotic and abiotic makeup; that is, their indigenous, norm- and value-laden map of reality.

The earliest (and most of the subsequent) systematic research on indigenous environmental knowledge was conducted by anthropologists. This occurred because of the turn toward a field-based orientation in the disciple. Anthropologists began adopting an extended fieldwork approach in the early 20th century. This development is generally credited to the Polish anthropologist Bronislaw Malinowski, who advocated that to understand another society it was necessary to live among its people in their traditional setting, learn their language, engage them in constant dialogue and increasingly informed interviewing, and participate to the degree possible in their everyday activities. While other anthropologists had spent time in the field before him, as Wax (1972, p. 3) asserts, Malinowski's research among the Trobriand Islanders of the western Pacific during the years 1916–1918 "yielded a series of epochal volumes which revolutionized the content and practice of anthropology."

As part of this transition, anthropologists came to distinguish inside and outsider viewpoints, labeling them the emic and etic points of view, respectively. These terms were introduced by the linguist Pike (1967), who derived them from the suffixes of the linguistic concepts "phonemic" and "phonetic," the former referring to any sound unit of significance in a particular language and the latter to the system of notations scholars use to represent these vocal sounds (McCutcheon 1999). An earlier but related advance by anthropology was the embrace of cultural relativism as a foundational axiom. This is the idea that the beliefs, values, norms, and behaviors of the people anthropologists study in the field should be understood in terms of their own culture, rather than be judged against criteria developed by another, different culture—a position often attributed to the writings of Franz Boaz (1887).

Concerned with understanding the emic perspective, and aware of the critical role of language in this endeavor, some anthropologists began to inquire about indigenous labeling and classification of plants and animals in the local environment, and discovered extensive assemblages of environment knowledge among indigenous people. Exemplary is the work of the British/Australian social anthropologist Ralph Bulmer. Based on his ethnographic research among the Karam, a horticultural tribe living in Kaironk Valley in highland New Guinea, Bulmer (1974, p. 12) found "a vast knowledge concerning the integration of the plant and animal communities—of the topographic, soil and climatic conditions required by wild as well as cultivated plants, of the kinds of plants and their parts which provide food or refuge for different kinds of animals, of which animals prey upon which other animals, of the role of birds and mammals in the propagation and dispersal of certain plants." Similarly, the American anthropologist Douglas Oliver (1989, pp. 275–276), writing in detail about the food and food-getting activities of indigenous island-dwellers in the Pacific, concluded that:

> behind their various practices of gardening, gathering, fishing, and hunting were rich accumulations of factual knowledge about the plants and animals they produced or otherwise obtained—knowledge not just about the individual items they produced or obtained, [but] necessarily, about the ways those items were interrelated with other aspects of the physical environment . . . [Moreover] it would be a mistake to conclude that Islanders' knowledge of their plant and animal universe was limited to it relevance to eating, or to apparel, or shelter, or ornamentation, or medicine, or to other practical uses alone . . . [They] did not close their eyes and ears to the living things around them simply if the latter served no practical value.

The latter point is also emphasized by Van Der Ploeg & Van Weerd (2010, p. 127) in noting that in their research among the Agta of the Northern Sierra Madre Natural Park area of the Philippines, they found that "[i]ndigenous knowledge of birds is not limited to economically important species, as is often assumed. Agta hunters are familiar with most discernible species."

While anthropologists like these recognized that the cultures of different societies label and group the elements of nature differently, as exemplified by the many more words for types of snow among the Inuit people of Canada than are to found in standard English, as asserted by Boaz (1911), they took it for granted that material components of the environment existed independent of culture. In short, "'[c]ultures' may differ, but nature does not . . . People see the world in different ways, but the world is still the world" (Heywood 2017). At the turn of the 21st century, however, some anthropologists began to question this bedrock assumption—a shift in perspective that has been labeled the "ontological turn" (Viveiros de Castro 2004; Pedersen 2011; Holbraad 2012). This approach, while critiquing the etic perspective (which subordinates the insider perspective to the conceptions of the outside observer), and asserting that anthropology has not been sufficiently culturally relativist in its accounts of other peoples, "proposes that worlds, as well as worldviews, may vary" (Heywood 2017). Heywood (2017) provides a clarifying example. If, during fieldwork, a member of the group an anthropologist is studying says that that the tree they are pointing to is really a spirit, should the anthropologist record this as a cultural belief? To do so, Heywood maintains, violates the individual's view that the object in question *is* a spirit, not that they believe it to be a spirit: "Calling it a belief, as a number of anthropologists writing before the ontological turn have pointed out, is both to mislabel it and to call it mistaken without actually saying so" (Heywood 2017). I encountered this problem during research in Haiti when a man I was talking to, a *houngan sur pwen* (lower-level male priest) in the Vodun tradition, told me that at night he turns into a bicycle and spies on his enemies. To interpret this statement as a belief, from the vantage of the ontological turn, when to my interlocker it was a statement of fact, could only be achieved by imposing my understanding of reality on to his understanding of reality.

The ontological turn, if it gained wide adherence within the discipline, would constitute a radical paradigm shift on the scale of a scientific revolution, including with regard to how we think about nature, interspecies interaction, and the human place in the world. But, as a fair number of anthropologists have pointed out, there are grave problems with it. Bessire & Bond (2014, p. 446) argue that:

> ontological anthropology is incapable of accounting for those disruptive beings [e.g., mining, petroleum, and logging company executives and other polluting elites] and things [e.g., industrial toxins, rising sea levels and planetary temperatures, disease vectors] that travel between ontologies. Today, it is not only pollution but also logging, mining, agriculture, and oil extraction that routinely impinge on the premier sites of ontology. Ontological anthropology avoids recognizing such confrontations, in part, by pressing all analysis of materiality ever further into sacred materials

An indigenous interlocutor may know a tree is a spirit, but that will not stop a logging company from cutting it down, slicing it into boards, and shipping it overseas to build structures in a wealthy country. Nor will it change the fact that the loss of such spirits/trees fuels climate change, which is now causing droughts and killing large stretches of rainforest. Moreover, are we to accept as a credible reality the statement by Donald Trump—one echoed by many like-minded politicians and conservative think tank writers and spokespersons who are members of the tribe of climate change deniers—that "Global warming is a total, and very expensive, hoax!" (quoted in Schulman 2017)? While advocates of the ontological turn are correct in demanding that we be mindful in writing about the realities of other peoples as well as the limits of all truth claims, there are severe costs to not striving for an objective understanding of the environment.

Another example that supports this conclusion involves the Eskimo people of Alaska. There is a body of ethnographic and other studies suggesting that the Eskimo have acquired traditional knowledge that they use to avoid diseases, such as botulism and trichinosis, that could be transmitted by unsafe handling of foraged meat, and otherwise to cope with the challenges of the Arctic environment. All of this knowledge, however, was inadequate to handle the discovery of a radioactive nuclear waste dump that had been buried beneath the ground in northwest Alaska by the

Atomic Energy Commission in the 1950s. People expressed uncertainty, confusion, and worry (Cassady 2007) knowing that the dump site was crossed by caribou harvested by Eskimo hunters. In fact, Cassady (2007, p. 91) notes that it was impossible for her to express "the magnitude of the vulnerability, resentment and confusion" that circulated in local communities. But despite this concern, the people Caasady interviewed expressed problematic ideas such as the belief that you can cook radiation out of contaminated meat or remove it by aging the meat. In short, there are dangers in treating indigenous understandings of the world as if they are unobjective and do not matter, but there are also dangers in assuming that there are not wider forces at work beyond local or self-interested understandings. Moreover, the involvement of communities and the utilization of local environmental knowledge have been shown to expand scientific understanding in a number of studies, including one on fish contamination in Brazil (Silvano & Begossi 2016) and one on changes in marine environments in the Canadian Arctic (Berkes *et al.* 2007). But, as Caron-Beaudoin & Armstrong (2019, p. 59) state, "there is a remarkable lack of meaningful consultation" regarding industrial development with affected indigenous communities.

2.3 Modern ecology

As noted, the modern science of ecology is one of the key fields concerned with developing an objective understanding of the complex interconnections that make up the environment. Of no less interest are the limits of interconnection. As the urban ecologist Liam Heneghan (2015), using the classic "butterfly effect" illustration of connectedness, states: "if the dominoes line-up and the circumstances are just so, a butterfly's wing beat over the Pacific may hurl a typhoon against its shores, but more often than not such lepidopterous catastrophes do not come to pass." Still, a lesson of ecology is that small changes in initial conditions can, at a certain point, lead to drastic outcomes—an issue taken up in Chapter 10.

Ecology's immediate scholastic roots lie in the efforts of 19th-century biologists to understand the number of species, their distribution around the planet, and the nature of relations among them. Questions about such issues motivated wide-eyed and curious naturalists of the era like Henry Walter Bates, Alfred Russel Wallace, and Charles Darwin (Fig. 2.2) to venture out and witness the species of the world in their local habitats. During his nearly 5-year voyage as ship's naturalist aboard the HMS *Beagle*, Darwin filled many notebooks and letters home with careful observations on animals, plants, and geology. His record of the distribution of species in the various places he visited in South America, including the Galápagos Islands, would provide him with the insights he used for the rest of his life in framing his understanding of biological evolution.

Over the years since Darwin and his peers carried out their groundbreaking studies,

Fig. 2.2 Charles Darwin. *Source:* National Portrait Gallery.

ecology has developed several core concepts to help frame this understanding.

2.3.1 Ecosystems

Central to the conceptual repertoire of ecology is the notion that the natural biotic/abiotic world is organized into overlapping ecosystems. While it is useful in building an understanding of the environment to demarcate and even label ecosystems, like all natural systems they actually "are open systems without clear boundaries" (Langmuir & Broecker 2012, p. 6). A tide pool, for example, can be described as an intertidal ecosystem consisting of animals (e.g., sea anemones, starfish, crabs, sponges, fish, mussels, barnacles, limpets), plants (e.g., seaweed, sea grass, rockweed, sea palm, sea lettuce), water, rocks, and sand (Fig. 2.3). However, tide pools are not isolated from the ebb and flow of the ocean, the gases of the atmosphere, or the rays of the sun.

The term "ecosystem" is used specifically to refer to a community of directly interacting living organisms and nonliving elements such as air, water, rocks, and soil. In an ecosystem, interactions occur among organisms (across all phyla) and between organisms and other environment components. Central to the relationships among these ecosystem components is the flow of matter and energy. A food chain, for example, involves the flow of matter and energy from autotrophs (e.g., plants) to heterotrophs (e.g., herbivores and carnivores) and, eventually, to decomposers—a flow that occurs in all ecosystems.

2.3.1.1 Ecosystem synergy

The biotic and abiotic components of an ecosystem are synergistic and reinforcing. Acreman *et al.* (2011) provide a good illustration of these kinds of interactions based on research on Somerset Levels and Moors (SLM), a coastal plain and wetlands located in the

Fig. 2.3 Tide pool. *Source:* rreeths/Pixabay.

county of Somerset in southwest England. The SLM contains a rich biodiversity of invertebrates, plants, migratory and local birds, fish, amphibians, reptiles, and several mammalian species. Archeological remains indicate human presence since the Paleolithic. The area has been mined for its rich peat soils of decomposing organic material—used for fuel and fertilizer—at least since Roman times, but intensely so since the industrialization of extraction in the 1950s. The result is considerable ecosystem damage and peat loss.

Freshwater is ever-present in the SLM because of its location near the outlet of a river basin, low-lying topography, permeable peat soils, underlying aquafer, and rainfall. The resulting wet environment "supports bird [and other animal] life that maintains biological diversity, attracts tourists, protects archaeological artefacts and reduces CO_2 emissions; raising water levels to or above the ground leads to net greenhouse gas uptake by the wetland" (Acreman *et al.* 2011, p. 1543). In light of climate change, it is notable that peat is the largest and most efficient terrestrial store of CO_2. On average, peat wetlands sequester 10 times more CO_2 per acre than any other ecosystem. When peat is mined, however, it is exposed to air. Carbon contained within it (from decomposed biota) combines with oxygen in the air to produce CO_2, which is emitted into the atmosphere (Dunn & Freeman 2011). In this way, peatlands can be transformed by human actions from a CO_2 sink (or storage site) to a CO_2 source; that is, from a resource that restricts global warming to a source that drives it.

Another example of ecosystem synergy is provided by black bears (*Ursus americana*). These omnivores eat various animals (including insects), though most of their diet is based on diverse plant foods. Their lives revolve around food acquisition, reproduction and offspring care, and hibernation. They have a keen sense of smell that enables them to locate various food sources. Pregnant black bears give birth to their young in dens, and mother bears lactate for about 3 months while living there (Alt 1989). During this period, females do not eat or drink, yet they produce more than 850 ounces of high-fat milk. After emergence from the den—a part of the abiotic environment critical to bear survival—the composition of their milk changes as new milk production is based on the nutrients available in the diet rather than on fat stores (Oftedal *et al.* 1993; Iverson *et al.* 2001). Thus, black bears must consume large quantities of fat-rich food before denning but may have more varied diets at other times of the year, depending on the season.

Research on black bears in New Mexico (Costello *et al.* 2001) found that key food items taken from the environment and used for energy and cell construction, as well as to produce and feed cubs, were flowering plants, grasses, juniper berries, squawroot, acorns, beechnuts, gooseberries, huckleberries, and blueberries. Blueberries, for example, provide energy in the form of sugar (15 grams for each cup consumed), as well vitamins and dietary fiber, while acorns are high in fats and carbohydrates. Notably, black bears are able to extract protein from flowering plants and grasses (Eagle & Pelton 1980). They are highly efficient berry-eaters, capable of consuming up to 30 000 a day if they are available in their range; they can double their body weight in years when fruits are plentiful. They harvest berries rapidly, using their sensitive and flexible lips, and swallow them whole (Landriault *et al.* 2018). Research in Massachusetts (McDonald & Fuller 2005) found that postdenning milk produced for cubs during the spring was higher in fat (26.7% versus 18.2%) when acorns were abundant as compared with lean acorn-production years. These plants acquire CO_2 from the atmosphere and nutrients like nitrogen and phosphorous from the soil to build cells, grow, and produce leaves, nuts, and fruit. Bears were found to also consume a significant number of ants during part of the year. Both the bears and the primarily plant foods they consumed, as well as the ants, release CO_2 back into the atmosphere as their bodies carry out cellular respiration. The waste excreted by bears provides energy and building material for bacteria and fungi living in the soil. These decomposers, in turn, release molecules like nitrogen back into the soil and atmosphere. The same occurs when bears and plants die. These resources are then available for plants and the cycling of energy and matter continues. Bears also disperse the seeds of the plants they consume. Some berry seeds pass through their digestive tracts unbroken and are able to germinate where they land. Each summer, black bears spread the seeds of their favorite berries throughout their home ranges, enabling berry movement to new patches and future bear food supplies.

These two examples, of course, are simplified illustrations intended to show some of the interconnections and synergies that make up an ecosystem. They are useful because, in the case of black bears, the arrival of European colonialists in North America led to a drastic drop in their numbers and range. Habitat loss (e.g., deforestation) and extensive hunting and persecution resulted in local extinctions and the disappearance of bears across significant sectors of their traditional range by the early 1900s. Subsequently, organized recolonization by conservation officers using bears from growing neighboring populations has been successful in many areas. The lessons of this example suggest the potential for species loss and habitat destruction, as well as the possibility of ecosystem restoration. Restoration of damaged peatlands is a more daunting challenge because of the time it requires: peat only forms at the rate of about 0.04 inches per year, so a 30-foot-deep peat bed takes 9000 years to form (Chalker-Scott 2014). While degraded peatland ecosystems have been restored using sphagnum moss and mulch, CO_2 continues to be released for a number of years through bacterial respiration during decomposition of the new organic matter.

2.3.1.2 Conceptualizing ecosystems

As an analytic concept, "ecosystem" was first used by the outspoken British botanist (and strong advocate of Freudian psychology) Arthur Tansley in 1935 (New Phytologist Foundation n.d.). Tansley, known as the father of British ecology, was the first president of the British Ecological Society and served as the initial editor of its *Journal of Ecology*. He developed the concept of the ecosystem to draw attention to traceable transfers and flows of energy (e.g., from the sun) and matter (e.g., nutrients, water) within specific geographic spaces. Ecosystems vary in size and complexity, and some scientists even view Earth as a single ecosystem. Ecosystem diversity references all the different habitats that exist on Earth, such as tropical or temperate forests, hot and cold deserts, wetlands, rivers, lakes and ponds, mountains and mountain valleys, caves, islands, shorelines, undersea kelp forests, coral reefs, hot springs, and deep-ocean vents, plus many more. Matter and energy flow not only within ecosystems but between them as well (e.g., a river carries mater from land-based to water-based ecosystems), highlighting ecosystem interconnection.

Ecosystems are identifiable because they tend to maintain a natural balance among their component parts over time; that is, there is a state of dynamic equilibrium across their community of component organisms and a stability in the numbers of each species within them. This stability makes them recognizable even if identifiable changes occur (e.g., a beach ecosystem before and after a major storm). Ecological researchers have tried to develop general predictions about ecosystem community structure based on abiotic and biotic factors. Ecosystems, however, are not really static. They are impacted by various factors, including evolution, invasive species, salt intrusion, pollutants, overhunting, natural disasters, and global warming. Intrusive change in an ecosystem can be illustrated with the case of Walden Pond in Concord, Massachusetts, made famous by the environmentalist author Henry David Thoreau (2016). Walden Pond is what is called a kettle hole, a sediment-bottomed shallow body of water. Created by glacial retreat, it is anchored by a beach and a surrounding wooded area. The detailed journal notes about the flora and fauna, weather, and seasons kept by Thoreau while living at Walden Pond from 1852 to 1858 have been used by modern scientists (Primack 2014; Stager *et al.* 2018) to show that because of climate change and other human influences, levels of phytoplankton have increased there, reducing water clarity, and ice on the surface now breaks up 2 weeks earlier than during his stay.

It has been found that ecosystems with higher biodiversity (a greater array of different lifeforms and genetic variations within species) tend to be more stable and to have greater resistance and resilience when faced with disturbances and disruptive events—an idea originally suggested by Darwin (Peterson *et al.* 1998). It is not easy in the short run to specify whether an ecosystem is disturbed (and could regain balance over time) or becoming a different, perhaps more degraded or simplified biotic/abiotic arrangement. Over longer periods of observation, however, it is evident that simplification occurs.

Ecosystems are different than habitats—another environmental term frequently used in this book. Habitats are the places where specific organisms live. Consequently, they exist everywhere on the planet, because organisms live everywhere. The Grand Canyon, for example, consists of multiple habitats, with cacti and warm desert scrub species living just above the Colorado River corridor, pinyon pine and junipers living above the desert scrub up to about 6000 feet above it, and abundant ponderosa pine living at even higher elevations. Other plant species and various animals also live in each of these habitats. Habitats exist in built environments as well, such as the tracks of the New York City subway, a habitat for the brown rat (*Rattus norvegicus*), as reflected in some of the popular names of this species (e.g., sewer rat, street rat). While ecosystem disruption refers to the loss of critical relationships, habitat disruption involves the loss of resources necessary for a species to live in a particular locale.

2.3.2 Biodiversity and the multitude of species

Another key concept of modern ecology that already has been mentioned several times is biodiversity. Earth is covered in organisms of diverse size, shape, and nature, and has been for billions of years, but all of them, both living and extinct, are related. Collectively, they form an extremely complex family tree. This is called the monophyly of all known life, meaning all the organisms on Earth are descended from a universal common ancestor (Theobald 2010). From those shared roots have grown kelp and elephants, amoebae and T. rex, sequoias and humans. Of special concern in this book is the influence of humans on global biodiversity and the diversity of life within regional or local ecosystems.

Drawing on botanical records and colonial diaries, the richness of biodiversity in a regional ecology, and its constriction at human hands, is aptly portrayed in Thomas Rainer & Claudia West's (2016, p. 13) *Planting in a Post-Wild World: Designing Plant Communities for Resilient Landscapes*:

> Imagine for a moment what it must have been like for the first European colonists arriving on the shores of [Massachusetts] . . . By all accounts, the landscape they encountered was a place teeming with diversity . . . Hundreds of species of birds flew over the coastline; tens of thousands of different plants covered the forests, and billions of oysters and clams filled the estuaries . . . Just beyond the coastal plain, chestnut trees—some nine stories tall—accounted for fully half of the canopy of the Piedmont. These giants showered the ground with their mast, sustaining black bears, deer, turkey, and other creatures.

After the Europeans arrived, "forest composition changed markedly in response to human disturbance and there was a sharp increase in rates of vegetation change" as communities grew, timber cutting expanded, and deforestation accelerated (Fuller *et al.* 1998, p. 76). There are now 427 native plant and animal species that are endangered, threatened, or of concern in Massachusetts (MassWildlife 2017). At greatest risk for extinction is the Plymouth red-bellied turtle (*Pseudemys rubriventris bangsi*). While Rainer & West's (2016) account erases the presence, before Europeans arrived, of humans on the shores and in the forests of what would come to be called Massachusetts, and how these people's lives changed markedly in response to European disturbance, it does suggest the fundamental diminishment of New World diversity wrought during and since the colonial era.

The question of how many species live on Earth is a fundamental one in science. Just under 2 million are currently identified, and the pace of discovery and naming is greater now than at any time during taxonomy's 250-year history. Yet we know there is a sizeable gap between the number of living species we know about and the actual number that exist. To help quantify global biodiversity, Mora *et al.* (2011) developed the hypothesis that the scientific assignment of species to higher-level taxa such as phylum, class, order, family, and genus follows a consistent and predictable pattern that can be used to estimate the total number of species in a taxonomic group. They validated this idea using well-studied taxa and then applied their findings to all lifeforms. They concluded that there are between 5.3 and 10 million species globally, not including species of bacteria and archaea that lack a membrane-bound nucleus. Of this total, they calculate that between 2 and 2.4 million are aquatic. In other words, over 80 percent of existing species on land and over 91 percent in the ocean remain to be identified.

Lest the smallest of lifeforms be ignored, researchers from the University of Queensland in Australia (Parks *et al.* 2017) carried out a study in which they sequenced all of the DNA in samples drawn from environmental and nonhuman gastrointestinal sources in order to focus on microbial populations from under-sampled lineages. They reconstructed the individual genomes of 7280 new bacteria and 623 new archaea. Of these, about a third were new to science, warranting the creation of 17 new bacterial phylums and three new achaeal ones. These newly discovered organisms add to the nearly 70 000 bacterial and archaeal genomes held in public repositories.

But the discoveries continue. In 2014, dental researchers (He *et al.* 2015) detected a very tiny bacterium, known as TM7x, that lives on the surface membrane of a larger bacterium, *Actinomyces odontolyticus* (or XH001), and parasitizes it. Both the host and the parasite live in human saliva. After attaching itself to its victim, the newly discovered bacterium draws nutrients out of the host. Eventually, holes form in the host's surface membrane and its internal contents drain out, killing it. There is evidence that TM7x succeeds by overcoming and suppressing the immune response of XH001. Organisms like TM7x are referred to as microbial dark matter because biologists are unable to culture them in labs, which hinders their discovery and classification.

Parasitism, like that practiced by TM7x, is an extremely common animal adaptation—one of the most common—and yet, with the exception of insects that parasitize other insects (e.g., the samurai wasp, *Trissolcus japonicas*, parasitizes the eggs of the brown marmorated stink bug, *Halyomorpha halys*), parasitic species commonly have been overlooked in the construction of local ecologies and food webs (Lafferty *et al.* 2006). An important reason for this is that parasites can be difficult to quantify using the standard methods of ecological research. Moreover, there is a problematic assumption built into the theoretical models of trophic relations: that bigger animals eat smaller ones. Parasites, however, commonly eat species that are larger than themselves. Many parasites, in fact, have complex multiple-host life cycles that are deeply interwoven in food webs. This marvelous complexity is shown clearly in the case of trematodes (flatworms), such as *Euhaplorchis californiensis*, which lives in the Carpinteria Salt Marsh near Santa Barbara, CA. Kevin Lafferty, co-director of the Parasite Ecology Group at the University of California, Santa Barbara, has studied these parasites for several decades (Hechinger & Lafferty 2005). His research began with a study of the California horn snail (*Cerithidea californica*), which inhabits the muddy bottom of the salt marsh. Lafferty found that about half of the horn snails he collected contained trematodes. These parasites eat the snail's reproductive organ, making it infertile, and remain inside it for the rest of its life, feeding on it while producing trematode larvae and releasing them into the brackish waters of the mudflat. These larvae swim around until they can penetrate the gills of the small California killifish (*Fundulus parvipinnis*), from where they swim inward and attach themselves in large numbers to the fish's brain. There, they form cysts that manipulate serotonin and dopamine metabolism, causing the fish to swim in ways that render it more noticeable and hence susceptible to predation by birds such as herons and egrets. Inside the intestine of these predator birds, the trematodes mature to adult status and release eggs, which are dispersed in bird droppings into the salt marsh, where they hatch and infect horn snails, reinitiating this multi-host life cycle involving (and requiring) mollusks, fish, and birds.

Garnering more attention than parasites in aquatic environments are amphibians. Notably, about 25 percent of all known amphibian species were discovered only in recent years; by comparison, since 1993, the number of previously unknown mammal species that have been identified increased by 10 percent. There are three amphibians orders: Anura (frogs and toads), Caudata (newts and salamanders), and Gymnophiona (legless caecilians). The AmphibiaWeb (2019) database contains 8034 amphibian species, of which 7091 are frogs and toads, 731 are newts and salamanders, and 212 are caecilians. Since 1985, the total number of recognized amphibian species has increased by over 60 percent.

While humans have traversed all terrestrial and many aquatic environments over the centuries, thousands of new plant and animal species are discovered every year. The Arizona State University International Institute for Species Exploration reported 16 969 new species (not including previously undocumented species of microbials) in the year 2006 alone (Stutz 2009). This was not an atypical year (International Institute for Species Exploration 2018). The Institute for Biodiversity Science and Sustainability at the California Academy of Sciences supports annual expeditions around the globe in search of new species. Academy scientists described 229 new plants and animals in 2018, including 120 wasps, 34 sea slugs, 28 ants, 19 fish (including a dazzling neon-colored fish from deep beneath Brazilian waters), 1 sea horse (which was about the size of a jelly bean), 7 flowering plants, and 1 moss (California Academy of Sciences 2018). Each of these remarkable finds further enriches our understanding of the complexities and fragilities of life on Earth.

Authenticating a new species is not easy work, as it requires awareness of a vast body of species descriptions dating to 1758. Validated species discovered in recent years includes a broad array of hundreds of new kinds of bats, a rodent species that was thought to have gone extinct millions of years ago, a pygmy deer in Bhutan, a macaque in the Himalayan foothills, a white titi monkey in Brazil, and a pygmy sloth in Panama. New species discovered in 2018 include the baffling beetle (*Nymphister kronaueri*), the heterotrophic flower (*Sciaphila sugimotoi*), the now extinct marsupial lion (*Wakaleo schouteni*), Swire's snailfish (*Pseudoliparis swirei*), and the Tapanuli orangutan (*Pongo tapanuliensis*) (College of Environmental Science and Forestry 2018). The latter, a native of Batang Toru, Sumatra, is considered the most imperiled great ape in the world. It is estimated that there are only 800 Tapanuli orangutans now living in a fragmented habitat that is spread over about

250 000 acres of forested lands. Insects, like the baffling beetle and a new type of exploding ant (*Colobopsis explodens*)—so named because a specialized group of its workers can rupture their abdomens to release a toxic liquid to deter invaders (Laciny *et al.* 2018)—represent more than half of all newly found species.

The discovery of new species is currently occurring at about twice the overall historic average because of the number of researchers in the field, the availability of funding, and technological advances. Additionally, a number of biodiversity hotspots, where species variety is at its peak, are now known. Stutz (2009) notes that "these are found in Madagascar, India, Indonesia, South America, and Southeast Asia. Along the Mekong River watershed . . . more than a thousand new species have been discovered over the last 10 years—an average of two new species each week—and the pace of discovery appears to be unabated. Running 3000 miles through China, Myanmar, Laos, Thailand, Cambodia, and Vietnam, the Mekong and its surrounding mountains and forests were always known to hold extraordinary reserves of biodiversity, more species per unit area than any river other than the Amazon." Additionally, the tropical forests of the Annamite Mountains along the border of Laos and Vietnam have proven to be a lost world that persisted through the last ice age. These poorly trafficked places have been made even more inaccessible by political instability and war.

When new species are identified and described, they must be named, using what is called a Latin binomial (e.g., *Homo sapiens*), though it is not always actually in Latin. A thick, 18-chapter book, the *International Code of Zoological Nomenclature* (International Commission on Zoological Nomenclature 2012), specifies how new species are named. Rules include: 1) the name must be unique and not in use; 2) the name should not give offence on any grounds; and 3) the discoverer cannot name the species after her- or himself. None of the rules bar name-givers from using humor, however. For example, there are five species of the small round fungus beetle that are named *Gelae baen* (to sound like "jelly bean"), *Gelae balae* ("jelly belly"), *Gelae donut* ("jelly doughnut"), *Gelae fish* ("jelly fish"), and *Gelae rol* ("jelly roll"). The right to inject humor explains the naming of a newly discovered Panamanian caecilian that spends most of its life underground: upon winning a contest to raise money for the Rainforest Trust, which allowed the winner to name the new species, Aidan Bell, co-founder of the British sustainable building materials company EnviroBuild, selected the name *Dermophis donaldtrumpi*, after the U.S. president. Bell felt that the amphibian's tendency to keep its head buried in the ground was similar to Donald Trump's do-nothing approach to climate change. Unfortunately, *Dermophis donaldtrumpi* is in line to be added to the vulnerable species list as it is "in danger of becoming extinct as a direct result of its namesake's climate policies," Bell said in a statement released to the press (quoted in Guy 2018).

Ironically, in the midst of the current sixth global extinction of species, when biodiversity is in rapid retreat, science is experiencing a golden age of new species discovery, some examples of which may go extinct not long after they are first identified.

2.3.2.1 Extinction

Newly discovered species cannot replace those that are being lost as a result of human-caused ecocrises. These loses are occurring at all scales from local to global. The ecosystem effect of a species extinction is not necessarily tied to its relative abundance. A keystone species (Paine 1966, 1969), like the top predator in an ecosystem food chain, may be relatively rare but play a major role in influencing properties like the pathways of energy and material flows that sustain the system. Within specific ecosystems, the extinction of a species may have limited effect because "(a) ecosystems may have multiple species that carry out similar functional roles, (b) some species may contribute relatively little to ecosystem properties, or (c) properties may be primarily controlled by abiotic environmental conditions" (Hooper *et al.* 2005, p. 4). Conversely, loss of a single species can have dramatic effects, such as substantial changes in overall community biomass and nutrient cycling (Symstad *et al.* 1998). Keystone species are critical to an ecosystem because they have low functional redundancy. This means that if the species were to be eliminated, there would be no other species available that could really fill its ecological niche. Consequently, the ecosystem would be forced to change, allowing invasive species to populate and alter the habitat or leading to other consequences due to changes in remaining species populations.

An example of the latter can be seen by looking at Yellowstone National Park as an ecosystem. When the U.S. government designated almost 3500 acres for the park in 1872, there were hundreds of wolves roaming the area, preying especially on large herds of elk and bison. In 1914, fearing the impact of the wolves on those herds, but also on privately owned livestock grazed on public land, Congress appropriated funds to destroy wolves, prairie dogs, and other animals injurious to agriculture and animal husbandry. Park Service hunters carried out these orders, and by 1926 wolves were virtually eliminated from Yellowstone. Loss of this keystone predator started a top-down trophic cascade. No longer preyed on by the wolves, elk populations exploded and began competing for food resources such as grasses, sedges, and reeds. As a result, these plants did not have time or space to grow. Overgrazing had direct influences on other species, such as fish, beaver, and songbirds, which relied on the plants and their products for survival. Moreover, stream banks began eroding as wetland plants failed to anchor soil and sediments, while lake and river temperatures increased as trees and shrubs failed to provide shaded areas. In the end, it was realized that the complex web of relations that sustained the Yellowstone ecosystem required the presence of wolves, and beginning in the 1990s, in response to a call that began with the biologist Aldo Leopald, the U.S. government began reintroducing them (from Canada). The results have been a shrinking of elk populations, improvements in vegetative cover, and recovery of beaver and songbird populations. According to Doug Smith, a wildlife biologist in charge of the Yellowstone Wolf Project, "[i]t is like kicking a pebble down a mountain slope where conditions were just right that a falling pebble could trigger an avalanche of [positive] change" (quoted in Farquhar 2019).

Unfortunately, this kind of turnaround is not the norm. A recent statistical analysis of 51 bird species currently recorded on the International Union for Conservation of Nature (IUCN) Red List as critically endangered, for example, concluded that eight should be reclassified as already extinct or very close to extinction (Butchart *et al.* 2018). Stuart Butchart, the principle investigator on the study, explains that: "Ninety per cent of bird extinctions in recent centuries have been of species on islands . . . However, our results confirm that there is a growing wave of extinctions sweeping across the continents, driven mainly by habitat degradation and disappearance from unsustainable agriculture and logging" (BirdLife International 2018).

Aside from the loss of unique lifeforms, extinctions like these are important because when they occur, they alter key processes needed for the sustainability of Earth's ecosystems (Loreau *et al.* 2002; Wardle *et al.* 2011). Biodiversity loss can undercut the viability of ecosystems by diminishing their ability to respond to environmental change and disruption.

2.3.2.2 The species quandary

It bears noting that despite its centrality to biology and all other disciplines concerned with living and extinct organisms, the species concept remains a difficult issue. There is no fully agreed upon definition of its precise meaning and no universally shared understanding of its parameters, a dilemma known in biology as "the species problem" (Hey 2006; Mayr 1957; Wheeler & Meier 2000). A primary source of this dilemma is biological diversity.

A definition of species that is useful for categorizing similarities and differences among organisms that reproduce sexually, for example, is not useful for organisms like bacteria that reproduce asexually. To Mayr (1942), a leading evolutionary biologist of the 20th century, a species is composed of organisms that can reproduce fertile offspring with one another and that are reproductively isolated from other populations. Many researchers, but far from all (e.g., Wheeler & Meier 2000), find Mayr's (1942) definition to be the most useful for their efforts to grapple with the biodiversity of organisms, past and present. However, as stressed in this book, environments are changing, and as a consequence, species are moving. The geographic barriers that once divided identified species and minimized their chances of interbreeding are collapsing. Across the globe, new hybrid species are being documented among toads, sharks, butterflies, bears, and trout, the offspring of interbreeding of species newly brought together by climate change. Warmer temperatures have facilitated grizzly bear and polar bear encounters, leading to hybrids called either pizzlys or grolar bears. Among birds, a similar development has occurred among golden-winged warblers and blue-winged warblers. One unexpected consequence of human impact on ecosystems is thus changes in planetary speciation.

2.3.2.3 The Convention on Biological Diversity

This discussion draws attention to the incredible diversity of life in modern ecology. Biodiversity is formally defined in the Convention on Biological Diversity as "the variability among living organisms from all sources including, inter alia, terrestrial, marine and other aquatic ecosystems and the ecological complexities of which they are part; this includes diversity within species, between species and of ecosystems" (Reid *et al.* 1993). The Convention is a treaty signed by 168 government leaders at the United Nations Conference on Environment and Development on June 5, 1992, a historic meeting known as the Rio Earth Summit. It asserts that biological diversity is about more than plants, animals, fungi, and microorganisms, the variability within and across species, the communities organisms form, and habitats and ecosystems. Additionally, and importantly, it is about the human need for food, water, clean air, medicines, shelter, and a healthy environment (Table 2.1). According to the Convention, "[b]iodiversity provides a large number of goods and services that sustain our lives" (Convention on Biological Diversity 2018).

Inherently, biodiversity is multidimensional, encompassing many aspects of the variability of life on Earth, including the taxonomic classification of species, impacts of species on one another and on the abiotic environment, the evolutionary relatedness of species, the genetic make-up of organisms, and landscape characteristics.

Table 2.1 Diverse benefits from nature. *Source:* Modified from World Wildlife Fund, Living Planet Report 2018. Retrieved from: https://www.worldwildlife.org/pages/living-planet-report-2018.

Cultural resources
• Emotional and physical health
• Recreation
• Esthetic values
• Spiritual values
Provisions
• Food
• Water
• Medicinal resources
• Oxygen
• Material resources
Processes
• Nutrient cycling
• Photosynthesis
• Soil formation
Regulations
• Air quality
• Climate and extreme events
• Water
• Erosion
• Pollination

Despite their importance to ecological health, there remain significant gaps in available biodiversity data. The kinds of knowledge that are limited include the time depth of information about particular habitats, the actual composition of species in an ecosystem, the number of individuals within local populations of organisms, and changes in the areal range in which species are found.

2.3.2.4 Bioadversity

Biodiversity "is changing across the world and this change is mostly negative" (Turak *et al.* 2017) from the standpoint of health across species (Butchart *et al.* 2010), a pattern known as bioadversity. A growing body of research, for example, indicates that among many species there are shifting habitat ranges, seasonal activities, life cycle events (e.g., the time of emergence of leaves and flowers or the arrival dates of migratory birds), predator–prey interactions, and migratory patterns (Forcada & Hoffman 2014; Zhang *et al.* 2014). Change in these and other features is not new. Since they appeared on the planet, plants and animals have reacted to environmental changes by adapting in various ways (e.g., changing color, body shape, behavior), migrating to new areas with different conditions, or, if unable to adapt successfully, going extinct. Which of these possibilities lie ahead for the current inhabitants of Earth, including ourselves? According to Nogués-Bravo *et al.* (2018), "How individual species and entire ecosystems will respond to future climate change are among the most pressing questions facing ecologists." But the issue is not just climate change—the same applies to species and entire ecosystem responses to other adverse anthropogenic environmental modifications.

With regard to human health, maintaining biodiversity is of vital importance. Human medicines often are first discovered in wild-dwelling species. So too are various pathogens and disease vectors, which can only be brought under control by studying their behaviors and habitats. As Chivian (1997, p. 8) maintains:

> The study of species and biodiversity may be the best means we have for recognizing future dangers to human health from global environmental degradation . . . [W]e must focus much greater attention on biodiversity loss, which looms as a slowly evolving, potential medical emergency of unprecedented proportions, still largely unappreciated by policymakers and the public.

A species of considerable interest with regard to medicine development is the great white shark (*Carcharodon carcharias*), an animal that usually is far more terrifying in our imaginations than it is in reality. Found as a top predator in the coastal surface waters of all the major oceans, great whites face multiple anthropogenic challenges (e.g., accidental deaths on fishing longlines, poaching for the shark fin soup market, sport fishing for shark jaw trophies, ocean pollution). The scale of these threats has led the IUCN to list them as a vulnerable species and the Convention on International Trade in Endangered Species of Wild Fauna and Flora (CITES) as an Appendix II species (i.e., a species for which trade must be controlled in order to avoid human utilization that is incompatible with its survival). Medical interest in great whites stems from the fact that they are known to be super-healers. During their long evolutionary history, they have developed a genetic tool-kit that enhances several key processes involved in wound healing. These include sequences of DNA that code for rapid damage response, including amplified blood-clotting agents and scaffolding proteins that serve as the foundation for new tissue. Not only do great whites have uniquely adapted versions of these genes, but it appears that they have them in greater supply than other species, including us. Moreover, in theory, sharks and other large-bodied, long-lived animals with many more body cells than humans should have a greater chance of developing oncogenic mutations. Great whites, however, do not have a higher incidence of cancer than humans, and this may be because of clusters of genes that protect the integrity of their genetic code itself, a trait known in biology as genome stability. "Understanding how these genes might be inoculating these animals from cancer could be a huge benefit to humans," according to Michael Stanhope (quoted in Molenti 2019), an evolutionary biologist at Cornell University who co-led a project to map the great white genome (Marra *et al.* 2019).

Because "global biological diversity is declining in the face of numerous pressures" (Wetzel *et al.* 2017, p. 78), and this loss has grave implications, since the early 1970s biodiversity has emerged as a critical and heated scientific and political issue. In the case of sharks, their potential relevance to human health goes beyond the issue of cancer and includes their ability to heal quickly from wounds and resist infections. By sequencing the sharks' genome, researchers are trying to identify which genes contribute to these abilities and translate this finding into advances in wound-healing human medicines. Already, a skin-graft device called the Integra Omnigraft Dermal Regeneration Matrix (Omnigraft) has been approved by the U.S. Food and Drug Administration (FDA 2016). This material, which uses a combination of silicone, cow collagen, and shark cartilage, is used to help heal certain life-threatening burns and diabetic foot ulcers. The medical potential of the shark genome also extends to a possible new way of treating pulmonary fibrosis, a lung disease that occurs when tissue becomes damaged and scarred, making it difficult for the lungs to work properly and causing progressive shortness of breath. A new drug called AD-114, which mimics shark antibodies, is being developing in Australia. It belongs to a class of therapies called i-bodies, which are proteins that combine the features of small molecules and antibodies. In animal testing, AD-114 has been found to only bind with scarred lung cells while ignoring healthy lung tissue (Pulmonary Fibrosis News 2016).

At The Hague in 2002, the U.N. member states that signed on to the Convention on Biological Diversity agreed to a series of goals aimed at significantly reducing the rate of loss of animal and plant populations at the global, national, and regional levels—goals that have not been achieved. Blocking achievement are multiple threats to biodiversity (European Commission 2004), including:

- Unceasing expansion of industrial production, including many new chemicals,
- Promotion of hyper-consumerism.
- Production of toxic waste and its release into the environment.
- Global increase in vehicular mobility.
- Stratospheric ozone depletion.
- Human population growth, leading to increasing demands for living space, food, and other resources. (The global human population has grown from approximately 1.65 billion in 1900 to an estimated 6.3 billion today, and is increasing by over 75 million a year. In 40 years, the United Nations predicts a world population of 9 billion).
- Urban expansion and commercial agriculture and forestry encroachments on habitats.
- Extension of roads, rail lines, airports, and electricity networks that fragment habitats and frighten away some species.
- Overexploitation of natural resources, including excessive hunting, collecting, and trade in species and parts of species.
- Pollution and its adverse effects on the health of animals and plants.
- Environmental disasters, such as oil spills that devastate marine and coastal fauna and flora.
- Climate change, which impacts ecosystems on the land and in the seas, causes sea level rise, enhances the intensity and frequency of wild fires, contributes to species decline and extinction, and significantly impacts health.
- Invasion of alien species, often introduced intentionally or unintentionally by humans, that thrive and overwhelm endemic species.
- Overfishing, which is depleting global fish stocks in lakes and oceans.

Not only does this wide array of environmentally adverse factors threaten biodiversity, the loss of which is a direct threat to health, but it also creates the conditions for co-occurring and interacting ecocrises. As Noyes & Lema (2015) observe:

> a large percentage of terrestrial and aquatic species are facing elevated risk of extinction with climate change acting as a major driving force that is worsened by interactions with other stressors including habitat fragmentation and modification, over-exploitation and harvesting, eutrophication, invading species, infectious disease, and chemical pollution.

Another expression of ecocrises interaction is seen in the case of honeybees, which play a critical role in human food production. Research indicates that fruit, vegetable, and seed/nut production for 87 of the leading global food crops is dependent upon animal pollinators like bees (Klein *et al.* 2007). Historically, beehives in North America endured a die-off rate of about 15 percent in the period between October and April because of the cold temperatures and lack of winter forage. Rates of loss, however, have almost doubled to 28 percent since the early 2000s as a result of the combined and interacting effects of stress, pests, diseases, and pesticide exposure (Goodrich 2017). In fact, in 2014, the United Nations reported that 40 percent of the plant pollinator species that are key to the world's food supply are facing extinction. A member of the team that produced this assessment, Berry Brosi, assistant professor of environmental sciences and bee researcher at Emory University in Atlanta, stated: "If pollinator declines continue at this rate it will have serious implications not just for human food security and economics but also for biodiversity and the health of ecosystems in general . . . When we lose even one pollinator species from an ecosystem, it can degrade the functioning of the system overall . . . Studies have shown this relationship between biodiversity of pollinators and both agricultural productivity and plant reproduction in wild ecosystems . . . Nutritionally, the pollinator declines will likely have the biggest impact on the poorest people of the world" (quoted in Clark-Emory 2016). Beyond nutritional issues, pollinators are needed for crops that are the source of biofuels (e.g., canola and palm oils), fibers (e.g., cotton and linen), forage for livestock, and construction materials like wood.

2.3.3 Regional and planet-wide natural interconnecting structures

While Earth consists of multiple ecosystems, there are other systems that crosscut and connect it, including jet streams and the hydraulic cycle. Jet streams are powerful, fast-moving rivers of weather-producing wind that meanders eastward 5–10 miles above the planet in the part of the atmosphere known as the tropopause (Moran 2012; Zimmermann 2013)—the boundary between the troposphere (where planetary weather is formed) and the stratosphere (which is weather-free). They form in places where large colder and warmer bodies of air meet, as a result of atmospheric heating caused by solar radiation and the rotation of Earth. They move at about 275 miles per hour. Usually, there are two jet streams: a subtropical stream traveling in the Southern Hemisphere that transports atmospheric moisture into storm systems and a more powerful polar-front stream encircling the Northern Hemisphere. The northern stream varies in its location between polar air from the far north and more temperate air found south of the polar regions. During summer in the Northern Hemisphere, a third stream blows toward the west in tropical high-altitudes areas. This reverse air stream is associated with the heating of Asia and may play a role in the formation of summer monsoons in the Indian Ocean. It is believed that jet streams were at least partially responsible for the intense drought that caused the Dust Bowl in the U.S. Midwest during the 1930s. Notably, climate scientists hypothesize that the jet streams will gradually as a result of global warming. The current decline in Arctic sea ice, reduced snow cover, and other climate changes are projected to increasingly make the Arctic warm faster than other areas of the planet; indeed, this has already been documented. This change diminishes the temperature gradient that pushes jet stream winds, with important consequences for weather stability and predictability. Such changes will have telling effects on ecosystems across the planet.

The water cycle is the constant movement of water on, in, and above Earth (U.S. Geological Survey 2016). On the planet's surface, water appears as rivers and streams, ponds, swamps, dew, and oceans, and in frozen form as land and sea ice, all of which are connected through their flows, by precipitation, and by water coursing under the ground in vast quantities in spaces between rocks and subsurface soils and in aquifers. Some precipitation makes its way downward into groundwater. There is also ice found underground in caverns and ice tubes. Fresh water (about half of which is as salty as sea water) is even found in subterranean pools 600 feet under the Atlantic Ocean. Above the planet, the atmosphere is a superhighway of water vapor that moves around the globe. When, during its movement, water vapor encounters cooler temperatures, it condenses to form clouds, which are pushed around by strong winds until the vapor falls to Earth as rain or snow. About 90 percent of the water vapor in the atmosphere is produced by evaporation from water bodies, while the remainder comes from the release of moisture by plants—a process known as transpiration. The oceans are the source of 90 percent of water evaporation that flows into the planetary water cycle. While they are the primary location of water storage on Earth—currently estimated to hold 321 000 000 cubic miles—they vary in volume over time due to climate change, as we see today with rising sea levels due to land ice melt and the increased mobility of molecules in warmer water. Within the oceans, currents like the Gulf Stream move immense amounts of water around the world.

The interconnection of global flows like the jet stream and the water cycle with climate make these systems an important part of the story of the changing ecology on Earth and the development and interface of ecocrises, including loss of biodiversity and climate impact on health.

2.3.4 Human-dominated ecosystems

All organisms, past and present, modify their ecosystems, and "studies on lakes, coral reefs, oceans, forests and arid lands have shown that smooth [environmental] change can be interrupted by sudden drastic switches to a contrasting state" (Scheffer *et al.* 2001). In the case of humans, however, the extent of ecosystem alteration can be both rapid and far-reaching (Fig. 2.4). At the local or regional scale, human-dominated ecosystems include agricultural fields, created and maintained pasturelands, planted forests, urban areas, populated islands, and similar settings where the effects of human presence are sweeping. It is even possible to talk of all of Earth as a human-dominated ecosystem, as there are no terrestrial ecosystems and almost none underwater that are untouched by human activity (Vitousek *et al.* 1997). When humans clear lands for agriculture, send airplanes across the skies, fish the oceans, operate manufacturing plants, or produce and use cement, we reconfigure the major biogeochemical cycles of Earth, including those of CO_2, nitrogen, and sulfur.

The concentration of CO_2 in the atmosphere has steadily increased as a result of human activities, especially since the Industrial Revolution, but starting long before that with alterations like clearing and planting. Even indigenous populations cleared forested areas through burning to increase grassy pastures in which to find and hunt grazing wild animals. Human alterations of the hydrological cycle can change regional weather patterns and even longer-term climate. Implementing irrigation systems raises atmospheric humidity in semiarid locations, which

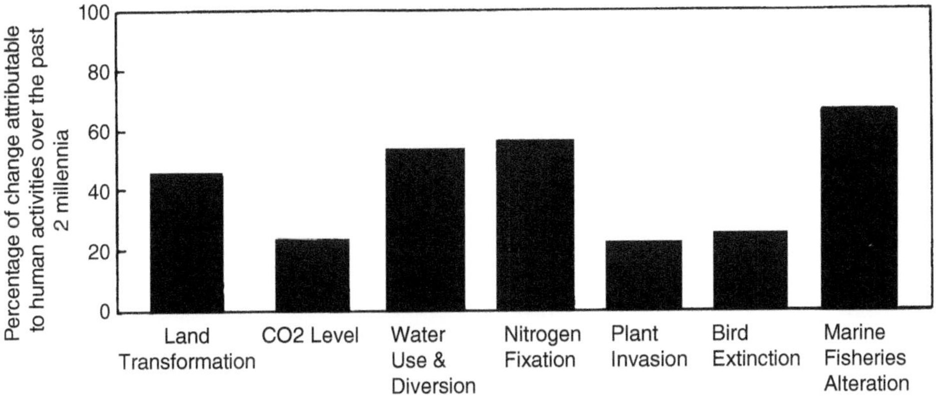

Fig. 2.4 Human alteration of key components of Earth ecosystems. *Source:* Modified from Vitousek *et al.* (1997).

promotes the frequency of rain and thunder storms. Humans have also altered the global nitrogen cycle. A vital biosynthesizing element for living organisms, nitrogen is present in abundance in the atmosphere but must be combined with carbon, hydrogen, or oxygen to form ammonia or other molecules in order to be useable by most lifeforms, a process known as fixation. Human activity has substantially altered the global nitrogen cycle through both fertilizer production and fossil fuel combustion. The global sulfur cycle, meanwhile, has been changed by the acquisition and burning of fossil fuels. Unprecedented increases in the amount of sulfur in both the atmosphere and the oceans is now occurring. Sulfur dioxide emitted by the fossil fuel industry reacts with water vapor in the atmosphere to form acid rain, which is damaging to forests, fish, and other aquatic animals. Oceans, too, feel the impact of acid rain, especially in coastal areas, which are becoming more acidic and being stripped of the carbonate ions needed by shell-making sea life. The ability of human technologies to change multiple biogeochemical cycles in these ways creates the potential for radical alterations of Earth.

2.3.5 Human ecology

Human ecology is the subfield of ecology that is specifically concerned with understanding relationships between people and their environment, including the built environment of human construction (e.g., an urban neighborhood, a riverine village, or a fishing camp). The presence of humans on Earth is the zigzagging, undirected product of evolutionary history, played out under the influence of complex and changing environmental and climatic conditions.

In human ecology, as in the broader field, the environment is conceptualized as an ecosystem. Thus, in human ecology research, a single farm can be studied as an ecosystem, as can New York City. Notably, humans are never the only organisms that inhabit built environments. Some nonhuman species, like dogs and zoo animals, occupy human-constructed environments because we bring them there. Others, like termites, coyotes, and brown rats, are uninvited but find such settlements contain abundant desirable resources. Of special note to human health in built environments is the role of animal disease vectors like mosquitoes and rats in the spread of infectious pathogens as a result of interacting ecocrises like climate change and flooding.

Some wild species, like the house sparrow (*Passer domesticus*), are so anthrodependent that if humans abandon a location, the local population goes extinct. The house sparrow is also quite sensitive to changes in human landscapes. In some urban areas, especially in the United Kingdom and Western Europe, their numbers are now declining because of changes in human ecology such as gentrification, which reduces the number of nesting sites and food availability.

Human social systems are a dominant feature of human-impacted ecosystems. As Marten (2001, p. 1) states:

> Although humans are part of the ecosystem, it is useful to think of human–environment interaction as interaction between the human social system and the rest of the ecosystem . . . The social system is a central concept in human ecology because human activities that impact on ecosystems are strongly influenced by the society in which people live. Values and knowledge . . . shape the way that we process and interpret information and translate it into action.

Human activities in an environment, such as driving a car, building a new factory, cutting down trees in a rainforest, drilling for oil in a deep-water site, launching a fleet of factory stern trawler fishing and processing vessels, or planting a field of corn, trigger a chain of effects that reverberate between social systems and ecosystems (Fig. 2.5). Marten (2001, p. 2) cites the following example of this echo process. People in many parts of India traditionally have used wood extracted from the environment for cooking fuel. This was not a great threat to the balance and regenerative capacity of the forest environment until the human population expanded rapidly beginning in the 1950s. As a result:

> Many forests have disappeared in recent years because people have cut so many trees and bushes for cooking fuel. Now there are not enough trees and bushes to provide all the fuel that people need. People have responded . . . by having their children search for anything that can be burned, such as twigs, crop residues . . . and cow dung. Fuel collection makes children even more valuable to their families, so parents have more children. The resulting

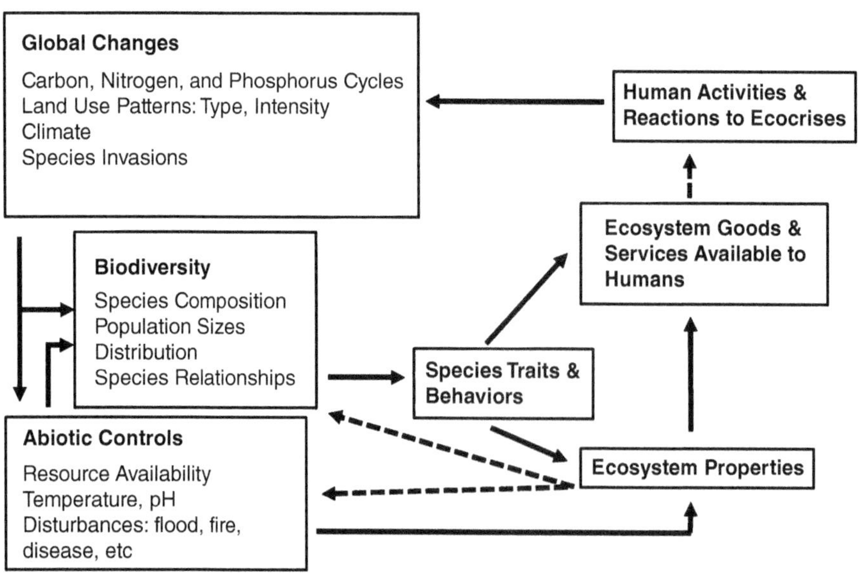

Fig. 2.5 Feedbacks between human activities and Earth properties leading to global change. *Source:* Modified from Hooper *et al.* (2005).

increase in population leads to more demand for fuel. (Marten 2001)

More recently, India's fertility rate has stabilized, with the average number of children born to a woman during her lifetime dropping from 5.9 in 1951 to 2.3 in 2011 (Nagajan 2016). But India's population remains quite large and the damage to the environment has been done, not only by small farmers but also by large planation corporations, beginning with colonial British commercial forestry operations. Forest loss has impacted many species and contributed to a growing water crisis in the country's villages and cities. This poses threats to health, particularly among poorer families.

As this last comment indicates, a central feature in contemporary human societies that shapes the way humans interact with the environment is social inequality. Health anthropologists have developed what they call an ecobiopolitical model for comprehending the complexities of societal–environmental interaction in the production of health and health inequality within society. This approach is informed by a synthetic and holistic exploration of the linkages that connect power and social structures, societal/environmental relations, and health and the environment. The term "ecobiopolitics" draws needed attention to the complex entwinements across conceptual domains that underlie the making of human health and disease.

In the ethnographic book *Flammable: Environmental Suffering in an Argentine Shantytown*, for example, the authors Javier Auyero and Debora Swistun (2009) describe a highly and multiply polluted slum area in Buenos Aires. Called Villa Inflamable, this shantytown is surrounded by a petrochemical complex, leather tanneries, various other factories, open-air garbage dumps, and the contaminated waters of the Río de la Plata, a river named for a ship explosion that occurred just offshore. The people of the community also face an indifferent state bureaucracy and duplicitous corporations controlled by elite polluters. The shanty and the surface water around it have high concentrations of arsenic (a potent, potentially lethal poison), cadmium (a carcinogen), chrome (which causes dermatitis, an ulceration of the skin), mercury (a neurotoxin, which damages all body systems), cyanide (which causes weakness, giddiness, headaches, vertigo, confusion, and heart stoppage), and phenol (which is corrosive to the eyes, the skin, and the respiratory tract, and causes burns), and blood samples drawn from some of the local inhabitants show startlingly high levels of lead. Residents suffer from diarrhea, respiratory problems, skin diseases, cancers, allergies, and anemia.

The people who live in the shanty endure what Auyero & Swistun (2009) call "environmental suffering," a term derived from the health social science concept of social suffering, which refers to experiences of group misery caused by occupying an oppressed or marginalized social ranking in a hierarchical social system. Environmental suffering can be defined as social suffering mediated by the environment, in this case among people living and working in a polluted or toxic environment that is the product of anthropogenic activities by dominant groups in the wider society. Social inequality commonly produces environmental inequality, environmental suffering, and health inequality.

Being in such an environment, people's lived experience can be highly stressful because of the direct effects of the illnesses of family members, the constant threat of such illnesses to household economic viability, and the persistent uncertainty about what is happening. A heavily polluted environment is a stressful environment, and the burden of living there is twofold: 1) direct health effects of pollutants on the body; and 2) the stress of being in a constant state of threat.

With multiple toxins in the land, air, and water, the people of Villa Inflamable are at constant risk of the perils of ecocrises interaction (e.g., interaction between hydrocarbons in the polluted air and runoff of tannery chemicals in the polluted water). Toxicological interactions among environmental pollutants that increase adverse health effects have been described (Krishnan & Brodeur 1994). Such interactions of two or more chemical contaminants may occur simultaneously or sequentially (involving interaction among toxins stored over time in body tissues). Other adverse ecocrises of diverse kinds also threaten human health and well-being, as detailed in subsequent chapters.

References

Acreman, M., Harding, R., Lloyd, C., McNamara, N., Mountford, J., Mould, D. *et al.* (2011). Trade-off in ecosystem services of the Somerset Levels and Moors wetlands. *Hydrological Sciences Journal* 56(8): 1543–1565.

Alt, L. (1989). Reproductive Biology of Female Black Bears and Early Growth and Development of Cubs in Northeastern Pennsylvania. Ph.D. dissertation. Morgantown, WV: West Virginia University.

AmphibiaWeb. (2019). Species by the numbers. https://amphibiaweb.org/amphibian/speciesnums.html.

Auyero, J. & Swistun, D. (2009). *Flammable: Environmental Suffering in an Argentine Shantytown*. New York: Oxford University Press.

Berkes, F., Berkes, M. & Fast, H. (2007). Collaborative integrated management in Canada's north: the role of local and traditional knowledge and community-based monitoring. *Coastal Management* 35: 143–162.

Bessire, L. & Bond, D. (2014). Ontological anthropology and the deferral of critique. *American Ethnologist* 41(3): 440–456.

BirdLife International. (2018). Spix's Macaw heads list of first bird extinctions confirmed this decade. https://www.birdlife.org/worldwide/news/spixs-macaw-heads-list-first-bird-extinctions-set-be-confirmed-decade.

Boaz, F. (1887). Museums of ethnology and their classification. *Science* 9: 589.

Boaz, F. (1911). *Handbook of American Indian Languages*, vol 1. Washington, D.C.: Bureau of American Ethnology.

Brosius, J., Lovelace, G. & Martin, G. (1986). Ethnoecology: an approach to understanding traditional agricultural knowledge. In: G. Marten (Ed.), *Traditional Agriculture in Southeast Asia: A Human Ecological Perspective*, pp. 187–198. Boulder, CO: Westview Press.

Bulmer, R. (1974). Folk biology in the New Guinea highlands. *Social Sciences Information* 13: 9–28.

Butchart, S., Walpole, M., Collen, B., van Strien, A., Scharlemann, J., Almond, R. *et al.* (2010). Global biodiversity: indicators of recent declines. *Science* 328(5982): 1164–1168.

Butchart, S., Lowe, S., Martin, R., Symes, A., Westrip, J. & Wheatley, H. (2018). Which bird species have gone extinct? A novel quantitative classification approach. *Biological Conservation* 227: 9–18.

California Academy of Sciences. (2018). Academy scientists describe 229 species in 2018. https://www.calacademy.org/press/releases/academy-scientists-describe-229-species-in-2018.

Caron-Beaudoin, É. & Armstrong, C. (2019). Biomonitoring and ethnobiology: approaches to fill gaps in indigenous public and environmental health. *Journal of Ethnobiology* 39(1): 50–64.

Cassady, J. (2007). A tundra of sickness: the uneasy relationship between toxic waste, tek, and cultural survival. *Arctic Anthropology* 44(1): 87–97.

Centre for Climate Adaptation. (2018). Climate change post: coastal erosion in Greece. https://www.climatechangepost.com/greece/coastal-erosion/.

Chalker-Scott, L. (2014). *The Myth of Permanent Peatlands*. Pullman, WA: Washington State University: Extension Urban Horticulture.

Chivian, E. (1997). Global environmental degradation and biodiversity loss: implications for human health. In: J. Rosenthal & F. Grifo (Eds.), *Biodiversity and Human Health*, pp. 7–38. Washington, D.C.: Island Press.

Clark-Emory, C. (2016). U.N. report warns 40% of pollinators face extinction. *Futurity*. https://www.futurity.org/bees-pollinators-extinction-1112572-2/.

College of Environmental Science and Forestry. (2018). Top ten new species. http://www.esf.edu/top10/.

Convention on Biological Diversity. (2018). Sustaining life on Earth. https://www.cbd.int/convention/guide/.

Costello, C., Jones, D., Hammond, K., Inman, R., Inman, K., Thompson, B. *et al.* (2001). A study of black bear ecology in New Mexico. Final Report. Federal Aid in Wildlife Restoration Project W-131-R. New Mexico Game and Fish.

Dunn, C. & Freeman, C. (2011). Peatlands: our greatest source of carbon credits? *Carbon Management* 2(3): 289–301.

Durham, W. (1976) Resource competition and human aggression, part I: a review of primitive war. *Quarterly Review of Biology* 51(3): 385–415.

Eagle, T. & Pelton, M. (1980). Seasonal nutrition of black bears in the Great Smoky Mountains National Park. In: *Bears: Their Biology and Management, Vol. 5, A Selection of Papers from the Fifth International Conference on Bear Research and Management*, pp. 94–101. Madison, WI: International Association for Bear Research and Management.

Egerton, F. (2001). A history of the ecological sciences: early Greek origins. *Bulletin of the Ecological Society of America* 82(1): 93–97.

Ehrlich, A. & Ehrlich, P. (1980). Ecoscience: the Greeks and Romans did it, too! *Mother Earth News*. https://www.motherearthnews.com/nature-and-environment/greeks-and-romans-zmaz80mjzraw#ixzz33TFsXNbu.

Eriksen, C. & Adams, M. (2010). Indigenous environmental knowledge. In B. Warf (Ed.), *Encyclopedia of Geography*, pp. 1–3. London: SAGE Publications.

European Commission. (2004). Biodiversity loss: facts and figures. http://europa.eu/rapid/press-release_MEMO-04-27_en.htm.

Farquhar, B. (2019). Wolf reintroduction changes ecosystem in Yellowstone. *My Yellowstone*. https://www.yellowstonepark.com/things-to-do/wolf-reintroduction-changes-ecosystem.

FDA. (2016). FDA approves Integra Omnigraft Dermal Regeneration Matrix to treat diabetic foot ulcers. U.S. Food and Drug Administration. https://www.fda.gov/news-events/press-announcements/fda-approves-integra-omnigraft-dermal-regeneration-matrix-treat-diabetic-foot-ulcers.

Forcada, J. & Hoffman, J. (2014). Climate change selects for heterozygosity in a declining fur seal population. *Nature* 511: 462–465.

Frängsmyr, T., Lindroth, S., Erikson, G. & Broberg, G. (1983). *Linnaeus, the man and his work*. Berkeley, CA: University of California Press.

Fuller, J., Foster, D., McLachlan, J. & Drake, N. (1998). Impact of human activity on regional forest composition and dynamics in central New England. *Ecosystems* 1: 76–95.

Giam, X. (2017). Global biodiversity loss from tropical deforestation. *Proceedings of the National Academy of Sciences of the U.S.A.* 114(23): 5775–5777.

Goodrich, B. (2017). *The California Almond Pollination Market: Contracts and Honeybee Colony Health. Dissertation*. Davis, CA: University of California Davis.

Guy, J. (2018). Blind, burrowing amphibian named after Donald Trump. *CNN*. https://www.cnn.com/2018/12/19/world/burrowing-amphibian-trump-scli-intl/index.html.

Hardesty, D. (1977). *Ecological anthropology*. New York: John Wiley and Sons.

He, X., McLean, J., Edlund, A., Yooseph, S., Hall, A., Liu, S.-Y., *et al.* (2015). Cultivation of a human-associated TM7 phylotype reveals a reduced genome and epibiotic parasitic lifestyle. *Proceedings of the National Academy of Sciences of the U.S.A.* 112(1): 244–249.

Hechinger, R. & Lafferty, K. (2005). Host diversity begets parasite diversity: bird final hosts and trematodes in snail intermediate hosts. *Proceedings of the Royal Society B: Biological Sciences* 272(1567).

Heneghan, L. (2015). The butterfly of doom: everything connected? 10 things wrong with environmental thinking. http://10thingswrongwithenvironmentalthought.blogspot.com/2011/10/everything-connected.html.

Hey, J. (2006). On the failure of modern species concepts. *Trends in Ecology & Evolution* 21(8): 447–450.

Heywood, P. (2017). Ontological turn, the. *The Cambridge Encyclopedia of Anthropology*. http://www.anthroencyclopedia.com/entry/ontological-turn.

Hodell, D., Curtis, J. & Brenner, M. (1995). Possible role of climate in the collapse of Classic Maya civilization. *Nature* 375: 391–394.

Holbraad, M. (2012). *Truth in Motion: The Recursive Anthropology of Cuban Divination*. Chicago, IL: University of Chicago Press.

Hooper, D., Chapin, F., Ewel, J., Hector, A., Inchausti, P., Lavorel, S. *et al.* (2005). Effects of biodiversity on ecosystem functioning: a consensus of current knowledge. *Ecological Monographs* 75(1): 3–35.

Hughes, J. (1975). Ecology in ancient Greece. *Inquiry* 18(2): 115–125.

Hughes, J. (1985). Theophrastus as ecologist. *Environmental Review* 9(4): 296–306.

International Institute for Species Exploration. (2018). ESF top 10 new species. https://www.esf.edu/species/.

International Commission on Zoological Nomenclature. (2012). International Code of Zoological Nomenclature, 4 edn. https://www.iczn.org/the-code/the-international-code-of-zoological-nomenclature/the-code-online/.

Iverson, S., McDonald, J. Jr. & Smith, L. (2001). Changes in the diet of free-ranging black bears in years of contrasting food availability revealed through milk fatty acids. *Canadian Journal of Zoology* 79: 2268–2279.

Klein, A.-M., Vaissière, B., Cane, J., Steffan-Dewenter, I., Cunningham, S., Kremen, C. & Tscharntke, T. (2007). Importance of pollinators in changing landscapes for world crops. *Proceedings of the Royal Society B* 274(1608).

Krishnan, K. & Brodeur, J. (1994). Toxic interactions among environmental pollutants: corroborating laboratory observations with human experience. *Environmental Health Perspectives* 102(Suppl. 9): 11–17.

Laciny, A., Zettel, H., Kopchinskiy, A., Pretzer, C., Pal, A., Kamariah, A. *et al.* (2018). *Colobopsis explodens* sp. n., model species for studies on "exploding ants" (Hymenoptera, Formicidae), with biological notes and first illustrations of males of the *Colobopsis cylindrica* group. *ZooKeys* 751: 1–40.

Lafferty, K., Dobson, A. & Kuris, A. (2006). Parasites dominate food web links. *Proceedings of the National Academy of Sciences of the U.S.A.* 103(30): 11 211–11 216.

Landriault, L., Obbard, M. & Rettie, W. (2018). A black bears quest for food. *Wise About Bears*. http://wiseaboutbears.org/black-bears/the-bear-family/a-bears-quest-for-food/.

Langmuir, C. & Broecker, W. (2012). *How to Build a Habitable Planet: The Story of Earth from the Big Bang to Humankind*. Princeton, NJ: Princeton University Press.

Long, A. Huby, P. & Fortenbaugh, W. (1985). *Theophrastus of Eresus: On His Life and Work*. New Brunswick, NJ: Transaction Publishers.

Loreau, M., Naeem, S. & Inchausti, P. (Eds.). (2002). *Biodiversity and Ecosystem Functioning: Synthesis and Perspectives*. Oxford: Oxford University Press.

Marra, N., Stanhope, M., Jue, N., Wang, M., Sun, Q. & Bitar, P. (2019). White shark genome reveals ancient elasmobranch adaptations associated with wound healing and the maintenance of genome stability. *Proceedings of the National Academy of Sciences of the U.S.A.* 116(10): 4446–4455.

Marten, G. (2001). *Human Ecology*. London: Earthscan.

MassWildlife. (2017). List of endangered, threatened, and special concern species. https://www.mass.gov/info-details/list-of-endangered-threatened-and-special-concern-species.

Mayr, E. (1942). *Systematics and the Origin of Species from the Viewpoint of a Zoologist*. New York: Columbia University Press.

Mayr, E. (1957). *The Species Problem.* Washington, D.C.: American Association for the Advancement of Science.

McCutcheon, R. (1999). Theoretical background: Insides, outsides, and the scholar of religion. In: R. McCutcheon (Ed.), *The Insider/Outsider Problem in the Study of Religion*, pp. 15–22. London: Cassell Academic.

McDonald, J. & Fuller, T. (2005). Effects of spring acorn availability on black bear diet, milk composition, and cub survival. *Journal of Mammalogy* 86(5): 1022–1028.

Molenti, M. (2019). Dun dun. Duun duuun! The great white shark genome is here. *Wired.* https://www.wired.com/story/dun-dun-duun-duuun-the-great-white-shark-genome-is-here/.

Mora, C., Tittensor, D., Adl, S., Simpson, A. & Worm, B. (2011). How many species are there on earth and in the ocean? *PLoS Biology* 9(8): e1001127.

Moran, J. (2012). *Weather Studies: Introduction to Atmospheric Science*, 5 edn. Boston, MA: American Meteorological Society.

Nagajan, R. (2016). The myth of India's population explosion. *Times of India.* https://blogs.timesofindia.indiatimes.com/staying-alive/the-myth-of-indias-population-explosion/.

New Phytologist Foundation. (n.d.). Sir Arthur George Tansley, 1871–1955. https://www.newphytologist.org/trust/tansley/about.

Nogués-Bravo, D., Rodríguez-Sánchez, F., Orsini, L., de Boer, E., Jansson, R., Morlon, H. *et al.* (2018). Cracking the code of biodiversity responses to past climate change. *Trends in Ecology and Evolution* 33(10): 765–776.

Noyes, P. & Lema, S. (2015). Forecasting the impacts of chemical pollution and climate change interactions on the health of wildlife. *Current Zoology* 61(4): 669–689.

Oftedal, O., Alt, G., Widdowson, E. & Jakubasz, M. (1993). Nutrition and growth of suckling black bears (*Ursus americanus*) during their mothers' winter fast. *British Journal of Nutrition* 70: 5–79.

Oliver, D. (1989). Food and food getting: the islands. In: D. Oliver (Ed.), *Oceania: The Native Cultures of the Pacific Islands*, vol 1. Honolulu, HI: University of Hawaii Press.

Paine, R. (1966). Food web complexity and species diversity. *American Naturalist* 100(910): 65–75.

Paine, R. (1969). A note on trophic complexity and community stability. *American Naturalist* 103(929): 91–93.

Paine, R. (2005). Cross environment talk in ecology: fact or fantasy? *Marine Ecology Progress Series* 304: 280–282.

Parks, D., Rinke, C., Chuvochina, M., Pierre-Alain Chaumeil, P.-A. & Woodcroft, B. (2017). Recovery of nearly 8000 metagenome-assembled genomes substantially expands the tree of life. *Nature Microbiology* 2: 1533–1542.

Pedersen, M. (2011). *Not quite shamans: spirit worlds and political lives in Northern Mongolia.* Ithaca, NY: Cornell University Press.

Peterson, G., Allen, C. & Hollimh, C. (1998). Ecological resilience, biodiversity, and scale. *Ecosystems* 1: 6–18.

Pike, K. (1967). *Language in Relation to a Unified Theory of the Structure of Human Behavior*, 2 edn. The Hague: Mouton.

Primack, R. (2014). *Walden Warming: Climate Change comes to Thoreau's Woods.* Chicago, IL: University of Chicago Press.

Pulmonary Fibrosis News. (2016). AD-114 for pulmonary fibrosis. https://pulmonaryfibrosisnews.com/ad-114-for-pulmonary-fibrosis/.

Rainer, T. & West, C. (2016). *Planting in a Post-Wild World: Designing Plant Communities for Resilient Landscapes.* Portland, OR: Timberwood Press.

Reid, W., Laird, S., Meyer, C., Gámez, R., Sittenfeld, A., Janzen, D., *et al.* (1993). *Biodiversity Prospecting: Using Resources for Sustainable Development.* Washington, D.C.: World Resource Institute.

Ross, W. & Forbes, F. (Trans.). (1967). *Theophrastus, Metaphysics.* Hildesheim: Georg Olms.

Scheffer, M., Carpenter, S., Foley, J., Folke, C. & Walker, B. (2001). Catastrophic shifts in ecosystems. *Nature* 413: 591–596.

Schulman, J. (2017). A timeline of every ridiculous thing trump has said about climate change.*ABC.* https://gma.yahoo.com/timeline-every-ridiculous-thing-trump-131002181.html.

Schulze, E.-D., Beck, E. & Muller-Hohenstein, K. (2002). *Plant Ecology*. New York: Springer.

Sears, P. (1964). Ecology: a subversive subject. *BioScience* 14(7): 11–13.

Silvano, R. & Begossi, A. (2016). From ethnobiology to ecotoxicology: Fishers' knowledge on trophic levels as indicator of bioaccumulation in tropical marine and freshwater fishes. *Ecosystems* 19: 1310–1324.

Stager, J., Wiltse, B., Hubeny, J., Yankowsky, E., Nardelli, D. & Primack, R. (2018). Climate variability and cultural eutrophication at Walden Pond (Massachusetts, USA) during the last 1800 years. *PLoS ONE* 13(4): e0191755.

Stutz, B. (2009). Finding new species: the golden age of discovery. YaleEnvironment[360]. https://e360.yale.edu/features/finding_new_species_the_golden_age_of_discovery.

Symstad, A., Tilman, D., Willson, J. & Knops, J. (1998). Species loss and ecosystem functioning: effects of species identity and community composition. *Oikos* 81(2): 389–397.

Theobald, D. (2010). A formal test of the theory of universal common ancestry. *Nature* 465: 219–222.

Thoreau, H. (2016). *Walden Pond; Or, Life in the Woods*. Eastford, CT: Martino Fine.

Turak, E., Brazill-Boasta, J., Cooney, T., Drielsma, M., DelaCruza, J., Dunkerley, G. *et al.* (2017). Using the essential biodiversity variables framework to measure biodiversity change at national scale. *Biological Conservation* 213(Part B): 264–271.

U.S. Geological Survey. (2016). Groundwater flow and the water cycle. https://water.usgs.gov/edu/watercyclegwdischarge.html.

Van Der Ploeg, J. & Van Weerd, M. (2010). Agta bird names: an ethno-ornithological survey in the Northern Sierra Madre Natural Park, Philippines. *Forktail* 26: 127–131.

Vayda, A. & Rappaport, R. (1968). Ecology: cultural and noncultural. In: J. Clifton (Ed.), *Introduction to Cultural Anthropology*, pp. 476–497. Boston, MA: Houghton Mifflin.

Viveiros de Castro, E. (2004). Exchanging perspectives: the transformation of objects into subjects in Amerindian ontologies. *Common Knowledge* 10: 463–484.

Wardle, D., Bardgett, R., Callaway, R. & Van der Putten, W. (2011). Terrestrial ecosystem responses to species gains and losses. *Science* 332(6035): 1273–1277.

Wax, M. (1972). Tenting with Malinowski. *American Sociological Review* 37(1): 1–13.

Webster, J., Brook, G., Railsback, B., Cheng, H., Edwards, L., Alexandere, C. & Reeder. P. (2007). Stalagmite evidence from Belize indicating significant droughts at the time of Preclassic Abandonment, the Maya Hiatus, and the Classic Maya collapse. *Palaeogeography, Palaeoclimatology, Palaeoecology* 250: 1–17.

Wetzel, F., Bingham, H., Groom, Q., Haasee, P., Kõljalgg, U., Kuhlmann, M. *et al.* (2017). Unlocking biodiversity data: prioritization and filling the gaps in biodiversity observation data in Europe. *Biological Convention* 221: 78–85.

Wheeler, Q. & Meier, R. (2000). *Species Concepts and Phylogenetic Theory: A Debate*. New York: Columbia University Press.

Vitousek, P., Mooney, H., Jane Lubchenco, J. & Melillo, J. (1997). Human domination of Earth's ecosystems. *Science* 277(5325): 494–499.

Zhang, X., Lei, Y., Zhihai, M., Kneeshaw, D. & Peng, C. (2014). Insect induced tree mortality of boreal forests in eastern Canada under a changing climate. *Ecology and Evolution* 4(12): 2384–2394.

Zimmermann, K. (2013). What is a jet stream? *LiveScience*. https://www.livescience.com/27825-jet-stream.html.

Zorgdrager, N. (2008). Linnaeus as ethnographer of Sami culture. *TijdSchrift voor Skandinavistiek* 29: 45–76.

3

The social and technological making of environmental crises

> *If we choose to take on this challenge, it appears that we can slow the rate of change substantially, giving us time to develop mechanisms so that the cost to society and the damage to ecosystems can be minimized. We could alternatively close our eyes, hope for the best, and pay the cost when the bill comes due.*
>
> Michael B. McElroy (quoted in Revkin 2008)

3.1 Earth is now a different place

The purpose of this chapter is to historically contextualize my primary argument concerning the growing risks of ecocrises interaction by examining the powerful, planet-wide effects of the rise of capitalism, the emergence and expansion of the Industrial Revolution, and the appearance and rise to power of a class of wealthy elites that are the primary drivers of catastrophic environmental pollution and ecosystem disruption. While change is an inherent feature of ecosystems, so too is balance. Since the rise of capitalism as a stalwartly productivist commodity/profit world economy, extraction of resources and the environmental release of the waste products of the industrial process have radically altered the planet and the lives of its inhabitants. The extent of this change has led environmentalist author Bill McKibben (2010, p. 2) to assert that "The world hasn't ended, but the world as we know it has—even if we don't quite know it yet. Earth is now a different place. A different planet." Why has this radical transformation occurred? Because humans, and in particular the world economic system they have created, have become a "geomorphic force with significant implications for their own health" (McKibben 2010, p. 2). Environmental and social scientists highlight our ever-increasing impact on the planet by referring to the geologic epoch we live in as the "Anthropocene" and our current global environmental trajectory as the "Hothouse Earth" pathway. How did we come to be on this road that appears headed in a perilous direction?

3.2 The *longue durée* and the rise and development of capitalism

Longue durée is a French expression that refers to the adoption of a long-term, broad view of history rather than a focus on short-term events and individual actors (like rulers and their specific deeds), as is common in discussion of the current environmental dilemma. This chapter begins with a *longue durée* perspective on the rise of capitalism and its development into an environmentally impactful global economic system that is pushing us toward crisis. Taking such a deep-rooted approach allows a fuller, more holistic understanding of how what was at one time a relatively insignificant species found on one

Ecosystem Crises Interactions: Human Health and the Changing Environment, First Edition. Merrill Singer.
© 2021 John Wiley & Sons, Inc. Published 2021 by John Wiley & Sons, Inc.

continent became a global species capable of instigating radical shifts on the whole planet and harmful ecocrises that threaten multiple species' survival.

3.2.1 Toward environment crises: critical turning points in human history

A number of momentous turning points have marked human history. These were not simple events that occurred at a single moment in time or on a precise date, but complex processes that unfolded over many years or even centuries. They were linked causally and overlapped temporally. As their full impact resonated across Earth, the human world was irrevocably changed, as was our planetary home.

The first turning point, of course, was the actual evolutionary emergence of the hominin line (i.e., modern humans, extinct human species, and all our immediate extinct ancestor species), and, eventually, as part of that development, the appearance on the planet of our species, *Homo sapiens*. Earth has existed for 4.5 billion years, while life on the planet began about 3.5 billion years ago. By contrast, modern humans and our immediate evolutionary ancestors have a history of only about 7 million years, and our species has only been around for 300 000 years, based on archeological findings of a skull, face, and jaw bones in Jebel Irhoud, on the Atlantic coast of Morocco (Richter *et al.* 2017). The interpretation of these findings is not that *Homo sapiens* originated in North Africa, but rather that the earliest members of our species evolved across Africa.

Before that, however, there were dramatic planetary events that predated humans but contributed to the possibility of the development of our species and the ecosystems of the world, such as photosynthesis, the oxidation of Earth (achieved by early bacteria), the greening of the continents with multicellular eukaryotes (i.e., plant and animal life), and the appearance of mammals (Cowen 2013). Since they first appeared, Earth has presented living organisms with endless challenges. All organisms through time have had to contend with shifting geography, climate, atmosphere, water and land environments, and predator species. The process of evolution has enabled them to meet these challenges, and life to continue. But change is bidirectional. Organisms also modify local and global environments, as exemplified by the original bacterial oxidation of Earth's atmosphere and the current oxygenation by phytoplankton, one-celled plants that live at the ocean surface. Eons of interplay between a changing Earth and its evolving organisms gave birth to humans and every other living and past species—a process likely to continue in a post-human world.

Understanding of the evolution of our species continues to expand with new fossil and artifact discoveries by archeologists and paleoanthropologists. Most scientists currently recognize between 15 and 20 different species of early humans. They disagree, sometimes vehemently, about how these species are related, which are in our direct line of descent, and which are branch lines that died out and left no living descendants (e.g., Wood & Collard 1999; Conroy & Pontzer 2012). Identified but now extinct species determined through careful fossil analysis to be closely related to us and to share our genus include *Homo heidelbergensis*, *Homo floresiensis*, *Homo naledi*, *Homo ergaster*, *Homo neanderthalensis*, *Homo erectus*, and *Homo habilis*. *Homo sapiens denisova* is an extinct subspecies of humans from Asia, one known to have interbred with Neanderthals (Slon *et al.* 2018). The latter, the closest of our extinct human relatives, resided across Europe and western Asia, reaching as far north as Britain and as far east as Uzbekistan, as well as parts of the Middle East. There is some debate as to whether they represent a distinct species or a subspecies of *Homo sapiens*. Researchers disagree about the point of divergence between the Neanderthal and modern human lineages, with estimates ranging between 750 000 and 400 000 years ago (Rieux 2014). Biological analysis indicates that Neanderthal DNA is

99.7 percent undistinguishable from modern human DNA and that most modern humans (other than people indigenous to Africa) have 1–4 percent Neanderthal DNA in their genome as a result of various periods of interbreeding (Green *et al.* 2010). It is not certain why Neanderthals went extinct some 30000 years ago, but several scholars theorize that global warming was a factor as they were physically and socially adapted to cold weather (Hodgkins *et al.* 2016).

At that point in time, early in human prehistory, our ancestors' behavior was not a driver of climate or environmental change. Human numbers were too few and our technologies, consisting of stone, bone, or wooden tools energized by human muscle power, were not capable of leaving a deep footprint on the environment. Ironically, these early limitations on our disruptive potential succumbed over the centuries because of the very success of our species in confronting the pitfalls of the world around us. The reverse impact, that of nonanthropogenic climate change on human evolution, however, is certainly quite plausible, as discussed in Chapter 4. Notably, the first simple stone tools used by our ancestors appeared in the archeological record in association with the fossils of the early bigger-brained members of our genus (Fig. 3.1). The landscapes inhabited by these early Homo groupings were transitioning from wooded areas to open grasslands as the climate was becoming drier. This shift in environmental scenery may be part of the reason early humans evolved away from living in trees and toward our modern form of bipedal locomotion on the ground. This freed our hands from a primary role in locomotion and created the potential for their use for other purposes. One of those, it turned out, was tool use, which was associated with the cerebral expansion vital to hand/tool coordination and the social communication and coordination that proved useful in surviving in a riskier environment on the ground (McGee & deMenocal 2017). None of these changes had to happen, and there was no progressive pathway in our development as a species; rather, our course involved various jigs and jags, and multiple dead ends (extinctions), as members of our genus coped (or failed to cope) with the challenges of changing environmental conditions. From the perspective of our broader family tree, that of Old World tailless anthropoid primates, Nengo (2018, p. 4) comments:

> Humans are one of many in a once lush and bushy tree of apes, now severely pruned, not the end production of some stairway to heaven. Humans, like chimpanzees, gorillas, and gibbons are the few survivors of a surprisingly diverse lineage.

Fig. 3.1 Early stone tools. *Source:* holmespj/Pixabay.

In light of the reference to the pruning of our hominid tree (occupied by all modern and extinct great ape species, including modern humans, chimpanzees, gorillas, and orangutans, as well as all of their immediate fossil ancestor relatives), it bears noting that very few branches survive today. Of all the species of hominids that ever lived, only eight are still occupants of Earth, and all but one of these, humans, are rare or endangered (at our own hands). Of vital concern is that the pruning end and we too avoid becoming a fossil species (at our own hands).

It is now firmly established that the birthplace of modern humans was in Africa. The second turning point also began there, and involved the peopling of the other continents. Researchers believe that early humans first migrated out of Africa into Asia probably between 2 million and 1.8 million years ago. Europe was penetrated somewhat later, between 1.5 million and 1 million years ago. Migration to other parts of the world, including islands that could only be reached after the development of watercraft, occurred much later. Thus, people did not arrive in Australia until 60 000 years ago. Debate continues over the appearance of humans in the Americas, though it generally is thought to have begun during the last 30 000 years. Through this radiation of human populations, we became a global species, and with the establishment of research stations in Antarctica, people pretty much can be found everywhere on the planet, even, with the inclusion of seagoing vessels and underwater divers, on and in the oceans. We have, of course, even pushed off of our planet during half a dozen crewed forays to the moon.

Notably, the initial thrust of humans out of Africa may well have been prompted by global warming that "rendered large parts of North, East, and West Africa unsuitable for hominin occupation" (Carto *et al.* 2009, p. 139). In addition to these "push" factors, there may have been climate-related "pull" factors at play as well. Research by Tierney *et al.* (2017) "suggests that warm and wet conditions from 120 000 to 90 000 y[ears] ago could have facilitated early waves of human migration toward the Levant and Arabia, as supported by fossil and lithic evidence." As this discussion of human evolution in and migration out of Africa confirms, we, like all species, are the products of interface with the environment.

The third turning point in human history was the development of farming and food production. Prior to this, human communities supported themselves by foraging plant and animal wild food sources. Diets based on gathering plants and shellfish, scavenging, hunting, and fishing varied by location and resource availability. Anthropologists working in Africa, Asia, and South America have conducted multiple ethnographic studies of existing and recent foraging societies that possess comparatively simple technological toolkits which appear similar to those found in the archeological record for all human communities that existed prior to about 12 000 years ago. While extrapolation of information from living societies to those of the past is not without great challenges, the ethnographic record offers a valuable tool for contemplating the nature of ancient lifeways. It suggests that for most of the history of our genus, a central feature of social life was equalitarianism. While there have always been individual differences among people in terms of size, intelligence, skill, health, and other traits, our ancient ancestors likely did not dwell in hierarchical social worlds characterized by status or wealth inequalities. Rather, in most environments, they lived most of the time in small, mobile groups of no more than 25–50 people, with whom they sustained face-to-face relationships across the various activities of daily life. Sharing of resources was the norm and social inequality was seen as a threat to needed group cooperation. Teamwork reduced the risk of starvation, attacks by predators, and even extinction. People did not possess domesticated work animals and relied on their own labor power to produce needed items. Mobility was unavoidable because of the movement of game animals and the eventual decline of local plant-based food sources. The size and quantity of material goods people possessed were limited because of their lack of beasts of burden and the demands of geographic mobility. Collective, consensual decision-making among adults and individual autonomy were the norm. As Kaplan *et al.* (1985) report, based on their ethnographic research among the Aché of eastern Paraguay, even the most skilled hunters received no political or material benefits. Further, Turnbull (1965) observes that among the Mbuti of the Democratic Republic of the Congo, the best hunters tend to keep a low profile in group gatherings in order to avoid public derision. Rather than retaining control over the food they foraged, people shared it to ensure that everyone was fed.

One issue of debate is the status of women in these kinds of communities. Researchers agree that a gendered division of labor is typical in egalitarian societies, with men filling the primary roles in hunting and women in the gathering of edible plants and childcare. But the separation of the domestic and public spheres of life among foragers narrow, and women can rely on their kinship ties to elude oppressive relationships with their husbands. Moreover, ethnographic accounts make it clear that women exercised control over their personal lives and daily activities and would no more avoid killing a vulnerable animal they encountered while gathering than men would ignore a rich vegetable resource they came across while hunting (Leacock 1992). Based on a thorough review of the ethnographic evidence, Lee (2018) concludes that forging societies past and present have tended to have balanced gender relations.

As Sahlins (1972) argues, a large body of ethnographic work has shown that in foraging societies, people have a considerable amount of leisure time and their lives are not impoverished or unelaborated. Most of the time, their needs are readily satisfied through their skillful use of even simple technologies combined with their considerable storehouse of local environmental knowledge and access to productive resources that is not hindered by private property rules. Thus, within the confines of their limited wants and needs, life under most circumstances allowed for a kind of prosperity among early human communities.

These patterns, however, began to change, slowly at first but eventually quite rapidly, with the development of farming and food production. Ultimately, this shift in food acquisition ushered in dramatic alteration of human social organization, population size and density, and human–environmental relationships. Why did farming develop? The classic answer has been that the discovery of how to grow food about 12 000 ago combined with the realization of the benefits of having control over this vital resource prompted people to launch the first green revolution. Anthropologists and other scientists who study this issue no longer find this simple explanation of ever advancing human enlightenment and rational decision-making convincing, however.

An alternative approach draws on the idea that dramatic change is a response, on the one hand, to emergent necessity, and, on the other, to newly achieved potential. Human communities in various places may have selected to increase their control over the wild plants and animals they knew well from their foraging lifestyles because of the effects of a changing climate (Gupta 2004). One plausible hypothesis involves Earth's environmental transition from the conditions brought on by the colder climate characteristic of the geotemporal epoch known as the Pleistocene to the warmer, wetter, and stabilizing trends characteristic of the subsequent Holocene. Beginning approximately 11 650 years ago, at the end of the last glacial period, climatic changes associated with the Holocene—as revealed by rising CO_2 levels found in ice and ocean-core samples—marked the changeover between these two epochs. In all but the most marginal environments at this time, "amelioration of the climate was followed immediately by the beginnings of plant-intensive resource-use strategies in some areas," and "eventually agriculture became the dominant strategy" (Richardson *et al.* 2001, p. 387).

Domestication appears to have occurred independently in over half a dozen different sites in subtropical and tropical settings in southwestern and southern Asia, northern and central Africa, and Central and South America. In each location, people developed their own assemblage of specific domesticated plants and animals. From these initial birthplaces, domestication spread out to other areas at the expense of local foraging lifestyles. In this interpretation, based on numerous archeological findings, the "increase in the CO_2 content of the atmosphere and increases in rainfall, rather abruptly changed Earth from a regime where agriculture was impossible everywhere to one where it was possible in various places" (Richardson *et al.* 2001, p. 404). Social groups that adopted plant-rich subsistence systems would have been able to out-compete groups that retained a less productive foraging subsistence strategy. Overall, the Holocene, with its equable climate, "was characterized by a persistent, but regionally highly variable, tendency toward subsistence intensification" (Richardson *et al.* 2001, p. 404). Prior to domestication, all vertebrate life on Earth was wildlife; by contrast, today wildlife accounts for only 3 percent of the terrestrial faunal biomass, with human beings, our livestock, and our pets making up the remaining 97 percent.

The fourth turning point in human history was the rise of urban centers, city-states, and eventually expansive empires like the massive Mongol empire of the 13th and 14th centuries. These social and political economic systems were fully dependent on floral and faunal domestication. The intensification of food production required the creation of permanent settlements to house full-time farmers and hydraulic system labor, while allowing the appearance of other specialists like soldiers, who could protect and expand planted acreage. Over time, social differentiation based on labor facilitated the institutionalization of status differences within the social body. The advent of elite social status became possible with the consolidation of advantage across various arenas of social life and production. Prospective elites could attempt "to manipulate the organization of technology to . . . intensify production" while seeking to control the distribution of the surplus food and other resources made possible by domestication (Nassaney 1992, p. 116). Surpluses could be deployed by emergent nobility to gain degrees of power with the recruitment of salaried abettors to enforce their claims within society and over conquered neighboring groups. To mark their special status, elites began to display symbols of authority, such as the wearing of fancy green-colored clothing and elaborate jewelry among the ancient Mayan (Fig. 3.2). Separation of aristocratic families from those of commoners in varying types of status hierarchies entailed transition away from the egalitarianism of foragers toward the inegalitarian class formation typical among city dwellers. In Mayan society, the majority of people toiled on the land. High above them in the social hierarchy was the nobility, known as *almehenob*, including the royal family. The division between these two social classes was rigid, with nobles claiming honored descent through elite family lineages and intermediary status between the gods and people.

In these kinds of societies, the group-wide egalitarian relationships that had been the norm for much of prior human history were jettisoned as the material and status gaps between social classes widened. Inequality came to cut across all spheres of life, and even found expression in differing death rituals and funerary practices. Economically, city-states and empires adopted a tributary mode of production (Gates 1996). The economies of these types of societies were constructed around the extraction of surplus goods (i.e., those beyond the minimum needed to survive) and labor from their immediate producers by political or military rulers. Scaffolding the payment of

Fig. 3.2 Chichen Itza, a pre-Columbian city built by the Mayan people. *Source:* jarmoluk/Pixabay.

tribute was a significant level of power disparity that enabled the development of political systems organized around social domination. Often, the ruling elite in such systems maintained control over the primary means of production, such as watering systems for agriculture. The Imperial governments of historic China (1368–1683), for example, oversaw far-flung tributary structures consisting of immense flood control, canal, irrigation, and road building and maintenance projects. Material tribute paid to the state by commoners primarily consisted of agricultural products, especially grains, but also coal, copper, salt, weapons, porcelain, and bricks. When possible, state-enforced "rights of extraction" were used by government officials to "to squeeze commoners for extraordinary extraction" (Gates 1996, p. 26).

Badges of social status were important in China (Mark 2016). Only nobles and royalty were allowed to wear silk clothing. Commoners, including those who processed silk into clothing for the wealthy, wore outfits made of hemp. Nonetheless, during the Sui dynasty, the emperor declared that there was too great a similarity between the clothing of commoners and that of the upper classes and that henceforth commoners could only wear blue or black garments. Distinctions between the classes extended to hairstyles as well. While all women had long hair, only elite women could afford to wear hairpins of ivory or precious metals. Class also structured space. Thus, while taverns were popular places of social interaction among men of all classes, members of the upper classes would not go to lower-class pubs, and members of the lower classes were not permitted to enter upper-class taverns.

Critical to the social organization of tributary states, in sum, was the establishment of distinct and unequal social classes with separate spaces and lifestyles, widening gaps in material wealth, and differential power to deploy labor and control surplus production. These advantages of the elite included control of state policies and actions. The imposition not only of social inequality but also of control over the products of farming, manufacture, and other industry was doubly reinforced by direct military might and the dissemination of a cultural ideology that rationalized hierarchy as natural and of spiritual origin (e.g., god-king beliefs). Evidently, this kind of detailed cultural elaboration and subterfuge was necessary to get people to (often willingly) do what the elites desired.

The fifth turning point was the development of capitalism. Its history lies in the shadow of the collapse of the best known and biggest of the western tributary states, the Roman Empire. In passing, because of the importance of environmental factors, it warrants noting the enduring disputation over the fall of Rome as an imperial entity. Various arguments have been put forward to explain this momentous event, including the burdens of being in a constant and costly state of war with neighboring peoples, pandemic diseases (e.g., small pox, bubonic plague), rebellion by subjugated populations, invasions by external enemies, a declining population, and excessive urbanization and expensive monument construction (Tainter 2006). An alternative climate-based explanation comes from analysis of tree ring data (Büntgen *et al.* 2011; McCormick *et al.* 2012). Paleoclimate research indicates that the rise and flourishing of the Roman Empire coincided with, and was probably immensely facilitated by, several centuries of stable, warm, moist summers that favored agricultural production, road construction, and military expansion. This period is known as the "Roman Climate Optimum." By contrast, the collapse of Rome was marked by extended droughts and much colder weather. This research affirms "climate forcing as one agent of distinct episodes of societal crisis" (Büntgen *et al.* 2011, p. 578).

In Western Europe, the downfall of the Roman Empire's unifying political and military edifices led to fragmentation and a period of local tributary production and smaller-scale fiefdoms. These are the hallmarks of an era between the 9th and 15th centuries commonly known as feudalism, though this concept has been a point of argument among historians for years (Brown 1974). While the actual social, legal, and organizational arrangements varied across Europe, the general pattern involved claimed ownership of arable land by local nobles—comprising initially of families that had been installed or recognized by Rome—sections of which were granted to local landlords in exchange for protection from rival fiefdoms, various forms of assistance, and the performance of military service. The landlords, in turn, divided their property among peasant farmers, occupants of the lowest rungs of the social ladder, who were allowed to live on it in exchange for their labor. In England, for example, the key ranks in the feudal pyramid were the king, who claimed all land in the kingdom, his vassals, who were granted parcels of land after pledging homage and fealty to him, and peasants, who cultivated the land and were provided access to small fields to maintain their own subsistence (Fleming 2001).

Over time, the foundations of the feudal mode began to shift. Three factors that promoted this geopolitical change were: 1) the growth in Italy of seaports and a class of merchants who developed the techniques of modern capitalist business (Reinert & Fredona 2017); 2) the return of agreeable climatic conditions that permitted agricultural surpluses sufficient to fund expansionist ventures (Diaz & Hughes 1994); and 3) rebellions by oppressed peasants, which, when added to the overseas trade of emergent global merchants, triggered a search for new sources of wealth beyond the boundaries of Europe (Wolf 1981). This drive for wealth launched the era of European global reach and military-enforced colonialism, the incubator of the capitalist world-system. Pre-industrial merchant capitalism, the product of these changes, involved control of labor and trade designed to produce wealth extracted from other lands. The arrangement was sufficiently successful to enable the ruling class of Europe "to live above its means, to invest beyond its savings" (Braudel 1974, p. 268).

The full transition from mercantile to industrial capitalism involved the conversion of mercantile wealth acquired through colonialism into workable capital (i.e., financial assets used to generate additional wealth through production and investment). Critical in this transition were the textile factories that developed in England during the 18th century. Previously, textile production was a cottage industry carried out in small workshops or in the homes of individual spinners, weavers, and dyers. Using fortunes tied to the influx of colonial wealth, affluent individuals began to invest in the acquisition of land, the construction of large textile factories, the purchase of new, labor-intensive machinery, and the hiring of workers—all with the goal of becoming even richer. Richard Arkwright, for example, with the help of well-heeled investors, built a number of textile factories, beginning in the village of Cromford in the East Midlands, early in the development of industrial capitalism. In the 1770s, his six-story factory, the first of many, employed nearly 600 people, including numerous women and small children. A working day of 12 hours was the norm, and injurious industrial accidents happened frequently. Trueman (2018) comments that while a "visitor to Arkwright's Cromford factory described the building as 'magnificent' . . . conditions inside for a worker were less than magnificent." Arkwright prospered enormously from his own factories, from licenses for his production patents, and from the acquisition of shares in over 100 other companies. He commissioned the building of Willersley Castle on 60 acres of land in Cromford as a home for his family and a visible expression of his success as an industrialist and entrepreneur. His wealth and prominence led to his being knighted in 1786 (Unwin 1968).

Profit produced by selling factory goods was used to significantly enhance the material lives of factory owners, meet costs, and invest in unending expansion and, when necessary, the social control of resistant or rebellious workers outraged by intensifying social inequality. Under emergent industrial capitalism, those who did not own factories or other means of production were forced to sell their labor if they wished to survive. But individuals who previously had worked the land or labored as artisans were repelled by the unforgiving conditions of privately owned factories, with their inexorable demands for lengthy hours of toil, often prison-like working environments, and the disciplining actions of profit-conscious supervisors.

Prior to the Industrial Revolution, textiles in England were made from wool, but in the early years of capitalism the rapidly developing machinery of the new industrial age supported a transition to the use of cotton. Cotton hastened the time of production and increased the profits of factory owners, which motivated the invention of even faster, more productive technologies. Water power, used to drive giant water wheels, energized the new equipment. Chapman (2013, p. 19) notes that by 1835, the steam engine "had become the predominant form of power in [almost] every cotton town in the North of England." Before long, coal, which was abundant in England, emerged as the leading energy source, turning the wheels of industry and transport (Wrigley 1988). The transition from water to steam did not occur because it was cheaper, or even more powerful. Instead, it offered factory proprietors improved access to the labor force, as well as additional opportunities to exploit their workers. Water-driven machines required a site near fast-flowing rivers or waterfalls, but potential factory workers in adequate numbers did not necessarily live close to these natural features. As a result, housing had to be built—as at Arkwright's factory in Cromford—at some cost to the owner. Moreover, in more sparsely populated areas, when workers complained about unfair conditions or wages, they were difficult to replace. Unlike flowing water, coal was a mobile energy source. It allowed factory owners to locate their plants in more populated areas, including cities, with growing masses of potential replacement workers hungry for jobs. Urban areas offered police and employable thugs to constrain worker unrest. From the perspective of the *nouveau riche* cotton-mill owners, the higher cost of coal was offset by the expanded power it gave them over their workers. Coal rapidly became the fuel driving capitalist economic growth and expansion into new areas of production, and triggered the full economic conversion to industrial fossil fuel use. This transition was not a natural evolution of ever-improving human technology, but rather was propelled by issues of profit and power. The painful inequalities historically built into industrialization are further seen in the fact that English textile factories received much of their cotton from the American South during a period when slavery provided the labor power for cotton plantations. The enslavement of Africans, in other words, played no small role in the birth of the modern industrial age. Further, coal miners were poorly paid, and mine owners often hired women and children at a much lower wage than men.

The use of coal also led to a massive rise in air pollution. In fact, one of the grievances of early British industrial workers living in densely crowded areas with a concentration of factories concerned the horrible air quality caused by coal smoke (Akatsu 2015). These workers and their families had to contend with noxious clouds of industrial pollutants that caused extensive respiratory diseases and rising death rates. Soot from the factories blackened the buildings around them and even their clothing. Hatton (2017) reports that "[a]s early as the 1850s, higher coal intensity was associated with higher death rates from respiratory diseases, especially among the old and the very young." Though it was not known at the time, the factory smokestacks were also emitting greenhouse gases, which were heating the planet.

The sixth turning point in human history was the global institutionalization of colonialism and of the capitalist world economic system. Colonialism had profound impacts. Local economies were restructured "to meet the needs of industrializing Europe by producing raw materials" (Carmody 2011, p. 1) as ruling class European prosperity soared. The organization of colonial extraction varied by location and colonial power but in all forms had dramatic impact around the globe. In West Africa, for example, the British colonial regime provoked a massive increase of agricultural production for export to Europe without imposing land ownership by non-Africans. By contrast, other areas of the continent, from Kenya to South Africa, witnessed the establishment of European-owned plantations and mines (Smith 1979). The most egregious and oppressive component was the slave trade, which ripped Africans from their homes and converted them and their descendants into sellable commodities who produced wealth for their owners while dwelling in abject poverty and suffering. Ideologically, the slave trade was self-servingly justified as a fair playing out of alleged innate differences between whites and Africans. As a social system whose mission was to create a new world order to the advantage of European countries (Mitchell 1988), colonialism left an indelible legacy of inequality around the globe. Reflecting a pattern of ecological imperialism (Crosby 1986), it also left a parallel legacy of restructured environments (e.g., denuded forestlands, reconfigured landscapes designed for agriculture). On the one hand, this was a product of the wealth extracted from the colonies, which funded the technological and institutional development of the colonial powers. On the other, it derived from the largely unquestioned assumption within Europe that colonists had the right to reconfigure other lands and other lives.

The global spread of capitalism, and the relations of production it brought into being, ultimately forged a world economic system. Wallerstein (1976, p. 229) argues that "a world-system is a social system, one that has boundaries, structures, member groups, rules of legitimation, and coherence." Capitalism allowed "constant economic expansion of the world-system, albeit a very skewed distribution of its rewards" (Wallerstein 1976, p. 230). While the tributary mode of production funded political empires that traversed wide regions under a governmental structure, capitalism subsumed a multiplicity of political entities within a single economy that today encompasses the entire world.

At the economic hub of the world-system are technologically developed "core" nations that have differing but relatively stable state structures: countries like the United States and Japan. These dominate the global market and use it to control and exploit underdeveloped "peripheral" societies, which tend to have weaker or even failed state structures and widespread and often severe levels of poverty. Core nations are class-divided societies with varying degrees of internal social inequality. Even quite wealthy core nations like the United States and United Kingdom have significant levels of wealth, status, and resource inequality. Situated between the hub and the periphery of this economic structure, in Wallerstein's (1976) analysis, are "semiperipheral" societies, such as Mexico, Brazil, and India, which have more developed economies than peripheral ones, and may even have rapidly expanding economic sectors, but also are burdened with steep social inequalities and high levels of poverty. Over time, a region may change its structural role in the world-economy, moving for example from core to semiperiphery or from periphery to semiperiphery, but the basic pattern of disparities within the system persist.

Inherent in its architecture, the world economic system is marked by unequal exchange involving a systematic flow of wealth from the periphery and semiperiphery to the core. This facilitates the accumulation of capital in the core, which is deployed in the expansion of global resource extraction and production and their accompanying escalation of ecosystem simplification and environmental disruption. In the poignant words of O'Dempsey & Munslow (2006, p. 54):

> Africa's canvas of complex humanitarian emergencies is a landscape of exploitation and degradation, wherein the portraits of millions of displaced, impoverished, vulnerable and powerless people are barely discernable to the rapacious eyes of the mighty, dazzled by the shimmer of oil, the sparkle of diamonds, and the glitter of the gilded frame.

Within the world-system, there is also a counterflow of environmental hazards running against the primary movement of resources. This entails the movement of refuse from the core to the periphery, such as the dumping of garbage, radioactive substances, and the worn-out electronic goods known or "e-waste" (Lipman 2002). Strikingly unequal access to the conditions that foster health across the core, semiperiphery, and periphery is another feature of the world-system, a pattern that is replicated within each region of the global complex (Baer *et al.* 2013; Singer & Erickson 2013).

The final turning point in human history, thus far, which extends the trajectory of earlier developments, is the full triumph of neoliberal corporate globalism. All people alive today inhabit a world produced by this potent turn. Because of it, "globalization" and "the global marketplace" are rapidly becoming household terms in the core industrial consumer societies of the world-system. The former is defined as "the closer integration of countries and peoples facilitated by cheaper transport and communications and freer flow of goods, services, capital and people across borders" (O'Dempsey & Munslow 2006). The latter entails access to and sale of goods all over the globe. For example, a garment corporation headquartered in New York City might obtain the material resources it needs to manufacture goods from the Philippines, Nigeria, and Brazil, ship them to a factory in China for assembly onboard container vessels owned by a company based in Denmark, and send the finished products by train to warehouses that supply retail outlets in Italy. This vast web is the global marketplace. It connects Zhang Baobao—a young garment worker who left his family's rural home to find work in a large industrial city in southwest China (Catherall 2017)—to Sidonie Sawyer (2015), who travels from Miami to Paris once a year to buy clothing. While Zhang and Sidonie are not likely to ever meet, their lives are directly and consequentially connected.

In the words of historian Patrick Iber (2018):

> The products we rely on—whether for food, fuel, communication—are woven together in supply chains that have the complexity of medieval tapestries. Some people can make unimaginable fortunes in microseconds, while others still scratch a living out of the dry ground. Eight men, it is calculated, hold as much wealth as the poorest half of the planet: 3.6 billion people. A global system underlies this vastly unequal distribution of wealth and power. But what to call this system? What is its name? It is not enough to call it capitalism, since capitalism has been around for hundreds of years and transforms itself with great dexterity

His answer: neoliberalism, a form of capitalist political economy in which the global system is structured to ensure rights for the owners of capital and to keep their market dominance safe. Neoliberalism has prompted bilateral trade agreements and international treaties that support the power to enforce contracts, keep trade flowing, and prevent the seizure of property by national governments, approaches intended to safeguard the international power of capital even at the price of democracy, fairness, and the health and welfare of most people (Kim *et al.* 2000). When economic problems emerge, like the inability to repay development loans to wealthy nations or to the international lending institutions they command, neoliberalism demands that they be resolved through fiscal structural adjustments at the expense of the poor. Commonly, these austerity measures take the form of government cuts to price controls on basic commodities, civil service employment, public services, and safety-net welfare assistance. If neoliberalism had a slogan, it would be "more markets, less government" (Rodrik 2017).

From the standpoint of environmental sustainability, an issue of direct concern to this book, neoliberalism has emphasized government deregulation. Underlying this is the idea that regulation restricts markets, profits, and economic growth. This assertion is epitomized in Ronald Reagan's inauguration-speech claim that "Government is not the solution to our problem; government is the problem" (Reagan 1981). Moreover, as Thomas Eriksen (2018) points out, neoliberal corporate globalization has broken the chain of responsibility. While in the bygone days of capitalist entrepreneurs like Richard Arkwright and Henry Ford capitalists were easily identifiable, "Today, the situation is far more complicated since ownership structures are transnational, corporate, and complex . . . and [the elite] are further away and less approachable than before."

Several decades ago, in an assessment of international relations, Susan Strange (1991, p. 246) raised two critical questions: "What is power in the world system/international political economy? And who has it?" Her answers emphasize the need to assess the power of corporations to "influence the elements of justice and freedom" within nation-states. Critics of the clout of multinational corporations argue that neoliberal corporate globalization has become a form of neocolonialism or corporatocracy: rule by multinational corporations. As Kim and colleagues (Millen *et al.* 2000, p. 384) note, "[y]esteryear's colonialism laid the foundation of today's neoliberalism." The quintessence of neocolonialism, says Nkrumah (1965), "is that the State . . . is, in theory, independent and has all the outward trappings of international sovereignty . . . [but in] reality its economic system and thus its political policy is directed from outside." Growth in the global power of multinational corporations, in short, has come at the expense of the erosion of the political sovereignty of nation-states. This power is expressed in various ways, including in the pressure corporations apply on poorer nations to have business- rather than worker-friendly labor laws and weak environmental protection regulations (Li & Zhou 2017). State control over multinational corporations is at its weakest in resource-rich but economically poor countries. In Nigeria, for example, the result has been "human rights and environmental abuses by business corporations . . . including transnational companies" (ICJ 2012, pp. 1–2). This paradox, where countries have an abundance of natural resources but suffer from extensive poverty, a lack of democracy, and worse development outcomes than countries with fewer such resources, has been called "the resource curse." In the southeastern Niger Delta region of Nigeria, minimally regulated crude oil extraction by multinational corporations like Shell, Mobil, and Chevron has led to multiple oil spills which have caused serious environmental and social harm, including pollution of the river basin and surrounding land areas, destruction of village subsistence farming and fishing sites, and expropriation of land from local residents. This kind of manipulation of a nation-state's decision-making authority to serve private interests is labeled "state capture," or more precisely in this case, "corporate state capture." Just as oil companies have polluted the environment of the Niger Delta, so have they polluted the politics of Nigeria (Bribena 2011).

An urgent question with clear implications for the relationship between the environment and health is whether neoliberal corporate globalism is still ascending or the social, economic, and other crises it has created have pushed it into retreat. Certainly, there are signs that opposition and doubt have emerged.

One base of opposition is the grassroots global justice movement, which is critical of the expansion of multinational corporate power and advocates for worker and human rights, fair trade, and environmental sustainability. The movement gained attention through a series of large street demonstrations and alternative conferences in response to international meetings of the Group of Eight (G8) Industrialized Nations (later G7), the World Trade Organization, the International Monetary Fund (IMF), and the World Bank (Klein 1999). A second expression of opposition is the appearance even within international promoters of neoliberalism of a critique of aspects of the program. While asserting that "[t]here is much to cheer in the neoliberal agenda," Ostry *et al.* (2016) of the IMF admit "there are aspects of the neoliberal agenda that have not delivered as expected." Two of the shortcomings of neoliberalism from an economic perspective are: 1) it is hard to demonstrate that it has produced economic growth for many countries; and 2) it promotes unemployment and enhances economic inequality.

Finally, there has been a rise during the second decade of the 21st century of an economic nationalist sentiment, as seen in the pronouncements and tariff threats and actions of Donald Trump as president of the United States. Even before Trump's election, in an editorial entitled "The End of Globalization?", the *Washington Post* (2015) remarked: "The seemingly limitless expansion of international trade that begin in the 1980s has slowed in recent years and may never recover its former momentum." During the U.S. presidential campaign of 2016, Trump declared: "We will no longer surrender this country or its people to the false song of globalism" (New York Times 2016). In another campaign speech, he argued:

> Our politicians have aggressively pursued a policy of globalization—moving our jobs, our wealth and our factories to Mexico and overseas. Globalization has made the financial elite who

> donate to politicians very wealthy. But it has left millions of our workers with nothing but poverty and heartache. (Politico 2016)

While initially criticism of globalism came from those on the political left, former president Trump and others represent a backlash that has emerged from the right. This response is not only found in the United States, as Trump's election provided "momentum to right-wing anti-globalisation throughout the world, and especially Europe" (Macgregor-Bowles & Bowles 2017, p. 141). The new brand of anti-globalism, however, "is not just about economic globalisation. It has involved ethno-nationalist and anti-immigrant components, for example among supporters of Trump" (Horner *et al.* 2018). To the ultra-right, the term "globalist" has long been used as a cover-term for Jews as part of its nativist anti-Semitic mythologies (Neiwert 2017).

But was Trump or his administration really anti-globalist? The Trump organization as a business entity has never operated in accordance with his anti-globalist oratory. Over the years, Trump has built a multinational business with holdings, investments, and other dealings in various foreign markets around the globe. Moreover, the Trump Organization outsources the production of Trump brand name merchandise to overseas companies and imports materials from multiple foreign producers for construction and manufacturing operations in the United States (Scoffen & Tan 2016; Hill 2017). In terms of his personal business activities, in short, Trump never acted like an anti-globalist, but rather as a quite enthusiastic international player. Similarly, while as president Trump stressed his "America First" agenda in speeches and other declarations, he affirmed in his presentation to the World Economic Forum in Davos, Switzerland in 2018 that "America is open for business" (Horsley & Naylor 2018). Additionally, a report issued by Good Jobs Nation (Faraday 2018), based on an analysis of contracts issued by the federal government, "found no evidence that the Trump administration has done anything to halt, or even slow, the outsourcing of American jobs by federal contractors." Rather, during Trump's presidency, contractor outsourcing has grown to record levels. As Campoy (2018) suggests, "Trump's trade philosophy—which appears to be mostly based on the principle of squeezing partners to extract privileges—harks back to the first wave of globalization [at the turn of the 20th century]. That first huge wave in the global exchange of goods took place amid stiff barriers [i.e., tariffs] erected by imperial powers hell-bent on conquering new markets while protecting their own." In short, Trump appears to engage in a brand of what might be called "anti-globalization globalization" that consists of public anti-globalist political rhetoric and active pro-globalist economics.

3.3 Environmental neoliberalism and the polluting elites

A classic book in the literature of sociology is *The Power Elites* by C. Wright Mills (1956). It sparked a realization of the importance of studying up and inspired other analyses like C. William Domhoff's *Who Rules America: The Triumph of the Corporate Rich* (2014) and Peter Phillips' *Giants: The Global Power Elites* (2018b). These works assess the supremacy of the corporate master class, the managers of global capital who represent one percent of the world's population but control more than half of its total wealth. Year after year, the global power elites "accumulate greater concentration of wealth and are embedded in an unrelenting guest for more" (Phillips 2018b, p. 34).

One sector of the power elites is of special concern to this book: the global polluting elites and climate changers (Kenner 2019). These are the very wealthy executives who direct the production processes that cause many of the estimated 9 million human deaths each year from pollution (Landrigan *et al.* 2018), 3 million of which result from outdoor air pollution alone (WHO 2016). They are the heads and primary owners (directly or as the largest stock holders) of fossil fuel, mining, other manufacturing, and agribusiness companies, their net worth, luxury lifestyle, and extensive political influence resting on wealth that comes from polluting activities. Every year, *Forbes* publishes its list of the 75 most powerful people in the world. Exemplary is Darren W. Woods, number 34 on the *Forbes* list, and Chairman of the Board, President, and Chief Executive Officer of ExxonMobil since 2017 (Ewalt 2018). Woods began working at ExxonMobil in 1992 as a planning analyst after completing a bachelor's degree in electrical engineering at Texas A&M University, followed by an MBA from Northwestern's Kellogg School of Management. Prior to becoming CEO, Woods directed the refining and chemical divisions of the company, the source of the majority of ExxonMobil's $7.8 billion net income. Before his promotion, Woods' estimated net worth was at least $31.5 million dollars (as of May 2015). His total annual compensation (as of 2019) is $18 777 800—more than anyone else at ExxonMobil (Wallmine 2019).

Woods' office in the executive wing of ExxonMobil's headquarters outside Dallas is nicknamed the God Pod. It is from here that he has promoted a company agenda of new drilling and refinery expansions, from the Permian Basin shale fields of West Texas to the deep waters off the coast of Guyana, with major production in Angola, Canada, Russia, Qatar, and the United Arab Emirates. Under Woods, the company announced it had discovered a significant oil field off the coast of Nigeria, began drilling in Liberia, increased investments in gas production and exports in Papua New Guinea, and expanded its refineries in Singapore and the Netherlands. All of this growth is being driven by hefty increases in both capital development and exploration spending. Woods' goal in the midst of global warming is a determined expansion of fossil fuel production (Krauss 2017).

Moreover, under his direction, in 2018, ExxonMobil spent $11 million on lobbying policy-makers and made over $2 million in campaign contributions (Center for Responsive Politics 2019). Notably, 27 out of 42 ExxonMobil lobbyists in 2017–2018 previously held government jobs, highlighting the revolving door that links U.S. office holders and private companies with a keen interest in government fossil fuel decision-making. ExxonMobil historically "has enjoyed a close relationship with Congress, successfully lobbying to gain commercial access to federal lands as well as the rollback of several Environmental Protection Agency initiatives deemed unfriendly to the oil industry" (Center for Responsive Politics 2019).

According to Richard Heede's (2014) analysis, ExxonMobil was responsible for over 3 percent of the world's total fossil fuel emissions between 1854 and 2010. Only Chevron accounts for more historic emissions, at 3.5 percent. In its Toxic 100 Air Polluters Index for 2018, the Political Economy Research Institute (PERI 2019) at the University of Massachusetts, Amherst listed ExxonMobil in ninth place. In 2014 alone, the company released over 38 million metric tons of CO_2 equivalent (which is a measurement that incorporates CO_2 as well as methane and other greenhouse gases). Hand-in-hand with its role in polluting has been the company's aggressive stance on denying and obscuring the role of fossil fuel use in driving climate change. This pattern has characterized ExxonMobil's practice during Woods' entire career, and as described in Chapter 8, continues to this day.

The impact of polluting elites on the world is considerable. As the Lancet Commission on Pollution and Health summarizes, "ambient air pollution, chemical pollution, and soil pollution—the forms of pollution produced by industry, mining, electricity generation, mechanised agriculture, and petroleum-powered vehicles—are all on the rise" (Landrigan *et al.* 2018, p. 462). Polluting elites are responsible for the release of dangerous and unnatural solid, gas, and liquid contaminants that replace or damage natural ecosystem components and have the ability to cause adverse system imbalance while threatening the health of flora and fauna, including humans. Increasingly since the beginning of the Industrial Revolution, pollution has been one of the most serious problems facing Earthlings of all species. Of particular concern here is corporate pollution and the super-rich who control polluting industries, including manufacturing, mining, and transportation. It is estimated that the rich "pollute the world fifty times more than the poor; but the latter are more likely to die of pollution" (Hart 2002, p. 27).

A particularly vivid example of the corporate polluting of the environment involves the St. Lawrence River in northern New York State. The St. Lawrence, one of the largest rivers in the world, provides a gateway between the Atlantic Ocean and the Great Lakes. Jacques Cartier, a Breton explorer sent in 1534 by King Francis I of France, was the first European to describe and map the area. He called it "La Grand Rivière" and claimed it for France despite its already being the home of indigenous people. The region he visited is covered by deciduous, mixed, and coniferous forests and populated by common north woodlands mammals like deer, muskrat, and bear, fish like pike and smallmouth bass, and many birds, including massive seasonal migrations of ducks and geese navigating the Atlantic Migratory Fly-Way. The area is home to several endangered species, including the Indiana bat (*Myotis sodalis*) and Blanding's turtle (*Emys blandingii*).

The primary Indian people Cartier encountered in his voyages to the New World, the St. Lawrence Iroquoians, disappeared by the end of the 16th century, for reasons that are still in dispute among archeologists (possibilities include intertribal conflict spurred by the European fur trade, European diseases, and absorption into the Mohawk and Onondaga tribal groups) (Trigger 1996). Beginning in the mid-18th century, Mohawk families settled in the region, a place they probably already knew well from earlier trading journeys. Their first small settlement was named the Saint Regis after the Catholic mission established there in 1755. Now known as Akwesasne, meaning "where the partridge drums," this settlement, which stretches along the shores and islands of the St. Lawrence, today occupies about 4400 sq. miles in the U.S.–Canada border area shared by New York State and the provinces of Ontario and Quebec and is home to over 12 000 people (Bonaparte 2008).

During the 1950s, various industries, including Reynolds Metals, General Motors, and Alcoa, established facilities on the banks of the St. Lawrence because of the low-cost energy available from the system of hydroelectric power dams constructed as part of the St. Lawrence Seaway (Singer & Hodge 2016). All of these factories made extensive use of polychlorinated biphenyls (PCBs), primarily produced by the Monsanto Chemical Company, dumping thousands of pounds of them into the river. PCBs are a set of anthropogenic organic chemicals consisting of atoms of carbon, hydrogen, and chlorine. They have multiple industrial applications in insulating fluids in capacitors, transformers, and other electrical equipment, as well as in plasticizers in paint. Their appeal in industrial production stems from their nonflammability, chemical stability, high boiling point, and electrical insulating properties. They also do not quickly break down once in the environment. They can remain for protracted periods, cycling among air, water, and soil. Further, they can bioaccumulate in animals, including fish, leading to exposure for people who ingest them (Lacetti 1993).

PCB exposure has been shown to have significant adverse health effects. Studies in humans indicate potential carcinogenic and noncarcinogenic adverse effects, including cognitive and motor development disorders (Berghuis *et al.* 2018), neurologic and psychiatric disorders (Coxon *et al.* 2018), hypertension and possibly cardiovascular disease (Raffetti *et al.* 2018), and diabetes (Silverstone *et al.* 2012). Identified environmental and health risks associated with PCBs led to their production being banned in the United States in 1979. PCBs were never manufactured in Canada.

In addition, aluminum refining in the St. Lawrence area produced various other toxic substances, such as fluoride, which were emitted into the air through unfiltered smokestacks.

These companies, and their owners and directors, were the polluting elites of the St. Lawrence River. Ultimately, General Motors was found by the Environmental Protection Agency (EPA 1990) to be the largest PCB polluter, with 10 dump sites along the St. Lawrence within the company's 210-acre riverfront property. It also built open pits to dispose of PCBs and other toxic chemicals (e.g., volatile organic compounds) and discharged PCB-contaminated waste into the river.

The area was designated as the St. Lawrence River Area of Concern (AOC) under the 1987 Great Lakes Water Quality Agreement between the United States and Canada. Environmental degradations listed in the AOC included loss of fish and wildlife habitat, impairment of fish and wildlife populations, fish tumors or other deformities, bird or animal deformities or reproduction problems, and demise of phytoplankton and zooplankton populations

In June 2009, during the automotive industry crisis, General Motors filed for bankruptcy. When it emerged unscathed, it was freed by the U.S. government of any legal obligation for its polluted properties at discarded plant sites (Higgins 2009). Some areas along the St. Lawrence were designated Superfund sites, which ranked them among the nation's most hazardous hot spots, including the 1600-acre former Reynolds Metals facility and the Alcoa factory sites. The EPA dredged 20 200 pounds of PCBs from the riverbed adjacent to Reynolds' Alcoa East facility, while Alcoa removed 8000 pounds from the Alcoa West site. The latter, enough to fill 33 railroad cars, were mislabeled as nonhazardous and shipped to mega-landfills in Alabama (Taylor 2014). Necessary long-term cleanup activities continue at the St. Lawrence site today (EPA 2012a, 2012b, 2018). Paul O'Neill, CEO of Alcoa from 1987 to 1999, referred to the polluting of the St. Lawrence by his company as an "unfortunate episode" (Sack 1991). While Alcoa was fined for its despoiling of the St. Lawrence, no elite polluters at the company, or at General Motors or Reynolds Metals, went to jail. As Greider (1984) observes, "A street hoodlum who assaults someone with a gun or a knife will normally get locked up for his crime, but the swift sword of justice hesitates when a business executive assaults an entire community with poisonous chemicals."

Tellingly, it has been revealed that Monsanto knew the PCBs it produced were "harmful and pervasive in the environment, and kept selling them in spite of that fact," according to Bill Sherman, Washington assistant attorney general; he adds, "They knew the dangers, but hid them from the public in order to profit" (quoted in Chow 2017). This assertion is based on Monsanto's own documents. It reflects a common pattern of knowing corporate production of "killer commodities" that put public health in jeopardy (Singer & Baer 2008).

The Akwesasne have experienced diverse environmental and health consequences because of their prolonged exposure to PCBs. Animals hunted for food and pelts have become scarce. Mohawk fisherman have found that the catfish, carp, and trout caught in their nets suffer from open skin lesions. Francis Jock, a Mowhawk fisherman, told a reporter that PCBs "got into everything connected to this river" (quoted in Schneider 1994). At the initiative of community members, the Akwesasne Task Force on the Environment (1996) was formed in the mid-1990s "to conserve, preserve, protect, and restore the environment, natural and cultural resources within the territory of Akwesasne." In partnership with university investigators, it undertook research to examine the health effects of PCB exposure among the Akwesasne (Santiago-Rivera *et al.* 1998).

Examination of the relation between the consumption of local fish and concentrations of PCBs in the milk of 97 nursing Mohawk mothers found a significant correlation between estimated lifetime fish consumption and PCB content (Fitzgerald *et al.* 1998). Research by Schell *et al.* (2008) demonstrated a reduction in thyroid function among Akwesasne Mohawk youth directly related to their blood level of PCBs. Attention and impulsivity problems among Akwesasne Mohawk males aged 17 to 21 also showed a significant positive relationship with blood PCB levels (Behforooz *et al.* 2017). Adolescent long-term memory and comprehension suffered "subtle negative effects of PCB exposure" (Newman *et al.* 2006, p. 439). Awkwesasne men exposed to PCBs were found to have lower testosterone levels (Goncharov *et al.* 2009). Other, longer-term consequences such as asthma and diabetes continue to mount in the Akwesasne population because of the enduring presence of industrial pollutants and other contaminants on their lands (Aminov *et al.* 2016).

Love Canal, New York offers another quite vivid example of the corporate polluting of the environment with significant health consequences. The story begins in 1892 when William T. Love, an entrepreneur from the Western Railroad Corporation, proposed digging a six-mile-long canal to connect the upper and lower Niagara River. His goal was to terminate the water of the upper Niagara in an artificial waterfall with a 280-foot drop into the lower Niagara as a source of cheap hydro-power. He proceeded to obtain the necessary land, hire workers, and start digging. However, a serious downturn in the national economy, the Panic of 1893, drove away Love's financial backing and the project was abandoned, leaving behind a partially dug mile-long section of canal.

In 1920, the land was sold at public auction and became a municipal and chemical disposal site until 1953. The main company that dumped there was Hooker Chemical Corporation, an eventual subsidiary of Occidental Petroleum. The City of Niagara also used Love Canal as a dump site for municipal garbage, and there are reports that the U.S. Army used it as a disposal site for the Manhattan Project (a World War II initiative that produced the first nuclear weapons) and other chemical warfare waste materials (Center for Health, Environment & Justice 2015). By 1953, it is estimated that about 21 000 tons of chemicals, many of them quite toxic, had filled the canal. No longer of use, Hooker workers piled 20 feet of dirt on top of it.

Facing local population growth, officials of the City of Niagara Falls decided to purchase the site in order to construct new neighborhoods and schools. Hooker was hesitant to sell due to safety concerns, but the city expropriated the property and built the 99th Street School and several neighborhoods on and around the buried canal. People moved in and began raising their families and sending their children to the new school. They were never given any warning or information that the property was located near a hazardous chemical waste dump (Haiman 2015).

Things began to go wrong. People complained of foul odors and reported that puddles of oil were forming on the ground and that strange colored liquids were seeping into their basements. Children started developing chemical burns and rashes on their hands and faces after playing outside. By the 1970s, local rates of various health problems began to go up. After years of receiving citizen complaints, the city and county hired a consulting firm to investigate. In 1976, the Calspan Corporation completed a study of the canal area and reported finding toxic chemical residues in the air and sump pumps of a high percentage of homes at the southern end of the canal. Calspan personnel also found 55-gallon drums just beneath or on the surface and high levels of PCBs in the storm sewer system. Very little was done, however, to respond to these findings, and the complaints of residents continued. In 1978, the New York State Department of Health conducted air and soil tests in basements and undertook a health study of the 239 families living near the canal. This research found an increase in reproductive problems among women, including miscarriages, still births, and crib deaths, and high levels of chemical contaminants in the soil and air. Rates of nervous breakdowns, hyperactivity, epilepsy, and urinary tract disorders also were elevated. Ultimately, in 1979, the EPA declared the area around Love Canal an emergency disaster area.

In 2008, on the 30th anniversary of the closing of the 99th Street School, Lois Marie Gibbs angrily reiterated to reporters her complaint about the government's handling of the disaster that brought her national attention as a community leader:

> Thirty years later, I'm still fighting the Love Canal battle . . . It's a constant fight. Thirty years ago we had to fight to learn why our kids were getting sick, why they were developing rare illnesses. We had to fight to win evacuation for those of us living amongst toxic chemicals that threatened harm to our bodies. We had to fight for years to win a medical fund to help us with our medical bills. We continue to fight to get the state of New York and every other state to stop allowing schools to be built on toxic and hazardous sites like the 99th Street School was. And now in 2008, we have to fight again to get the New York Department of Health to fully study and present findings about the real results from Love Canal—not just partial data. We don't want a whitewash, we want the truth. (quoted in PR Newswire 2008)

Three years after the 99th Street School was closed, Gibbs launched the Center for Health, Environment & Justice to help other communities and families threatened by conditions similar to those at Love Canal. According to the Center, "Love Canal awoke a community to the unpleasant and unfortunate realization of how toxic wastes affect our lives, and destroy our environment. Residents at Love Canal always believed that the government would automatically protect them. They were wrong; in some cases dead wrong!" (Center for Health, Environment & Justice 2015).

3.4 The Anthropocene or the Capitolocene?

In the previous chapters, the current geotemporal epoch was referred to as the Anthropocene, reflecting the dramatic influence of humans on Earth's atmospheric, geologic, hydrologic, biospheric, and other system processes. In light of the issues explored in the present discussion, some social and environmental scientists wonder if this is the most accurate label for the current era. An alternative designation that has been proposed by scholars who study the socioeconomic factors driving our current precarity is "Capitalocene." The crux of their argument is the question of culpability. They point out that "Anthropocene" suggests that collectively humans as an undifferentiated whole are equally involved in the unprecedented reconfiguring of Earth. In the edited volume *Anthropocene or Capitalocene? Nature, History, and the Crisis of Capitalism* (Moore 2016), advocates of the need to bring capitalism into the discussion stress that the environment is not just a context in which human relationships transpire; its shifting form, and its kairos, are direct reflections of the structure of changing political-economic social relations through time, most dramatically since the rise of capitalism. The real problem, in short, is capitalism. From this perspective, the Capitalocene, the age of capital, is "a multispecies assemblage, a world-ecology of capital, power, and nature" (Moore 2016, p. xi). Historically, during the life course of capitalism, "successive state–capital–science complexes . . . map, identify, quantify and otherwise" make the environment legible and useful to capitalist production, with increasingly dismal environmental consequences (Moore 2017, p. 237). Capitalism does not produce the environment; it co-produces it with natural forces. Humanity as a whole, however, if it is ever again to inhabit an equitable and sustainable home, does have responsibility—namely, to bring a better world into being.

Contributors to Moore's book raise various issues with the term "Anthropocene." Justin McBrien (2016) points out that the accumulation of capital is the accretion of potential extinctions of world-ecologies. He uses the term "Necrocene" to label the era of capitalist-caused extinctions of species and depletions of natural resources. Elmar Alvater (2016) adds that in the Capitalocene, nature may be romantically idealized as a reaction to the environmental assaults of the Industrial Revolution, but in its treatment it continues to be converted into a capitalist asset. For Eileen Crist (2016), the term "Anthropocene" is anthropocentric. It implies that ours is a brilliant if disruptive species fully distinguished from other Earthlings and worthy of a demarcating label. She maintains that what we need to strive for is an ecozoic era in which humans and other animals form a "union in diversity" (Crist 2016, p. 141). To achieve such a world, Ian Angus (2016, p. 216) argues, the "climate change vandals much be stopped. We need to work with everyone who is willing to join in fighting climate change in general, and the fossil fuel industry specifically." By adopting the term "Anthropocene," says Baer (2017), earth scientists and others have failed to acknowledge that "global capitalism is the elephant in the room when it comes to ecological and climatic crises"—a point also highlighted by Newell (2011).

One sign of the elephant is deadly land and water grabs by transnational and domestic companies seeking to gain access to profitable resources and productive locations. These drives for control of real estate and water rights have led to large-scale displacements of local residents. Resisters and indigenous environmental defenders in various parts of the world are killed every year protecting their land, forests, and waterways against the expansion of commercial agriculture, mining companies, and logging interests. This unreported aspect of the Capitalocene continues unembedded. Global Witness (2014), an environment and human rights organization, reports that in the years 2011 and 2012:

> indigenous leader Jimmy Liguyon was shot dead in front of his family in the Philippines, reportedly for refusing to sign away his land to miners. José Cláudio Ribeiro da Silva and his wife Maria were assassinated by masked gunmen in Brazil . . . after having denounced the encroachment of loggers into their forest reserve . . . [I]n Cambodia, 14-year-old Hen Chantha was shot dead by military police who were evicting her village to make way for a rubber plantation.

In 2017, another 207 environmental activists were murdered, making it the deadliest year on record according to Global Witness (2017). Agribusiness had the most deaths associated with it, with a reported 46 activists killed in disputes over large-scale agriculture projects. Next most deadly were the oil and mining industries, which historically have been particularly dangerous for environmental activists. The majority (almost 60 percent) of the assassinations of environmental activists in 2017 occurred in Latin America, followed by the Philippines. Brazil was an especial risky place, with an estimated 57 murders. Additionally, Ben Leather, the lead investigator on the study for Global Witness, notes that "There are a whole range of tactics used to silence [environmental] defenders and the killings only represent the sharpest edge of that" (quoted in Zachos 2018). One of these tactics is the issuing of death threats. Maria do Socorro Costa da Silva, for example, lives in an indigenous community in the Amazonian state of Pará next to the Tocantins River. She and other activists have waged a struggle against water pollution and other environmental damage caused by the largest aluminum processing plant in the Amazon. Waste from the plant and related facilities, they assert, has caused diarrhea and vomiting and poisoned fish and local produce. According to da Silva, who has been diagnosed with cancer, "[e]very day we drink the water, every day we die a little bit. This is not just now, it has been happening for years" (quoted in Watts 2018). The Oslo-based Norsk Hydro owns 92 percent of the aluminum refinery (the world's largest outside China), as well as 51 percent of the Hydro Albras aluminum factory. In response to their activism, da Silva and her fellow activists have been threatened with bodily harm. In December 2017 and March 2018, Fernando Perreira and Paulo Sérgio Nascimento, two leaders of the community activist group Cainquiama, were shot dead (Phillips 2018a; Watts 2018). Says da Silva, "Fernando died, Paulo Sergio died, but the truth is that it [the industrial plant] is slowly killing all of us" (quoted in Watts 2018).

3.5 The future of Eaarth

The future of Earth as we know it, or "Eaarth" as Bill McKibben titles our already different planet, is contingent. A dark, worst-case, deadly Earth scenario envisions a future in which major cities like New York and Amsterdam sink below a rising ocean, famine is widespread because of unpredictable or inadequate precipitation, species extinctions and loss of biodiversity run rampant, safe drinking water becomes scarce, the oceans are acidic, wildfires are intense and the fire season ever longer, the air is increasingly toxic, deadly heat waves are common and costly, and disease vectors spread around the globe, inflicting additional suffering and loss of life. In McKibben's words, "If not hell then a place with a similar temperature" (quoted in Loria 2017).

By no means do most environmental scientists predict such a macabre near future. The majority, however, recognize that ecocrises are already taking their toll, and that it is increasing day by day. The greater the number and the impacts of ecocrises, the higher the potential for interactions to occur that magnify their overall health, social, and environmental harms. Stopping further advance down the bleak pathway of cascading ecocrises is consequently, without doubt, absolutely vital to human health and well-being and the survival of untold other species.

References

Akatsu, M. (2015). The problem of air pollution during the industrial revolution: a reconsideration of the enactment of the Smoke Nuisance Abatement Act of 1821. In: S. Sugiyama (Ed.), *Economic History of Energy and Environment*, pp. 85–112. Tokyo: Springer.

Akwesasne Task Force on the Environment. (1996). *Protocol for Review of Environmental and Scientific Research Proposals*. Hogansburg, NY: Akwesasne Task Force on the Environment.

Alvater, E. (2016). The capitalocene, or, geoengineering against capitalism's planetary boundaries. In: Moore, J, (Ed.), *Anthropocene or Capitalocene? Nature, History, and the Crisis of Capitalism*, pp. 138–152. Oakland, CA: PM Press.

Aminov, Z., Haase, R., Rej, R., Schymura, M., Santiago-Rivera, A., Morse, G. *et al.* (2016). Diabetes prevalence in relation to serum concentrations of polychlorinated biphenyl (PCB) congener groups and three chlorinated pesticides in a Native American population. *Environmental Health Perspectives* 124: 1376–1383.

Angus, I. (2016). *Facing the Anthropocene: Fossil Capitalism and the Crisis of the Earth System*. New York: Monthly Review Press.

Baer, H. (2017). Anthropocene or Capitalocene? Two political ecological perspectives. *Human Ecology* 45(3): 433–435.

Baer, H., Singer, M. & Susser, I. (2013). *Medical Anthropology and the World System: A Critical Perspective*. Santa Barbara, CA: Praeger.

Behforooz, B., Newman, J., Gallo, M., Akwesasne Task Force on the Environment & Schell, L. (2017). PCBs and measures of attention and impulsivity on a continuous performance task of young adults. *Neurotoxicology and Teratology* 64: 29–36.

Berghuis, S., Van Braeckel, K., Sauer, P., & Bos, A. (2018). Prenatal exposure to persistent organic pollutants and cognition and motor performance in adolescence. *Environment International* 29(121): 13–22.

Bonaparte, D. (2008). *Creation & Confederation: The Living History of the Iroquois*. Akwesasne, NY: Wampum Chronicles.

Braudel, F. (1974). *Capitalism and Material Life, 1400–1800*. New York: Harper Torchbooks.

Bribena, E. (2011). Nigeria, oil and gas exploration and the Niger Delta Question: a study in corporate social responsibility. Thesis submitted for the degree of Doctor of the Law. Potchefstroom: North-West University.

Brown, E. (1974). *The Tyranny of a Construct: Feudalism and Historians of Medieval Europe*. Oxford: Oxford University Press.

Büntgen, U., Tegel, W., Nicolussi, K, McCormick, M., Frank, D. Trouet, V. *et al.* (2011). 2500 years of European climate variability and human susceptibility. *Science* 331(6017): 578–582.

Campoy, A. (2018). Trump is a globalist. Just a chaotic one. *Quartz*. https://qz.com/1246283/trump-is-a-globalist-just-a-chaotic-one/.

Carmody, P. (2011). *The New Scramble for Africa*. Cambridge: Polity Press.

Carto, S., Weaver, A., Hetherington, R., Lam, Y. & Wiebe, E. (2009). Out of Africa and into an ice age: on the role of global climate change in the late Pleistocene migration of early modern humans out of Africa. *Journal of Human Evolution* 56(2): 139–151.

Catherall, S. (2017). Who made my clothes? Inside Cotton On's Chinese factories. *Life & Style*. https://www.stuff.co.nz/life-style/fashion/93382961/Who-made-my-clothes-Inside-Cotton-Ons-Chinese-factories.

Center for Health, Environment & Justice (2015). Love Canal: A Guidebook. http://chej.org/publications/love-canal-a-guidebook/.

Center for Responsive Politics. (2019). Client Profile: Exxon Mobil. https://www.opensecrets.org/lobby/clientsum.php?id=d000000129&year=2018.

Chapman, S. (2013). The cost of power in the industrial revolution in Britain: the case of the textile industry. *Midland History* 1(2): 1–24.

Chow, L. (2017). Poison papers: Monsanto knew pcbs were toxic for years but sold them anyway. *EcoWatch*. https://www.ecowatch.com/monsanto-pcbs-2470957270.html.

Conroy, G. & Pontzer, H. (2012). *Reconstructing Human Origins: A Modern Synthesis*. New York: W.W. Norton.

Cowen, R. (2013). *History of Life*. Oxford: Blackwell Scientific.

Coxon, T., Goldstein, L. & Odhiambo B. (2018). Analysis of spatial distribution of trace metals, PCB, and PAH and their potential impact on human health in Virginian counties and independent cities, USA. *Environmental Geochemistry and Health* 41: 783–801.

Crist, E. (2016). On the poverty of our nomenclature. In: Moore, J. (Ed.), *Anthropocene or Capitalocene? Nature, History, and the Crisis of Capitalism*, pp. 14–33. Oakland, CA: PM Press.

Crosby, A. (1986). *Ecological Imperialism: The Biological Expansion of Europe, 900–1900*. Cambridge: Cambridge University Press.

Diaz, H. & Hughes, M. (1994). *The Medieval Warm Period*. Boston, MA: Kluwer Academic.

Domhoff, W. (2014). *Who Rules America: The Triumph of the Corporate Rich*. New York: McGraw Hill.

EPA. (1990). Superfund record of decision: General Motors/Central Foundry Division, NY. Environmental Protection Agency. https://cumulis.epa.gov/supercpad/SiteProfiles/index.cfm?fuseaction=second.ars&id=0201644&doc=Y&colid=63986&region=02&type=AR.

EPA. (2012a). Reynolds Metals Company. Region 2 Superfund. Environmental Protection Agency. https://cumulis.epa.gov/supercpad/cursites/csitinfo.cfm?id=1000383.

EPA. (2012b). Aluminum Company of America. Region 2 Superfund. Environmental Protection

Agency. https://cumulis.epa.gov/supercpad/cursites/cscdocument.cfm?id=0700087.

EPA. (2018). Construction work scheduled to resume in Massena, N.Y. in preparation for cleanup of the Grasse River Superfund Site. Environmental Protection Agency. https://archive.epa.gov/epa/newsreleases/construction-work-scheduled-resume-massena-ny-preparation-cleanup-grasse-river.html.

Eriksen, T. (2018). How globalization has broken the chain of responsibility. *Sapiens*. https://www.sapiens.org/culture/globalization-downfall-gladstone-australia/.

Ewalt, D. (2018). The world's most powerful people 2018. *Forbes*. https://www.forbes.com/sites/davidewalt/2018/05/08/the-worlds-most-powerful-people-2018/#706065a56c47.

Faraday, G. (2018). Trump's First Year: Outsourcing by Federal Contractors Worsens. Good Jobs Nation. https://actionnetwork.org/user_files/user_files/000/019/433/original/GOOD_JOBS_NATION_Trump's_First_Year_-_Outsourcing_by_Federal_Contractors_Worsens.pdf.

Fitzgerald, E., Hwang, S., Bush, B., Cook, K. & Worswick, P. (1998). Fish consumption and breast milk PCB concentrations among Mohawk women at Akwesasne. *American Journal of Epidemiology* 148(2): 164–172.

Fleming, P. (2001). *Family and Household in Medieval England*. New York: Palgrave.

Gates. H. (1996). *China's Motor: A Thousand Years of Petty Capitalism*. Cornell, NY: Cornel University Press.

Global Witness. (2014). Deadly environment: the dramatic rise in killings of environmental and land defenders. https://www.globalwitness.org/en/campaigns/environmental-activists/deadly-environment/.

Global Witness. (2017). At what cost: irresponsible business and the murder of land and environmental defenders in 2017. https://www.globalwitness.org/en/campaigns/environmental-activists/at-what-cost/.

Goncharov, A., Rej, R., Negoita, S., Schymura, M., Santiago-Rivera, A., Morse, G. *et al.* (2009). Lower serum testosterone associated with elevated polychlorinated biphenyl concentrations in Native American men. *Environmental Health Perspectives* 117(9): 1454–1460.

Green, R., Krause, J., Briggs, A., Maricic, T., Stenzel, U., Kircher, M. *et al.* (2010). A draft sequence of the neandertal genome. *Science* 328(5879): 710–722.

Greider, W. (1984). Rx for the environment. *Rolling Stone*. https://www.rollingstone.com/culture/culture-news/rx-for-the-environment-62616.

Gupta, A. (2004). Origin of agriculture and domestication of plants and animals linked to early Holocene climate amelioration. *Current Science* 87(1): 54–59.

Haiman, A. (2015). The story of Love Canal. *A Birding Naturalist*. https://abirdingnaturalist.wordpress.com/2015/05/03/the-story-of-love-canal/.

Hart, K. (2002). World society as an old regime. In C. Shore & S. Nugent (Eds.), *Elite Cultures: Anthropological Perspectives*, pp. 22–41. New York: Routledge.

Hatton, T. (2017). Air pollution in Victorian-era Britain—its effects on health now revealed. *The Conversation*. https://theconversation.com/air-pollution-in-victorian-era-britain-its-effects-on-health-now-revealed-87208.

Heede, R. (2014). Tracing anthropogenic carbon dioxide and methane emissions to fossil fuel and cement producers, 1854–2010. *Climate Change* 122(1–2): 229–241.

Higgins, T. (2009). GM gets to dump its polluted sites. The Center for Auto Safety. https://www.autosafety.org/gm-gets-dump-its-polluted-sites/.

Hill, J. (2017). A look inside Trump's global business interests. *ABC News*. https://abcnews.go.com/Politics/inside-trumps-global-business-interests/story?id=44416694.

Hodgkins, J., Marean, C., Turq, A., Sandgathe, D., McPherron, S. & Dibble, H. (2016). Climate-mediated shifts in Neandertal subsistence behaviors at Pech de l'Azé IV and Roc de Marsal (Dordogne Valley, France). *Journal of Human Evolution* 96: 1–18.

Horner, R., Haberly, D., Schindler, S. & Aoyama, Y. (2018). How anti-globalisation switched from a left to a right-wing issue—and where it will go next. *The Conversation*. https://theconversation.com/how-anti-globalisation-switched-from-a-left-to-a-right-wing-issue-and-where-it-will-go-next-90587.

Horsley, S. & Naylor, B. (2018). Trump to Davos: "America is open for business." *NPR*. https://www.npr.org/2018/01/26/580319938/trump-to-davos-america-is-open-for-business.

Iber, P. (2018). Worlds apart: how neoliberalism shapes the global economy and limits the power of democracies. *New Republic*. https://newrepublic.com/article/147810/worlds-apart-neoliberalism-shapes-global-economy.

ICJ. (2012). Access to Justice: Human Rights Abuses Involving Corporations—Nigeria. Geneva: International Commission of Jurists.

Kaplan, H., Hill, K., Cadeliña, R., Hayden, B., Hyndman, D., Preston, R. *et al*. (1985). Food sharing among Ache foragers: tests of explanatory hypotheses. *Current Anthropology* 26: 223–245.

Kenner, D. (2019). *Carbon Inequality: The Role of the Richest in Climate Change*. London: Routledge.

Kim, J., Millen, J., Irwin, A. & Gershman, J. (Eds.). (2000). *Dying for Growth: Global Inequality and the Health of the Poor*. Monroe, MA: Common Courage Press.

Klein, N. (1999). *No Logo: Taking Aim at the Brand Bullies*. Toronto, ON: Random House Canada.

Krauss, C. (2017). Darren Woods, Exxon's new chief, begins to make his mark. *New York Times*. https://www.nytimes.com/2017/03/01/business/energy-environment/darren-woods-exxon-mobil-investors.html.

Lacetti, G. (1993). A Public Health Assessment: General Motors/Central Foundry Division, Massena, St. Lawrence County, NY. Albany, NY: New York State Department of Health.

Landrigan, P., Fuller, R., Acosta, N., Adeyi, O., Arnold, R., Basu, N. *et al*. (2018). The Lancet Commission on Pollution and Health. *Lancet* 391: 462–512.

Leacock, E. (1992). Women's status in egalitarian society: implications for social evolution. *Current Anthropology* 33(1): 225–259.

Lee, R. (2018). Hunter-gatherers and human evolution: new light on old debates. *Annual Review of Anthropology* 47: 513–531.

Li, X. & Zhou, Y (2017). Offshoring pollution while offshoring production? *Strategic Management Journal* 38(11): 2310–2329.

Lipman, Z. (2002). A dirty dilemma: hazardous waste trade. *Harvard International Review* 23(4): 67–71.

Loria, K. (2017). The planet's worst-case climate scenario: "If not hell then a place with a similar temperature." *Business Insider*. https://www.businessinsider.com/climate-change-worst-case-scenario-bill-mckibben-2017-8.

Macgregor-Bowles, I. & Bowles, D. (2017). Trump, Brexit, right-wing anti-globalisation, and an uncertain future for public health. *AIMS Public Health* 4(2): 139–148.

Mark, E. (2016). Daily life in ancient China. *Ancient History Encyclopedia*. https://www.ancient.eu/article/890/daily-life-in-ancient-china/.

McBrien, J. (2016). Accumulating extinction planetary catastrophism in the Necrocene. In: J. Moore (Ed.), *Anthropocene or Capitalocene? Nature, History, and the Crisis of Capitalism*, pp. 116–137. Oakland, CA: PM Press.

McCormick, M., Büntgen, U., Cane, M., Cook, E., Harper, K., Huybers, P. *et al*. (2012). Climate change during and after the Roman Empire: reconstructing the past from scientific and historical evidence. *Journal of Interdisciplinary History* 43(2): 169–220.

McGee, D. & deMenocal, P. (2017). *The African Humid Period Recorded in Multi-proxy Data: Climatic Signatures and Cultural Consequences*. Oxford: Oxford Research Encyclopedia of Climate Science.

McKibben, B. (2010). *Eaarth: Making a Life on a Tough New Planet*. New York: Henry Holt.

Millen, J., Irwin, A. & Kim, J. (2000). Conclusion: pessimism of the intellect, optimism of the will. In: J. Kim, J. Millen, A. Irwin & J. Gershman (Eds.), *Dying for Growth:*

Global Inequality and the Health of the Poor, pp. 382–390. Monroe, MA: Common Courage Press.

Mills, C. (1956). *The Power Elites*. New York: Oxford University Press.

Mitchell, T. (1988). *Colonising Egypt*. Cambridge: Cambridge University Press.

Moore, J, (Ed.). (2016). *Anthropocene or Capitalocene? Nature, History, and the Crisis of Capitalism*. Oakland, CA: PM Press.

Moore, J. (2017). The Capitalocene Part II: accumulation by appropriation and the centrality of unpaid work/energy. *Journal of Peasant Studies* 45(2): 237–279.

Nassaney, M. (1992). Communal societies and the emergence of elites in the prehistoric American Southeast. *Archeological Papers of the American Anthropological Association* 3(1): 111–143.

Neiwert, D. (2017). *Alt-America: The Rise of the Radical Right in the Age of Trump*. New York: Verso.

Nengo, I. (2018). Great ape haters. *Anthropology News* 59(5): 3–5.

Newell, P. (2011). The elephant in the room: capitalism and global environmental change. *Global Environmental Change* 21: 4–6.

Newman, J., Aucompaugh, A., Schell, L., Denham M, DeCaprio, A., Gallo, M. *et al.* (2006). PCBs and cognitive functioning of Mohawk adolescents. *Neurotoxicology and Teratology* 28(4): 439–445.

New York Times. (2016). Transcript: Donald Trump's foreign policy speech. https://www.nytimes.com/2016/04/28/us/politics/transcript-trump-foreign-policy.html.

Nkrumah, K. (1965). *Neo-Colonialism, the Last Stage of Imperialism*. London: Thomas Nelson & Sons.

O'Dempsey, T. & Munslow, B. (2006). Globalisation, complex humanitarian emergencies and health. *Annals of Tropical Medicine & Parasitology* 100(5–6): 501–515.

Ostry, J., Lougani, P. & Fuceri, D. (2016). Neoliberalism: oversold? *Finance and Development* 53(2).

Phillips, D. (2018a). Pollution, illness, threats and murder: is this Amazon factory the link? *Guardian*. https://www.theguardian.com/world/2018/mar/16/brazil-pollution-amazon-aluminium-plant-norwegian

Phillips, P. (2018b). *Giants: The Global Power Elites*. New York: Seven Stories Press.

PERI. (2019). Toxic 100 Air Polluters Index. Political Economy Research Institute. https://www.peri.umass.edu/toxic-100-air-polluters-index-2018-report-based-on-2015-data.

Politico. (2016). Full transcript: Donald Trump's jobs plan speech. https://www.politico.com/story/2016/06/full-transcript-trump-job-plan-speech-224891.

PR Newswire. (2008). Thirty years after Love Canal more questions than answers remain—do the children of Love Canal count? https://search-proquest-com.ezproxy.lib.uconn.edu/docview/448604343?accountid=14518&rfr_id=info%3Axri%2Fsid%3Aprimo.

Raffetti, E., Donato, F., Speziani, F., Scarcella, C., Gaia, A. & Magoni, M. (2018). Polychlorinated biphenyls (PCBs) exposure and cardiovascular, endocrine and metabolic diseases: a population-based cohort study in a North Italian highly polluted area. *Environment International* 120: 215–222.

Reagan, R. (1981). Inaugural address. The American Presideny Project. http://www.presidency.ucsb.edu/ws/index.php?pid=43130.

Reinert, S. & Fredona, R. (2017). Merchants and the origins of capitalism. Working Paper 18-021. Cambridge: Harvard Business School.

Revkin, A. (2008). 1988–2008: climate then and now. *New York Times Blog*. https://dotearth.blogs.nytimes.com/2008/06/23/1988-2008-climate-then-and-now/.

Richardson, P., Boyd, R. & Bettinger, R. (2001). Was agriculture impossible during the Pleistocene but mandatory during the Holocene? A climate change hypothesis. *American Antiquity* 66(3): 387–411.

Richter, D., Grün, R., Joannes-Boyau, R., Steele, T., Amani, F., Rué, M. *et al.* (2017). The age of the hominin fossils from Jebel Irhoud, Morocco, and the origins of the Middle Stone Age. *Nature* 546: 293–296.

Rieux, A. (2014). Improved calibration of the human mitochondrial clock using ancient genomes. *Molecular Biology and Evolution* 31(10): 2780–2792.

Rodrik, D. (2017). The fatal flaw of neoliberalism: it's bad economics. *Guardian*. https://www.theguardian.com/news/2017/nov/14/the-fatal-flaw-of-neoliberalism-its-bad-economics.

Sack, K. (1991). Alcoa to pay New York State $7.5 million in waste fines. *New York Times*. https://www.nytimes.com/1991/07/12/nyregion/alcoa-to-pay-new-york-state-7.5-million-in-waste-fines.html.

Sahlins, M. (1972). *Stone Age Economics*. Chicago, IL: Aldine.

Santiago-Rivera, A., Morse, G. & Hunt, A. (1998). Building a community-based research partnership: lessons from the Mohawk Nation of Akwesasne. *Journal of Community Psychology* 26(2): 163–174.

Sawyer, S. (2015). How, when and where to shop for bargains in Paris. *Huffington Post*. https://www.huffingtonpost.com/sidonie-sawyer/how-when-and-where-to-sho_b_7148596.html.

Schell, L., Ravenscroft, J., Cole, M., Jacobs, A., Newman, J. & The Akwesasne Task Force on the Environment (2008). Health disparities and toxicant exposure of Akwesasne Mohawk Young Adults: a partnership approach to research. *Environmental Health Perspectives* 113(12): 1826–1832.

Schneider, K. (1994). Efforts revive river but not Mohawk life. *New York Times*. https://www.nytimes.com/1994/06/06/us/efforts-revive-river-but-not-mohawk-life.html.

Scoffen, K. & Tan, S. (2016). A decade of Trump's imports. *Washington Post*. https://www.washingtonpost.com/graphics/politics/trump-products/.

Silverstone, A., Rosenbaum. P., Weinstock, R., Bartell, S., Foushee, H., Shelton, C. & Pavuk, M. (2012). Polychlorinated biphenyl (PCB) exposure and diabetes: results from the Anniston Community Health Survey. *Environmental Health Perspectives* 120(5): 727–732.

Singer, M. & Baer, H. (Eds.). (2008). *Killer Commodities: Public Health and the Corporate Production of Harm*. Malden, MA: AltaMira/Rowman Littlefield.

Singer, M. & Erickson, P. (2013). *Global Health: An Anthropological Perspective*. Long Grove, IL: Waveland Press.

Singer, M. & Hodge, G. (2016). Ecobiopolitics in the making of Native American reservation health inequities. In: M. Singer (Ed.), *A Companion to the Anthropology of Environmental Health*, pp. 193–215. San Francisco, CA: Wiley-Blackwell.

Slon, V., Mafessoni, F., Vernot, B., de Filippo, C., Grote, S., Viola, B. *et al.* (2018). The genome of the offspring of a Neanderthal mother and a Denisovan father. *Nature* 561: 113–116.

Smith, S. (1979). Colonialism in economic theory: the experience of Nigeria. *Journal of Development Studies* 15(3): 38–59.

Strange, S. (1991). Big business and the state. *Millennium Journal of International Studies* 20(2): 245–250.

Tainter, J. (2006). Archeology of overshoot and collapse. *Annual Review of Anthropology* 35: 59–74.

Taylor, D. (2014). *Toxic Communities: Environmental Racism, Industrial Pollution, and Residential Mobility*. New York: New York University Press.

Tierney, J., deMenocal, P. & Zander, P. (2017). A climatic context for the out-of-Africa migration. *Geology* 45(11): 1023–1026.

Trigger, B. (1996). The original Iroquoians: Huron, Petun, and neutral. In: E. Rogers & D. Smith (Eds.), *Aboriginal Ontario: Historical Perspectives on the First Nations*, pp. 41–63. Toronto, ON: Dundurn Press.

Trueman, C. (2018). The cotton industry and the Industrial Revolution. *The History Learning Site*. https://www.historylearningsite.co.uk/britain-1700-to-1900/industrial-revolution/the-cotton-industry-and-the-industrial-revolution/.

Turnbull, C. (1965). The Mbuti Pygmies: An Ethnographic Survey. *American Museum of Natural History Anthropological Papers* 50(3).

Unwin, R. (1968). *Samuel Oldknow and the Arkwrights, the Industrial Revolution in Marple and Stockport, New York*. New York: Augustus Kelly.

Wallerstein, I. (1976). *The Modern World-System: Capitalist Agriculture and the Origins of the European World-Economy in the Sixteenth Century*. New York: Academic Press.

Wallmine. (2019). Darren Woods net worth. https://wallmine.com/people/33491/darren-w-woods.

Washington Post. (2015). The end of globalization? https://www.washingtonpost.com/opinions/the-end-of-globalization/2015/09/20/b48b9f60-5e2f-11e5-9757-e49273f05f65_story.html.

Watts, J. (2018). "They should be put in prison": battling Brazil's huge alumina plant. *Guardian*. https://www.theguardian.com/environment/2018/jul/21/land-grab-corruption-pollution-amazon-rainforest-brazil-maria-do-soccoro-silva.

Wood, B. & Collard, M. (1999). The human genus. *Science* 284(5411): 65–71.

WHO. (2016). Ambient Air Pollution: A Global Assessment of Exposure and Burden of Disease. Geneva: World Health Organization.

Wolf, E. (1981). *Europe and the People Without History*. Berkeley, CA: University of California Press.

Wrigley, E. (1988). *Continuity, Chance and Change*. Cambridge: Cambridge University Press.

Zachos, E. (2018). Why 2017 was the deadliest year for environmental activists. *National Geographic*. https://www.nationalgeographic.com/environment/2018/07/environmental-defenders-death-report/.

4

Engaging catastrophe

The number of extreme catastrophes that have more than negligible probability of occurring in this century is alarmingly great, and their variability startling.
Richard Posner (2004, p. 21)

4.1 Introduction to a dismal theme

In its classical meaning, in drama or poetry, "catastrophe" refers to closure, an awaited resolution of tensions, a finale. Aristotle is known for favoring shocking endings in tragic drama. In his volume *Poetics*, he maintained that the catastrophe should excite pity and fear. Like almost all words, "catastrophe" took on new connotations over time and across contexts. It was adopted in geology, as embodied in catastrophism theory. This way of understanding Earth's geological history asserted that our planet has been shaped over time by multiple sudden and violent events, like massive floods and shattering volcanic eruptions, which sometimes were global in scope and impact. For some catastrophic thinkers, the Biblical account of Noah's devastating flood exemplified these world-shaping occurrences, while others avoided theological interpretations and focused on naturalist accounts of major change-causing events in nature.

In the contemporary vernacular, the term "catastrophe" transports elements of Aristotelian dramaturgy and catastrophism theory into a usage that is synonymous with horrid disaster, usually with a distinctly human-centered focus. Events are catastrophic because of their effect directly or indirectly on us, the apex species in our self-interested assessment. The deeply engrained cultural theme of catastrophe has also been said to have "a distinguished career within literature and the arts" (Dole *et al.* 2016, p. 5), such as Albert Camus' *The Plague* or Daniel Defoe's *A Journal of the Plague Year*, but also in philosophy, which some have read as "an extended reflection on catastrophe" (Dole *et al.* 2016, p. 4). The theme is also prominent in theology and in politics. In the former, catastrophe often is portrayed as either divine retribution for human sin or part of a divine plan that is beyond our mortal capacity to comprehend. In the latter, there has emerged the idea that progress is often only possible through the catastrophic destruction of the old order of society. As the 18th-century French revolutionary Maximilien Robespierre is alleged to have said, "omelets are not made without breaking eggs." In her book, *The Culture of Disaster*, Marie-Hélène Huet (2012) argues that since the Enlightenment, Western culture has been haunted by a sense of impending disaster. In the view of British sociologist Anthony Giddens (Giddens & Pierson 1998, p. 209), this cultural theme is an expression of our nervous relationship with modernity; we are, he asserts, "increasingly preoccupied with the future (and also with safety)" in a world that

Ecosystem Crises Interactions: Human Health and the Changing Environment, First Edition. Merrill Singer.
© 2021 John Wiley & Sons, Inc. Published 2021 by John Wiley & Sons, Inc.

feels intrinsically risky. This concern about living in a risky world accelerated in the aftermath of the April 1986 nuclear disaster in Chernobyl in what was then part of the Soviet Union. The reactor explosion and release of radioactive material prompted German sociologist Ulrich Beck to write his influential book, *Risk Society: Towards a New Modernity* (1992), in which he proposes that environmental risk—which, like Chernobyl, endures across geographic boundaries and generations—has become the predominant product of modern industrial society.

In this light, in a world of mounting climate change and numerous ecocrises, from the toxic befouling of the air we depend on for survival to the unceasing pile-up of very long-lived nuclear waste, it is appropriate to ask: are we on the edge of catastrophe? This chapter considers catastrophe as a cultural theme, as a biological event with regard to the extinction of species, and as a social process within the context of societal collapse. Also examined is the rise of a social narrative about the environment and its cornucopia of species being in crisis.

4.2 Prepping for doomsday

There is a dystopian genre of science fiction writing known as apocalyptic fiction that explores the demise of our planet, including, often, reflection on the human role in provoking global ruin. While novels of doom are not new, the genre has attracted many 21st-century writers, including Margaret Atwood, Cormac McCarthy, and Suzanne Collins. Though the tone of this literature has varied over time, in the view of historian Jill Lepore (2017), we are living in a golden age of particularly pessimistic apocalyptic fiction. The genre has become the "fiction of submission, the fiction of an untrusting, lonely, and sullen [era] . . . the fiction of helplessness and hopelessness" (Lepore 2017). In cinema, a similar pattern, involving "a curious increase in the number and popularity of [dystopian] films" (Schmidt 2014), is described by movie critics. If Carlyon (2014) is right in arguing that "[s]cience fiction is at its best when it is using the future to talk about the present, and sci-fi films always reflect what is happening in the world at the time they were made," what does this say about how people are feeling in our current critical environmental moment?

The discipline of social psychology has developed a perspective on this issue. In psychology, a specific type of thanatophobia (the intense fear of death) is sometimes called "doomsday phobia," a pressing dread of impending global demise. Some have linked this emotional condition to the rising popularity of what is called "doomsday prepping," the effort to get mentally, physically, educationally, logistically, and materially prepared for an ominous, fast-arriving, and ugly future. It is estimated that there are now millions of people who are zealously preparing for the collapse of society as we know it. Doomsday prepping is even attracting the interest of quite affluent sectors of society "in Silicon Valley and New York City, among technology executives, hedge-fund managers, and others in their economic cohort" (Osnos 2017). Sectors of the wealthy digital elite, writes Jason Rhodes (2018), are embracing a doomsday perspective on the future "because that ideology mirrors their politics and their sociological views of people," namely a dystopian view of the masses.

Companies—labeled by some as "doom boom" opportunists (O'Connor 2015)—have begun to recognize that doomsday fears are good for business, as preppers constitute an accessible market niche anxious to acquire survival products like storable food and water, air filtration masks, survival books, hand-cranked radios, weapons, specialty and conventional tools, and, in the case of rich preppers, fully furnished underground bunkers and refuge property in places like New Zealand. Online distributors such as ReadyStore sell freeze-dried foods that are guaranteed to last for 20–30 years, providing their customers "with insurance for decades to come," as well as a wide array of survival gear including solar-powered radios, plant seeds, and survival training books (ReadyStore 2019). Other companies, like the Combined Universal Martial Applications Survival School in Chicago, train individuals on how to survive after SHTF ("shit-hits-the-fan" in prepper-speak) by building makeshift traps to catch their own food, starting fires with a metallic rod and Vaseline-soaked cotton ball, planning for "bugging in" (surviving at home during a catastrophe) and "bugging out" (leaving for a prepared safe place), assembling a GO (a "get-out" bag, filled with enough emergency supplies to last at least 72 hours), identifying "back-alley" routes out of the city, building a lean-to in the woods, and related survival tactics (Vaughn 2014). According to Yahoo Finance, "Given their numbers and their willingness to spend money, catering to doomsday preppers has become a multibillion dollar business" (Ogg 2013). As an advertising copywriter points out, companies that can identify "ways to provide preppers with information or new products to help them survive a coming catastrophe will gain a vocal and loyal customer base" (Spooner 2018). Marketing firms have developed focused strategies to assist the sellers of prep merchandise to reach their target market. Best ROI Lists (2019), an email and online marketing company in Boynton Beach, FL, for example, advises clients that "[i]f there's one core quality that preppers value above all else, it's trust. They need to trust in the items and supplies that they purchase in order to guarantee the safety of both themselves and their loved ones in the event of the unthinkable." If a prep sell can establish trust, they will "have access to a dedicated, passionate community of like-minded individuals who are all ready to engage, share and purchase at a moment's notice." This is made tricky, however, because preppers "are also a skeptical bunch . . . They are not a forgiving bunch if they learn claims made in your marketing are not true, so be prepared when you market to such a sophisticated audience."

The COVID-19 pandemic, which led to the emptying of supermarket shelves and the limited availability of basic commodities, has accelerated the spread of prepper activity. According to Bradley Garrett (2020), who spent 3 years conducting ethnographic research with nearly 100 preppers from six countries, to committed preppers "the ongoing COVID-19 pandemic is merely a 'mid-level' event—a warm up for what is to come. They anticipated and prepared for a disaster just like this one, and—unlike most of us—say they weren't caught by surprise." What the pandemic has done, Garrett asserts, is "increase people's dread—and therefore willingness—to normalise prepping practices." As a result, the companies that sell prepper worst-case-scenario supplies also ran low on their stocks. Knibbs (2020) reports that prepackaged kits, from bug-out bags to food storage bundles, began "filling up shoppers' carts across the country, as people [sat] in their living rooms, look at the internet, and [fretting]."

Super-rich preppers represent a special assemblage within the wider doomsday shopper crowd. Antonio García Martínez, for example, author of the bestselling tell-all autobiography *Chaos Monkeys* (2016), about life in the tech bubble of Silicon Valley, and a former Facebook product manager, has bought five wooded acres on an island in the Pacific Northwest and had generators, solar panels, furniture, other building and living supplies, and thousands of rounds of ammunition delivered there, asserting that "Within 30 years half of humanity could not have a job. It could get ugly, there could be revolution . . . You may not believe it but its coming . . . In the post-America, the 5.56 millimeter round will be the currency of the new America. I guarantee you" (quoted in Bartlett 2019). When Martínez began telling his peers in the Bay Area about his "project," many told him about their own preparations. In Martínez's view: "I think people who are particularly attuned to the levers by which society actually works understand that we are skating on really thin cultural ice right now" (quoted in Osnos 2017).

Another wealthy Silicon Valley prepper is Tim Chang, a managing director at Mayfield Fund, a venture-capital firm. He affirms that "[t]here's a bunch of us in the Valley. We meet up and have these financial-hacking dinners and talk about backup plans people are doing. It runs the gamut from a lot of people stocking up on Bitcoin and cryptocurrency, to figuring out how to get second passports if they need it, to having vacation homes in other countries that could be escape havens" (quoted in Osnos 2017). Chang's strategy has been to buy up real estate, both to generate passive income and in order to have safe havens to flee to in case of an emergency. He says, "I kind of have this terror scenario: 'Oh, my God, if there is a civil war or a giant earthquake that cleaves off part of California, we want to be ready'" (quoted in Osnos 2017).

The addition of an ever-wider array of preppers has pushed the trend into mainstream culture. In February 2012, National Geographic Channel launched a reality show called *Doomsday Preppers*, which, along with several spinoff programs like *Doomsday Castle*, features preppers (and prepping families) showing off their plans for evaluation and scoring by a panel of judges from the consulting company Practical Preppers, who provide recommendations for improvements. The program treats prepping as a kind of extreme competitive sport populated by colorful but very serious players. The premiere episode attracted more than 4 million viewers, making it one of the most popular shows in the channel's history. A 2012 survey commissioned by National Geographic found that more than 62 percent of Americans think the world will experience a major catastrophe in less than 20 years. The gravest threats according to respondents are earthquakes (64 percent), hurricanes (63 percent), terrorism (55 percent), and financial collapse (51 percent) (Cision PR Newswire 2012).

A vital question about preppers is: "are they prudent and smart people, or are they suffering from Doomsday Phobia?" (Pinkerton 2017). Additionally, it must be asked whether the privatized, individualized, masculinized, commodity-driven, hobby-like orientation of many preppers constitutes a meaningful societal response to cascading and interacting anthropogenic climate/environment catastrophes.

Existing science suggests that Earth as a planet is not at great risk from calamitous anthropogenic impacts. The planet will survive whether there are humans on it or not. Even life will survive, given its propensity to rise intact—if in new forms—from past mass extinctions. It is not likely that the most disastrous anthropogenic ecocrises will render the entire planet an uninhabitable place. As with past catastrophic events, some species will manage to persevere and usher in the next phase of biological evolution. But, Earth as a domicile for humans and the lifeforms with which we are most deeply entwined is another matter. As discussed in previous chapters, the gravest threat our environmental home faces is from multiple ecocrises that degrade ecological life support systems and ecosystem processes. Researchers have documented many past examples of these crises in considerable detail, including—if humans were around at the time as a species—both the role of our behaviors in their development and their effects on our physical and social health. Further, they have begun to address the ways in which ecocrises can spread across geographical boundaries and scales to become more destructive events and, as specifically examined in this book, interact to cause intersystem calamities and globally distributed risks.

Given that ecocrises have global effects, put billions of people at risk, and are promoted by the economic and environmental behaviors of multinational companies and the super-rich, what is most needed is not asocial or individualized planning for disaster survival by those who can afford it but coordinated social responses that address inherent injustices in the production and experience of peril and are aimed at preventing calamity and achieving a sustainable future. One thing we can be sure of, both because there is a long history of it and because of the power of human activities: environmental conditions will change.

4.3 The record of past radical environmental change

When we think of our planet as an entity in the universe, commonly what comes to mind is a nearly spherical heavenly body traveling in elliptical orbit around the sun. In geosystem science, Earth is conceived of not just as a bounded object hurling through defined space but as a system made up of several interacting subsystems: the geosphere (solid parts of Earth), hydrosphere (water in all forms located on, under, and above the surface of Earth), atmosphere (gases surrounding Earth), and biosphere (regions of Earth inhabited by living organisms). Earth's subsystems may remain relatively stable for long periods, change gradually over time, or change quite abruptly. All of these options have consequences for living organisms, for ecosystems, and for biodiversity. Because they are interconnected, alterations in one subsystem can cause changes to other subsystems or the whole system in various ways.

Both the fossil and the archeological records suggest that since the first evolution of life some 3.8 billion years ago there have been periods of both gradual and more abrupt change, as reflected in the state of Earth's subsystems. Abrupt and consequential planetary changes are evident not only in the distant past, long before the appearance of our early ancestors, but also in the more recent past, since the appearance of modern humans. Both of these temporal frames are considered in this section.

4.3.1 Planetary change and mass extinction

In the fossil record of the distant past, there are inscribed mass extinctions that eliminated large portions of existing lifeforms. Species extinctions occur continuously over time (Fig. 4.1), as does the evolution of new species. In mass extinctions, the appearance rates of

Fig. 4.1 Dodo displayed at Oxford University Museum of Natural History. *Source:* 134213/ Pixabay.

new species are notably lower per specified time interval than corresponding species disappearance rates. While researchers can describe five mass extinction events in the past, when more than 50–75 percent of species disappeared, there remains uncertainty whether there were earlier ones (before 540 million years ago) that are not recorded in fossil assemblages because the organisms involved lacked preservable hard body parts. Mass microbial extinctions likely occurred but are not (yet) detectable. To be recorded, in other words, extinction events must be visible to human—or, more precisely, scientific—ways of knowing about the environment. For example, not that long ago, the ecological upheavals associated with mass fauna extinctions were not believed to have shaped terrestrial plant evolution. This understanding began to change with the introduction of high-resolution regional paleoecological studies (McElwain & Punyasena 2007).

Each of Earth's mass extinctions "set the stage for new ecosystems with very different assemblages of organisms" compared to those that preceded them (Langmuir & Broecker 2012, p. 511). Understanding successive past ecosystems is the concern of paleoecology researchers. They seek to infer the adaptive responses of extinct communities of organisms to their habitats and to disruptions of those habitats based on the fossil record. They read the fossil record as "an interplay between evolutionary trends and environmental fluctuations" (Valentine 1973, p. 466).

The causes of mass extinction have been extensively investigated and hotly disputed among researchers. Diverse environmental upheavals have been suggested or linked to intensified extinction rates during specific temporal epochs, including global warming, global cooling, anoxic events involving significant marine loss of oxygen, asteroid impacts, volcano eruptions releasing massive amounts of atmospheric particles and basalt lava, biolide (meteor) explosions in the atmosphere, and plate tectonic shifts. The interaction of disruptive environmental events has also been suggested. Based on their research, Arens & West (2008) find that:

> the highest frequency of intervals with elevated extinction occurred when continental flood basalt volcanism and bolide impact co-occurred. In contrast, neither volcanism nor impact alone yielded statistically elevated extinction frequencies . . . Impacts, like marine anoxic incursions, are pulse disturbances that are sudden and catastrophic, and cause extensive mortality. Volcanism, like climate and sea level change, is a press disturbance that alters community composition by placing multi-generational stress on ecosystems.

In the view of Smithsonian paleontologist Doug Erwin (quoted in Brannen 2017b), following a cataclysmic event "[t]here's a very rapid collapse of the ecosystem during . . . mass extinctions." Critical to the collapse, in Erwin's view, is the faltering of food chains, which underlines the internal fragility of such systems despite their general resilience.

The five past mass extinction events, some of which encompass two or more sequential extinction episodes or pulses, are identified by the epoch in which they occurred:

- **Ordovician–Silurian Extinction (440 million years ago).** During this event, numerous small marine organisms, the dominant biota on the planet, died out. It is estimated that 85 percent of marine species were lost. Significant climate change, which triggered the spread of a massive ice sheet and caused sea level decline, is believed by some to have been the cause. While several other factors may have been involved, and ecocrises interaction may have played a role, "when the deteriorating climate intensified, causing widespread glaciation, a tipping point was reached resulting in catastrophe" (Harper *et al.* 2014). An additional factor proposed by paleontologist Thijs Vandenbroucke and coworkers (2015) is that metal poisoning (from lead, arsenic, and iron), unleashed by depleted oxygen in the oceans, contributed to the dramatic loss of ocean life. Their findings on the appearance of ancient sea animal fossils are consistent with the damage done to more modern marine life on exposure to toxic levels of metal. Species that were widely disseminated fared better than those with more local distribution (Sheeman & Coorough 1990).
- **Devonian Extinction (365 million years ago).** This event devastated tropical marine ecosystems and wiped out 70–80 percent of lifeforms, especially reef-dwelling species. Nonmarine organisms were also impacted. A marked drop in species diversity is identifiable in the fossil record (McGhee 1988). The event had two pulses, 15 million years apart. The second led to "the restructuring of vertebrate ecosystems worldwide" and provided the foundation of modern vertebrate biodiversity (Sallen & Coates 2010). Tetrapods (four-limbed organisms, including amphibians, reptiles, and mammals), actinopterygians (ray-finned fish, which now represent almost all existing fish species), and chondrichthyans (which includes sharks and rays), all of which were uncommon during the Devonian, experienced a subsequent massive diversification. Volcanic eruptions, climate change, and ocean anoxia (lack of oxygen) caused by biochemical runoff into the marine environment from new kinds of plants have all been suggested as

the engines driving this mass extinction. Based on lithological and geochemical findings, Barish (2017, p. 863) believes that the Devonian mass extinction was caused by a combination of climate and volcanic forces producing "emissions of harmful chemical compounds and aerosols [that caused] greenhouse warming; darkening of the atmosphere, which prevented photosynthesis; and stagnation of oceans and development of anoxia." Based on a broad review of the published research, House (2002) finds consistent evidence of climatic change as one clear driver of this extinction event, but no consistent evidence of volcanic eruption or tectonic plate activity as general causes. Analysis by Carmichael *et al.* (2016) of shallow ocean sedimentary geology and trace element geochemistry supports the argument that the extinction was caused by increasing ocean anoxia associated with excess plant nutrient runoff under conditions of climate stress.

- **Permian–Triassic Extinction (250 million years ago).** This event is also known as the Great Dying, because 95 percent of all lifeforms went extinct (Benton & Twitchell 2003). Marine life, like sea scorpions and trilobites, suffered heavy species losses, as did 70 percent of terrestrial vertebrates. Researchers have posited diverse causes for this event, including global warming triggered by a massive release of CO_2 during colossal volcanic eruptions in Siberia. These eruptions disgorged enough basalt lava to cover 400 million acres to a depth, in places, of almost 2 miles, accompanied by a massive release of CO_2 (Reichow *et al.* 2002). As a result, the chemical and temperature balance of the atmosphere and oceans suffered a catastrophic breakdown. High-resolution oxygen isotope data support the view that global warming during the main pulse of extinction "was one of the causes of the collapse of the marine and terrestrial ecosystems" (Joachimski *et al.* 2012). Research on tropical temperatures at the time support this conclusion (Cui & Kump 2015). Penn *et al.* (2018) simulated Permian–Triassic greenhouse gas-driven global warming using an Earth system model of climate change. Earth system models integrate the interactions of atmosphere, ocean, land, ice, and biosphere to estimate the state of regional and global climate under a wide range of conditions. Increases in atmospheric greenhouse gas levels simulated by this model raised near-surface ocean temperatures by more than 9 °F while depleting global marine oxygen levels by almost 80 percent. The researchers concluded that "The combined physiological stresses of ocean warming and O2 loss largely account for the spatial pattern and magnitude of extinction observed in the fossil record of the 'Great Dying'. These results highlight the future extinction risk arising from a depletion of the ocean's aerobic capacity that is already under way" (Penn *et al.* 2018).
- **Triassic–Jurassic Extinction (210 million years ago).** This event consisted of the demise of many marine reptile species, some large amphibians, numerous reef-building organisms, and large numbers of nautiloid mollusks. By contrast, dinosaurs, crocodiles, turtles, mammals, and fish were not heavily affected. Paleobotanical and geochemical research suggests that a sudden rise in atmospheric CO_2 and ocean acidity caused by volcanic release and methane produced by the decomposition of organic matter buried on the ocean bottom were the entwined causes of this mass extinction (Schaller *et al.* 2011; Greene *et al.* 2012). Volcanic evidence has been found in what is known as the Central Atlantic magmatic province, Earth's largest continental location of basalt (Pálfy & Zajzon 2012). Deposits occurred while Earth's current landmasses were still one supercontinent called Pangea.
- **Cretaceous–Tertiary Extinction (65.5 million years ago).** This event is the best known of the past mass extinctions because almost all species of dinosaurs disappeared, though those that were the ancestors of modern birds survived. An issue of keen interest is

why the birds did not disappear with the other dinosaurs. One possible reason concerns body size. The largest dinosaurs that have been unearthed weighed about 20 million times more than the smallest, indicating this faunal group had exceptional physical diversity (Benson *et al.* 2018). Maniraptorans, the feathered dinosaurs, for example, evolved into larger and smaller subgroups as they dispersed across ecosystems. Species on the smaller branches of the maniraptorian tree—those with a body weight under 4 ounces—survived, while larger maniraptors perished along with all other dinosaurs (Benson *et al.* 2014). One line of thought (Field *et al.* 2018), supported by studies of the microfossils of ancient pollen, is that changing environmental conditions led to the devastation of forests, as well as of the flying birds that were dependent on them. Ground-dwelling birds, by contrast, survived in new floral landscapes dominated by ferns. From these survivors, new taxa of birds capable of flight later evolved to occupy the new forests that eventually developed. One fossil that fits this interpretation has been nicknamed Lori (actual scientific name *Hesperornithoides miessleri*). A chicken-size carnivore, Lori was unearthed in a layer of approximately 150 million-year-old rocks called the Morrison Formation, which covers a vast ribbon of the western United States and is centered in Wyoming and Colorado (Hartman *et al.* 2019). The fossil has a large sickle-shaped claw on each foot, wings, feathers, and other bird-like features, but is clearly terrestrial. Animated debate continues concerning whether the Cretaceous–Tertiary mass extinction was caused by global climate change unleashed by extensive volcanic eruption or a large extraterrestrial asteroid impact at Chicxulub beneath the Yucatán Peninsula in Mexico (Alvarez *et al.* 1980; Gerta *et al.* 2018). Some research suggests it was a product of both: that the asteroid that left a huge crater in Mexico delivered the final blow to the resilience of ecosystems already weakened by the volcanic eruptions (Petersen, Dutton & Lohmann 2016). In light of the contemporary threat of climate change, when the Smithsonian National Museum of Natural History reopened its Hall of Fossils in 2019 after a 4-year renovation, a central theme was the impact of ancient planetary heating and the associated rise in the amount of carbon dioxide in the atmosphere on dinosaur life. One of the effects was a shift toward dwarfism: a decrease in size generation by generation. According to Smithsonian curator and paleobotanist Scott Wing, "It could be that the warming itself led the smaller individuals to be more successful" (quoted in Stein 2019). But it is also known from contemporary experiments that increasing the amount of carbon dioxide in the atmosphere can cause plants to make less protein. Thus, as ancient plants became less nutritious for the animals that ate them, smaller species prevailed.

While there remains much to be learned about mass extinction events of the past, climate change is now believed to have played a role—in many cases, the decisive role—in all of these sweeping losses of biodiversity. An additional lesson is the importance of ecocrises interactions. The interaction of two or more climate/environmental factors can tip the scales toward biological disaster. A third insight gained from the study of past mass extinctions is that "ecosystems, although remarkably resilient, really do have breaking points . . . [a]nd that history should be considered extremely sobering" (David Jablonski, quoted in Boissoneault 2018). Such lessons are vital in responding to the sixth mass extinction that many believe is currently underway in the Anthropocene/Capitoloscene (Brannen 2017b).

4.3.2 The sixth mass extinction?

Researchers from different academic disciplines and environmental activists have begun to entertain the possibility that we are both witnessing and causing a sixth mass extinction of species (Leakey & Lewin 1992). Some voices deny this is occurring, however. One dissenter is Doug Erwin, who studies the Permian–Triassic mass extinction and the processes of evolutionary novelty that emerge following such extinctions. In his view, those engaged in the popular narrative about the sixth mass exaction are making facile comparisons between what happened in the past and what is occurring now (Brannen 2017a). He argues that mass extinctions parallel the pattern of bankruptcy described by the pitiable character Mike Campbell in Hemingway's classic novel *The Sun Also Rises*. When asked by his acquaintance Bill Gordon how he went bankrupt, Mike replies, "Gradually and then suddenly" (Hemmingway 2006, p. 141). In other words, mass extinctions begin slowly, but at some point a tipping point is passed and the rate of die-offs suddenly picks up and cascades full force until the number of species disappearances is catastrophic.

Analyses of various kinds have led others to quite different conclusions, however. Focusing on biodiversity loss over the last 300 years, Dirzo & Raven (2003) argue that species diversity is not evenly distributed across ecosystems, with tropical environments constituting diversity hot spots. As noted in Chapter 2, key hot spots are found in Madagascar, India, Indonesia, South America, and South East Asia. A species naturally found in only one location, such as a hot spot, is termed an endemism. Currently, Earth's biodiversity hot spots are facing rapid habitat destruction and other ecocrises, putting many endemisms at risk of extinction and collectively paving the way for a mass-level extinction.

Another approach to assessing the meaning of current extinction levels is to focus on specific kinds of organisms that appear to be at grave risk. For Wake & Vredenburg (2008), a focal group today is the amphibians. It is estimated that one third of amphibian species are currently threatened with extinction, and many others face increasing risk. The cause is ecocrises interaction, and for Wake & Vredenburg (2008, p. 1472) the lesson is clear:

> Multiple factors acting synergistically are contributing to the loss of amphibians. But we can be sure that behind all of these activities is one weedy species, Homo sapiens . . . A primary message from the amphibians, other organisms, and environments, such as the oceans, is that little time remains to stave off mass extinctions, if it is possible at all.

Extinction rates at present are rising because of a plethora of direct and indirect—often synergic—factors. As Strona & Bradshaw (2018) affirm, "it is becoming increasingly evident how biotic interactions, in addition to permitting the emergence and maintenance of diversity, also build up complex networks through which the loss of one species can make more species disappear (a process known as 'co-extinction'), and possibly bring entire systems to an unexpected, sudden regime shift, or even total collapse." This dilemma appears to be playing out in the case of the resident orca populations (*Orcinus orca*) of the inland waters of Washington State and British Columbia, an area known as the Salish Sea. Due to the overfishing of wild salmon stocks, the warming temperature of breeding rivers driven by climate change, noise pollution from ships, and other anthropogenic changes, the Chinook (King) salmon (*Oncorhynchus tshawytscha*) population is declining. This has consequences for the orcas, which feed primarily (about 85 percent of total intake) on Chinook. There is fear that we are on a pathway to co-extinction of both species (Wild Orca 2014).

As our understanding of the role of ecological interactions in shaping ecosystems advances, it is becoming clear that co-extinction is a product of the unraveling of ecological interdependencies and the bottom-up cascading (from lower to higher trophic levels) of extinction ultimately involving several higher-level consumers. Even species with the greatest resilience can be expected to "fall victim to the synergies among extinction drivers . . . as extreme stresses drive biological communities to collapse" (Strona & Bradshaw 2018). As a result, it appears that our current extinction estimates need to be recalibrated upward to reflect species co-extinctions (Koh *et al.* 2004).

One of the problems with assessing whether we are in the midst of a mass extinction is that fewer than 3 percent of the approximately 2 million existing species we know about and have given names to have been properly appraised for their extinction status by the International Union for Conservation of Nature (IUCN) (Vié *et al.* 2009). Assessment has tended to focus on cases where threat is noticeable, or even on opportunistically gaining access to new information.

Another problem is in establishing the best metric to use in evaluating rates of extinction. An approach that has gained favor in recent decades uses the metric of extinctions per million species-years (E/MY) (Pimm *et al.* 1995). Of concern are both the rate (number of species extinctions over a selected unit of time) and the magnitude (the percentage of species alive in a previous time that have gone extinct) of loss. Barnosky *et al.* (2011) argue that "to document where the current extinction episode lies on the mass extinction scale . . . requires us to know both whether current extinction rates are above background rates (and if so, how far above) and how closely historic and projected biodiversity losses approach 75% of the Earth's species." Based on what we know about current extinction levels among mammals, amphibians, birds, and reptiles, is the current rate-magnitude of species loss on par with the rate-magnitude of recognized mass extinctions? These researchers believe the answer is yes. How many more years will it take for current extinction rates to produce species losses equivalent to past mass extinction events? According to the calculations of Barnosky *et al.* (2011), the answer is between 240 and 540 years—a mere drop in the bucket of geological time.

Moreover, Barnosky *et al.* (2011) focus only on mammals, amphibians, birds, and reptiles. What of other kinds of organisms? As Lister & Garcia (2018) report, "[a]rthropods, invertebrates including insects that have external skeletons, are declining at an alarming rate." Significant loss of arthropods threatens species loss up the food chain. Lister & Garcia's (2018) research in Puerto Rico's lush Luquillo rainforest adds to the growing recognition of this significant biomass loss (e.g., Dirzo *et al.* 2014; Habel *et al.* 2016). They found insect biomass losses of 98 and 78 percent for ground-foraging and canopy-dwelling arthropods, respectively, over a 36-year period, with annual losses of 2.7 and 2.2 percent (Lister & Garcia 2018). Similarly, in Germany, Hallman *et al.* (2017) measured total insect biomass (using large, tent-like net structures to trap the insects) over a 27-year period in 63 nature protection areas, in order to infer the status and biomass trend of local entomofauna. They found a seasonal decline of 76 percent and a midsummer decline of 82 percent in flying insect biomass over the course of the study. This marked decline occurred regardless of habitat type. It represents an average 2.8 percent loss in insect biomass per year in habitats subject to rather low levels of human disturbance. Over almost three decades, the study found a steady declining trend. Meanwhile, in the Northeastern United States, University of Connecticut professor David Wagner (2012) reports a decline in numbers of moths, especially larger moths—a change he attributes to the combined effects of habitat destruction, loss of early successional habitats and afforestation, coastal development, overgrazing by deer, and climate change. As University of Delaware entomologist Doug Tallamy notes, "You have total ecosystem collapse if you lose your insects . . . How much worse can it get than that?" (quoted in Borenstein 2018).

The threat to insect biodiversity is global. Some have referred to it as an arthropocalypse. A comprehensive review of 73 historical reports of insect declines from across the planet (Sánchez-Bayo & Wyckhuys 2019) identifies worrisome rates that suggest the potential extinction of 40 percent of the world's insect species within the next few decades. In terrestrial ecosystems, Lepidoptera (butterflies and moths), Hymenoptera (bees, ant, and wasps), and Coleoptera (beetles) appear to be the taxa most affected. In the less popularly known aquatic environments, four major taxa, Odonata (dragonflies and damselflies), Plecoptera (stone flies), Trichoptera (caddisflies), and Ephemeroptera (mayflies), have already lost a notable proportion of their species. While some of the affected groups are highly specialized to live in certain econiches, others are common and generalist species that dwell in diverse settings. Sánchez-Bayo & Wyckhuys (2019) show that habitat loss due to conversion of land areas for intensive agriculture and urbanization and exposure to agrochemical pollutants are the main drivers of decline, in interaction with climate change. The latter "is particularly important in tropical regions, but only affects a minority of species in colder climes and mountain settings of temperate zones" (Sánchez-Bayo & Wyckhuys 2019, p. 8). Overall, they conclude insect biodiversity around the globe is in a "dreadful state" (Sánchez-Bayo & Wyckhuys 2019, p. 24).

Insects are not only disappearing from nature, but also from high school and college Introduction to Biology textbooks. An analysis by Gangwani & Landin (2018) shows that the amount of content dedicated to insects has decreased by more than 75 percent in textbooks published since 2000 compared to those published before 1960. In books published from 1900 to 1920, the number of pages focused on insects averaged 9 percent, while in those published from 2000 to 2017, it represented only 0.59 percent. Gangwani & Landin (2018) maintain that their findings:

> imply a decline in awareness of the human–insect relationship. Even insects' crucial presence in ecosystems is presented in a detached manner, as if humans are not also a part of those ecosystems. This may affect human cognizance of the importance of insects—a troubling prospect, because human activities are directly and indirectly contributing to sharp declines in insect abundance and diversity all over the world.

According to the World Wildlife Fund's Living Planet Report 2018 (WWF 2018), because of the way we take nature for granted, the total population of more than 4000 mammal, bird, fish, reptile, and amphibian species declined rapidly in the 44 years between 1970 and 2014. The primary driver of this, in the view of various researchers, conservationists, and environmental activists, was anthropogenic ecocrisis synergies of the sort discussed in this book. Rather than debating what is the gravest threat to species-sustaining ecosystems, climate change or biodiversity loss, it is argued here that both in complex interaction with each other and many other anthropogenic disruptions, overexploitations, contaminations, and transformations of habitats are equally important.

Nor is significant contemporary species loss limited to animals. The first global analysis of plant extinction (Humphreys *et al.* 2019) found that the number of known seed-bearing plant extinctions was more than four times that on the IUCN Red List of Threatened Species, though 50 species included on the Red List have subsequently been rediscovered (but may remain at grave risk of extinction). Moreover, plants are disappearing faster than in the past. Recently described plants are becoming extinct at almost twice the rate of those that were described before 1900, though at a rate lower than for vertebrates. Paralleling the situation with animals, this extinction is not random. The study found that plant species have gone extinct since 1900 at a rate around 500 times higher than the natural plant extinction rate. The primary driver of this is human activity. As part of the global analysis, in searching for living examples of an unusual-looking grass species, *Sartidia perrieri*, originally described in 1914, the research team looked in the areas where it had been seen in the past, in the central region of Madagascar, and found only grazing cattle and rice cultivation. In hunting in difficult terrain in Tanzania for species of spiny wild aubergines (relatives of the common eggplant), team members were startled to find that the "coastal forest [was] completely destroyed" (Maria Vorontsova, quoted in Carrington 2019). Between 1990 and 2007, coastal forest cover in the country decreased by more than one third, and this loss has continued ever since. Agricultural expansion, charcoal production, and logging for timber and firewood are the most important anthropogenic causes (Burgess *et al.* 2018).

Geographically, the primary plant extinction areas have been either tropical or Mediterranean climates, and many are islands. Islands are particularly vulnerable extinction hot spots because they represent unique habitats for endemic species and are subject to the rapid spread of invasive species that outcompete the native flora. Exemplary are the Chile sandalwood (*Santalum fernandezianum*), which lived on only one group of Pacific islands, and the St. Helena olive tree (*Nesiota elliptica*), which is found exclusively on the remote South Atlantic island that gives it its name (made famous as the place of Napoleon Bonaparte's exile and death). Hawaii has suffered more plant losses than anywhere else on Earth with 79 extinctions, but other hot spots are Australia, Brazil, India, Mauritius, and Madagascar.

No families of plants are disproportionately at risk for extinction (the group at highest risk is woody perennials), which differentiates them from animals, suggesting differences in the prevailing extinction processes for these two kingdoms.

4.3.3 Planetary change in the archeological record

Human ecodynamics is the relationship among climate, environment, and sociocultural systems. This entails both how climatic and environmental conditions influence human lifeways and how human behaviors influence climate and the environment. Within anthropological archeology, there has been a steady growth in the level of focus on these issues over the last two decades, though the turning point was in 1956 with the convening by the Wenner-Gren Foundation for Anthropological Research of an international symposium entitled Man's Role in Changing the Face of the Earth. On one hand, this environmental turn in archeology has spawned numerous studies of the extent of human modification of the environment since the late Pleistocene. A central premise of this work is that we do not simply occupy or adapt to environmental niches, but are niche constructors (Boivin *et al.* 2015). However, our capacity to remake the world as we see fit has its limitations, as does our ability to insulate ourselves from powerful environmental/climatic shifts and intense events like floods, wildfires, mudslides, earthquakes, tornados, and hurricanes. On the other hand, the archeological record is an important resource for understanding global change in historic context. This information is derived from analyses of proxy data consisting of culturally modified artifacts and ecofacts—such things as plant remains, animal bones and teeth, shells, grains of pollen, pieces of wood, charcoal, lithics, soil, and sediments excavated from human occupation—and activity sites ranging from transitory spear blade knapping spots to cities. With the help of an increasingly large array of dating techniques, archeologists are able to suggest the temporal provenience of their finds (Kirch 2005).

Much of archeological research focuses on the local level, based on research at a specific site, and may reflect local conditions that deviate from broader regional or climatic/environmental trends at the global level. One strength of archeology, however, is its ability to reveal the ways that past human communities coped with ecological challenges. Archeologists have come to recognize that "environmental change is a significantly more important factor in influencing human activity in the landscape" than was understood in the past (Turney *et al.* 2006, p. 37).

One expression of this significance is seen in what are called "rapid change events" (RCCs) (Mayewski *et al.* 2004). These events are called "rapid" because of their swift onset (within 10–20-year periods), but once in process they may endure for several hundred years. They are "considered one of the main environmental factors causing socioeconomic and cultural changes, migrations, and even collapses" (Berger *et al.* 2016, p. 1847) because they trigger abrupt shifts in vegetation composition and environmental conditions. Because of their sudden commencement, they ominously challenge societal resilience and coping mechanisms.

The occurrence of RCCs is recorded in an array of proxy data—natural indicators of past climates, including tree rings, glacial ice cores, fossil pollen, ocean and lake sediments, and corals—used by paleoclimatologists to reconstruct past climate conditions (Fig. 4.2). The widths, densities, and isotopic compositions of the rings formed each year inside trees, for example, reflect variations in climate. Similarly, the sediments that collect on ocean bottoms contain billions of tons of dead plankton and other marine organisms, dust blown from terrestrial zones, and materials carried by rivers. Analysis of long cores drilled into the ocean floor can be used to interpret past climates. The types and concentrations of microorganisms found in such cores record past changes in the temperature and composition of the ocean. These kinds of proxy data allow paleoclimatologists to extend their temporal archive of climate information back in time for thousands of years.

Weninger *et al.* (2009) used such data to assess the impact of RCCs on prehistoric communities in the Eastern Mediterranean region. Of prime utility in this analysis were marine bottom sentiments from the Aegean Sea off the coast of Crete. The researchers found that variations in the sea surface temperature recorded in Aegean cores resulted from the rapid passage of extremely cold polar/continental air masses sweeping down from the Balkans. In the region of concern, the domestication of wild grain plants coincided with an abrupt increase in annual rainfall about 10 000 years ago. This development is reflected in an increase in settlements culturally organized around farming found in the archeological record. As part of this turn toward food source control, archeozoologic data indicate that the domestication of goats (from the wild bezoar ibex) occurred within

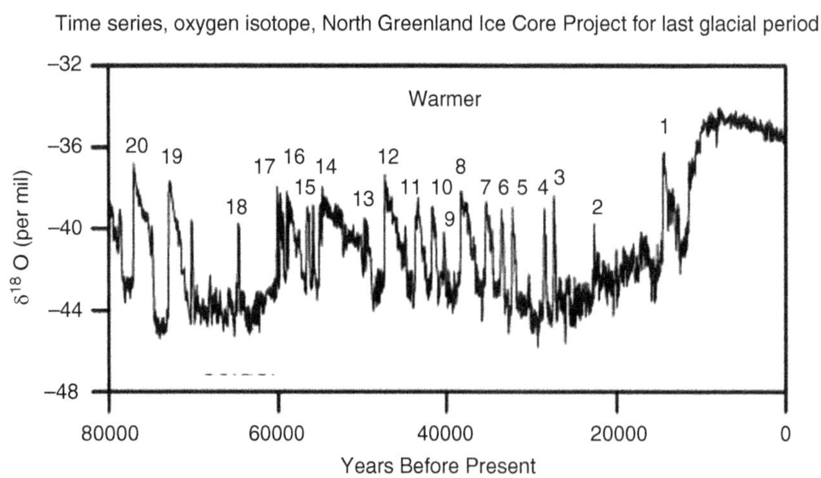

Fig. 4.2 Time series for an oxygen isotope from the North Greenland Ice Core Project. *Source:* Modified from National Centers for Environmental Conditions, NOAA.

the same time zone. An important archeological site revealing early goat domestication is in the Zagros Mountains of Iran at Ganj Dareh (Zeder & Hesse 2000). Of note, this site was only occupied for one or two centuries and then abandoned. Another occupation site that developed during the same period is 'Ain Ghazal, which is located on the bank of the Zarqa River in what is now metropolitan Amman, the capital of modern Jordan. The site was discovered in 1974 during a road construction project. Founded about 10 200 years ago, the subsistence economy of 'Ain Ghazal was based on the cultivation of barley, a type of wheat, and legumes on arable land, as well as goat herding. The remains of bones indicates that residents hunted deer, gazelle, cattle, horses, and pigs. They also foraged for various wild plants. The settlement, consisting of rectangular mud-brick houses, continued to grow until it spread to both banks of the river and encompassed an area of about 10–12 acres. Excavated at the site were several batches of carefully crafted human statues and busts made from marsh reeds, wrapped in coiled layers of twine and plastered with lime, with clear facial features (Simmons *et al.* 1988; Grissom 2000). Animal figures also were found. After its initial period of settlement, there was a sudden jump, about 9500 years ago, in the population to about 3000 people and a tripling of the acreage covered by the site. Then, about 500 years later, the area of occupation shrunk in size again. It continued to shrink thereafter until only a tiny farming community remained. Ultimately, 'Ain Ghazal, after 1700 years of occupation, was completely deserted and covered over by rubble and gravel slide deposits. Based on their examination of the archeological data from 'Ain Ghazal, in conjunction with climatic proxy data, Weninger *et al.* (2009, p. 49) conclude that "RCC-deterioration may well have been a major factor underlying social change, but if so, always reacting within a wide regional spectrum of social, cultural, economic and religious factors." They also recognize "the existence in the Near East of other important climatic and environmental factors, besides RCC" and that "some of these factors appear to interact with the RCC mechanism" (Weninger *et al.* 2009, p. 49).

The ancient name Mesopotamia (Greek for "land between the rivers") labels the crescent-shaped region surrounding the Tigris and Euphrates rivers, stretching from the Persian Gulf to the Eastern Mediterranean, parts of which are now occupied by the modern countries of Iraq, Kuwait, Saudi Arabia, Syria, and Turkey. These rivers and the highly fertile lands around them provided the foundation that enabled food cultivation and sedentarization to develop, and eventually led to the emergence of local, politically independent city-states reliant on the farming of barley and wheat. By 4500 years ago, new cities were arising across this terrain, which is often referred to today as the Fertile Crescent. Armed hostilities between these parallel entities over water and land rights were common. Approximately 4300 years ago, the standing army of the city of Akkad, located on the western bank of the Euphrates, succeeded in overtaking the city-states of northern and southern Mesopotamia and uniting them into a tributary empire. The invention of the composite bow gave the Akkadian army a great advantage over its rivals. While the precise boundaries are in dispute, at its height the Akkadian Empire had some degree of control over lands and trade routes stretching from the coast of the Mediterranean in the west to what is now Iran in the east. Repeated military campaigns enforced the tributary flow of wealth to Akkad. Trusted administrators were installed in positions of power in many of the conquered cities. Within the lands of the empire, there were two human-dominated ecosystems: a rain-fed agricultural zone located in the northern sector and an irrigated alluvial plain in the southern one. Culturally, the language, clothing patterns, and religious beliefs and practices of Akkad infiltrated local ways of life across the Fertile Crescent. The Akkadian Empire transitioned through five dynastic rulers, beginning with Sargon, who had the longest reign of the Akkadian kings, before collapsing after less than 150 years (Weiss *et al.* 1993).

Archeological research at Tell Leilan in northeastern Syria sheds some light on the demise of the Akkadian Empire. Though founded before its rise, the city-state of Shekhna came under Akkadian imperial control but was subsequently abandoned (though it was later reoccupied by the post-Akkadian Assyrian Empire). Excavation shows that the site was buried by 3 feet of sediment coincident with a catastrophic drought and the drying of the Tigris and Euphrates floodplains. In this sediment, researchers found a volcanic ash layer rich in glass shards, suggesting a volcanic origin of rapid climatic changes that occurred during the collapse of the empire. Courty & Weiss (1994, p. 109), who excavated the site, conclude that both Tell Leilan and regional "soil data comprise a proxy record for . . . climate change and provide one high resolution signal to link environmental and societal phenomena" across a wide region.

Findings from archeological research at 'Ain Ghazal in Jordan and Tell Leilan in Syria, as well as research on the ancient Mayans of Central America (discussed in Chapter 2), indicate the significant effects of abrupt planetary change on human societies. These societal collapses and the linkage of them and of mass species extinctions to systems change underline the necessity of understanding Earth as we contemplate our continued survival on it. The role of climate change in societal collapse, however, is far from the only issue of interest. In the view of Weiberg & Finné (2008), identifying and describing "instances of resilience and persistence in ancient societies during periods of climate stress are necessary as counter weights to simplified collapse archaeology." Understanding how humans perceive and react to climate change at different time scales has been described as one of the 25 grand challenges for archeology (Kintigh *et al.* 2014). Ironically, as archeologists intensify their interest in past climates and develop increasingly powerful research methods to study them, climate change itself is actively undermining these efforts. In many places, for example, coastal sites are rapidly being eroded by sea level rise. In Greenland and throughout the Arctic, permafrost thawing is leading to the rapid loss of well-preserved climate archives in organic middens. Increasing temperatures and drought stress in the Southwestern United States are driving wildfires that damage archeological sites located in forests. Globally, the archeology of climate change is being put at risk by climate change itself (Guedes *et al.* 2016).

It is not clear if the magnified focus on climate and the growing body of findings on this connection within archeology have permeated the public perspective on contemporary climate change. One approach toward disseminating this information is expressed in the volume *Public Archaeology and Climate Change* (Dawson *et al.* 2017), which promotes new approaches from around the world to both studying and managing archeological sites that are threatened by climate change. Specifically, it describes efforts, learned from the natural heritage sector, to engage communities and employ "citizen science" initiatives. By involving local communities in projects and making data accessible, archeologists are seeking to generate public discussions on climate change and inform policy-makers.

4.4 Popular concern with the environment

Popular concern with the environment unavoidably permeates human history, in the sense that we are dependent on the resources of the environment and cautious about the risks it presents. Inside a cave on the island of Sulawesi in Indonesia, there is a rock painting of a babirusa (*Babyrousa* sp.), colloquially called the "pig-deer" because of the large upward incurving tusks of the male of the species, that was rendered at least 35 400 years ago, making it "among the earliest dated figurative depictions worldwide, if not the earliest one" (Aubert *et al.* 2014, p. 223). Other animals represented on the walls of caves in the area include the miniature water buffalo (*Anoa* sp.) and the Celebes warty pig (*Sus celebensis*). Similarly, in Arnhem Land and the Kimberley in northern Australia, extensive ancient rock art includes paintings of various living animals (e.g., fish, monitor lizards, turtles, possums, and wallabies), as well as images of now extinct megafauna like the Tasmanian tiger (Chaloupka 1993; Morwood 2002). In Europe, in places like the famous Lascaux Cave complex in southwestern France, there are multiple images of horses, cattle, bison, lions, and an extinct species of deer, as well as one picture of a bear and one of a rhinoceros, all dating to 17 300 years ago (Whitley 2009) (Fig. 4.3). While it is impossible to know precisely what these ancient animal images meant to their creators, they certainly suggest a keen interest in the environment, and, very likely, in the food sources they coveted and the potential predators and dangerous animals they feared. Concern about the environment as being in crisis is of much more recent origin but is documented in the rise of preservationist and other environmental movements.

4.4.1 History of the environmental movement

The origin of the contemporary global environmental movement can be traced to mid-19th-century Europe and the artistic and intellectual perspective of Romanticism. A reaction to the objective, rational approach of emergent naturalist science and the sweeping material changes occurring in industrial capitalist society, this worldview emphasized an emotional appreciation and embrace of nature and wild places as pure and spiritual sources of renewal. Among Romance thinkers, the natural world was seen as inherently good, while human society was thoroughly corrupt (McMichael & Crews 1997). The emergence of this movement inspired still popular authors and poets such as William Wordsworth, Samuel Taylor Coleridge, John Keats, and Percy Bysshe Shelly and painters such as Francisco Goya, John William Waterhouse, and Eugène Delacroix. Romanticism emphasized the individual, the

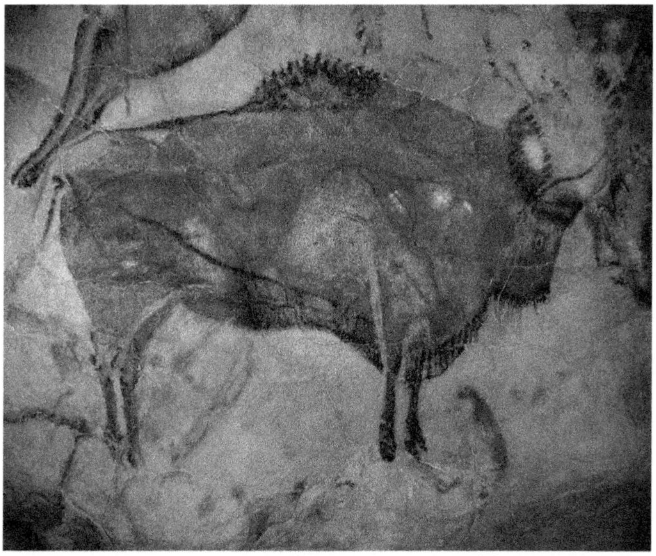

Fig. 4.3 Bison cave painting, Altamira Cave, Santillana del Mar, Cantabria, Spain. *Source:* janeb13/Pixabay.

subjective, the irrational, the experiential, the imaginative, the spontaneous, the visionary, the transcendental, and the beauty of nature.

Beginning in the late 1800s, interest in the natural sciences began to grow, with emergent fields like botany, geology, and meteorology embracing an empirical approach to the investigation of nature. In the arts and humanities, this focus on nature, as expressed by Alexander von Humboldt, an influential German proponent of Romantic philosophy, promoted the existence of a "natural unity and feeling of harmonious accord pervading the universe" (Bohn 1847). As Thubron (2015) points out, "[T]he transcendentalism in much of Humboldt's writing deeply affected . . . the English Romantics." It is likely that Humboldt influenced renowned landscape artists like John Constable, whose painting "The Hay Wain," completed in 1821, has been described as "a triumph of the ecological viewpoint" (Rees 1982, p. 260). Humboldt's extensive writings also were read and absorbed by other influential figures, including Charles Darwin, who noted that they inspired his voyages aboard the *Beagle*, and environmentalists like George Perkins, Henry David Thoreau, and John Muir (Wulf 2015). For this audience, Humboldt's ideas about the ecological threat of unrestrained human activities made a lasting impression (Grove 1995).

Romanticism can be seen at least in part as a backlash against the polluting effects of the Industrial Revolution. Inherent in grand social change was an ethic of controlling the forces and resources of nature to promote production. At the same time, nature was utilized increasingly as a trash receptacle for industrial waste, which was pumped into the air, emptied into the rivers, and spilled on to the land. The cities of the nascent industrial era were dark, dank, and polluted. Alexis de Tocqueville (1958, p. 7), a French diplomat and historian who visited Manchester, England in 1835, observed, "[a]sort of black smoke covers the city." Henry Mayhew (1968), a magazine editor and social researcher, added that in London in 1850, "[it] was impossible to tell where the sky ended and the city began; and as you peered into the thick haze you could, after a time, make out the dusky figures of tall factory chimneys plumed with black smoke; while spires and turrets seemed to hang midway between you and the earth, as if poised in the thick grey air." While soot-blackened skies, buildings, and clothing were evident to see, the health effects of the toxic air on the residents of early industrial cities were not assessed at the time. Recently, researchers have collected data on coal consumption by industry to estimate annual coal burning in each district of Victorian-era Britain (Beach & Hanlon 2017). Coal-use intensity was found to be highest in the Midlands, in the North of England, and in South Wales. The highest rates of use were associated with the highest rates of infant mortality, an association known as the urban mortality penalty. Raising local industrial coal use by 1 standard deviation from the mean was found to be associated with an increase in infant mortality of approximately 6–8 percent and in under-5 mortality of 8–15 percent (Beach & Hanlon 2017). Exposure to burned coal fumes contributed to repeated respiratory illnesses, slower growth during childhood, and shorter adult stature (Hatton 2017). Also contributing to infant mortality was infectious disease caused by unclean water and improper sewage disposal (Kesztenbaum & Rosenthal 2011). As the industrial cities became all but unlivable, people had ever increasing motivation to contemplate more desirable alternatives.

A "back-to-nature" theme developed among the Romantics as so-called wild environments came to be envisioned as places for the spiritual renewal of the battered human soul (Oosthoek 2015). This attitude was taken up by countryside or nature tourism enthusiasts, who, as Squire (1988, p. 237) comments, "fostered impressions of an idyllic, untamed eden" and inspired "public fascination with [a] mythologized and emotional portrait" of nature. Among the influential nature tourist writers was poet William Wordsworth, who presented visiting natural places as having the power to elevate consciousness.

Dismayed by the environment changes ushered in by factory production, people began to call on the government to protect wild spaces, and early species conservation groups like The Selborne League (founded in 1885), the Society for the Protection of Birds (1889), the National Trust for Places of Historic Interest or Natural Beauty (1894), and the Society for the Promotion of Nature Reserves (1912) appeared across England. The Society for the Protection of Birds, for example, was first organized as the Plumage League by a group of affluent women, led by the philanthropist Emily Williamson, who pledged on moral grounds not to wear the ornamental feathered hats that had gained favor as high-fashion apparel in the late Victorian era (Clarke 2004). This profitable clothing trend had led to the near extermination of the Great Crested Grebe and other birds as the commercial trade in feathers expanded to become a multinational industry designed to fill elite fashion demand. In 1891, the Society merged with Fur, Fin and Feather Folk, founded by Eliza Phillips and Catherine Hall, to expand its number of chapters and enhance its efforts to lobby influential people on behalf of bird protection. The organization was granted a Royal Charter in 1904 and became the Royal Society for the Protection of Birds. The RSPB remains an active environmental protection advocate today.

In Germany, the contemporary environmental movement, as well as the county's robust environmental protection and sustainable industrial development policies, also stem from the Romantic era. In art, painters like Caspar David Friedrich produced works that are considered representative masterpieces of Romanticism. One historic thrust of German environmental thinking emerged from 19th-century agrarian romanticism, which acclaimed that the peasantry and the rural lifestyle represented the foundation of the German people. Advocates of this view believed that children from the city should be sent to work on farms to build their character and determination. Under the slogan "blood and soil," this orientation laid a foundation for Nazi ideology and violent antisemitism based on the idea that Germans were firmly rooted in the land while Jews were nomadic (Grunberger 1971). Nazi doctrine included a "back to the land" theme, asserting that city life, under Jewish influence, was demoralizing and bred weakness. Protecting German forests emerged as a Nazi objective, at least in part based on the idea that pre-Roman German tribes had been forest-dwellers and that wooded areas rejuvenated an asserted ancient warrior spirit in Aryan people (Williams 2007). The Nazis often are portrayed as being particularly environmentally conscious because:

> they passed elaborate nature and animal protection laws. However, as much of the recent scholarship on this topic demonstrates, Nazi environmental protection policies were often just rhetorical claims that found little application in actual practice . . . At the same time, we should be cautious not to let the Nazi period dominate [the field of German environmental history] (as it has done in many other subfields of German history) because there is much more to German environmental history than the relatively short-lived, albeit tremendously consequential, Nazi period. (Branz *et al.* 2009)

Long prior to the Nazi brand of narrow nationalist "environmentalism," Germany, in the early 19th century, gave birth to the vibrant life reform movement, which advocated an array of nonmainstream ideas from alternative medicine to organic food to nudism, but which included a back-to-nature component. Though most participants were left-leaning or apolitical, there were ultra-conservative and even antisemitic Nazi-precursor sectors in the broad-based movement that included hundreds of different groups. Many contemporary

German environmental have historic ties to the life reform movement.

Across the Atlantic in the United States, the environmental movement began to take shape when naturalist John Muir convinced Congress to establish Yosemite National Park in California in 1890. Muir's nature and advocacy writings in popular periodicals, produced in a powerful poetic style, also were persuasive in the creation of both the Grand Canyon and Sequoia national parks. About the Sierra Madre Mountains of California, for example, he wrote: "Its marvelous beauty, displayed in such striking and alluring forms, *woos* the admiring wanderer on and on, higher and higher, charmed and enchanted. Benevolent, solemn, fateful, pervaded with divine light, every landscape glows like a countenance hallowed in eternal repose . . . throbbing and pulsing with the heartbeats of God" (Hatch 2012, p. 126).

Muir came to environmentalism after he was temporarily blinded during an explosion in the factory where he worked in Wisconsin. Building on his childhood interest in botany, he responded to the accident by embracing a nature tourism lifestyle, drawing sketches of the changing landscape he saw as he walked from Indiana to Florida. Eventually making his way to San Francisco, Muir first visited the Yosemite Valley in 1868, and soon found work herding sheep so that he could stay in the area. In 1870, he visited Yosemite with Joseph LeConte, a geologist, who said of Muir that he "gazes and gazes and cannot get his fill . . . Plants and flowers and forests, and sky and clouds and mountains seem actually to haunt his imagination" (quoted in National Park Service 2015).

Muir (1912, p. 256), like others before him, experienced emersion in nature as therapeutic:

> The making of gardens and parks goes on with civilization all over the world, and they increase both in size and number as their value is recognized. Everybody needs beauty as well as bread, places to play in and pray in, where Nature may heal and cheer and give strength to body and soul alike.

In an increasingly fast-paced and urbanized world, at the core of Muir's approach as a naturalist, as expressed in the philosophy of the Sierra Club, the environmental and outdoor recreation organization he co-founded in 1892, was the dual idea that nature should be both protected and enjoyed. He embraced the creation of national parks as a mechanism whereby people could "experience nature" through participatory engagement but in which nature could be protected from the harms of prolonged human presence.

Unavoidably, Muir's understanding of nature was not natural, in the sense that it was socially constructed. Reflecting the conventional attitudes of his day, Muir believed that the indigenous people he encountered during his Yosemite sojourns were not capable of appreciating the true beauty of the environment and therefore had "no right place in the landscape" (Muir 1894, p. 93). Such attitudes, which Jacoby (2006, p. 3) refers to as part of the "elite discourse about conservation," rationalized the removal of Native Americans living in areas that were selected as national parks (Spence 1999).

While Muir focused on landscape preservation, other early American environmentalists set their sights on saving endangered species like the American bison (*Bison bison*). Europeans first gained sight of this impressive animal in 1521 when Hernán Cortés and his fellow invaders reached the capital of the Aztecs and observed captured bison in Montezuma's zoo. During the early 1800s, as Americans migrated westward, tens of millions of bison widely roamed the prairies of North America (Morris 2015). It is estimated that in 1858, the year Theodore Roosevelt was born, the population densities for people and bison in the United States were about 10 and 17 per square mile, respectively, while by 1900, when Roosevelt was 42 and the new vice president of the country, there were 25 people per square mile and very few bison left standing (Aune & Plumb 2016). The formidable herbivores, the largest of the wild fauna of North America, were on the brink of joining six other species of bison that already were extinct (Hornaday 1889).

The hides of the downed animals were cut from their carcasses, prepared, and shipped eastward, and then on to Germany for processing into bison leather, the preferred material for making belts for industrial machinery. In his collection of letters and observation notes, published under the title *Illustrations of the Manners and Customs and Conditions of the North American Indians*, the eminent painter and author George Catlin (1886) reported that in 1832, 150 000–200 000 bison robes were marketed annually. This translates to the yearly killing of 2–3 million bison. The U.S. government sanctioned this carnage, and even paid a bounty for skulls as a way to undercut the food supply of resistant Native American populations. The increasingly wealthy and powerful railroad industry urged that bison herds be culled, if not outright eliminated, to keep them from blocking tracks and damaging locomotives during collisions. Shooters were hired to ride the rails and take down as many bison as they could. Horns were used by the makers of umbrellas and fans. Bones were used by sugar refineries (to filter impurities from sugar cane juice), made into fertilizer, used as material for button making, and turned into bone china. Pyramids of bison bones piled up at railroad stations awaiting shipment to factories.

Foreseeing the rapidly approaching disappearance of the bison around them, in the late 1800s a small number of ranchers collected the scattered remnants of the remaining herds in an effort to save the species. One of these was Ernest Harold Baynes, the manager of a private bison herd in New Hampshire, who came to play a national role in bison conservation. The largest private owner may have been Michael Pablo of Montana, who in 1906 amassed a herd of about 700 animals on Blackfoot tribal land, where he had grazing rights through his mother (Hewitt 1919). Elk Island National Park, near Edmonton in western Canada, became active in bison conservation in 1907 when it bought several hundred animals from Pablo. Other bison were collected to put on public display, such as a small herd maintained in a park and fairgrounds known as Bismarck Grove in eastern Kansas. Ironically, grazing land was provided by the Union Pacific Railroad, which adopted the image of a bison as its company symbol.

A key figure in the bison rescue effort was William Hornaday, a taxidermist by training, naturalist by passion, and, eventually, zoo director by employment (Carter 2018). In 1800, Hornaday won the top prize in a taxidermy contest for his display entitled "A Fight in the Tree-Tops" depicting two male Bornean orangutans brawling over a female. The display changed the direction of taxidermy from merely stuffing animal hides for exhibit to seeking scientific accuracy in true-to-life mounted scenes. Two years later, Hornaday was appointed the Chief Taxidermist for the Smithsonian Institution in Washington, D.C. In this role, he travelled in 1886 to Montana with the goal of shooting enough bison to assemble a collection for the Smithsonian. It took him more than 2 months to kill 22 bison, a clear sign that the animal had become a rare item on the prairie. Several years later, he compiled the first bison census, concluding that there just over 800 remaining in North America. Subsequently, he directed a successful effort to establish a zoo in Washington, D.C., and in 1896 he became the first director of the Bronx Zoo in New York City. Based on his own observations and encouragement from Ernest Baynes, Hornaday concluded that it might be possible to save the bison from extinction. At the Bronx Zoo, he created a bison exhibit with a breeding pair and two calves shipped from the Black Hills of South Dakota.

In 1905, the American Bison Society was formally launched as an organized initiative to prevent the extinction of the American bison through a breeding program and to use animals raised at the Bronx Zoo to populate new "wild" bison reverses elsewhere in the country, including the Wichita Mountain Preserve in Oklahoma, the National Bison Range in Montana, Wind Cave National Park in South Dakota, and the Niobrara Wildlife Preserve in Nebraska. Theodore Roosevelt, who had himself shot several bison while hunting, agreed to serve as honorary president of the Society. Roosevelt knew Hornaday through the Boone and Crockett, a nonprofit organization he had founded in 1887 made up of well-off trophy hunters who recognized that overharvesting and lack of conservation efforts would drive the game they wished to purse into extinction.

These scattered efforts to save bison from disappearing ultimately succeeded—an occurrence that has been widely touted as a major conservation accomplishment. Bison numbers rebounded and herds multiplied to the point that bison meat became readily available in supermarkets. Based on sales in restaurants and retail stores and online, in 2018 the bison industry was valued at $350 million. The same year, the bison population of North America was approximately 400 000, though over 90 percent of the animals are raised on private ranches for commercial use (National Bison Association 2018). In other words, while bison are not at risk of species extinction, they continue to suffer grave risk of ecological extinction. The grasslands that were their traditional habitat have been fragmented and converted for use in agricultural and livestock production or paved over for urban development. As the IUCN (2018) notes:

> There has been a modest increases in the number of conservation herds and individuals in populations managed for species conservation and ecological restoration, however, all mature individuals occur within active management programs which if ceased would result in the species qualifying for a threatened status. About 97 percent of the continental population is managed for private captive commercial propagation; very few of these herds are managed primarily for species conservation and none is managed in the public interest for conservation.

Rescue of the bison population from extinction raises the question of whether it is possible to go one step further and attempt the genetic de-extinction of lost species. This would entail extracting ancient DNA from the tissue or bone remains of an extinct species to use in genetically programing a living relative to produce a new species with as many traits of the extinct one as possible. The technology that makes this idea plausible is called the Crispr-Cas9 system, which, in effect, is a biotechnological cut-and-paste tool that enables gene editing: a type of bioengineering in which selected DNA is inserted, eliminated, altered, or substituted in the genome of a living organism. Considerable media interest (e.g., Hedgecock 2017) has focused on the work of Harvard geneticist George Church (TEDx Talks 2013) and his colleagues, who are interested in the de-extinction of the woolly mammoth (*Mammuthus primigenius*) using genetic material extracted from the extensive collection of mammoth finds to recover the species' complete nuclear genome (Miller *et al.* 2008). Despite this attention, a more likely first de-extinction, if one ever occurs, will probably involve the passenger pigeon (*Ectopistes migratorius*).

A fast-flying nomadic wanderer that specialized on a diet of mast acorns, the passenger pigeon was at one time perhaps the most numerous bird on the planet, with an estimated population of 3–5 billion. The last member of the species died in captivity at the Cincinnati Zoo in 1914. Named Martha, she became a national symbol among species conservationists. Using unavoidably degraded tissue taken from the toe pads of museum-housed passenger pigeon specimens, researchers have gained partial insight into the bird's genome. Other needed information was found in the

DNA of the band-tailed pigeon (which also has been heavily hunted), its closest living relative. The goal is to create a new lineage of birds born of band-tailed pigeons that carry genes for distinctive features (e.g., red breast and legs in males, special tail feathers, living in dense groups) of the extinct passenger pigeon. While the project still has a long way to go, plans have already been developed for how to begin testing the release of the old/new bird "back" into the wild by the year 2032 (Revive & Restore n.d.).

One of the roadblocks is the deforestation that has occurred in North America since dense flocks of migrating passenger pigeons darkened its skies. Precolonial acorn- and other nut-producing forests are now urban areas and farmland. Indeed, besides the enormous technical challenges and costs (which could be used to protect living species from extinction), human radical transformation of the environment is the elephant in the de-extinction room. Nonetheless, the IUCN has published a set of guidelines for the synthetic biology of de-extinction practice (Novack 2018). Concern about extinction and the popularity of the idea in the public environmental consciousness will no doubt influence the course of this area of scientific work.

4.4.2 Environmental crisis and the media

How are environmental problems and crises shaped as public issues? Many have looked to the mass media as a primary influence on public knowledge and attitudes about the kinds of ecocrises examined in this book. Anders Hansen (1991, p. 44) argues that to understand the degree and kind of coverage of the environment in the mass media, it is necessary to examine "the importance of cultural resonances in the privileging of some issues over others." In other words, to avoid attributing undue influence to the media in the development of public concerns, it is necessary to consider interaction among various institutions (e.g., science, religion, government, education) in the process of meaning construction in society. Consequently, Hansen is critical of lineal approaches that assume that environmental information flows from a source institution (e.g., a government announcement, release of new scientific research findings, the statement of a prominent religious figure) through the mass media to the wider public. The role of hegemony—the pervasive spread, replication, and normalization of the perspectives of the wealthy and most powerful across societal institutions—also must be considered in assessing which environmental messages are most consistently and loudly echoed through the media (Hall 1996).

In an early appraisal of the relationship between the level of media focus on the environment and the extent of public concern about environmental issues, Funkhouser (1973) found a high level of correlation. He did not conclude, however, that the media drives the public's actual level of concern so much as that it influences public awareness about the state of particular conditions in the environment. People may be concerned about the risks of polluted air without this focus becoming generalized into a global unease with the overall health of the environment or even having much influence on attitudes about other environmental issues like acid rain or wildlife protection. Moreover, some environmental and environmental health issues, like earthquakes and floods, have a sudden, acute, and immediate life-threatening onset that insures media attention—though it does not guarantee how the event will be framed in terms of the assessment of causality (e.g., the human role). Others, like climate change, have a slower, less immediately obvious onset and expression that allows media attention to wax and wane or be subject to well-funded denial campaigns by vested private interests. As Hansen (2010, p. 96) observes: "[n]ews is largely event-focused and event-driven, and it is this inherent event orientation which is an important factor in determining which environmental issues get news coverage and which don't." Additionally, private media ownership and the relationship of owners to powerful political and economic interests affect the framing of environmental reporting (Boykoff & Yulsman 2013). Various researchers have examined the ways in which economic pressures and ownership structures have influenced environmental news production (Carvalho 2005).

With a growth in the awareness that there are multiple sources of meaning-making in society—if not always a parallel recognition that these are not equally empowered—the actual role of the media in the establishment of public priorities has become more complicated and a topic of both contestation and periodic rethinking. Consistently, however, it has been shown that the social authority status of an environmental information source is significant and that environmental activists are not treated as primary definers of environmental issues by the media (Corbett 1998; Taylor *et al.* 2000). This pattern has come to be known as the media's authority-order bias. Because of this structural bias, the media "play an inherent role in representing certain interpretations of the biosphere" (Luedecke & Boykoff 2017).

Further, as the divide between points of view in the population has both widened and become more politically charged, and the same has occurred in the channels of mass media communication—which have significantly multiplied since the rise of the digital age—it has become clear that speaking in a singular frame is problematic because there are diverse publics and medias (Lester 2010). Now, people can more easily select media outlets that resonate with their preexisting points of view, a development that modulates public concern about pressing environmental issues.

The nature of the tenacious environmental issues we face is discussed in the following chapter by way of a series of detailed case studies.

References

Alvarez, L., Alvarez, W., Asaro, F. & Michel, H. Extraterrestrial cause for the Cretaceous–Tertiary extinction. *Science* 208: 1095–1108.

Arens, N. & West, I. (2008) Press-pulse: a general theory of mass extinction? *Paleobiology* 34(4): 456–471.

Aubert, M., Brumm, A., Ramli, M., Sutikna, T., Saptomo, E., Hakim, B. *et al.* (2014). Pleistocene cave art from Sulawesi, Indonesia. *Nature* 514: 223–227.

Aune, K. & Plumb, G. (2016). The president who saved the American bison. *Scientific American* https://blogs.scientificamerican.com/guest-blog/the-president-who-saved-the-american-bison/.

Barish, M. (2017). Causes of the great mass extinction of marine organisms in the Late Devonian. *Oceanology* 56(6): 863–875.

Barnosky, A., Matzke, N., Tomiya, S., Wogan, G., Swartz, B. Quental, T. *et al.* (2011). Has the Earth's sixth mass extinction already arrived? *Nature* 471: 51–57.

Bartlett, J. (2019). Why one man left Silicon Valley and set up a survival camp. *BBC News.* https://www.bbc.com/news/av/world-us-canada-40814714/why-one-man-left-silicon-valley-and-set-up-a-survival-camp.

Beach, B. & Hanlon, W. (2017). Coal smoke and mortality in early industrial economy. *Economic Journal* 128: 2652–2675.

Beck, U. (1992) *Risk Society: Towards a New Modernity*. New Delhi: Sage.

Benson, R., Campione, N., Carrano, M., Mannion, P., Sullivan, C. Upchurch, P. *et al.* (2014). Rates of dinosaur body mass evolution indicate 170 million years of sustained ecological innovation on the avian stem lineage. *PLoS Biology* 12(5): e1001853.

Benson, R., Hunt, G., Carrano, M. & Campione, N. (2018). Cope's rule and the adaptive landscape of dinosaur body size evolution. *Palaeontology* 61(5): 13–48.

Benton, M. & Twitchell, R. (2003). How to kill (almost) all life: the end-Permian extinction event. *Trends in Ecology & Evolution* 18(7): 358–365.

Berger, J.-F., Lespez, L., Kuzucuoglu, C., Glais, A., Hourani, F., Barra, A. & Guilaine, J. (2016). Interactions between climate change and human activities. *Climate of the Past* 12: 1847–1877.

Best ROI Lists. (2019). Prepper marketing is based on trust, not just message. https://bestroilists.com/prepper-marketing/.

Bohn, H. (1847). *Cosmos*, vol 4. London: Henry George Bohn.

Boissoneault, L. (2018). How the ancestors of birds survived the dino-killing asteroid. *Smithsonian Magazine*. https://www.smithsonianmag.com/science-nature/how-ancestors-birds-survived-dino-killing-asteroid-180969167/.

Boivin, N., Zeder, M., Fuller, D., Crowther, A., Larson, G., Erlandson, J. *et al.* (2015). Ecological consequences of human niche construction: examining long-term anthropogenic shaping of global species distributions. *Proceedings of the National Academy of Sciences of the U.S.A.* 113(23): 6388–6396.

Borenstein, S. (2018). Bye bye bugs? Scientists fear non-pest insects are declining. *Associated Press*. https://www.apnews.com/eccd411638994db6b5388c052c00dd3b.

Boykoff, M. & Yulsman, T. (2013). Political economy, media, and climate change: sinews of modern life. Wiley Interdisciplinary Reviews. *Climate Change* 4(5): 359–371.

Brannen, P. (2017a). Earth Is not in the midst of a sixth mass extinction. *The Atlantic*. https://www.theatlantic.com/science/archive/2017/06/the-ends-of-the-world/529545/.

Brannen, P. (2017b). *The Ends of the World: Volcanic Apocalypses, Lethal Oceans, and Our Quest to Understand Earth's Past Mass Extinctions*. New York: Ecco Press.

Brantz, D., Bissibl, B., Warde, P., Winiwarter, V., Zeller, T. & Lekan, T. (2009). The nature of German environmental history. *German History* 27(1): 145–149.

Burgess, N., Malugu, I. & Sumbi, P. (2018). The fate of unique species in Tanzania's coastal forests hangs in the balance. *The Conversation*. https://theconversation.com/the-fate-of-unique-species-in-tanzanias-coastal-forests-hangs-in-the-balance-102281.

Carlyon, R. (2014). The end Is nigh: the rising popularity of the apocalypse in modern cinema. *Screen Robot*. https://screenrobot.com/end-nigh-rising-popularity-apocalypse-modern-cinema/.

Carmichael, S., Waters, J., Batchelor, C., Coleman, D., Suttner, T., Kido, E. *et al.* (2016). Climate instability and tipping points in the Late Devonian: detection of the Hangenberg Event in an open oceanic island arc in the Central Asian Orogenic Belt. *Gondwana Research* 32: 213–231.

Carrington, D. (2019). "Frightening" number of plant extinctions found in global survey. *Guardian*. https://www.theguardian.com/environment/2019/jun/10/frightening-number-of-plant-extinctions-found-in-global-survey.

Carter, D. (2018). The early days: rangers' role in bison restoration. National Bison Association. https://bisoncentral.com/genetic-item/the-early-days-ranchers-role-in-bison-restoration/.

Carvalho, A. (2005). Representing the politics of the greenhouse effect. *Critical Discourse Studies* 2(1): 1–29.

Catlin, G. (1886). *Illustrations of the Manners and Customs and Conditions of the North American Indians*. London: H.G. Bohn.

Chaloupka, G. (1993). *Journey in Time, the World's Longest Continuing Art Tradition: The 50000-Year Story of the Australian Aboriginal Rock Art of Arnhem Land*. Auckland: Reed.

Cision PR Newswire. (2012). If it's the end of the world as we know it. . . New survey reveals Americans are not prepared! National Geographic Channel releases survey timed to new series doomsday preppers. https://www.

prnewswire.com/news-releases/if-its-the-end-of-the-world-as-we-know-it-new-survey-reveals-americans-are-not-prepared-national-geographic-channel-releases-survey-timed-to-new-series-doomsday-preppers-138865199.html.

Clarke, R. (2004). Pioneers of conservation: the Selborne Society and the Royal SPB. http://eprints.bbk.ac.uk/8254/1/Clarke_2004_Pioneers_of_Conservation%2BFront Cover1pps.pdf.

Corbett, J. (1998). The environment as theme and package on a local television newscast. *Science Communication* 19(3): 222–237.

Courty, M.-A. & Weiss, H. (1994). The scenario of environmental degradation in the Tell Leilan region, NE Syria, during the late third millennium abrupt climate change. In: H. Nüzhet Dalfes, G. Kukla, & H. Weiss (Eds.), *Third Millennium BC Climate Change and Old World Collapse*, pp. 107–148. Berlin: Springer-Verlag.

Cui, Y. & Kump, L. (2015). Global warming and the end-Permian extinction event: proxy and modeling perspectives. *Earth-Science Reviews* 149: 5–22.

Dawson, T., Nimura, C., López-Romero, E. & Daire, M.-Y. (2017). *Public Archaeology and Climate Change*. Oxford: Oxbow Books.

de Tocqueville, A. (1958). *Journeys to England and Ireland*. New Haven, CT: Yale University Press.

Dirzo, R. & Raven, P. (2003). Global state of biodiversity and loss. *Annual Review of Environmental Resources* 28: 137–167.

Dirzo, R., Young, H., Galetti, M., Ceballos, G., Isaac, N. & Collen, B. (2014). Defaunation in the Anthropocene. *Science* 345(6195): 401–406.

Dole, C., Hayashi, R., Poe, A. Sarat, A. & Wolfson, B. (2016). When is catastrophe: an introduction. In: C. Dole, R. Hayashi, A. Poe, A. Sarat & B. Wolfson (Eds.), *The Time of Catastrophe: Multidisciplinary Approaches to the Age of Catastrophe*, pp. 1–18. London: Routledge.

Field, D., Bercovici, A., Berv, J., Dunn, R., Fastovsky, D., Lyson, T. *et al.* (2018). Early evolution of modern birds structured by global forest collapse at the end-cretaceous mass extinction. *Current Biology* 28(11): 1825–1831.

Funkhouser, G. (1973). The issues of the sixties: an exploratory study in the dynamics of public opinion. *Public Opinion Quarterly* 50: 533–538.

Gangwani, K. & Landin, J. (2018). The decline of insect representation in biology textbooks over time. *American Entomologist* 64(4): 252–257.

Garrett, B. (2020). Living with bunker builders: doomsday prepping in the age of coronavirus. *The Conversation*. https://theconversation.com/living-with-bunker-builders-doomsday-prepping-in-the-age-of-coronavirus-136635.

Gerta, K., Mateo, P., Punekar, J., Khozyem, H., Gertsch, B., Spangenberg, J. *et al.* (2018). Environmental changes during the Cretaceous–Paleogene mass extinction and Paleocene–Eocene thermal maximum: implications for the Anthropocene. *Gondwana Research* 56: 69–89.

Giddens, A. & Pierson, C. (1998). *Making Sense of Modernity: Conversations with Anthony Giddens*. Stanford, CA: Stanford University Press.

Greene, S., Martindale, R., Ritterbush, K., Bottjer, D., Corsetti, F. & Berelson, W. (2012). Recognising ocean acidification in deep time: an evaluation of the evidence for acidification across the Triassic–Jurassic boundary. *Earth-Science Reviews* 113: 72–93.

Grissom, C. (2000). Neolithic statues from 'Ain Ghazal: construction and form. *American Journal of Archaeology* 104(1): 25–45.

Grove, R. (1995). *Green Imperialism: Colonial Expansion, Tropical Island Edens and the Origins of Environmentalism 1600–1860*. Cambridge: Cambridge University Press.

Grunberger, R. (1971). *The 12-Year Reich: A Social History of Nazi Germany 1933–1945*. New York: Holt, Rinehart, and Winston.

Guedes, J., Crabtree, S., Bocinsky, R. & Kohler, T. (2016). Twenty-first century approaches to ancient problems: climate and society. *Proceedings of the National Academy of Sciences of the U.S.A.* 113(51): 14483–14491.

Habel, C., Segerer, A., Ulrich, W., Torchyk, O., Weisser, W. & Schmitt, T. (2016). Butterfly community shifts over two centuries. *Conservation Biology* 20(4): 754–762.

Hall, S. (1996). The problem of ideology: Marxism without guarantees. In D. Morley & K.-H. Chen (Eds.), *Stuart Hall: Critical Dialogues in Cultural Studies*, pp. 25–46. London: Routledge.

Hallman, C., Sorg, M., Jongejans, E., Siepel, H., Hofland, N., Schwan, H. *et al.* (2017). More than 75 percent decline over 27 years in total flying insect biomass in protected areas. *PLoS One* 12(10): e0185809.

Hansen, A. (1991). The media and social construction of the environment. *Media, Culture and Society* 13: 443–458.

Hansen, A. (2010). *Environment, Media and Communication*. Abigdon: Routledge.

Harper, D., Hammarlund, E. & Rasmussen, C. (2014). End Ordovician extinctions: a coincidence of causes. *Gondwana Research* 25(4): 1294–1307.

Hartman, S., Mortimer, M., Wahl, R., Lomax. D., Lippincott, J. & Lovelace, D. (2019). A new paravian dinosaur from the Late Jurassic of North America supports a late acquisition of avian flight. *PeerJ* 7: e7247.

Hatch, S. (2012). *The Contemplative John Muir: Spiritual Quotations from the Great American Naturalist*. Morrisville, NC: Lulu Press.

Hatton, T. (2017). Air pollution in Victorian-era Britain—its effects on health now revealed. *The Conversation*. https://theconversation.com/air-pollution-in-victorian-era-britain-its-effects-on-health-now-revealed-87208.

Hedgecock, S. (2017). What the woolly mammoth de-extinction project actually means. *Forbes*. https://www.forbes.com/sites/sarahhedgecock/2017/03/01/what-the-woolly-mammoth-de-extinction-project-actually-means/#2ef4fe674a61.

Hemmingway, E. (2006). *The Sun Also Rises*. New York: Scribner.

Hewitt, C. (1919). The coming back of the bison. *Natural History*. http://www.naturalhistorymag.com/picks-from-the-past/101507/the-coming-back-of-the-bison?page=3.

Hornaday, W. (1889). *The Extermination of the American Bison*. Washington, D.C.: National Museum.

House, M. (2002). Strength, timing, setting and cause of mid-Palaeozoic extinctions. *Palaeogeography, Palaeoclimatology, Palaeoecology* 181(1–3): 5–25.

Huet, M.-H. (2012). *The Culture of Disaster*. Chicago, IL: University of Chicago Press.

Humphreys, A., Govaerts, R., Ficinski, S., Lughadha, E. & Vorontsova, M. (2019). Global dataset shows geography and life form predict modern plant extinction and rediscovery. *Nature Ecology & Evolution* 3: 1043–1047.

IUCN. (2018). American bison. International Union for Conservation of Nature Red List of Threatened Species. http://www.iucnredlist.org/details/2815/0.

Jacoby, K. (2006). *Crimes Against Nature: Squatters, Poachers, Thieves, and the Hidden History of American Conservation*. Berkeley, CA: University of California Press.

Joachimski, M., Lai, X., Shen, S., Jiang, H., Luo, G., Chen, B. *et al.* (2012). Climate warming in the latest Permian and the Permian–Triassic mass extinction. *Geology* 40(3): 195–198.

Kesztenbaum, L. & Rosenthal, J.L. (2011). The health cost of living in a city: the case of France at the end of the 19th century. *Explorations in Economic History* 48(2): 207–225.

Kintigh, K., Altschul, M., Beaudry, R., Drennan, A., Kinzig, T., Kohler, W. *et al.* (2014). Grand challenges for archaeology. *American Antiquity* 79(1): 5–24.

Kirch, P. (2005). Archeology and global change: the Holocene record. *Annual Review of Environment and Resources* 30: 409–444.

Knibbs, K. (2020). The prepping industry wasn't prepared for the coronavirus. *Wired*. https://www.wired.com/story/prepping-industry-coronavirus/.

Koh, L., Dunn, R., Sodhi, N., Colwell, R. Proctor, H. & Smith, V. (2004). Species coextinctions and the biodiversity crisis. *Science* 305(5690): 1632–1634.

Langmuir, C. & Broecker, W. (2012). *How to Build a Habitable Planet*. Princeton, NJ: Princeton University Press.

Leakey, R. & Lewin, R. (1992). *The Sixth Extinction: Patterns of Life and the Future of Humankind*. New York: Doubleday.

Lepore, L. (2017). A golden age for dystopian fiction. *New Yorker*. https://www.newyorker.com/magazine/2017/06/05/a-golden-age-for-dystopian-fiction.

Lester, L. (2010). *Media and Environment: Conflict, Politics and the News*. Cambridge: Polity Press.

Lister, B. & Garcia, A. (2018). Climate-driven declines in arthropod abundance restructure a rainforest food web. *Proceedings of the National Academy of Sciences of the U.S.A.* 115(44): E10397–E10406.

Luedecke, G. & Boykoff, M. (2017). Media communication on environmental issues. In: D. Richardson, N. Castree, M. Goodchild, A. Kobayashi, W. Liu & R. Marston (Eds.), *The International Encyclopedia of Geography*. Hoboken, NJ: John Wiley & Sons.

Martínez, A. (2016). *Chaos Monkeys: Obscene Fortune and Random Failure in Silicon Valley*. New York: Harper.

Mayewski, P., Rohling, E., Stager, J., Karlé, N., Maasch, K., Meeker, L., *et al.* (2004). Holocene climate variability. *Quaternary Research* 62: 243–255.

Mayhew, H. (1968). *The Morning Chronicle Survey of Labour and the Poor: The Metropolitan Districts*, vol. 4. New York: Dover Publications.

McElwain, J. & Punyasena, S. (2007). Mass extinction events and the plant fossil record. *Trends in Ecology & Evolution* 22(10): 548–557.

McGhee, G. (1988). The Late Devonian Extinction Event: evidence for abrupt ecosystem collapse. *Paleobiology* 14(3): 250–257.

McMichael, G. & Crews, F. (Eds.). (1997). *Anthology of American literature: Colonial through Romantic*, vol. 1, 6 edn. Upper Saddle River, NJ: Prentice Hall College Division.

Miller, W., Drautz, D., Ratan, A., Pusey, B., Qi, J., Lesk. A., *et al.* (2008). Sequencing the nuclear genome of the extinct woolly mammoth. *Nature* 456(7220): 387–390.

Morris, D. (2015). *Bison*. London: Reaktion Books.

Morwood, M. (2002). *Visions from the Past: The Archaeology of Australian Aboriginal Art*. Crows Nest, NSW: Allen & Unwin.

Muir, J. (1894). *The Mountains of California*. New York: The Century Co.

Muir, J. (1912). *The Yosemite*. New York: The Century Co.

National Bison Association. (2018). Current status. https://bisoncentral.com/current-status/.

National Park Service. (2015). Muir's influences. https://www.nps.gov/yose/learn/historyculture/muir-influences.htm.

Novack, B. (2018). De-extinction. *Genes* 9(11): 548.

O'Connor, R. (2015). Preppers: meet the paranoid Americans awaiting the apocalypse. *The Week*. https://theweek.com/articles/565576/preppers-meet-paranoid-americans-awaiting-apocalypse.

Ogg, J. (2013). Industries making the most money on doomsday preppers. *Yahoo Finance*. https://finance.yahoo.com/news/industries-making-most-money-doomsday-103557775.html.

Oosthoek, K. (2015). Romanticism and nature. Environmental History Resources. https://www.eh-resources.org/romanticism-and-nature/.

Osnos, E. (2017). Doomsday prep for the super-rich. *New Yorker.* https://www.newyorker.com/magazine/2017/01/30/doomsday-prep-for-the-super-rich.

Pálfy, J. & Zajzon, N. (2012). Environmental changes across the Triassic–Jurassic boundary and coeval volcanism inferred from elemental geochemistry and mineralogy in the Kendlbachgraben section (Northern Calcareous Alps, Austria). *Earth and Planetary Science Letters* 395–396: 121–134.

Penn, J., Deutsch, C., Payne, J. & Sperling, E. (2018). Temperature-dependent hypoxia explains biogeography and severity of end-Permian marine mass extinction. *Science* 362(6419): eaat1327.

Petersen, S., Dutton, A. & Lohmann, K. (2016). End-Cretaceous extinction in Antarctica linked to both Deccan volcanism and meteorite impact via climate change. *Nature Communications* 7: 12079.

Pimm, S., Russell, G., Gittleman, J. & Brooks, T. (1995). The future of biodiversity. *Science* 269: 347–350.

Pinkerton, P. (2017). Doomsday preppers—smart thinking or doomsday phobia. *Outdoor Revival*. https://www.outdoorrevival.com/instant-articles/doomsday-preppers-smart-thinking-doomsday-phobia.html/2.

Posner, R. (2004). *Catastrophe: Risk and Response*. Oxford: Oxford University Press.

ReadyStore (2019). Homepage. www.thereadystore.com.

Rees, R. (1982). Constable, Turner and views of nature in the nineteenth century. *Geographic Review* 72(3): 253–269.

Reichow, M., Saunders, A., White, R., Pringle, M., Al'Mukhamedov, A., Medvedev, A. & Kirda, N. (2002). 40Ar/39Ar dates from the West Siberian Basin: Siberian flood basalt province doubled. *Science* 296(5574): 1846–1849.

Revive & Restore. (n.d.). De-extinction defined. https://reviverestore.org/passenger-pigeon-de-extinction/.

Rhodes, J. (2018). The real reason tech billionaires are prepping for doomsday. *Salon*. https://www.salon.com/2018/05/12/the-real-reason-tech-billionaires-are-prepping-for-doomsday/.

Sallen, L. & Coates, M. (2010). End-Devonian extinction and a bottleneck in the early evolution of modern jawed vertebrates. *Proceedings of the National Academy of Sciences of the U.S.A*. 107(22): 10131–10135.

Sánchez-Bayo, F. & Wyckhuys, K. (2019). Worldwide decline of the entomofauna: a review of its drivers. *Biological Conservation* 232: 8–27.

Schaller, S., Bottjer, D., Corsetti, F., Berelson, W. & Zonneveld, J. (2011). A subseafloor carbonate factory across the Triassic–Jurassic transition. *Geology* 40(11): 1043–1046.

Schmidt, C. (2014). Why are dystopian films on the rise again? *JSTOR*. https://daily.jstor.org/why-are-dystopian-films-on-the-rise-again/.

Sheeman, P. & Coorough, P. (1990). Brachiopod zoogeography across the Ordovician–Silurian extinction event. *Geological Society, London, Memoirs* 12: 181–187.

Simmons, A., Köhler-Rollefson, I., Rolleson, G., & Kafafi, Z. (1988). 'Ain Ghazal: a major Neolithic settlement in Central Jordan. *Science* 240(4848): 35–39.

Spence, M. (1999). *Dispossessing the Wilderness: Indian Removal and the Making of the National Parks*. Oxford: Oxford University Press.

Spooner, A. (2018). A case study in niche marketing: doomsday preppers. Point Content Marketing. http://pointcontentmarketing.com/a-case-study-in-niche-marketing-doomsday-preppers/.

Squire, S. (1988). Wordsworth and Lake District tourism: romantic reshaping of landscape. *Canadian Geographer* 32(3): 237–247.

Stein, V. (2019). Dinosaurs are back with a climate change warning at the Smithsonian. *PBS News Hour*. https://www.pbs.org/newshour/science/dinosaurs-return-to-the-smithsonian-with-a-climate-change-warning.

Strona, G. & Bradshaw. C. (2018). Co-extinctions annihilate planetary life during extreme environmental change. *Scientific Reports* 8: 16724.

Taylor, C., Lee, J. & Davie, W. (2000). Local press coverage of environmental conflict. *Journalism & Mass Communication Quarterly* 77(1): 175–192.

TEDx Talks. (2013). Hybridizing with extinct species: George Church at TEDxDeExtinction. https://www.youtube.com/watch?v=oTH_fmQo3Ok.

Thubron, C. (2015) "The Invention of Nature," by Andrea Wulf. *New York Times*. https://www.nytimes.com/2015/09/27/books/review/the-invention-of-nature-by-andrea-wulf.html?_r=0.

Turney, C., Baillie, M., Palmer, J. & Brown, D. (2006). Holocene climatic change and past

Irish societal response. *Journal of Archaeological Science* 33(1): 34–38.

Valentine, J. (1973). *Evolutionary Paleoecology of the Marine Biosphere*. Englewood Cliffs, NJ: Prentice-Hall.

Vandenbroucke, T., Emsbo, P., Munnecke, A., Nuns, N., Ludovic, D., Lepot, K. *et al.* (2015). Metal-induced malformations in early Palaeozoic plankton are harbingers of mass extinction. *Nature and Communication* 6: 7966.

Vaughn, C. (2014). How to survive the Chicago apocalypse. *Chicago Reader*. https://www.chicagoreader.com/Bleader/archives/2014/04/24/how-to-survive-the-chicago-apocalypse/.

Vié, J.-C., Hilton-Taylor, C. & Stuart, S. (Eds). (2009). Wildlife in a changing world—an analysis of the 2008 IUCN Red List of Threatened Species. International Union for Conservation of Nature. https://www.iucn.org/content/wildlife-a-changing-world-analysis-2008-iucn-red-list-threatened-speciestm.

Wagner, D. (2012). Moth decline in the Northeastern United States. *News of the Lepidopterists' Society* 54(2): 52–56.

Wake, D. & Vredenburg, V. (2008). Are we in the midst of the sixth mass extinction? A view from the world of amphibians. *Proceedings of the National Academy of Sciences of the U.S.A.* 105: 11466–11473.

Weiberg, E. & Finné, M. (2008). Resilience and persistence of ancient societies in the face of climate change: a case study from Late Bronze Age Peloponnese. *World Archaeology* 50: 584–602.

Weiss, H., Courty, M, Wetterstrom, W., Guichard, F., Senior L., Meadow, R. & Curnow, A. (1993). The genesis and collapse of third millennium North Mesopotamian civilization. *Science* 261(5124): 995–1004.

Weninger, B., Clare, L., Rohling, E., Bar-Yosef, O., Böhner, U., Budja, M. *et al.* (2009). The impact of rapid climate change on prehistoric societies during the Holocene in the Eastern Mediterranean. *Documenta Praehistorica* 36: 7–59.

Whitley, D. (2009). *Cave Paintings and the Human Spirit: The Origin of Creativity and Belief.* Amherst, NY: Prometheus Books.

Wild Orca. (2014). Orcas and salmon: making the connection. https://www.wildorca.org/orcas-and-salmon/.

Williams, J. (2007). *Turning to Nature in Germany: Hiking, Nudism, and Conservation, 1900–1940.* Stanford, CA: Stanford University Press.

Wulf, A. (2015). *The Invention of Nature: Alexander von Humboldt's New World.* New York: Vintage.

WWF. (2018). Living Planet Report 2018: Aiming Higher. https://www.wwf.org.uk/sites/default/files/2018-10/LPR2018_Full%20Report.pdf.

Zeder, M. & Hesse, B. (2000). The initial domestication of goats (*Capra hircus*) in the Zagros Mountains, 10000 years ago. *Science* 287: 2254–2257.

5

A home in peril: major contemporary environmental crises

There are, obviously, all kinds of ecological perils out there. We've overfished our seas, we've overcut our forests. Fresh water is beginning to run short, and species are disappearing at a rapid rate. You can come up with a long and troubling list, including the disturbing fact that most of the world's people are so poor they can barely summon the energy to care about the larger world.

Bill McKibben (2006)

5.1 Case studies in contemporary environmental crises

Many of the ecocrises enmeshed in adverse interactions of concern to this book because of their impacts on health have already been discussed. This chapter considers Earth today as we confront the patterns of sweeping change reshaping life on the planet. While the list of ecocrises is manifold, seven are provided close examination here: deforestation, acidification of oceans, eutrophication of estuarine and coastal waters, depletion of the ocean biome, pollution of the global water system, oil spills, and desertification. This closer look allows the identification of the potential for development of the harm-enhancing, one-two punch of ecocrises interaction.

5.2 Deforestation

Since the first tree evolved almost 400 million years ago, forests and their assemblages of species have come and gone. The earliest forests were swampy areas populated by tree-like giant club mosses, horse tails, and ferns. As these plants died and decomposed, they became an important source of the coal deposits we now use for energy, the consumption of which leads to greenhouse gas release, global warming, and resultant death of contemporary forests. Paleobotanists are familiar with various extinct species of trees, like the Sigillaria, a spore-bearing genus, or *Araucarioxylon arizonicum*, a conifer species whose fossilized remains can be viewed in the Petrified Forest National Park in Arizona. The tree-related question of most immediate importance is how many of the 60 000 tree species extant in the world today are at risk for extinction because of direct and indirect human actions. According to Botanical Gardens Conservation International (BGCI) in London, based on a 2017 compilation of data from 500 different botanical organizations around the globe, 300 species are critically endangered (Beech *et al.* 2017). High on this list is *Karomia gigas* in Tanzania, a member of the mint family (Lamiaceae) and a botanical relative of oregano, rosemary, and thyme (Clarke *et al.* 2011). It is one of the rarest plant species, with fewer than 100 individual trees remaining

Ecosystem Crises Interactions: Human Health and the Changing Environment, First Edition. Merrill Singer.
© 2021 John Wiley & Sons, Inc. Published 2021 by John Wiley & Sons, Inc.

in the wild (a small number of juvenile plants also reside under the careful watch of botanists at the Missouri Botanical Garden greenhouse). Its dwindling numbers are believed to be a consequence of habitat loss as forests have been cleared to open up land for agricultural use. Also threatening this species is wood cutting for fuel and timber; logging remains an ongoing threat. And the case of *Karomia gigas* is far from an isolated one. Biologist Roy Gereau, Director of Tanzania Botanical Research and Conservation Programme, has assembled a list of nearly 500 tree species from Tanzania that also are at serious risk (Missouri Botanical Garden 2018).

In addition to other anthropogenic threats, scientific research shows that global climate change is having "a significant impact on forested ecosystems, causing increases in tree mortality rates, and decreases in tree growth and health" (Brouwers *et al.* 2015). This is evident, among other forest hot spots, in Australia, a country that has endured intense droughts since the beginning of the 21st century, especially in the years 2003–2012. Parts of Australia are still confronted by drought stress, with rainfall measures indicating a decrease in precipitation levels dating to 1994. A growing number of researchers attribute the intensification of drought in Australia—a land nicknamed the Sunburnt Country after Dorothea Mackellar's poem, "My Country"—to global warming (Whetton *et al.* 1993; Mpelasoak *et al.* 2007).

In the southwestern section of Western Australia, 26 percent of mature forest trees died in response to the extreme drought and multiple heatwaves that hit the country in 2010/2011. Similar mass tree mortality befell the Mulga Lands bioregion in Queensland in eastern Australia during the 2003–2007 drought. In addition to the negative impact of these losses on other species that depend on trees, they also reduce the natural carbon storage process (capture of carbon from the atmosphere and its subsequent sequestration for many years) that occurs in forests.

In the midst of the prolonged drought, the 2019–2020 brushfire season in Australia was unprecedented. Several fires burned more than 46 million acres (72 000 square miles) and killed 34 people. It is estimated that more than a billion animals perished (Center for Disaster Philanthropy 2020). While firefighters were able to save the last grove of 200 prehistoric Wollemi pine trees, ravaged in the fires was the only stand of nightcap oak (*Eidothea hardeniana*), one of 20 extremely rare ancient plant species found in a small area in the Nightcap National Park in New South Wales. The nightcap oak dates back to the era when Australia and Antarctica formed a single land mass. Many of the few remaining trees were killed or badly damaged, though some showed signs of new growth once the fires were quelled (Ross 2020).

In 2008, BGCI found that 400 medicinal plants—those used in pharmaceutical and traditional disease treatments—are at risk of extinction (Hawkins 2008). The primary causes are overcollection and deforestation. Trees, such as the Pacific yew (the bark of which is used in the treatment of a number of types of cancer), are among the threatened species. In addition to being a danger to current medicines, loss of medicinal plant species imperils the discovery of future disease cures.

Researchers are finding that trees worldwide are much more sensitive to drought than they previously realized (Choat *et al.* 2012, 2018). Based on a global study of tree vulnerability, Choat and colleagues determined that even small changes in rainfall can cause high levels of tree mortality. They note that "[r]ecent evidence suggests rising global temperatures are already amplifying drought-induced forest change" (Choat *et al.* 2012, p. 752). There is concern that climate-related drought could effect rapid forest collapse, converting tropical forests from net carbon sinks into robust sources of atmospheric carbon during the 21st century (Phillips *et al.* 2009). In a spiraling feedback loop, loss of trees pushes global heating, which causes further loss of trees, as tree losses extend over time from species with weak drought resistance to those with stronger resistance capacity.

The primary focus thus far has been on the role of anthropogenic climate change on forest loss, but a second driving force is the cutting of trees for lumber and as part of land-clearing projects to make way for agriculture and livestock. This process has had significant impacts on the globe's largest rainforest countries: Brazil, the Democratic Republic of the Congo, and Indonesia. The degree of loss and the extent of remaining primary forest vary among these nations. Brazil's total area of loss is more than double that of Indonesia, and five times that of the Democratic Republic of the Congo. Indonesia has the least amount of remaining primary forest (Turubanova *et al.* 2018).

Significant forest loss is not limited to these three countries. Two states in eastern Australia, Queensland and New South Wales, for example, have been described as being "in the midst of a full-blown land-clearing crisis," with over 7 million acres of virgin forest slated to be bulldozed by 2030 (Slezak 2018). The World Wildlife Fund (WWF 2011a) describes eastern Australia as a biodiversity global land-clearing hot spot. Under pressure are both its tropical rainforests and temperate forests. Development of pasture for livestock is the dominant driver of deforestation, accounting for almost 90 percent of clearing in both primary and secondary forests and woodlands. Queensland is home to almost half of the meat cattle industry in Australia. Following a long history of land clear dating to the arrival of Europeans in 1787, concern about mounting deforestation led to legislative action at the turn of the 21st century. Passage of the Vegetation Management Act 1999 achieved a reduction in the pace of clearing, but policy changes in 2013 reversed this trend and sparked a subsequent increase (Neldner *et al.* 2017). The Statewide Landcover and Trees Study of 2015–16 found that about 300 000 acres of trees and shrubs were cleared between 2012 and 2016, with a 45 percent increase in clearing between 2014/15 and 2015/16 (DSITI 2017). Mature *Acacia aneura* forest, commonly known as mulga forest, has been bulldozed to harvest foliage to feed to livestock while opening up new land for pasture. In addition to the threat they present to the unique biodiversity of Australia, including iconic species like the koala, deforestation and rainfall runoff in the northeast have become significant contributors of sediment pollution, adversely affecting the Great Barrier Reef (Siriwardena *et al.* 2006; Cowie *et al.* 2007). As the Queensland Government (2017) reports:

> Strong scientific evidence indicates increased quantities of sediment are entering the Great Barrier Reef lagoon and grazing lands are a key source. Sediment discharged from rivers reduces sunlight available to sea grasses and corals and can smother coral growth. The resulting decline in water quality also increases the risk of serious long-term effects on reef health and decreases its resilience to pressures caused by climate change, ocean acidification and tropical cyclones.

As this statement indicates, the interaction of deforestation with other ecocrises can have significant detrimental environmental consequences (Guzmán & Jiménez 1992). In the case of damaging interaction with coral system pollution, human health is put at direct risk, as coral reefs are a source of protein and employment for an estimated half a billion or more people (EPA 2018). Corals are home to at least one-fourth of the total number of species living in marine environments. They are also the source of antibacterial, antitumor, and painkiller medicines. Forests too are critical to health. A growing body of scientific evidence indicates that the removal of tropical forests contributes to the spread of vector-borne infectious diseases, including malaria and leishmaniasis (Saccaro *et al.* 2015; Fornac *et al.* 2016).

After the Amazon, the next largest tropical rainforest is found in the Congo Basin, an area that covers parts of six African countries: Cameroon, the Central African Republic, the Democratic Republic of the Congo, Equatorial Guinea, Gabon, and the Republic of the Congo.

The Basin contains almost 800 000 square miles of forest, an area approximately the size of Mexico. It is known for its extensive biodiversity, including a number of iconic African species such as mountain gorillas, lowland gorillas, chimpanzees, and forest elephants. Research by Turubanova *et al.* (2018), using Earth-observing satellites and very high-resolution imagery, estimates rainforest loss between 2000 and 2014 at 41 million acres, with loss in the Democratic Republic of the Congo accounting for more than that in the other five countries combined. While small-scale forest clearing for either commercial agriculture or subsistence is the largest direct driver of forest loss in the region, agro-industry has been developing since 2004 and "large-scale agro-industrial clearing is likely to become a more significant contributor to forest loss in the future" (Turubanova *et al.* 2018). Clearing for the planting of palm oil trees is beginning in the Basin, but it has had nowhere near the same level of impact on primary forest as in Indonesia (Feintrenie 2014). Industrial selective logging is also an important factor, especially in the Republic of the Congo, Cameroon, and Gabon.

5.3 Acidification of the oceans

Like rainforests, coral reefs have high rates of biodiversity. Parallels between these terrestrial and marine ecosystems have led reef environments to be called the "rainforests of the sea." After walking for the first time on a coral island (Keeling Island) in the northeast Indian Ocean, Charles Darwin wrote in his diary that reefs "surely rank high among the wonderful objects of this world" (Rookmaaker 2006, p. 723). Unfortunately, also like rainforests, coral reefs are under significant anthropogenic threat from a multiple of sources, including overfishing, overcollection of reef invertebrates (e.g., sea urchins, lobsters, sea cucumbers, conchs), bomb fishing (with dynamite, fertilizer and fossil fuels, and other munitions), coral mining (for construction materials), industrial pollution (e.g., heavy metals, heated wastewater, fertilizers), sediment pollution, and acidification (Brown & Dunn 1988; Pet-Soede & Erdmann 1998; Hodgson 1999; Cesar 2000; Reichelt-Brushett & Harrison 2005). While few reefs face all of these threats, most if not all are suffering from at least one. In such cases, the potential for ecocrises interaction and a multiplication of damage weighs heavily on reef survival.

Coral reef construction depends on a symbiotic relationship between corals (known as polyps) and microscopic planktonic organisms called zooxanthellae. Zooxanthellae can swim freely in the ocean at several stages of their life cycle using their tails, called flagellum—appendages that not only propel them through the water but also function as sense organs, detecting chemicals and temperatures in the marine environment. A new coral colony can acquire zooxanthellae from the seawater around it or else receive them in other ways, such as during fertilization. Fertilization occurs during the mass synchronized release by polyps of eggs and sperm into the open water. These join together to form free-floating larvae, which settle on the ocean bottom, attach themselves to rocks or soil, and metamorphose into new polyps ready to add on to exiting reefs or start new coral colonies. Corals also reproduce asexually through budding.

Zooxanthellae live protected from predators in the coral tissues, where they capture sunlight and convert it through photosynthesis into the oxygen, glucose, carbon, and nitrogen needed for the coral host's metabolism, growth, and reproduction (Wang & Douglas 1998). Corals also use the nutrients supplied by their zooxanthellae to produce calcium carbonate, which is secreted to form the hard limestone reef structures in which the polyps live. They can grow from 0.04 to 0.4 inches per year, depending on the species and conditions in the surrounding environment. Zooxanthellae are the source of the celebrated palette of colors of many stony corals (those with hard skeletal structures). The intimate coral–zooxanthellae mutualist relationship allows for reef building in nutrient-poor tropical waters.

Stressful conditions can disrupt this symbiotic relationship and lead to the expulsion of zooxanthellae by the host. This process causes a decline in coral coloration, an alteration called bleaching (Fig. 5.1). The triggering of zooxanthellae expulsion may have evolved as a regulatory mechanism that assists corals in coping with environmental change. During thermal stress, for example, the photosynthetic capacity of zooxanthellae is damaged, and their expulsion limits a build-up of faulty cells. When polyps remain without zooxanthellae for too long or the magnitude of thermal stress is too great, however, continued bleaching leads to mass polyp death and the collapse of coral reef ecosystems (Fujise *et al.* 2014).

In addition to higher thermal temperatures, coral bleaching is also driven by ocean acidification. Acidification occurs as marine water absorbs CO_2 from the atmosphere. As CO_2 levels in the atmosphere increase due to anthropogenic activities, so too do levels in the surface layers of the ocean. CO_2 migrating from the atmosphere into the ocean reacts chemically with water molecules to form carbonic acid, an increase that leads to the acidification of seawater. In terms of its magnitude and rate, the current increase of CO_2 concentration in the oceans is "unparalleled in the evolutionary history of the past 20 million years" (Bollmann 2010, p. 36). Acidification is taking a severe toll on corals. As Lohr *et al.* (2007, p. 2976) note, the "frequency and intensity of coral bleaching have increased in the last two decades, leading to mass mortality of corals and a resulting reduction in the biodiversity of reefs . . . Bleaching has been observed in more than 50 countries and in the three major oceans."

Fig. 5.1 Bleached coral. *Source:* aliantha/123RF.

Richard Vevers has traveled around the world to photograph coral reefs. He visited Australia's Great Barrier Reef during its most devastating mass die-off. No dive has stunned him more, however, than one at the New Caledonia Barrier Reef, located about 1000 miles from the Great Barrier Reef, one early March, when he witnessed the corals turning snow white, lighting up fluorescently as their color left them. He notes, "[i]n the past people simply haven't gone to the right location at the right time . . . I was blown away . . . I've never seen something so beautiful, but it's dying . . . The soft corals were just decomposing—animals literally dripping off the rocks . . . The most horrifying part was that we just absolutely stank of rotting animals. That's when you really realize that reefs are made up of billions of animals" (quoted in Worland 2016).

Monitoring by the Australian Institute of Marine Science (AIMS) shows that since 2016, about half of the Great Barrier Reef has died from bleaching (Fig. 5.2). As a result, much of the marine ecosystem along the reef's north coast is now barren, and the central and south coasts have been hard hit as well. AIMS (2018) reports that "The geographic scale of recent bleaching means that breeding populations of corals have been decimated over large areas, reducing the potential sources of larvae to recolonise reefs over the next years. It is unprecedented in the 30+ year time series that all three regions of the GBR [Great Barrier Reef] have declined and that many reefs have now very low coral cover."

In experimental research carried out at the Great Barrier Reef using tank-dwelling crustose coralline algae (a rock-hard red algae that functions as an adhesive in coral reef ecosystems), staghorn corals (*Acropora intermedia*), and

Fig. 5.2 The Great Barrier Reef. *Source:* steinchen/Pixabay.

massive Porites (*Porites lobate*) coral, Anthony *et al.* (2008) used CO_2 dosing at three different levels and two different water temperatures to test the effects on coral bleaching and coral structure-building calcification. After 8 weeks, they found that higher-level CO_2 dosing led to 40–50 percent bleaching for the crustose coralline algae and staghorn corals. This effect was stronger than the effect of temperature. In the Porites corals, there was a strong synergy between CO_2 and temperature. While high temperature expanded the bleaching response by 10–20 percent in crustose coralline algae and staghorn corals, it amplified it up to 50 percent in Porites corals. High-CO_2 dosing also led to reductions of between 130 and 190 percent in calcification rates and to dramatic reductions in daily productivity (defined as hourly rates of photosynthesis minus respiration across a day) in the crustose coralline algae. These researchers conclude that "future predictions of bleaching in response to global warming must also take account of the additional effect of acidification and . . . any potential adaptation and acclimatization by coral reef organisms to thermal stress . . . may be offset or overridden by CO_2 effects" (Anthony *et al.* 2008, p. 17443).

As the case of the Great Barrier Reef, the world's largest coral reef system, suggests, the combined interactive effects of two primary anthropogenic ecocrises, ocean acidification and global warming, have put the world's coral reefs in significant danger of major diminishment and even total collapse. Magnifying and interacting with these linked threats to reef ecosystems is the anthropogenic introduction of sediment, nutrient, and industrial pollution into the waters of reef environments. Admixing of the components of this cacophony of threats, as already indicated, has direct boomerang effects on human nutritional health and socioeconomic well-being; indeed, on the health of countless aquatic and ocean-dependent species.

There is another threat to fish populations from ocean acidification: fish are losing their sense of smell. This is the finding of an experimental study by Cosima Porteus of the University of Exeter in the United Kingdom (Porteus *et al.* 2018). Fish have the capacity to smell their surroundings using receptors in their nasal cavity that are comparable to those found in the human nose. Porteus and colleagues designed an experiment using juvenile European sea bass in containers that held either current levels of CO_2 or the levels of ocean CO_2 predicted for the end of this century. They then added the scent of a sea bass predator at 2, 7, and 14 days after the fish were introduced to the containers. They found that the longer the sea bass were in high-CO_2 seawater, the less likely they were to respond to a predator smell compared to the fish in the containers with current levels of CO_2. The latter would tend to stop swimming and freeze in place when they smelled the predator's scent. They also ran experiments to test the impact of CO_2 on detecting the scent of food items, finding that fish were less likely to detect 8 out of 10 odors introduced in high-CO_2 seawater. Overcall, the ability of sea bass to smell critical components of their environment was reduced by up to half, and some smells had to be five times more concentrated before they were sensed. Also, the fish in the high-CO_2 environment had to be 42 percent closer to the source of a scent to detect it. This research is supported by similar findings on the impact of CO_2 on olfactory ability in other fish species (Ferrari *et al.* 2011; Williams *et al.* 2018). Notably, additional research shows that raising CO_2 levels affects fish vision (Chung *et al.* 2014) and response to sound (Simpson *et al.* 2011) as well. This body of work indicates that the overall sensory capacity of fish is compromised by elevated CO_2 levels, suggesting that ocean fish—a critical human food source—face multiple threats from climate change, as well as from overfishing.

5.4 Eutrophication of estuarine and coastal waters

With beautiful sunsets, a 19th-century spiral-staircase lighthouse, and luxurious resorts, Sanibel, one of the barrier islands in Lee County on the southwestern side of Florida, is a popular tourist destination that attracts visitors from around the world. It is widely acknowledged for its species-rich picturesque shell beaches and wildlife refuges that are home to over 220 species of birds, as well as bobcats, alligators, marsh rabbits, and river otters. The refuges provide migratory shelter for birds like ospreys, brown pelicans, herons, egrets, wood storks, and roseate spoonbills. But Sanibel and other coastal areas in Lee County also have become known for another environmental feature: large quantities of dead fish and other marine animals washed up on their shorelines. In the first 2 weeks of August 2018, for example, 3 million pounds of aquatic wildlife, including 600 000 pounds from Sanibel Island, were collected by county workers and burned at the local solid waste facility (Gillis 2018). On particularly bad days, workers were picking up six dead fish for every foot of beach they traversed. The carcass of a juvenile whale shark even washed up. But this is only a fraction of the demise. Most of the fish dying in the waters off Sanibel do not wash up on to beaches, they are found floating offshore or lie at the bottom of the Gulf of Mexico. The same is true of other marine animals, including hundreds of sea turtles, especially loggerhead and endangered Kemp's Ridley turtles. Also dying or showing grave signs of illness are fish-eating seabirds like cormorants.

The environmental change driving this mass die-off is eutrophication, the overenrichment of water by organic nutrients such as nitrogen and phosphorus. Overenrichment at Sanibel leads each year to a water condition known as "red tide," a descriptive label for the discolored coastal waters produced by a high concentration, or bloom, of a neurotoxin-producing microscopic algae called *Karenia brevis*. This unicellular dinoflagellate (two-tailed) microorganism generates brevetoxins (Baden 1989), poisonous compounds that affect vertebrate nervous systems by altering the flow of sodium ions in the body (Kirkpatrick *et al.* 2004), producing a depolarization of nerve cells that interferes with nerve transmission, causing a disruption of normal neurological processes. Other effects include immune depression, bronchial constriction, and the rupture or destruction of red blood cells. For herbivorous fish that consume the algae, the consequences can be fatal. And when these fish are consumed by other marine life, the toxins spread through the ocean food chain. Fish, filter-feeding mollusks, and seagrass are known to accumulate high concentrations of algal toxins in body tissues. Researchers have linked mass-mortality events involving endangered Florida manatees (*Trichechus manatus latirostris*) and bottlenose dolphins (*Tursiops truncates*) to dietary sources (Flewelling *et al.* 1995; Landsberg *et al.* 2009). Even after the bloom subsides, there is ongoing mortality. Testing of nesting female loggerheads a year after the 2013 bloom at Sanibel found that concentrations of brevetoxin in the blood were "significantly correlated with liver brevetoxin concentrations of dead-in-nest hatchlings and that brevetoxins could be related to a decreased reproductive success in this species" (Perrault *et al.* 2016).

Awareness of red tides in Florida and their role in marine deaths has a long history among native populations, a fact documented by Spanish explorers in the 16th century. While this algae is found in the United States across the coastal areas of the Gulf of Mexico as far west as Texas (and down the eastern coast of Mexico) and up the Atlantic coast to North Carolina, it is most prevalent along the southwest shores of Florida, where it usually lasts for about five months. The deadly 2017–18 event was twice as long, however. This extended intense presence of *Karenia brevis* reflects a longer-term pattern. Research by Brand & Compton (2007) revealed an approximate 13–18-fold increase in the abundance of the algae in this region in 1994–2002 compared to 1954–1963. Further, while in the latter period concentrated *Karenia brevis* occurred primarily in the fall months, during the former it was abundant also in the winter and spring. In 2006–07, there was a particularly problematic algae bloom in Florida that lasted for 17 months, leading to mortalities of bottom-dwelling crustacean communities, fish, turtles, birds, and marine mammals. This occurred because algae consume oxygen (primarily at night), and a large bloom can use up most or even all of that available, leading to anoxic dead zones devoid of oxygen-dependent life. Researchers believe that the active 2004 hurricane season may have played an important part in the 2005–06 bloom. Four hurricanes (Charley, Frances, Ivan, and Jeanne) crossed over Florida during August and September 2004, dumping over 25 inches of rain—almost twice the regular average for this region. Repeated heavy precipitation swept runoff nutrients on land into the ocean (Hu *et al.* 2006).

Within a bloom, every algae cell can replicate itself hundreds of thousands of times. During the 2017–18 red tide, researchers at the Sanibel-Captiva Conservation Foundation (SCCF) found concentrations of toxic algae of over 10 million cells per liter of seawater, and in some locations as high as 140 million cells per liter. This compares with usual *Karenia brevis* concentrations of 1000 cells per liter or lower (Wei-Haas 2018).

Surveys conducted in the United States and Europe show that 78 percent of the assessed continental U.S. coastal area and approximately 65 percent of Europe's Atlantic coast display symptoms of eutrophication (OSPAR 2003; Howarth & Ramakrishna 2005). Resulting harmful algae blooms (HABs) are not peculiar to one species of dinoflagellates, nor to the waters of the United State and Europe, nor even to oceans. There are approximately 5000 species of marine microscopic algae, of which about 300 cause "red tides" (though the actual hue of the discoloration varies), while about 80 produce potent toxins (Hallegraeff 2003). Available evidence suggests a global expansion in the intensity, duration, and biogeographic distribution of these events. Red tides on the Southern California coast are produced by an algae species known scientifically as *Lingulodinium polyedrum*. While this organism turns the ocean red by day, at night it emits a bright blue bioluminescent glow that has made it a popular attraction when it occurs every few years. Despite its ability to produce an entertaining light show, this widespread algae, like *Karenia brevis*, is harmful to wildlife and humans (McGowan *et al.* 2017). Deadly red tides caused by *Lingulodinium polyedrum* have been reported in San Diego, California, in

Spain, and in the Adriatic Sea off the coasts of Italy and Yugoslavia. The powerful toxin it produces, called yessotoxin, is linked to shellfish poisoning in humans (Paz *et al.* 2008). Some yessotoxins are 1000 times more powerful than cyanide.

In 2005, juveniles from the southern colony of right whales (*Eubalaena australis*) began washing up on the beach near Peninsula Valdes, a Patagonian nature reserve on the coast of Argentina. Most of the deaths (about 90 percent) were of calves under 3 months of age (Wilson *et al.* 2015). The protected bays found in the area provide important calving and nursery grounds for right whales, so named because they were designated as the right whales to hunt by 19th-century commercial whalers from the United States and Europe. Because of the mass slaughter that ensued, the global population crashed and the species hurled toward extinction. With the passage of protective international laws in the 1930s and the eventual decline of illegal hunting, the numbers rebounded, with a 5–10-fold population increase since the 1970s. However, the species is still scarce relative to its historical abundance, so the deaths in Argentina attracted global attention and pressure to identify the cause. The number of deaths has been striking. The National Oceanic and Atmospheric Administration (NOAA 2015a) reported that right whale deaths at Peninsula Valdes increased more than 10-fold from fewer than 6 per year before 2005 to 65 per year in the period 2005–2014. Various causal factors have been suggested, including a lack of adequate concentrations of the krill consumed by mother whales to build the blubber reserves needed to support their nursing calves' growth (a development associated with summer ocean warming and climate change), increased attacks by kelp gulls (which gouge out pieces of skin and blubber for food), and toxic algae blooms. Research by Wilson *et al.* (2015) supports the latter possibility, based on the timing of the baby whale deaths, examination of the bodies of dead whales for biotoxins produced by the diatom algae *Pseudo-nitzschia* spp. and the dinoflagellate *Alexandrium tamarense*, local shellfish harvesting closure dates, and satellite chlorophyll data. Their autopsies found trace levels of algae toxins in the tissues of some dead whales and fragments of diatom cell walls in their feces. Additionally, they noted that right whales were present at Peninsula Valdes during times when the local shellfish industry closed due to high frequencies of algae and biotoxins. Finally, there was a positive statistical relationship between monthly *Pseudo-nitzschia* densities and calf deaths. While uncertainty about the young right whale death continues, and research is ongoing, these initial findings suggest a role for local algae bloom.

The actual level of the global spread of HABs is illustrated by reviewing the history of blooms linked to two related algae species, *Alexandrium (Gonyaulax) tamarense* and *Alexandrium (Gonyaulax) catenella*. Before 1970, these organisms were only reported for the temperate waters of Europe and North America; 20 years later, they were documented in South Africa, Australia, New Zealand, India, Thailand, Brunei, Sabah, the Philippines, and Papua New Guinea (Hallegraeff *et al.* 2004).

Eutrophication in estuaries—partially enclosed environments of brackish water formed in places where river or stream flow connects to the open sea—can be devastating to local animals and plants. Data for U.S. estuaries suffering from long periods (multiple months) of low levels of dissolved oxygen, or hypoxia, show a steady increase from the 1980s through the 1990s (see Table 5.1). Nauset Estuary on Cape Cod, Massachusetts, for example—a Mid-Atlantic ecosystem comprising 1200 acres of inlets, salt marsh ponds, tidal channels, and open bays—suffers from recurrent HABs caused by the toxic saxitoxin-producing dinoflagellate *Alexandrium fundyense*. These events lead to a nearly annual closure (20 of 21 years from 1992 to 2012) of the estuary to shellfishing to avoid public health emergencies caused by human food poisoning. Such closures continue

Table 5.1 Percentage of U.S. estuaries with reports of low dissolved oxygen or hypoxia. *Source:* Modified from Rabalais *et al.* (2009).

Region	Percentage of estuaries		
	1980s	1990s	2000s
Atlantic Coast	72	143	184
Gulf of Mexico	66	84	91
Pacific Coast	21	26	55
Total U.S.	**159**	**253**	**330**

to occur (CapeCod.com 2018). Research at Nauset Estuary found warmer years produced earlier blooms (Ralston *et al.* 2014). Monitoring of Salt Pond within the estuary by the School of Marine Science and Technology at University of Massachusetts, Dartmouth found declining trends in water quality and oxygen levels, rising amounts of nitrogen, and increasing bloom intensity (Coastal Systems Group 2018). Across the world from Massachusetts, frequent HABs have become a major environmental issue in the Yangtze River estuary, the biggest such ecosystem in China (Chai *et al.* 2006). The areal extent of these blooms increased dramatically during the closing years of the 20th century and early years of the 21st, and a hypoxic zone has formed in the East China Sea close by (Chen *et al.* 2003). The dominant algae species involved in these HAB events is *Prorocentrum dentutam* (Shen *et al.* 2011), affirming that while the eutrophication of estuarine and coastal waters is a global crisis, the proximal causative agent—the primary dinoflagellate species involved—varies across the types and locations of threatened hydraulic ecosystems. Distal or ultimate causes also vary but tend to involve anthropogenic environmental changes and their interconnections, as discussed later.

Algae boom on lakes, including those that are primary sources of drinking water and water for other domestic uses, is also significant. In 2013, microcystin toxin was detected in the water treatment facility for Carroll Township, Ohio. Levels of the poisonous substance were at 3.5 times the safety threshold for drinking water, and six people suffered gastrointestinal illnesses during the outbreak. A year later, water from Lake Erie was shut off to 500 000 residents in Toledo, Ohio because of a bloom caused by the toxic *Mycrocystis aeruginosa* (Paerl *et al.* 2014), a so-called blue-green algae (though it is actually a cyanobacterium). The same kind of bloom on Lake Taihu caused a drinking water crisis in Wuxi, Jiangsu Province, China in 2007; as the lake was the sole water supply to the city, the bloom left approximately 2 million residents without drinking water for a week (Qin *et al.* 2010). A possible link between blue-green algal toxins, drinking water, and liver cancer has been identified in research in Florida (Fleming *et al.* 2002).

The primary cause of intense aquatic algae blooms is eutrophication, which is considered a form of environmental pollution. As mentioned in the discussion of corals, it is one of the threats to reef structures, because it feeds the growth rates of ephemeral seaweeds (a type of algae) at the expense of coral. The major nutrients that algae require to thrive are nitrogen and phosphorus, as well as smaller quantities of several minerals and vitamins. Unicellular algae is capable of rapidly increasing its rate of production and biomass accumulation when exposed to abundant nutrients. There has been considerable debate about the causes of nutrient enrichment, particularly whether it is primarily driven by natural forces (not involving human activity) or unnatural ones (involving human activity).

Abundant research documents the direct and indirect effects of human behavior on eutrophication. In the coastal ocean environment, eutrophication has primarily been driven by high quantities of human-produced nitrogen and phosphorous entering into water systems. Selman *et al.* (2008) identify agriculture, especially rainfall-driven runoff of animal manure and commercial fertilizers, as the primary contributor to eutrophication in the United States and European Union. Fertilizers applied to golf courses and lawn and landscaped areas also wind up in water systems (Howarth *et al.* 2000). Sewage and industrial wastes, which usually receive some level of treatment prior to discharge, are secondary sources, though in some locations they can be primary (Barile 2018). Additionally, both fossil fuel combustion by mobile (automobiles, buses, trucks, ships, airplanes, and off-road vehicle) and fixed (electric power generation, manufacturing facilities) sources and the vaporization of fertilizer release considerable amounts of nitrogen into the atmosphere (where it also drives climate change). Atmospheric nitrogen is redeposited on land and in bodies of water by wind, snow, and rain. By contrast, in Latin America, Asia, and Africa, where wastewater from sewage and industry often flows into bodies of water untreated, these sources frequently constitute the primary contributors to eutrophication. In rapidly developing nations in Africa and Latin America, however, power plants, industry, and transportation have become major sources of atmospheric nitrogen (Zhao *et al.* 2010). All of these anthropogenic causes are estimated to have about doubled the rate of formation of reactive nitrogen on the land surface of the planet between 1850 and 2000. As Howarth & Ramakrishna (2005, p. 299) highlight, this "rate of change is extraordinarily rapid."

The central human role in eutrophication is seen in the case of the inland Sacramento-San Joaquin River Delta in northern California. Located to the east of San Francisco on the western edge of the Central Valley, it is one of the world's few inverted river deltas. It constitutes a large estuary system that was once called the Everglades of the West. The two rivers that feed water into it capture runoff from the western flank of the Sierra Nevada Mountains and the southern portion of the Cascades, as well as part of the coastal range. Water quality in the delta has suffered under human pressure. Studies show the presence of toxic algae and of the kinds of chemicals that feed rampant algae growth (Kurobe *et al.* 2013). Many of these derive from fertilizers with high nitrogen and phosphorous content. Another chemical in the waters of the delta, ammonia, is a common byproduct of human urine and feces. Studies show that the Sacramento-region sewage treatment plant—which discharges treated wastewater produced by almost 1.5 million residents—is its largest single source. The ammonia load in the wastewater from Sacramento has grown rapidly as a result of rapid urban growth and an aging water treatment system (Weiser 2008). While the Sacramento regional sanitation district originally fought efforts to upgrade its system, it finally agreed in 2014 to build the EchoWater Project, which, when completed sometime between 2021 and 2023, will ensure wastewater discharges from the treatment plant finally meet state water quality requirements in relation to discharges of ammonia, nitrates, and other pollutants (Quinton 2016). Moreover, Environmental Protection Agency (EPA) data released by the California Sportfishing Protection Alliance (CSPA 2007) indicate that the city of Stockton violated its wastewater discharge permit 50 times in 2005, allowing ammonia to exceed allowed levels by between 23 and 1245 percent on different occasions.

Beginning in 1999, research on the delta identified the presence of bloom-causing *Microcystis aeruginosa*. This toxic cyanobacteria was found to be widely distributed throughout both fresh and brackish water in the delta water system. Microcystin, a liver toxin produced by cyanobacteria, had entered the food chain and was found in both total zooplankton and clam tissue (Lehman *et al.* 2005). Based on continued analysis, Lehman *et al.* (2010) found microcystins were present throughout the delta food chain. Higher concentrations occurred in striped bass than in their prey species. This suggests that toxins were accumulating at higher levels of the food chain. Examination of tissues in delta fish livers indicated that the health of two common fish in the estuary, striped bass and Mississippi silversides, was being affected by the tumor-promoting microcystins. In 2014, a Microcystis bloom contained the highest biomass and greatest toxin concentration, and had the earliest initiation and longest duration (8 months, compared to the usual 3–4 months), since the blooms began in 1999. It occurred during a period of extreme nutrient concentration in the delta, including a 20-year high in phosphorus concentration. It also coincided with high water temperatures. As 2014 was also the fourth driest year on record in the delta, and more broadly in California, Lehman *et al.* (2014, p. 94) concluded that "severe drought conditions can lead to a significant increase in the abundance of Microcystis and other cyanobacteria, as well as their associated toxins." They linked the intensified bloom of 2014 to global warming. Notably, a connection between climate change and toxic cyanobacteria blooms has been identified in research in other locations (e.g., Gkelis *et al.* 2014). Moreover, Paerl & Paul (2012) link synergistic interaction between human nutrient overenrichment and climate change in harmful loom-forming cyanobacteria. Warming promotes cyanobacterial proliferation as the growth rates of these organisms improve at relatively high water temperature (Butterwick *et al.* 2005; Kosten *et al.* 2011). Further, the intense storm promoted by climate changes contributes to fertilizer run-off into bodies of water. Some algal species, like *Lyngbya majuscule* (in marine environments) and *Lyngbya wollei* (in freshwater environments), are opportunistic invaders of disturbed water bodies, where they proliferate as dense floating mats following climate perturbations like hurricanes. As reviewed later, these ecocrises interactions contribute to negative health outcomes.

Eutrophication and algae bloom generate significant health risks. There are at least five recognized HAB-related illnesses in humans: ciguatera fish poisoning (CFP), paralytic shellfish poisoning (PSP), neurotoxic shellfish poisoning (NSP), amnesic shellfish poisoning (ASP), and diarrheic shellfish poisoning (DSP). While these clinical conditions share signs and symptoms, they are produced by exposure to different toxins. The two kinds of exposure that produce them are consumption of infected fish or shellfish and contact with aerosolized NSP toxins (Grattan *et al.* 2016b).

- **Ciguatera fish poisoning (CFP)** is caused by the consumption of coral reef fish that have accumulated a class of neurotoxins known as ciguatoxins in their bodies and roe. The risks of contracting this malady have been known for centuries in the tropical marine environments where it is found, though it is often misdiagnosed because of varying individual symptomology and the exportation of tropical fish to areas where it is less familiar to the public and clinical responders. The toxins are produced by the marine dinoflagellate *Gambierdiscus toxicus* and related Gambierdiscus species that live in coral reef ecosystems. Eutrophication, turbidity from hurricanes, dredging, coastal development, and warming sea surface temperatures are known to stimulate Gambierdiscus growth and result in higher cell densities (Parsons *et al.* 2010). Plant-eating fish consume these dinoflagellates and, through bioaccumulation, the toxin advances up through the food web as they in turn are consumed by carnivorous species. Hundreds of different fish are believed to be possible ciguatoxin vectors. Legrand (1998) has suggested that particularly toxic strains of Gambierdiscus dinoflagellates are responsible for ciguatera outbreaks. CFP is the most frequently reported seafood-related disease in the United States, with an estimated 15 000 cases annually (Pennotti *et al.* 2013), as well as the most common foodborne illness caused by finfish globally, with up to 100 000 (Lehane & Lewis 2000; Isbister & Kierna 2005). CFP is endemic in the islands of the Caribbean, Indo-Pacific, and Indian Ocean, but over time has shown a progressive expansion at a global scale. In French Polynesia, the sea snail *Tectus niloticus,* known commonly as a top shell, has been identified as a novel vector (Gatti *et al.* 2018). Initial symptoms usually develop within 12 hours after eating infected marine life. Commonly, the first indications of poisoning are moderate to severe gastrointestinal upset and abdominal cramps. Cardiovascular problems, involving both a slowing heart rate and low blood pressure, may also occur. In Pacific regions, there have been reports of rapid progression to respiratory distress, coma, and even death (Lange 1987). In an outbreak in Madagascar, over 500 people developed CFP following the consumption of shark meat, 98 of whom died (Habermehl *et al.* 1994). Within a few hours to up to 2 weeks after exposure, an array of neurological complaints have been reported, including: 1) leg pain and weakness; 2) tingling in the face and throat; 3) burning or tingling in the hands, arms, and legs; 4) dizziness, headache, blurred vision, and faintness; 5) nausea and vomiting; 6) metallic taste in the mouth; 7) loss of flexibility; and 8) reversal of normal thermal sensations, such that cold objects feel hot and, in some cases, hot objects are experienced as ice cold (Farstad & Chow 2001; Achaibar *et al.* 2007). While gastrointestinal and cardiac symptoms of CFP are generally temporary, the resolution of acute neurologic symptoms can persist steadily for weeks or months, or even years, and in some cases can return after periods of assumed recovery. Relapse has been associated with the consumption of alcohol, fish, and other foods.

- **Paralytic shellfish poisoning (PSP)**, a potentially lethal condition, is linked to the consumption of bivalve mollusks like oysters, mussels, scallops, and clams. Children who eat these seafoods are at particular risk because of several related marine neurotoxic alkaloids collectively referred to as saxitoxins (Rodrigue *et al.* 1990; Wiese *et al.* 2010). These damaging toxins primarily are produced by three genera, Alexandrium, Gymnodinium, and Pyrodinium, of dinoflagellate algae. Bivalve shellfish, marine snails like whelks and periwinkles, lobsters, crabs, and plankton-eating fish and their predators all act as saxitoxin vectors for human exposure. Temperate and tropical marine coasts are the most risky regions. In North America, this includes Alaska and the Pacific Northwest on one side of the continent and the St. Lawrence region of Canada on the other. PSP, however, has expanding global distribution, occurring from Chile to Japan to the United Kingdom. Cyanobacteria also produce saxitoxins that cause contaminations of drinking and recreational water resources. The initial, and in some cases only, symptom of PSP infection is numbness or tingling around the mouth and lips, which may spread to the neck, face, fingers, and toes. The onset of other complaints, including headache, abdominal pain, nausea, vomiting, dizziness, diarrhea, trouble swallowing, incoherent speech, and various neurologic symptoms, may follow (Grattan *et al.* 2016a).
- **Neurotoxic shellfish poisoning (NSP)** follows the consumption of bivalve mollusks infected with brevetoxins from *Karenia brevis* or other genus relatives that, most commonly, are recreationally harvested at the time of or after a red tide bloom (Fleming *et al.* 2011). Cases have been reported from Florida and North Carolina in the United States, as well as from Mexico and New Zealand. The largest recorded outbreak, involving more than 180 human cases, occurred in 1992–93 in New Zealand following the consumption of cockles, green shell mussels, and oysters. Symptoms include both gastrointestinal and, of greater alarm for sufferers, neurological indicators like partial limb paralysis, slurred speech, dizziness, malaise, body pain, and loss of coordination of body movements. NSP has not, however, been the cause of human fatalities, nor does is it produce later relapse once symptoms resolve. Wave action during red tides may cause aerosolization of the toxin, which is a special risk for people with asthma or other breathing disorders (Grattan *et al.* 2016b).
- **Amnesic shellfish poisoning (ASP)**, also known as domoic acid poisoning (DAP) because amnesia is not always present, was first diagnosed in 1987 on Prince Edward Island, Canada. In this outbreak, ASP caused three deaths in a total of 105 human cases

following the ingestion of blue mussels. Symptoms included abdominal cramps, vomiting, disorientation, and temporary or permanent loss of short-term memory. Lethal cases have been reported (Dhar *et al.* 2015). The causative neurotoxin, domoic acid, which can initiate neuronal cell rupture, has been found to be produced by the diatom species *Pseudo-nitzschia* (Van Egmond *et al.* 2004), which has worldwide distribution. Diatoms were one of the organisms that survived the mass Cretaceous–Tertiary Extinction

- **Diarrheic shellfish poisoning (DSP)** is caused by the consumption of contaminated bivalve shellfish. Shellfish are filter feeders that pump in water to filter out and consume algae and other food items. When they ingest biotoxin-producing algae, the toxin accumulates in their fatty tissues, though it does not harm the shellfish themselves. It does, however, impact their human consumers. The predominant DSP symptoms are acute diarrhea, nausea, vomiting, and abdominal pain, with complete recovery usually occurring within 3 days (Van Egmond *et al.* 2004). DSP toxins are produced by dinoflagellates of the genua *Dinophysis* spp. and *Prorocentrum*. The most affected areas are Europe and Japan (Yasumoto *et al.* 1980). DSP incidences, or at least the presence of DSP toxins, are increasing globally in marine waters from South America to New Zealand.

In sum, eutrophication has become a growing global ecocrisis produced by multiple interacting anthropogenic stressors and causing multiple threats to the health of various aquatic, airborne, and terrestrial species, including humans.

5.5 Depletion of the oceans

We call our planetary home Earth, a term we also use to label soil, the natural substance found on the planet's land surfaces. Yet we know that 71 percent of Earth's actual surface is ocean, not soil (Fig. 5.3). Were aquatic mammals capable of speech as we know it, it is certain they would not call the third planet from our sun Earth. We might imagine they would assert we live on Ocean. More importantly, whichever our species and wherever we live on the planet, we are all dependent on the oceans to sustain life. However, we live at a time when our oceans are highly vulnerable, and the risk has been growing steadily during the 20th and 21st centuries. Oceans, the cradle of life, are facing a dramatic depletion of their lifeforms. As stressed in a letter by 125 ocean scientists from 27 countries to the head of the World Trade Organization (WTO) in 2007:

> We have spent lifetimes dedicated to studying the oceans ecosystems and marine fisheries . . . There is no longer any question—we have reached a critical state. The world's ocean ecosystems are at a tipping point . . . There are only decades left before the damage we have inflicted on the oceans becomes permanent. We are at a crossroads. One road leads to a world with tremendously diminished marine life . . . The choices we make today will determine our path for the future. (Pauly *et al.* 2007)

Fig. 5.3 Blue Earth. *Source:* yasar simit/Adobe Stock Photos.

As with the other ecosystems examined thus far, the current dangers to the species that dwell in oceans, as well as those that depend upon them, are multiple and entwined and reflect human impact on the environment. The most significant problems include coral die-off and algae blooms, which have already been reviewed, but also overfishing, various forms of pollution, the oceanic effects of climate change on the distribution of marine species, the toxic chemical breakdown of floating plastic, and seafood safety.

Overfishing refers to the harvesting of a marine or freshwater fish stock, be it a single species or the populations inhabiting a particular underwater ecosystem, toward the point of unsustainability (Fitch 2015). Sustainable harvest meets the seafood needs of the present without compromising foreseeable future needs. Extreme overfishing is the decline of an aquatic population that is greater than the level of reproduction to the point of collapse or even extinction (Hilborn 2012). The consequences of overfishing are described by Andy Rosenberg, a scientist, environmental activist, and leader of a project called the History of Marine Animal Populations: "We see similar patterns of human impacts on the oceans pretty much everywhere, and in many cases [it leads to] real depletion" (quoted in Doyle 2009). This is not a new problem, having been discussed by the zoologist Walter Garstang of Oxford University and the International Council for the Exploration of the Sea at the turn of the 20th century (Sinclair 1997). Early concern with overfishing reflected a turn toward population thinking in fisheries science. The problem today is rapidly worsening, however (Clover 2006). The U.N. Food and Agriculture Organization (FAO 2016) estimates that over 30 percent of fish stocks are either fished to capacity or overfished.

Some species have suffered significant depopulation. One of these is the Canadian cod (*Gadus morhua*), the icon of the world fish crisis. Author Mark Kurlansky (1997) calls cod "the fish that changed the world" because of their role in the settlement of the eastern coast of North America. The Atlantic fisheries off the coast of Newfoundland and Labrador were once teeming with them. Greenish-brown to gray on their sides and back, with a lighter underbelly, they possess a barbel, or whisker-like projection, on their chin, which gives them the appearance of a catfish. Their backs are decorated by three prominent dorsal fins. There have been reports of cod as long as 6 feet and weighing over 200 pounds, though those typically caught by fishermen in recent times are much smaller (Kennedy 2019).

Cod bones, such as otoliths (an inner ear bone), are plentiful in pre-contact Native American middens, indicating cod were a vital component of coastal indigenous diets. This is understandable, because these groundfish were once so plentiful in the North Atlantic that people said you could almost walk across the water on their backs. Fish by the thousands would gather over a spot on the ocean bottom and form a spawning haystack that ran up the water column toward the surface. Schooners—sailing ships with two or more masts—could land 70 000 tons of cod a year in the Gulf of Maine in the 1860s (Doyle 2009). High fecundity and plentiful food availability fueled an abundant cod population, a cod-dominated ecosystem, a huge fishing industry, and a way of life (Kurlansky 1997). The harvest was heavy for several centuries, but it increased dramatically in the 1950s–1980s with the introduction of new fishing technology like larger, powerful trawlers equipped with radar, electronic navigation systems, and sonar. A mechanized fleet of ships was able to bring in 3000 tons of cod in a single haul by 2010. The northern cod biomass had fallen to 1 percent of its earlier level by 1992. Facing a historic collapse of unprecedented dimensions, on July 2 of that year the Canadian government declared a moratorium on northern cod fishing.

Throughout the region, but especially in the outports of Newfoundland, the cod fishery was the bedrock of people's sociocultural identity. For many families it also represented their primary source of livelihood as fishers, fish plant workers, fish sellers, fish transporters, or employees in related businesses. With the collapse, over 35 000 people living in over 400 coastal communities lost their jobs. The local economy nosedived, but the short-term impact was blunted by government support payments and through transition to crabbing and shrimping (Hamilton & Butler 2001). Since then, cod fish stocks have slowly improved, though they are still far from the glory days of abundance. According to a stock assessment of the Gulf of Maine undertaken by the NOAA (2017), cod are still overfished and levels are still below the agency's target biomass level. The same is true of cod in the Georges Bank between Cape Cod, Massachusetts and Cape Sable Island, Nova Scotia.

Another fish that has been deeply impacted by overfishing is the bluefin tuna. From a commercial standpoint, it is currently considered the most valuable and most prized fish species in the world. In 2013, Kiyoshi Kimura paid $1.7 million for a single 489-pound bluefin at the first auction of the new year at Toyko's Tsukiji wholesale fish market. While the jaw-dropping price was in no small part driven by the global publicity it brought to Kimura's Sushi Zanmai restaurant chain (motivation that also prompted his successful bid of $3.1 million for a 612-pound bluefin in 2019; Wingfield-Hayes 2019), it reflects the relatively recent skyrocketing demand for bluefin sushi (Corson 2007) (Fig. 5.4).

Fig. 5.4 Tuna sushi. *Source:* cathryn040/Pixabay.

Shaped like torpedoes with a pointy snout, retractable fins, high tailbeat frequency, and eyes set flush to their streamlined bodies, bluefins have been molded by natural selection for speed and continuous swimming. They dwell in the pelagic zone or open sea, as contrasted with water zones that are near the coast or around reefs. With the keenest vision of any bony fish, they use sight to spot and hunt their prey. They have speciated into four groups, the Atlantic (*Thunnus thynnus*), the Pacific (*Thunnus orientalis*), the Southern (*Thunnus maccoyii*), and the Longtail (*Thunnus tonggol*), though the latter is not considered a "true bluefin" by some because of a number of anatomical and developmental features it possesses (Graham & Dickson 2004).

Several traits of the bluefin make it highly vulnerable to overfishing: it only spawns at a late age, it reproduces when it attains a large body size, and it gathers in groups to procreate (Longo & Clark 2012). Sensitivity to ocean pollution may also contribute to breeding and migratory disruption, as Addis *et al.* (2009) have suggested with reference to the impact of mining on Atlantic bluefins in the Western Mediterranean. Other factors that have increased bluefin vulnerability are its size and its place as a predator at the top of the food chain. These are risky attributes given "the tendency for fisheries to first exploit species at higher trophic levels and subsequently move down the food web," as well as the "unique human propensity to cull the largest members of a population" first (Payne *et al.* 2016). As a result, "the global ocean has lost more than 90% of large predatory fishes" (Myers & Worm 2003, p. 282).

As the Pew Charitable Trusts, a nongovernment organization (NGO) that includes in its mission enhancement of fisheries management and ocean conservation, has cautioned, bluefin populations in both the Atlantic and the Pacific oceans are rapidly declining under the pressure of commercial overfishing (Nickerson 2016; Hobkins & Wheatley 2018). Part of the problem, as Pew emphasizes, is that "[c]alls by scientists and environmentalists to reduce fishing pressure have been ignored too often by politicians who put short-term economic and political gains ahead of long-term sustainability" (Mulvaney 2015, p. 2). Another part, and one that has a strong influence on policy directions, is fishing technology development, as it has been "driven by the increasing opportunities to profit from the sale of bluefin tuna on the global market" (Longo & Clack 2012). The global bluefin tuna industry has taken on the model of capitalist food production, as developed in industrial agriculture, including capital investment in production facilities and expansive technologies, adherence to a productivist ethic that asserts that growth is a necessity, opposition to regulation that restricts expansion and profit, pressing governments for industry subsidies, and enlargement of the market through mass advertising. For example, in 2015, the European Union agreed to a 60% increase in the total allowable bluefin catch in the Atlantic Ocean and Mediterranean Sea over the 3-year period 2015–2017. In 2017, the number of vessels legally authorized by the European Union to fish for bluefin was 782, with the quota for that year set at just under 13 500 tons. The following year, it permitted a jump to 1088 vessels and an expanded quota of 15 850 tons, indicating that expansion is continuing (European Commission 2017, 2018). The EU fishing fleet, which is government-subsidized, has acquired new technologies that allow it to effectively spot, track, and capture bluefin. The WWF (2011b, p. iii) reports that modern fishing fleets have improved their capture efficiency "by increasing the installed power of main engines and purse-seine net lifting cranes, extending vessel length, increasing purse-seine net efficiency, installing more efficient main seine winches and more powerful seine skiffs, and using tuna-spotting aircrafts and sonar detection to locate schools." Calculations carried out by Myers & Worm (2003, p. 282), biologists affiliated with Dalhousie University, indicate that there is "an 80% decline typically [occurring] within 15 years of industrialized exploitation" of a marine fish species.

The impact of the corporatization of the Atlantic bluefin harvest has been dramatic. One of the largest scientific collaborations ever conducted, the Census of Marine Life, was a decade-long (2000–2010) international program involving the participation of over 2500 researchers in 550 oceanic expeditions, as well as considerable time spent in labs and archives, in order to assemble existing information and fill in gaps in our knowledge of the scope of oceanic biodiversity. One such gap involved the reasons for the collapse of the North Atlantic bluefin population (MacKenzie & Myers 2007). The bluefin historically was a top predator in the North Atlantic marine ecosystem, feeding primarily on herring, mackerel, garfish, sprat, and squid. The Census found that migratory bluefins once "teemed in northern European waters (e.g., North Sea, Norwegian Sea) for a few months each summer until an industrialized fishery geared up in the 1920s and literally filled the floors of European market halls with them" (Census of Marine Life 2007). After World War II, the bluefin harvest exploded, as did processing facilities and demand. This caused dramatic declines and range contractions for bluefins off the coast of northern Europe. The species "was depleted in 40 years (1910–1950) and still remains rare there today" (Census of Marine Life 2011, p. 3). According to Brian MacKenzie of the University of Denmark, "We've shown bluefin tuna were here for a long time in high numbers. High fishing pressure preceded the species' virtual disappearance from the area and apparently played a key role but other factors under study might have compounded the fishery's demise—the catch of juvenile tuna in subsequent years, for example" (quoted in Census of Marine Life 2007).

Pacific bluefin may be facing a similar fate to their North Atlantic cousins. According to the Center for Biological Diversity (2016), a Tucson, Arizona-based nonprofit species conservation organization, "[u]nder pressure from overfishing, bluefin tuna populations have reached dangerously low levels, declining more than 97 percent since fishing [them] began." Moreover, the Center notes that almost all Pacific bluefin tuna being harvested today are caught before they have reproduced, leaving inadequate numbers old enough to breed and propagate the species. NOAA (2015b) data from 2014 show that the Pacific bluefin population that year produced the second-lowest number of young fish seen since 1952. Further, the biomass of spawning females, those aged 3–5 years (when they reach maturity) or older, is extremely low. Without young fish to mature into the future spawning stock, the population faces collapse.

On June 20, 2016, the National Marine Fisheries Service (NMFS) received a petition from the Center for Biological Diversity and 13 co-petitioners, including the Ocean Foundation, Earthjustice, Center for Food Safety, Defenders of Wildlife, Greenpeace, and Sierra Club, requesting that Pacific bluefin tuna be listed as endangered or threatened under the Endangered Species Act (ESA) throughout all or a significant part of its range. The NMFS concluded that the action might be warranted and initiated a review of the status of the bluefin. While acknowledging that important kinds of information were not available, including sufficient knowledge about the ability of bluefins to recover from population declines in the past, as well as adequate understanding of the full effects of ocean warming on bluefin food sources and spawning, the review team concluded that the short- and long-term risks for extinction were low (Pacific Bluefin Tuna Status Review Team 2017). The NMSF subsequently decided that ESA protections were not warranted. In response, Catherine Kilduff, an attorney with the Center for Biological Diversity, stated: "Pacific bluefin tuna will spiral toward extinction unless we protect them. The Endangered Species Act works, but not when the Trump administration ignores the plight of animals that need help . . . This disappointing decision makes it even more important for consumers and restaurateurs to boycott bluefin until the species recovers" (Center for Biological Diversity 2017). The Center then launched an ongoing consumer boycott and began seeking and receiving commitments from sushi restaurants to take part in this action.

Overall, while a substantial amount of uncertainty still exists in understanding the current status of the various bluefin groups, and parties disagree on the extent of the problem, in part because bluefins are a source of great profit for some and a vital and majestic component of fragile ecosystems for others, bluefin tuna have "many biological and economic traits that have led several fish stocks to extreme depletion" (Fromentin & Powers 2005). As Rosenberg (2003) observes, fish management policy-makers are quick to try and "capitalize on any apparent stock increase or marginal fishing opportunity but only slowly [respond to] apparent decreases in stock."

As shallow-water fish stocks have diminished around the world, the fishing industry has expanded its operations into deeper waters (Roberts 2000, 2019). The commercial targeting of deep-water fisheries has been made possible by advances in marine technologies, the result of which is a series of boom-and-bust harvestings of various species, including the orange roughy (*Hoplostethus atlanticus*) in the waters off New Zealand, the roundnose grenadier (*Coryphaenoides rupestris*) in the Northwest Atlantic, the marbled rockcod (*Notothenia rossii*) near Antarctica, and the armorhead (*Pseudopentoceros pectoralis*) near underwater mountains in the Pacific. On the high seas, beyond the jurisdiction of any nation, environmentally damaging deep-water trawling proceeds without regulation.

Overfishing is a paradoxical problem. In the short run, to the degree that its product is equitably distributed, it could enable the provision of protein and essential nutrients and minerals like omega-3 fatty acids, vitamins D and B2, calcium, phosphorus, iron, zinc, iodine, magnesium, and potassium to people facing hunger. Food is a critical issue in a world in which an estimated 124 million people in 51 countries are currently facing dangerous food insecurity and in need of urgent humanitarian action (Food Security Information Network 2018). The immediate dilemma is that as marine species are driven toward population collapse or even extinction, the so-called bounty of the sea does not evenhandedly reach those most in need. The United States, a very wealthy nation, imports about 80–90 percent of the seafood its residents consume. U.S. imports of edible seafood products in 2017 were valued at $21.5 billion, a rise of 10.5 percent over 2016 figures (NOAA 2018). By contrast, in a West African country like Senegal, dwindling coastal fish stocks pose a threat to livelihoods and food security. It is believed that fish account for approximately 75 percent of the animal protein intake in Senegal, most of it caught by local fishermen off the county's coast. Why are West Africa's once rich fish stocks declining? More than 400 foreign fishing vessels, mainly from China, Europe, and Russia, operate in West African waters. These ships include megatrawlers capable of catching up to 20 000 tons of fish a year. While Senegal has responded by ending its licensing of foreign industrial trawlers to catch pelagic fish, foreign trawlers from powerful nations often flout fishing regulations and continue to take fish illegally in waters reserved for Senegalese locals (Guilbert & Bacchi 2017). According to Voice of America (VOA 2017):

> Foreign fishing vessels, many from China, prowl the waters off West Africa every day. They capture millions of fish—catches that used to go to local boats. The fish are then shipped to China, Europe and the United States, satisfying a global demand for seafood and fueling a multibillion-dollar industry.

Notably, the largest source of U.S. seafood imports in 2017 was China. The top fish imports (based on volume) it receives from other nations are shrimp, freshwater fish, tuna, salmon, groundfish, crab, and squid.

The longer-term problem of overfishing is ocean depletion, a threat that is magnified by interaction with other ecocrises, including ocean pollution and climate change. Using temperature-dependent computer population models to assess the influence of warming on the productivity of 235 populations of 124 different fish and invertebrate species found in 38 ecoregions (e.g., the Celtic-Biscay Shelf, the North Sea, the Baltic Sea), Free *et al.* (2019) concluded that while some marine populations have responded positively (9 populations) to ocean warming, more have responded significantly negatively (19 populations). The type and magnitude of the responses were explained by four factors: ecoregion, taxonomy, life history (e.g., rate of growth to adulthood, lifespan), and exploitation history. Hindcasts (a way of testing a mathematical model by predicting past atmospheric and ocean response events and seeing if the prediction matches what actually happened) carried out by these researchers found that the maximum sustainable harvesting of the evaluated marine populations decreased by 4.1 percent from 1930 to 2010, with five heavily hit ecoregions experiencing losses ranging from 15 to 35 percent. In short, climate change, in conjunction with overfishing, has already had significant adverse impacts on productivity, and the problem will only worsen as the oceans continue to heat up (Blanchard *et al.* 2012; Barange *et al.* 2014; Moore *et al.* 2018). The nature of the ecocrises involved is clear: "overfishing has reduced the resilience of populations to climate change, and climate change will likely hinder efforts to rebuild overfished populations" (Free *et al.* 2019).

5.6 Pollution of waters

Jacques Yves Cousteau (1910–1997), award-winning documentary filmmaker, author, undersea adventurer, and co-inventor of the scuba tank, has been hailed as the greatest advocate for marine environmental protection of the 20th century. His television program,

The Undersea World of Jacques Cousteau, introduced millions of people to the majestic world concealed beneath the ocean surface. Though long a lover of ocean diving, his journey as a defender of the sea began in 1959 when a leading marine scientist told him that despite the high cost to humans and marine life the disposal of the mounting accumulation of radioactive nuclear waste in the ocean was an unavoidable necessity as there was nowhere else to put it. Hoping to serve as a voice for protecting the oceans, Cousteau responded by acquiring a repurposed World War II minesweeper named the *Calypso* and sailing around the world filming episodes for his program. Wherever he sailed, he observed that the products of human activity were having an adverse effect on marine life. He became increasingly concerned about human willingness to treat the sea as a "global sewer, the ultimate receptacle" of our discard, which eventually would be:

> glutted with all the pollution generated on Earth. Each time our population doubles, by some estimates, pollution multiplies six times—all of it into the fluids of life, all of it ending up in the sea. Out of sight, out of mind . . . In every area in which water is contaminated, so too are its creatures; after washing through the global sea and atmosphere, pollution percolates up the ladder of life. (Schiefelbein & Cousteau 2007)

In the last book he wrote, Cousteau lamented the toil of pesticides on penguins, the presence of PCBs in all fish species that had been tested, the discovery of dead beluga whales with more than 30 hazardous chemical pollutants in their body tissues, and many more examples of the widespread effects of marine pollution (Schiefelbein & Cousteau 2007).

As Cousteau realized, every year billions of pounds of physical trash and multiple kinds of industrial waste, as well as other varieties of pollutants, enter the world's oceans. They have diverse points of origin and varied routes of entry. Some contamination is local, such as the area around a sunken ship that had been carrying hazardous material. It is estimated that there are tens of thousands of shipwrecks on the coastal seabed, which together hold 3–25 million tons of oil. Other hazardous cargo includes munitions, chemicals, and radioactive waste. Research by Ndungu *et al.* (2017) on two World War II German ships that were sunk in Norwegian marine waters, for example, found that one was carrying a large quantity of highly toxic mercury, some of which was spread on the ocean bottom, while the other was hauling oil, which had leaked and contaminated coastal sediment with polycyclic aromatic hydrocarbons, compounds linked to cancer, cardiovascular disease, and unhealthy fetal development.

Some forms of ocean pollution become highly dispersed because of winds and ocean currents. Exemplary of the latter is plastic (Fig. 5.5). It is estimated that nearly 300 million tons of plastic are deposited in the oceans annually (Barnes *et al.* 2009). Ocean travelers have been witnessing this for decades. In 1968 and 1970, anthropologist Thor Heyerdahl (1971) and colleagues initiated two Atlantic Ocean crossings from Morocco to islands of the Caribbean aboard papyrus raft-ships named *Ra* and *Ra II*, after the ancient Egyptian sun god. While Heyerdahl's motivation for the voyages was to test his hypothesis that bygone mariners could have carried out successful trans-Atlantic voyages by sailing with the Canary Current, unexpectedly he also discovered visible pollution floating on the ocean surface on 6 days out of the 8 weeks of sailing in 1969 and on 40 of the first 43 days the following year.

Since the sailings of the *Ra* rafts, multiple studies have been carried out on ocean debris, especially floating plastic, the most abundant type (Ivar so Sul & Costa 2013). It is believed that at least 10 percent of manufactured plastics—from diverse products including fishing line, packaging, bottles, synthetic textiles, tires, paints, cosmetics and hygiene products (Fig. 5.6), and electronic equipment—ends up in the ocean (Barnes *et al.* 2009; Barboza

Fig. 5.5 Beached ocean plastic. *Source:* SandraAltherr/Pixabay.

Fig. 5.6 Seahorse and cotton swab floating in Indonesian waters. *Source:* Courtesy of Justin Hofman, photographer.

et al. 2018). Research using data from multi-vessel and aircraft surveys on what has been called the Great Pacific Garbage Patch (Lebreton *et al.* 2018), a major ocean plastic accumulation zone that has formed in the subtropical waters between California and Hawaii, concludes than there are at least 79 000 (with an estimated range between 45 000 and 129 000) tons of plastic floating inside an area of 618 000 square miles of ocean surface. This is 4–16 times higher than previous reports. Further, this research indicates that buoyant plastic in this zone is increasing exponentially and at a faster pace than in surrounding waters. Buoyant plastic is estimated to represent about 60 percent of the plastic load floating in the ocean, with the remainder being washed up on shore as an ever-increasing component of beach debris wrack lines or sinking to the sea floor and accumulating in bottom sediments and underwater canyons (Law 2017).

WE ARE GOLF (Dethier 2018), a coalition of professional golf's governing bodies and the golf industry, estimates that there are 23.8 million players of the game in the United States. These golfers spend $6 billion a year on golf equipment, apparel, and supplies, including plenty of golf balls. The demand for balls is high because so many are lost during play. It is estimated 1.2 billion balls are manufactured and 300 million go missing (into lakes, ponds, forests, wetlands, deserts, backyards, gardens, parking lots, cemeteries, rooftops, woodchuck holes, and the ocean) every year in the United States alone (Pennington 2010). Along the coast of central California there are a number of golf courses that use the ocean as a hazard or an out-of-bounds area. The most famous is Pebble Beach Golf Links, the site of the 2019 U.S. Open Championship. In 2017, Alex Weber, a high school student and scuba diving enthusiast, began to grow alarmed at the large numbers of golf balls she was finding on the ocean floor near Pebble Beach (Fig. 5.7). Seeking guidance, she contacted Maththew Savoca, a Stanford University scientist who studies marine plastic pollution. He advised her to begin counting the balls she found and removed from the ocean floor. From May 2016 to June 2018, 39602 golf balls were collected by individuals and organizations from intertidal and nearshore environments adjacent to or downriver from golf courses in Carmel. An additional 11079 were collected by the National Marine Sanctuary and the Pebble Beach Corporation, totaling approximately 2.5 tons of debris. Examination of decomposition patterns in the collected balls showed that about 62 pounds of synthetic material had been eroded away (Weber *et al.* 2019). While there are over 80 different types of golf balls with somewhat differing construction materials, commonly they are made of a butadiene rubber center surrounded by one or more rubber layers encased in a tough skin made from blends of ethylene and methacrylic acid (Scott 2017). The extent of the risk to oceans from remains uncertain.

On the other side of the North American continent, research has revealed the ubiquitous presence of microplastic pollution (i.e., less than 0.2 inches in length) in subsurface water in the Northeast Atlantic Ocean (Lusher *et al.* 2014). Microplastic was found in almost all of the water samples Lusher and colleagues collected from a continuous intake located on the forward starboard side of their research vessel. The fragmentation of plastic in microparticles follows known mechanical, chemical, and biological weathering processes once it reaches the marine environment and is exposed to sun, water, and other ocean conditions. Microplastic also enters the ocean through the runoff of micron-sized plastic particles used as exfoliates in cosmetics, produced by the ship-breaking industry, and used as industrial abrasives in sandblasting. The ocean accumulation of microplastic, which is occurring around the globe (Andrady 2011; Eriksen *et al.* 2014; Kanhai *et al.* 2017; Lavers & Bond 2017), is of great concern because it is difficult to remove from the environment and is consumed regularly by invertebrates and fish, and thereby enters and threatens health up and down the marine food chain, from zooplankton to whales. Microplastic has been found throughout the ocean water column, from the surface to deep-sea sediments (Van Cauwenberghe *et al.* 2013). In 2018, Hawaii's Kilauea volcano erupted, spewing streams of hot lava that crossed fields, blocked roads, and spilled into the ocean, where it burst into tiny shards of glass and rubble and created a new beach, Pohoiki. Based on samples taken from the newborn beach, scientists soon found that it was

Fig. 5.7 Harbor seal examining golf balls off the coast of California. *Source:* Courtesy of Alex Weber, photographer.

polluted with hundreds of tiny pieces of plastic washed up from the ocean (Gibbens 2019). After it is consumed by organisms, microplastic is distributed by the blood to body tissues, potentially resulting in several types of adverse effects, including "internal and/or external abrasions and ulcers; and blockages of the digestive tract, which can result in satiation, starvation and physical deterioration" (Wright *et al.* 2013, p. 487). This damage can lead to a decline in reproductive fitness, diminished predator avoidance, impaired feeding ability, and death. Microplastic ingestion has been observed in a range of animals consumed by humans, including fish, shellfish, and crustaceans. There remains the question of its effects on human health:

> The long-term impact of MPs [microplastics] on human health remains largely unknown since most studies to date have been limited to the impact on marine life consuming MPs directly. It is clear, however, that humans are exposed as these particles are found in our seafood and water supply, in the air we breathe, and in the dust in our home. How the damaging effects of MPs on marine, freshwater, and land ecosystems impact human health still needs to be discovered. (Kontrick 2018)

In a metastudy, Kieran Cox and colleagues (2019) carried out an analysis using 402 data points drawn from 26 studies to estimate the annual microplastic consumption in the United States. They concluded that it ranged from 39 000 to 52 000 plastic particles per annum, depending on age and sex; this increased to 74 000 and 121 000 when inhalation was added. Additionally, people who meet their recommended water intake through only bottled sources may be ingesting an extra 90 000 microplastics annually. These kinds of statistics have pushed many people to begin rethinking the future of plastics in human society.

5.7 Oil spills

On the night of April 20, 2010, at about 11 p.m. CDT, a surge of methane gas blasted through a concrete core recently installed on the BP-owned Deepwater Horizon drilling platform, floating 42 miles off the coast of Louisiana. The surge instigated an explosion and fire that took the lives of 11 rig workers, left another 17 seriously injured, and led to the sinking of the oilrig. Crude oil and gases, buried deep beneath the seafloor, began gushing from the source, located over 5000 feet beneath the water's surface. This tragic event is acknowledged as the worst oil spill in world history—though the far lesser known Taylor oil spill (see later) may have been bigger. By the time the well was finally capped 87 days after the explosion, an estimated 4–5 million barrels of oil had leaked into the Gulf. Winds, waves, and water currents spread it across the ocean's surface to form a 5000-square-mile slick, equivalent to the size of Connecticut, that endangered marine life, polluted beaches, islands, estuaries, and marshlands, and brought the regional fishing industry and people's livelihoods to a standstill (Shroder & Konrad 2011).

Prior to the Gulf oil disaster, the largest oil spill in the United States had been on March 24, 1989, when the *Exxon Valdez* struck Bligh Reef 25 miles from the city of Valdez, Alaska, rupturing the haul of the supertanker and spilling over 11 million gallons of thick crude oil into Prince William Sound. A storm soon dispersed the oil more than 470 miles to the southwest along the Kenai Peninsula, Kodiak Archipelago, and the Alaska Peninsula. Countless animals died from contact with the oil, and it has remained on coastal areas for decades. The National Wildlife Federation (Hubbell 2016) estimates that in the days immediately following the spill, 100 000–250 000 seabirds, 2800 sea otters, 12 river otters, 300 harbor seals, almost 250 bald eagles, 22 orcas, and innumerable fish and smaller organisms were killed. Loss of food sources led to additional animal deaths in the following days and weeks.

Though limited in number and scope, studies of the physical health consequences of oil spill exposure have documented both short- and long-term impacts. A review of the available research on spills from seven oil tankers found that the kinds of physical harm that had been inflected on various lifeforms included throat, skin, and eye irritations, respiratory problems, nausea, headaches and neuroendocrine function, and genotoxicity (Aguilera *et al.* 2010). For example, a study of acute health problems among people involved in cleanup efforts following the storm-caused 1997 spill by the Russian tanker *Nakhodka* in Anto, Japan found various acute symptoms emerged after the beginning of cleanup activities, including headache and symptoms of the eyes and throat (Morita *et al.* 1999). Longer-term studies suggest more serious sequelae. In 2007, the crude oil carrier *Hebei Spirit* was struck by a crane barge, had three of its oil tanks punctured, and began leaking near Mallipo Beach and wetlands in South Korea. A study of five major cancers and leukemia immediately before and 2 years after the spill found that the incidence rate of prostate cancer increased an annual average of almost 40 percent (Choi *et al.* 2018). Using longitudinal data from two surveys as part of the Gulf Long Term Follow-Up (GuLF) STUDY, Strelitz *et al.* (2018) assessed the occurrence across 5 years of congestive heart disease among Deepwater Horizon oil spill workers and community members. They found residential proximity to the spill and duration of participation in cleanup work were associated with a 29–43 percent higher hazard of experiencing heart disease events.

Equally important are the traumatic mental health impacts of such events. One source of this trauma is from personal loss of work and income, or from seeing the economic impacts suffered by loved ones. Another comes from witnessing the impact on the environment and wildlife. Palinkas (2012) observes that the mental health impacts of oil spills are similar to those of natural disasters, technological disasters like factory explosions, and even acts of terrorism. However, the acute or active impact phase of the dispersion of oil spills has a longer duration (sometimes months) than some other types of disasters, such as tornados and tsunamis. Moreover, in communities hit hard by spills, people are stressed by "increased social conflict and competition for limited resources . . . [the need to engage] in litigation to seek compensation for damages . . . a loss of social connection and a sense of separation from other people . . . increased uncertainty about long-term outcomes . . . and diminished trust in the ability of public officials and institutions to mitigate these outcomes or prevent future disasters" (Palinkas 2012, p. 205).

Studies of people living in areas impacted by oil spills have identified heightened levels of depression and anxiety, as well as post-traumatic stress. In a study of the mental health impacts of the *Exxon Valdez* spill 1 year after the disaster, Palinkas et al. (1993) examined the relationship between exposure to both the spill and subsequent cleanup efforts and the prevalence of generalized anxiety disorder (GAD), post-traumatic stress disorder (PTSD), and depressive symptoms in 13 Alaskan communities. Based on a community survey of almost 600 people, they found that 20 percent of men and 9 percent of women suffered from GAD and PTSD, while 17 and 14 percent had symptoms of depression. For all the mental health parameters they analyzed, exposed individuals showed scores several times higher than unexposed individuals. Gill & Picou (1998) monitored the longer-longer impacts of the *Exxon Valdez* spill on the affected communities using a 4-year longitudinal study design and an applied survey on social disruption and psychological stress. Their findings indicated the experience of chronic stress among community members. Expectations and desires to leave the affected community increased over the study period. Social disruption, however, declined, as did (at a slower pace) event-related psychological stress.

In a parental interview study of the prevalence of physical and mental health effects among over 1400 children living in communities in Gulf coastal areas that were adversely impacted by the Deepwater Horizon event, Abramson *et al.* (2013) found that half of parents reported that their children had a spill-related physical, environmental, or economic exposure. About 40 percent identified some type of health effect: 18 percent said their children experienced breathing problems, 15 percent that their children developed skin problems, 16 percent that their children suffered from visual problems, and 22 percent that their children had emotional or behavioral problems. Across the four communities included in the study, the team heard of significant issues related to children's health and well-being related to the spill. An array of other studies confirm the mental health risks created by oil spills that occur near to and reach populated areas (Grattan *et al.* 2011; Osofsky *et al.* 2011; Morris *et al.* 2013; Hansel *et al.* 2015).

Another longer-term consequence of oil spills appears to be an increased rate of serious chronic diseases among human populations. This connection was found by D'Andrea & Reddy (2018) in their assessment of the long-term adverse health effects of the 2010 Deepwater Horizon Gulf oil spill exposure on cleanup workers. This work involved a review of the medical records of 88 individuals who worked in oil spill cleanup matched with 44 unexposed individuals. Compared with the unexposed subjects, oil spill-exposed study participants had significantly reduced blood platelet counts. Their hemoglobin and hematocrit levels were increased significantly at both initial and 7-year follow-up medical exams. Additionally, they had significantly increased ALP, AST, and ALT levels (which are signs of liver or muscle damage) on both exams. Of note, at their 7-year follow-up, most of the oil spill-exposed workers had developed chronic rhinosinusitis and reactive airway dysfunction syndrome as new symptoms that were not reported during their initial visit. Additionally, more abnormalities were seen in pulmonary and cardiac functions in these participants. The researchers concluded that "intensive participation even for short periods (only few weeks) in the oil spill cleanup work activities has resulted in persistent long-term adverse health effects" (D'Andrea & Reddy 2018).

Larry Schweiger, speaking as president of the National Wildlife Federation, the largest conservation organization in the United States, cautions: "The impacts of oil spills continue long after the TV cameras have gone home. It is still possible to find oil on the shores of Prince William Sound that is nearly as toxic as it was a quarter-century ago" (quoted in Hubbell 2016). While many of the animal species that were mired in the oil leaked by the *Exxon Valdez* have recovered their pre-spill population numbers, some, like the Pacific herring, have not. These fish spawned in the waters of Prince William Sound shortly after the spill. Four years later, the herring fishery in the Sound collapsed. Interaction among multiple adverse factors, including damage from the spill, overfishing, and an increase in several fish diseases, has been hypothesized as the cause of the population's failure to rebound.

While the Deepwater Horizon oil spill received considerable public and media attention, the same cannot be said for the possibly larger Taylor spill that began in 2004 and is continuing to this day. It began when Hurricane Ivan, a fierce storm whose intensity has been linked to climate change (Emanuel 2005), destroyed a Taylor Energy MC20 oil platform with 28 wells located approximately 10 miles off the Gulf coast of Louisiana. Recognition that the spill was ongoing only occurred in 2010 when it was brought to light by environmentalists who were monitoring the nearby Deepwater Horizon catastrophe (Fears 2018). It is estimated that between 380 and 4500 gallons of crude oil leak into the ocean from the deep-water site every day (Mason *et al.* 2019). Taylor Energy long tried to keep this leak secret in order to shield its reputation, until it lost a lawsuit that exposed the event. The magnitude of the leak is 100–1000 times greater than the company's initial estimate, which set the amount of oil flowing into the ocean at less than 3 gallons per day. Once part of the forest of oil platforms in the now industrialized Gulf of Mexico, the Taylor site may continue to leak its toxic load through the end of the 21st century (BSEE 2015). Unable to fix the ongoing leak, Taylor Energy liquidated its oil and gas assets and ceased production and drilling in 2008. On its website, the company says it only exists now to respond to the continued spill. Though it maintains that any oil and gas that is still leaking at the site is coming from oil-soaked sediment and bacterial breakdown of the oil, this is contradicted by the assessment led by Mason *et al.* (2019), which concluded that the oil "is in fact coming from the reservoirs, from these oil pipes, and not from the remnant oil at the bottom of the ocean." Given the intention of the Trump administration to eliminate some of the Obama-era safety regulations for offshore drilling that were opposed by the oil industry, it is evident that lessons have not been learned from past oil spills and that the potential for new significant spills is increasing (Wamsley 2019).

5.8 Desertification

While Earth is a watery world, almost 10 percent of its surface (one third of the land surface) is desert. We use the word "desert" to label sparsely vegetated arid land with a very warm climate that annually receives less than 10 inches of sporadic rain. The largest area that fits this definition (i.e., a warm desert, as opposed to Antarctica, a cold desert) is the 3.5 million-square-mile zone known as the Sahara; the next largest, the Arabian Desert, is one fourth its size.

Deserts, however, do not have fixed sizes; over time, their dimensions change as a result, especially, of climate alterations, though at smaller timescales weather is also a factor as there are seasonal patterns of expansion and contraction (Kröpelin *et al.* 2008). In the case of the Sahara, research indicates that the desert has expanded significantly in the last century, due both to natural variability and anthropogenic climate change (Thomas & Nigam 2018). Between 1920 and 2013, it grew by about 8.5 percent, with most of the expansion occurring in the south.

Desertification is the progressive process of deterioration of fertile land into a place that is too dry to support human life and the food sources required to sustain it. In particularly fragile ecosystems, desertification can become irreversible as rising temperatures due to global warming diminish the annual level of rainfall. Additionally, desertification is a self-accelerating process that drives the desert edge across previously productive areas. In the Sahel, the midcontinent semiarid transition zone between the Sahara Desert to the north and the Sudanian Savanna to the south, for example, the desert is creeping southward at the rate of about 30 miles a year (Holthuijzen & Rugaimukamu 2011). In some parts of the Sahara, however, in the counterintuitive way that climate change impacts the planet, global heating may increase rainfall (because warmer air can hold more moisture).

Several factors propel desertification, including direct human activities, especially livestock overgrazing, and recurrent drought linked to climate change. A stark consequence of desertification is seen in Arauane, a village located in an arid area of Mali, in west Africa. In the past, community members remember, the village was in an oasis. But the desert spread and sand buried the local mosque up to its minaret. By the early 1990s, Arauane was surrounded by sand dunes on three sides that rose to a height of 65 feet, burying the village and forcing people to live in underground homes (SakhaliaNet 2014).

In 2000, then Secretary-General of the United Nations Kofi Annan called on member states to set up what came to be called the Millennium Ecosystem Assessment, an effort designed to assess and report on the consequences of ecosystem change—including desertification—for human well-being. The assessment report, which included a section on "Desertification Synthesis," was issued in 2005 (Adeel *et al.* 2005). It affirms that desertification is a global environmental issue, but that its magnitude and impacts vary significantly from place to place and change over time. Notably, desertification has become a global health issue because it puts over 1 billion people at risk for food and water insecurity, hunger, and starvation. In addition to diminishing food production, the report affirms that desertification limits access to clean drinking water and adequate sanitation, further damaging health. Infant mortality in 2005, for example, averaged over 60 per thousand live births in communities in dryland environments, compared to about 40 in coastal settings. Moreover, desertification causes reduction of ground vegetation cover, which increases the formation of aerosols and dust, contributing to illness including fever, coughing, and irritated eyes. Dust can be blown tremendous distances; that gusted off drylands in East Asia and the Sahara, for example, has been implicated in respiratory problems in North America and the Caribbean. Akpinar-Elci *et al.* (2015) found that Saharan dust, in conjunction with high seasonal humidity, facilitated the inhalation of particulate matter that exacerbated asthma among residents of the Caribbean island of Grenada. Similarly, on the island of Guadeloupe, inhaled Saharan dust has been associated with increased risk of visiting the health emergency department for children with asthma (Cadelis *et al.* 2014).

5.9 Concluding remarks

The ecocrisis case studies reviewed in this chapter reveal the extent of environmental degradation across diverse ecological zones and areas of the world, the significant contribution humans make to ecocrises, the importance of ecocrises interaction in exacerbating environmental problems, and the many health difficulties humans suffer because of environmental disruption. In this discussion, it has become clear that anthropogenic climate change is a powerful force in planetary health, interacting with and multiplying the harmful effects of diverse ecocrises. The various factors contributing to ocean depletion, for example, are enhanced by the heating of the oceans. This is seen in climate-related changes at the deep seafloor. Deep ocean areas, at depths below 650 feet, which are critical to the healthy functioning of ocean ecosystems (e.g., through nutrient production and cycling, and the provision of marine species nursery grounds), are changing as a result of global warming. Expected adverse modifications include "alterations in the distributions and health of open-ocean and deep-sea fish populations and commercially exploited stocks" and drops in the "embryonic survival rates of vulnerable deep-sea oviparous (egg-laying) elasmobranchs [e.g., sharks] that currently deposit their [egg] capsules at the seafloor in very narrow oceanographic niches with distinct O_2, salinity and temperature conditions" (Sweetman *et al.* 2017).

One ocean group not being harmed by climate change and ocean depletion, but rather benefitting from them, is the jellyfish. While rises in their population levels may not be as drastic as is asserted by some popular and academic science writers (Gershwin 2013), it is evident that anthropogenic activities are making the oceans particularly hospitable for jellyfish species (Berwald 2017). Given that jellyfish toxin "induces a local reaction of the skin, which can be painful and leave scaring," as well as a systemic reaction that "with some species is lethal," oceans filled with jellyfish are not desirable from the standpoint of human health (Badré 2014). Additionally, blooms of the moon jellyfish (*Aurelia*), which have been observed in coastal areas in Asia, Europe, Australia, and the United States, cause various negative impacts, such as clogging the cooling intakes of coastal power plants and interfering with fisheries, aquacultural operations, and tourism (Dong 2019). On the upside, a plentiful supply of jellyfish provides a food source for those marine animals that feed on them, such as leatherback turtles, penguins, albatross, tuna, and chum salmon (Hays *et al.* 2018). Even this silver lining has a dark cloud: leatherbacks mistake plastic bags dumped in the ocean for jellyfish and can die from their error.

The next chapter addresses specific health and other threats of ecocrises interaction.

References

Abramson, D., Peek, L., Redlener, I., Beedasy, J. & Aguilar, T. (2013). Children's Health After the Oil Spill: A Four-State Study, Findings from the Gulf Coast Population Impact (GCPI) Project. New York: National Center for Disaster Preparedness, Columbia University. https://www.academia.edu/11772469/Childrens_Health_after_the_Oil_Spill_A_Four_State_Study_Findings_from_the_Gulf_Coast_Population_Impact_GCPI_Project.

Achaibar, K., Moore, S. & Bain, P. (2007). Ciguatera poisoning. *Practical Neurology* 7(5): 316–322.

Addis, P., Locci, I. & Cau, A. (2009). Anthropogenic impacts on the bluefin tuna (*Thunnus thynnus*) trap fishery of Sardinia (Western Mediterranean). *Collected Volume of Scientific Papers ICCAT* 63: 174–185.

Adeel, Z., Safriel, U., Niemeijer, D. & White, R. (2005). Desertification Synthesis. Millennium Ecosystem Assessment. Washington, D.C.: World Resources Institute.

Aguilera, F., Méndez, J. Pásaro, E., & Laffon, B. (2010). Review on the effects of exposure to spilled oils on human health. *Journal of Applied Toxicology* 30: 291–301.

AIMS. (2018). Great Barrier Reef suffers multiple, regional-scale impacts. Australian Institute of Marine Science.

Akpinar-Elci, M., Martin, F., Behr, J. & Diaz, R. (2015). Saharan dust, climate variability, and asthma in Grenada, the Caribbean. *International Journal of Biometerology* 59(11): 1667–1671.

Andrady, A. (2011). Microplastics in the marine environment. *Marine Pollution Bulletin* 62: 1596–1605.

Anthony, K. Kline, D., Diaz-Pulido, G., Dove, S. & Hoegh-Guldberg, O. (2008). Ocean acidification causes bleaching and productivity loss in coral reef builders. *Proceedings of the National Academy of Sciences of the U.S.A.* 105(4): 17442–17446.

Baden, D. (1989). Brevetoxins: unique polyether dinoflagellate toxins. *FASEB Journal* 3(7): 1807–1817.

Badré, S. (2014). Bioactive toxins from stinging jellyfish. *Toxicon* 91: 114–125.

Barange, M., Merino, G., Blanchard, Scholtens, J., Harle, J., Allison, E. *et al.* (2014). Impacts of climate change on marine ecosystem production in societies dependent on fisheries. *Nature Climate Change* 4: 211–216.

Barboza, L., Vethaak, A., Lavorante, B. Lundebye, A.-K. & Guilhermino, L. (2018). Marine microplastic debris: an emerging issue for food security, food safety and human health. *Marine Pollution Bulletin* 133: 336–348.

Barile, P. (2018). Widespread sewage pollution of the Indian River Lagoon system, Florida (U.S.A.) resolved by spatial analyses of macroalgal biogeochemistry. *Marine Pollution Bulletin* 128: 557–574.

Barnes, D., Galgani, F., Thompson, R. & Barlaz, M. (2009) Accumulation and fragmentation of plastic debris in global environments. *Philosophical Transactions of the Royal Society of London Series B* 364: 1985–1998.

Beech, E., Rivers, M., Oldfield, S. & Smith, P. (2017). Global tree search: the first complete global database of tree species and country distributions. *Journal of Sustainable Forestry* 36(5): 454–489.

Berwald, J. (2017). *Spineless: The Science of Jellyfish and the Art of Growing a Backbone.* New York: Riverhead Books.

Blanchard, J., Jennings, S., Holmes, R., Harle, J., Merino, G., Allen, J. *et al.* (2012). Potential consequences of climate change for primary production and fish production in large marine ecosystems. *Philosophical Transactions of the Royal Society B* 367: 2979–2989.

Bollmann, M. (2010). World Ocean Review, 2010: Living with the Oceans. Hamburg: Maribus.

Brand, L. & Compton, A. (2007). Long-term increase in *Karenia brevis* abundance along the Southwest Florida coast. *Harmful Algae* 6(2): 232–252.

Brouwers, N., Hardy, G., Ruthrof, K., Matusick, G. & Zeppel, M. (2015). Climate change: trees under pressure. Ecological Society of Australia. https://www.ecolsoc.org.au/hot-topics/climate-change-trees-under-pressure.

Brown, B. & Dunn, R. (1988). The environmental impact of coral mining on coral reefs in the Maldives. *Environmental Conservation* 15(2): 159–166.

BSEE. (2015). Incident archive—Taylor Energy (Mississippi Canyon) oil spill—U.S. Coast Guard fact sheet. Bureau of Safety and Environmental Enforcement. https://www.bsee.gov/newsroom/library/incident-archive/taylor-energy-mississippi-canyon/fact-sheet.

Butterwick, C., Heaney, S. & Talling, J. (2005). Diversity in the influence of temperature on the growth rates of freshwater algae, and its ecological relevance. *Journal of Freshwater Biology* 50: 291–300.

Cadelis, G., Tourres, R. & Molinie J. (2014). Short-term effects of the particulate pollutants contained in Saharan dust on the visits of children to the emergency department due to asthmatic conditions in Guadeloupe (French Archipelago of the Caribbean). *PLoS One* 9(3): e91136.

CapeCod.com. (2018). Red tide forces Nauset estuary shellfishing closure. https://www.capecod.com/newscenter/red-tide-forces-nauset-estuary-shellfishing-closure/.

Census of Marine Life. (2007). Bluefin Tuna—Past and Present. http://www.coml.org/medres/tuna07/CoML_tuna_release_public.pdf.

Census of Marine Life. (2011). Scientific Results to Support the Sustainable Use and Conservation of Marine Life. New York: Census of Marine Life International Secretariat.

Center for Biological Diversity. (2016). Pacific bluefin tuna one step closer to Endangered Species Act protection: overfishing continues as international negotiations over reducing catch stall. https://www.biologicaldiversity.org/news/press_releases/2016/bluefin-tuna-10-07-2016.html.

Center for Biological Diversity. (2017). Trump administration denies pacific bluefin tuna Endangered Species Act protection: after 97 percent decline, species faces extinction without help. https://www.biologicaldiversity.org/news/press_releases/2017/pacific-bluefin-tuna-08-07-2017.php.

Center for Disaster Philanthropy. (2020). 2019–2020 Australian bushfires. https://disasterphilanthropy.org/disaster/2019-australian-wildfires/.

Cesar, H. (2000). Coral reefs: their functions, threats and economic value. In: H. Cesar (Ed.), *Collected Essays on the Economics of Coral Reefs*. Kalmar: CORDIO.

Chai, C., Yu, Z., Song, X. & Cao, X. (2006). The status and characteristics of eutrophication in the Yangtze River (Changjiang) Estuary and the adjacent East China Sea. *China Hydrobiologia* 563: 313–328.

Chen, C., Zhu, J., Beardsley, C. & Franks, P. (2003). Physical–biological sources for dense algal blooms near the Changjiang River. *Geophysical Research Letters* 30: 1515–1518.

Choat, B., Jansen, S., Brodribb, T., Cochard, H., Delzon, S., Bhaskar, R., *et al.* (2012). Global convergence in the vulnerability of forests to drought. *Nature* 491: 752–755.

Choat, B., Brodribb, T., Brodersen, C., Duursma, R., López, R. & Medlyn, B. (2018). Triggers of tree mortality under drought. *Nature* 558: 531–539.

Choi, K., Park, M., Ha, M., Hur, J., & Cheong, H. (2018). Cancer incidence trend in the Hebei Spirit oil spill area, from 1999 to 2014: an ecological study. *International Journal of Environmental Resarch and Public Health* 5(5): 1006.

Chung, W.-S., Marshall, N., Watson, S.-A., Munday, P. & Nilsson, G. (2014). Ocean acidification slows retinal function in a damselfish through interference with GABAA receptors. *Journal of Experimental Biology* 217(3): 323–326.

Clarke, G., Burgess, N., Mbago, F., Mligo, C., Mackinder, B. & Gereau, R. (2011). Two "extinct" trees rediscovered near Kilwa, Tanzania. *Journal of East African Natural History* 100(1&2): 133–140.

Cloverm, C. (2006). *The End of the Line: How Overfishing is Changing the World and What We Eat.* New York: The New Press.

Coastal Systems Group. (2018). *Town of Orleans Estuaries Water Quality Monitoring Database Development and Review.* New Bedford, MA: University of Massachusetts Dartmouth.

Corson, T. (2007). *The Story of Sushi: An Unlikely Saga of Raw Fish and Rice.* New York: HarperCollins.

Cowie, B., Thornton, C. & Radford, B. (2007). The Brigalow Catchment Study: I. Overview of a 40-year study of the effects of land clearing in the Brigalow bioregion of Australia. *Soil Research* 45(7): 479–495.

Cox, K., Covernton, G., Davies, H., Dower, J., Juanes, F. & Dudas, S. (2019). Human consumption of microplastics. *Environmental Science and Technology* 5312: 7068–7074.

CSPA. (2007). California Sportfishing Protection Alliance to sue city of Stockton for sewage spills and illegal discharges. Press release. California Sportfishing Protection Alliance.

D'Andrea, M. & Reddy, G. (2018). The development of long-term adverse health effects in oil spill cleanup workers of the Deepwater Horizon offshore drilling rig disaster. *Frontiers in Public Health* 6: 117.

Dethier, D. (2018). 15 numbers you need to know about the U.S. golf economy. Golf.com. https://www.golf.com/tour-news/2018/05/01/15-numbers-you-need-know-about-us-golf-economy.

Dhar, B., Cimarelli, L., Singh, K., Brandi, L., Brandi, A., Puccinelli, C. *et al.* (2015). Molecular detection of a potentially toxic diatom species. *International Journal of Environmental Research and Public Health* 12(5): 4921–4941.

Dong, Z. (2019). Blooms of the moon jellyfish aurelia: causes, consequences and controls. In C. Sheppard (Ed.), *World Seas: An Environmental Evaluation*, 2 edn, vol. III: *Ecological Issues and Environmental Impacts*, pp. 163–171. London: Academic Press.

Doyle, A. (2009). European over-fishing began 1000 years ago. *Reuters*. https://uk.reuters.com/article/uk-oceans/european-over-fishing-began-1000-years-ago-idUKTRE54N1Q520090524.

DSITI. (2017). Land Cover Change in Queensland 2015–16: A Statewide Landcover and Trees Study (SLATS) Report. Brisbane, QLD: Queensland Department of Science, Information Technology and Innovation. https://publications.qld.gov.au/dataset/4dbd1416-52b7-467a-8410-17a70ddf16bf/resource/60a7902d-7a9d-49a7-90b1-a54686fbcef5/download/edocs-5799682-v1-slatsreport2015-16.pdf.

Emanuel, K. (2005). Increasing destructiveness of tropical cyclones over the past 30 years. *Nature* 436: 686–688.

EPA. (2018). Basic information about coral reefs. Environmental Protection Agency. https://www.epa.gov/coral-reefs/basic-information-about-coral-reefs.

Eriksen, M., Lebreton, C., Carson, H., Thiel, M., Moore, C., Borerro, J. *et al.* (2014). Plastic pollution in the world's oceans: more than 5 trillion plastic pieces weighing over 250 000 tons afloat at sea. *PLoS One* 9: e111913.

European Commission. (2017). Bluefin tuna season 2017: E.U. determined to uphold high control standards in bluefin tuna fishery. https://ec.europa.eu/fisheries/bluefin-tuna-season-2017-eu-determined-uphold-high-control-standards-bluefin-tuna-fishery_en.

European Commission (2018). Bluefin tuna season 2018: E.U. determined to maintain control standards and guarantee recovery trend in eastern bluefin tuna stock. https://ec.europa.eu/fisheries/bluefin-tuna-season-2018-eu-determined-maintain-control-standards-and-guarantee-recovery-trend_en.

FAO. (2016). The State of World Fisheries and Aquaculture 2016. Rome: Food and Agriculture Organization of the United Nations.

Farstad, D. & Chow, T. (2001). A brief case report and review of ciguatera poisoning. *Wilderness and Environmental Medicine* 12: 263–269

Fears, D. (2018). A 14-year-long oil spill in the Gulf of Mexico verges on becoming one of the worst in U.S. history. *Washington Post*. https://www.washingtonpost.com/national/health-science/a-14-year-long-oil-spill-in-the-gulf-of-mexico-verges-on-becoming-one-of-the-worst-in-us-history/2018/10/20/f9a66fd0-9045-11e8-bcd5-9d911c784c38_story.html?utm_term=.1af351a9eb94.

Feintrenie, F. (2014). Agro-industrial plantations in Central Africa, risks and opportunities. *Biodiversity and Conservation* 23: 1577–1589.

Ferrari, M., Dixson, D., Munday, P., McCormick, M., Meekan, M., Sih, A. *et al.* (2011). Intrageneric variation in antipredator responses of coral reef fishes affected by ocean acidification: implications for climate change projections on marine communities. *Global Change Biology* 17: 2980–2986.

Fitch, L. (2015). Two fish, one fish, no fish: Alberta's fish crisis. *Wild Lands Advocate* 23(5): 14–18.

Fleming, L., Rivero, C., Burns, J., William, C., Bean, J., Shea, K. *et al.* (2002). Blue green algal (cyanobacterial) toxins, surface drinking water, and liver cancer in Florida. *Harmful Algae* 1(2): 157–168.

Fleming, L., Kirkpatrick, B., Backer, L., Walsh, C., Nierenberg, K., Clark, J. *et al.* (2011). Review of Florida red tide and human health effects. *Harmful Algae* 10: 224–233.

Flewelling, L., Naar, J., Abbott, J., Baden, D., Barros, N., Bossart, G. *et al.* (2005). Red tides and marine mammal mortalities: Unexpected brevetoxin vectors may account for deaths long after or remote from an algal bloom. *Nature* 435(7043): 755–756.

Food Security Information Network. (2018). 2018 Global Report on Food Crises. https://www.wfp.org/publications/global-report-food-crises-2018.

Fornac, K., Abidin, T., Alexander, N., Brock, P., Grigg, M., Murphy, A. *et al.* (2016). Association between landscape factors and spatial patterns of *Plasmodium knowlesi* infections in Sabah, Malaysia. *Emerging Infectious Diseases* 22(2): 201–209.

Free, C., Thorson, J., Pinsky, M., Oken, K., Wiedenmann, J. & Jensen, O. (2019). Impacts of historical warming on marine fisheries production. *Science* 363(6430): 979–983.

Fromentin, J.-M. & Powers, J. (2005). Atlantic bluefin tuna: population dynamics, ecology, fisheries and management. *Fish and Fisheries* 6(4): 281–306.

Fujise, L., Yamashita, H., Suzuki, G., Sasaki, K., Liao, L. & Koike, K. (2014). Moderate thermal stress causes active and immediate expulsion of photosynthetically damaged zooxanthellae (symbiodinium) from corals. *PLoS One* 9(12): e114321.

Gatti, C., Lonati, D., Darius, H., Zancan, A., Rué, M., Schicchi, A. *et al.* (2018). *Tectus niloticus* (Tegulidae, Gastropod) as a novel vector of ciguatera poisoning: clinical characterization and follow-up of a mass poisoning event in Nuku Hiva Island (French Polynesia). *Toxins* 10(3): 102.

Gershwin, L.-A. (2013). *Stung!: On Jellyfish Blooms and the Future of the Ocean*. Chicago, IL: University of Chicago,

Gibbens, S. (2019). Newborn Hawaii beach is already polluted with tiny pieces of plastic. *National Geographic*. https://www.nationalgeographic.com/environment/2019/05/newborn-hawaii-beach-already-polluted-with-plastic/.

Gill, D. & Picou, J. (1998). Technological disaster and chronic community stress. *Social & Natural Resources* 11: 795–815.

Gillis, C. (2018). Millions of pounds of dead fish have washed up on our beaches. This is what happens to them next. *News-Press*. https://www.news-press.com/story/news/2018/08/17/red-tide-blue-green-algae-fish-kill-sea-turtle-lake-okeechobee-caloosahatchee-river/978883002/.

Gkelis, S., Papadimitriou, T., Zaoutsos, N. & Leonardos, I. (2014). Anthropogenic and climate-induced change favors toxic cyanobacteria blooms: evidence from monitoring a highly eutrophic, urban Mediterranean lake. *Harmful Algae* 39: 322–333.

Graham, J. & Dickson, K. (2004). Tuna comparative physiology. *Journal of Experimental Biology* 207: 4015–4024.

Grattan, L., Roberts. S., Mahan, McLaughlin, P., Otwell, W. & Morris, J. (2011). The early psychological impacts of the Deepwater Horizon oil spill on Florida and Alabama communities. *Environmental Health Perspectives* 119(6): 838–843.

Grattan, L., Boushey, C., Tracy, K., Trainer, V., Roberts, S., Schluterman, N. *et al.* (2016a). The association between razor clam consumption and memory in the COASTAL cohort. *Harmful Algae* 57: 20–25.

Grattan, L., Holobaugh, S. & Morris, J. (2016b). Harmful algal blooms and public health. *Harmful Algae* 57(B): 2–8.

Guilbert, K. & Bacchi, U. (2017). All at sea: foreign fishing fleets drain West African waters. *Reuters*. https://www.reuters.com/article/africa-fish-overfishing-idUSL8N1I43AH.

Guzmán, H. & Jiménez, C. (1992). Contamination of coral reefs by heavy metals along the Caribbean coast of Central America (Costa Rica and Panama). *Marine Pollution Bulletin* 24(11): 554–561.

Habermehl, G., Krebs, H., Rasoanaivo, P. & Ramialiharisoa, A. (1994). Severe ciguatera poisoning in Madagascar: a case report. *Toxicon* 32(12): 1539–1542.

Hallegraeff, G. (2003). Harmful algal blooms: a global overview. In: G. Hallegraeff, D. Anderson & A. Cembella (Eds.), *Manual on Harmful Marine Microalgae*, pp. 25–50. Paris: UNESCO.

Hallegraeff, G., Anderson, D., Cembella, A. & Enevoldsen, H. (Eds.). (2004). Manual on Harmful Marine Microalgae, 2 rev. edn. Paris: UNESCO.

Hamilton, L. & Butler, M. (2001). Outport adaptations: social indicators through Newfoundland's cod crisis. *Research in Human Ecology* 8(2): 1–11.

Hansel, T., Osofsky, H., Osofsky, J. & Speier, A. (2015). Longer-term mental and behavioral health effects of the Deepwater Horizon Gulf oil spill. *Journal of Marine Science and Engineering* 3: 1260–1271.

Hawkins, B. (2008). *Plants for Life: Medicinal Plant Conservation and Botanic Gardens*. London: Botanic Gardens Conservation International.

Hays, G., Doyle, T. & Houghton, J. (2018). A paradigm shift in the trophic importance of jellyfish? *Trends in Ecology & Evolution* 33(11): 874–884.

Heyerdahl, T. (1971). Atlantic Ocean pollution and biota observed by the "Ra" expeditions. *Biological Conservation* 3(3): 164–167.

Hilborn, R. (2012). *Overfishing: What Everyone Needs to Know*. Oxford: Oxford University Press.

Hobkins, R. & Wheatley, T. (2018). U.S. moves to reverse protections for Atlantic bluefin tuna. Pew Charitable Trusts. https://www.pewtrusts.org/en/research-and-analysis/articles/2018/04/18/us-moves-to-reverse-protections-for-atlantic-bluefin-tuna.

Hodgson, G. (1999). A global assessment of human effects on coral reefs. *Marine Pollution Bulletin* 38(5): 345–355.

Holthuijzen, W. & Maximillian, J. (2011). Dry, hot, and brutal: desertification in the Sahel of Mali. *Journal of Sustainable Development in Africa* 13(7): 245–268.

Howarth, R. & Ramakrishna, K. (2005). Nutrient management. In: K. Chopra, R. Leemans, P. Kumar & H. Simons (Eds.), *Ecosystems and Human Wellbeing: Policy Responses*, pp. 295–311. Washington, D.C.: Island Press.

Howarth, R., Anderson, A., Cloern, J., Elfring, C., Hopkinson, C., Lapointe, B. *et al.* (2000). Nutrient pollution of coastal rivers, bays, and seas. *Issues in Ecology* 7: 1–7.

Hu, C., Muller-Karger, F. & Swarzenski, P. (2006). Hurricanes, submarine groundwater discharge, and Florida's red tides. *Geophysical Research Letters* 33: L11601,

Hubbell, D. (2016). Remembering Exxon Valdez. National Wildlife Federation. https://blog.nwf.org/2014/03/remembering-exxon-valdez/.

Isbister, G. & Kiernan, M. (2005). Neurotoxic marine poisoning. *Lancet Neurology* 4(4): 219–228.

Ivar so Sul, J. & Costa, M. (2013). The present and future of microplastic pollution in the marine environment. *Environmental Pollution* 185: 352–364.

Kanhai, D., Officer, R., Lyashevska, O., Thompson, R. & O'Connor, I. (2017). Microplastic abundance, distribution and composition along a latitudinal gradient in the Atlantic Ocean. *Marine Pollution Bulletin* 115(1–2): 307–314.

Kennedy, J. (2019). Brief history of cod fishing. Thought Company. https://www.thoughtco.com/brief-history-of-cod-fishing-2291538.

Kirkpatrick, B., Fleming, L., Squicciarini, D., Backer LC, Clark R, Abraham, W. *et al.* (2004). Literature review of Florida red tide: implications for human health effects. *Harmful Algae* 3(2): 99–115.

Kontrick, A. (2018). Microplastics and human health: our great future to think about now. *Journal of Medical Toxicology* 12(2): 117–119.

Kosten, S., Huszar, V., Becares, E., Costa, L., van Donk, E. & Hansson, L. (2011). Warmer climate boosts cyanobacterial dominance in shallow lakes. *Global Change Biology* 18(1): 118–126.

Kröpelin, S., Verschuren, D., Lézine, A., Eggermont, H., Cocquyt, C., Francus, P. *et al.* (2008). Climate-driven ecosystem succession in the Sahara: the past 6000 years. *Science* 320(5877): 765–768.

Kurlansky, M. (1997). *Cod: A Biography of the Fish that Changed the World*. New York: Penguin.

Kurobe, T., Baxa, D., Mioni, C., Kudela, R., Smythe, T., Waller, S. & The, S. (2013). Identification of harmful cyanobacteria in the Sacramento–San Joaquin Delta and Clear Lake, California by DNA barcoding. *Springerplus* 2: 491.

Landsberg, J., Flewelling, L. & Naar (2009). Karenia brevis red tides, brevetoxins in the food web, and impacts on natural resources: decadal advancements. *Harmful Algae* 8: 598–607.

Lange, W. (1987). Ciguatera toxicity. *American Family Physician* 3(4): 177–182.

Lavers, J. & Bond, A. (2017). Exceptional and rapid accumulation of anthropogenic debris on one of the world's most remote and pristine islands. *Proceedings of the National Academy of Sciences of the U.S.A.* 114: 6052–6055.

Law, K. (2017). Plastics in the marine environment. *Annual Review of Marine Science* 9: 205–229.

Lebreton, L., Slat, B., Ferrari, F., Sainte-Rose, B., Aitken, J. Marthouse, S., *et al.* (2018). Evidence that the Great Pacific Garbage Patch is rapidly accumulating plastic. *Scientific Reports* 8: 4666.

Legrand, A. (1998). Ciguatera toxins: origin, transfer through the food chain and toxicity to humans. In: B. Reguera, J. Blanco, M. Fernández & T. Wyatt (Eds.), *Harmful Algae*, pp. 39–43. Vigo: Xunta de Galicia.

Lehane, L. & Lewis, R. (2000). Ciguatera: recent advances but the risk remains. *International Journal of Food Microbiology* 61(2–3): 91–125.

Lehman, P., Boyer, G., Hall, C. Waller, S. & Gehtrts, W. (2005). Distribution and toxicity of a new colonial *Microcystis aeruginosa* bloom in the San Francisco Bay Estuary, California. *Hydrobiologia* 541(1): 87–99.

Lehman, P., Teh, S., Boyer, G., Nobriga, M., Bass, E. & Hogle, C. (2010). Initial impacts of *Microcystis aeruginosa* blooms on the aquatic food web in the San Francisco Estuary. *Hydrobiologia* 637(1): 229–248.

Lehman, P., Kurobe, T., Lesmeister, S., Baxa, D., Tung, A. & Teh, S. (2014). Impacts of the 2014 severe drought on the Microcystis bloom in San Francisco Estuary. *Harmful Algae* 63: 94–108.

Lohr, J., Munn, C. & Wilson, W. (2007). Characterization of a latent virus-like infection of symbiotic zooxanthellae. *Appled and Experimental Microbiology* 73(9): 2976–2981.

Longo, S. & Clark, B. (2012). The commodification of bluefin tuna: the historical transformation of the Mediterranean fishery. *Journal of Agrarian Change* 12(2–3): 204–226.

Lusher, A., Burke, A., O'Connor, I. & Officer, R. (2014). Microplastic pollution in the Northeast Atlantic Ocean: validated and opportunistic sampling. *Marine Pollution Bulletin* 88(1–2): 325–333.

MacKenzie, B. & Myers, R. (2007). The development of the northern European fishery for north Atlantic bluefin tuna *Thunnus thynnus* during 1900–1950. *Fisheries Research* 87(2–3): 338–239.

Mason, A., Taylor, J. & MacDonald, I. (2019). An integrated assessment of oil and gas release into the marine environment at the Former Taylor Energy MC20 Site. NOAA Technical Memorandum 260. Silver Spring, MD.: NOAA National Ocean Service, National Centers for Coastal Ocean Science.

McGowan, J., Deyle, E., Ye, H., Carter, M., Perrentti, C, Seger, K. *et al.* (2017). Predicting coastal algal blooms in southern California. *Ecology* 98(5): 1419–1433.

McKibben, B. (2006). State of the planet 2006. *Seed Magazine*. http://seedmagazine.com/content/article/state_of_the_planet_2006/.

Missouri Botanical Garden. (2018). Back from the brink: saving a species on the edge of extinction. https://discoverandshare.org/2018/10/18/back-from-the-brink-saving-a-species-on-the-edge-of-extinction/.

Moore, J., Fu, W., Primeau, F., Britten, G., Lindsay, K., Long, M., *et al.* (2018). Sustained climate warming drives declining marine biological productivity. *Science* 359: 1139–1143.

Morita, A., Kusaka, Y., Deguchi, Y., Moriuchi, A., Nakanaga, Y., Iki, M. *et al.* (1999). Acute health problems among the people engaged in the cleanup of the Nakhodka oil spill. *Environmental Research* 81(3): 185–194.

Morris, J., Grattan, L., Mayer, B. & Blackburn, J. (2013). Psychological responses and resilience of people and communities impacted by the Deepwater Horizon oil spill. *Transactions of the American Clinical and Climatological Association* 124: 191–201.

Mpelasoak, F., Jones, K. & Bates, B. (2007). Comparison of suitable drought indices for climate change impacts assessment over Australia towards resource management. *International Journal of Climatology* 28(10): 1283–1292.

Mulvaney, K. (2015). *Turning the Tide: Ending Overfishing in North-Western Europe*. Washington, D.C.: Pew Charitable Trusts.

Myer, R. & Worm, B. (2003). Rapid worldwide depletion of predatory fish communities. *Nature* 423: 280–283.

NOAA. (2015a). What was killing the young right whales? New research finds a suspect. National Oceanic and Atmospheric Administration. https://phys.org/news/2015-10-young-whales.html.

NOAA. (2015b). Pacific bluefin tuna. National Oceanic and Atmospheric Administration. https://www.fisheries.noaa.gov/species/pacific-bluefin-tuna.

NOAA. (2017). Operational assessment of 19 northeast groundfish stocks, updated through 2016. National Oceanic and Atmospheric Administration. https://repository.library.noaa.gov/view/noaa/16091.

NOAA. (2018). Imports and exports of fishery products annual summary, 2017. National Oceanic and Atmospheric Administration. https://www.st.nmfs.noaa.gov/Assets/commercial/trade/Trade2017.pdf.

Ndungu, K., Beylich, B., Staalstrøm, A., Øxnevad, S., Bergea, J. Braaten, H., *et al.* (2017). Petroleum oil and mercury pollution from shipwrecks in Norwegian coastal waters. *Science of the Total Environment* 593–594: 624–633.

Neldner, V., Laidlaw, M., McDonald, K., Mathieson, M., Melzer, R., Seaton, R. *et al.* (2017). Scientific Review of the Impacts of Land Clearing on Threatened Species in Queensland. Brisbane, QLD: Queensland Government.

Nickerson, A. (2016). New science puts decline of Pacific bluefin at 97.4 percent. Washington, D.C.: Pew Charitable Trusts. https://www.

pewtrusts.org/en/research-and-analysis/articles/2016/04/25/new-science-puts-decline-of-pacific-bluefin-at-974-percent.

Osofsky, H., Osofsky, J. & Hansel, T. (2011). Deepwater Horizon oil spill: mental health effects on residents in heavily affected areas. *Disasters Medicine and Public Health Preparedness* 5(4): 280–286.

OSPAR. (2003). OSPAR Integrated Report 2003 on the Eutrophication Status. London: OSPAR Commission.

Pacific Bluefin Tuna Status Review Team. (2017). Status review report of Pacific bluefin tuna (*Thunnus orientalis*). Report to National Marine Fisheries Service, West Coast Islands Regional Office. ftp://ftp.library.noaa.gov/noaa_documents.lib/NMFS/SWFSC/TM_NMFS_SWFSC/NOAA-TM-NMFS-SWFSC-587.pdf.

Paerl, H. & Paul, V. (2012). Climate change: links to global expansion of harmful cyanobacteria. *Water Research* 46(5): 1349–1363.

Paerl, H., Gardner, W., McCarthy, M., Peierls, B. & Wilhelm, S. (2014). Algal blooms: noteworthy nitrogen. *Science* 346(6206): 175.

Palinkas, L. (2012). A conceptual framework for understanding the mental health impacts of oil spills: lessons from the *Exxon Valdez* oil spill. *Psychiatry and Interpersonal Processes* 75(3): 203–222.

Palinkas, L., Petterson, J., Russell, J. & Downs, M. (1993). Community patterns of psychiatric disorders after the *Exxon Valdez* oil spill. *American Journal of Psychiatry* 150: 1517–1523.

Parsons, M., Settlemier, C. & Bienfang, P. (2010). A simple model capable of simulating the population dynamics of Gambierdiscus, the benthic dinoflagellate responsible for ciguatera fish poisoning. *Harmful Algae* 10(1–2): 71–80.

Pauly, D., Worm, B., Sumaila, U., Jackson, J., Rosenberg, A., Safina, C. *et al.* (2007). Letter to WTO Director-General Lamy. May 24, 2007. https://grist.files.wordpress.com/2007/06/scientists_letter_final_5_24_07.pdf.

Payne, J., Bush, A., Heim, N., Knope, M. & McCauley, D. (2016). Ecological selectivity of the emerging mass extinction in the oceans. *Science* 353(6305): 1284–1286.

Paz, B., Daranas, A., Norte, M., Riobo, P., Franco, J. & Fernández, J. (2008). Yessotoxins, a group of marine polyether toxins: an overview. *Marine Drugs* 6(2): 73–103.

Pennington, B. (2010). On par: the burden and boon of lost golf balls. *New York Times*. https://archive.nytimes.com/query.nytimes.com/gst/fullpage-9500E1DE163FF930A35756C0A9669D8B63.html.

Pennotti, R., Scallan, E., Backer, T. & Angulo, F. (2013). Ciguatera and scombroid fish poisoning in the United States. *Foodborne Pathology and Disease* 10(12):1059.

Perrault, J., Bauman, K., Greenan, T., Blum, P., Henry, M. & Walsh, C. (2016). Maternal transfer and sublethal immune system effects of brevetoxin exposure in nesting loggerhead sea turtles (*Caretta caretta*) from western Florida. *Aquatic Toxicology* 180: 131–140.

Pet-Soede, L. & Erdmann, M. (1998). An overview and comparison of destructive fishing practices in Indonesia. *SPC Live Reef Fish Bulletin* 4: 28–36.

Phillips, O., Aragão, L., Lewis, S., Fisher, J., Lloyd, J., López-González, G. *et al.* (2009). Drought sensitivity of the amazon rainforest. *Science* 323: 1344–1347.

Porteus, C., Hubbard, P., Webster, T., van Aerle, R., Canário, A., Santos, E. *et al.* (2018). Near-future CO2 levels impair the olfactory system of a marine fish, *Nature Climate Change* 8: 737–743.

Qin, B., Zhu, G., Gao, G., Zhang, Y., Li, W., Paerl, H. *et al.* (2010). A drinking water crisis in Lake Taihu, China: linkage to climatic variability and lake management. *Environmental Management* 45(1): 105–112.

Queensland Government. (2017). Reef protection regulations. https://www.qld.gov.au/environment/agriculture/sustainable-farming/reef/reef-regulations.

Quinton, A. (2016). Massive public works project will help clean Sacramento River. Capital Public Radio. http://www.capradio.org/articles/2016/02/16/largest-public-works-

project-in-sacramento-history-will-help-clean-sacramento-river/.
Rabalais, N., Turner, R., Díaz, R. & Justić, D. (2009). Global change and eutrophication of coastal waters. *ICES Journal of Marine Science* 66(7): 1528–1537.
Ralston, D., Keafer, B., Brosnahan, M. & Anderson, D. (2014). Temperature dependence of an estuarine harmful algal bloom: resolving interannual variability in bloom dynamics using a degree day approach. *Limnology & Oceanography* 59(4): 1112–1126.
Reichelt-Brushett, A. & Harrison, P. (2005). The effect of selected trace metals on the fertilization success of several scleractinian coral species. *Coral Reefs* 24(4): 524–534.
Roberts, J. (2000). Seabed photography, environmental assessment and evidence for deep-water trawling on the continental margin west of the Hebrides. *Hydrobiologia* 441(1): 173–183.
Roberts, J. (2019). Cold-water coral reefs. In: J. Cochran, H. Bokuniewicz & P. Yager (Eds.), *Encyclopedia of Earth Sciences*, 3 edn, vol. 1, pp. 675–687. Amsterdam: ScienceDirect.
Rodrigue, D., Etzel, R., Hall, S., de Porras, E., Velasquez,O., Tauxe, R., *et al.* (1990). Lethal paralytic shellfish poisoning in Guatemala. *American Journal of Tropical Medicine and Hygiene* 42(3): 267–271.
Rookmaaker, K. (Ed.). (2006). Darwin's *Beagle* diary (1831–1836). Darwin Online. http://darwin-online.org.uk/content/frameset?pageseq=1&itemID=EHBeagleDiary&viewtype=text.
Rosenberg, A. (2003). Managing to the margins: the overexplotation of fisheries. *Frontiers in Ecology and the Environment* 1(2): 102–106.
Ross, H. (2020). Precious nightcap oak survives the carnage of bushfire with show of uplifting resilience. *ABC News*. https://www.abc.net.au/news/2020-06-15/ancient-nightcap-oak-natures-great-survivor/12353034.
Saccaro, N., Mation, L. & Sakowski, P. (2015). Impacts of Deforestation on the Incidence of Diseases in the Brazilian Amazon. Rio de Janeiro: Institute for Applied Economic Research.
SakhaliaNet. (2014) Sahara territory: Mali ethnography. http://freevst.x10.mx/sahara/mali%20ethnography.htm.
Schiefelbein, S. & Cousteau, J. (2007). *The Human, the Orchid and the Octopus: Exploring and Conserving Our Natural World*. New York: Holtzbrinl.
Scott, A. (2017). What's inside golf balls, and can chemistry make them fly farther? *Chemical & Engineering News*. https://cen.acs.org/content/cen/articles/95/web/2017/06/s-inside-golf-balls-chemistry.html.
Selman, M., GreenHalgh, S., Diaz, R. & Sugg, Z. (2008). Eutrophication and hypoxia in coastal areas: a global assessment of the state of knowledge. WRI Policy Note, number 1. Washington, D.C.: World Resources Institute.
Shen, Li, Huiping, X., Guo, X. & Li, M. (2011). Characteristics of large-scale harmful algal blooms (HABs) in the Yangtze River Estuary and the adjacent East China Sea (ECS) from 2000 to 2010. *Journal of Environmental Protection* 2: 1285–1294.
Shroder, T. & Konrad, K. (2011). *Fire on the Horizon: The Untold Story of the Gulf Oil Disaster*. New York: HarperCollins.
Simpson, S., Munday, P., Wittenrich, M., Manassa, R., Dixson, D., Gagliano, M. *et al.* (2011). Ocean acidification erodes crucial auditory behavior in a marine fish. *Global Change Biology* 7: 917–920.
Sinclair, M. (1997). Prologue. Recruitment in fish populations: the paradigm shift generated by ICES Committee A. In: R. Chambers & E. Trippel (Eds.), *Early Life History and Recruitment in Fish Populations*, pp. 1–28. London: Chapman.
Siriwardena, L., Finlayson, B. & McMahon, T. (2006). The impact of land use change on catchment hydrology in large catchments: the Comet River, Central Queensland, Australia. *Journal of Hydrology* 326(1): 199–214.

Slezak, M. (2018). "Global deforestation hotspot": 3m hectares of Australian forest to be lost in 15 years. *Guardian*. https://www.theguardian.com/environment/2018/mar/05/global-deforestation-hotspot-3m-hectares-of-australian-forest-to-be-lost-in-15-years.

Strelitz, J., Keil, A., Richardson, D., Heiss, G., Gammon, M., Kwok. R. *et al.* (2018). Self-reported myocardial infarction and fatal coronary heart disease among oil spill workers and community members 5 years after Deepwater Horizon. *Environmental Research* 168: 70–79.

Sweetman, A., Thurber, A., Smith, C., Levin, L., Mora, C., Wei, C.-L. *et al.* (2017). Major impacts of climate change on deep-sea benthic ecosystems. *Elementa, Science of the Anthropocene* 5: 4.

Thomas, N. & Nigam, S. (2018). 20th-century climate change over Africa: seasonal hydroclimate trends and Sahara Desert expansion. *Journal of Climate* 31: 3349–3370.

Turubanova, S., Potapov, P., Tyukavina, A. & Hansen, M. (2018). Ongoing primary forest loss in Brazil, Democratic Republic of the Congo, and Indonesia. *Environmental Research Letters* 13: 074028.

Van Cauwenberghe, L., Vanreuse, A., Mees, J. & Janssen, C. (2013). Microplastic pollution in deep-sea sediments. *Environmental Pollution* 182: 495–499.

Van Egmond, H., Van Apeldoorn, M. & Speijers, G. (2004). Marine biotoxins. Rome: Food and Agriculture Organization of the United Nations.

VOA. (2017). Chinese overfishing threatens West African economies. *Voice of America*. https://www.voanews.com/a/chinese-overfishing-threatens-west-african-economies/3947764.html.

Wamsley, L. (2019). Trump administration moves to roll back offshore drilling safety regulations. *NPR*. https://www.npr.org/2019/05/03/720008093/trump-administration-moves-to-roll-back-offshore-drilling-safety-regulations.essed.

Wang, J. & Douglas, A. (1998). Nitrogen recycling or nitrogen conservation in an alga-invertebrate symbiosis? *Journal of Experimental Biology* 201: 2445–2453.

Weber, A., Weber, M. & Savoca, M. (2019). Quantifying marine debris associated with coastal golf courses. *Marine Pollution Bulletin* 140: 1–8.

Wei-Haas, M. (2018). Red tide is devastating Florida's sea life. Are humans to blame? *National Geographic*. https://www.nationalgeographic.com/environment/2018/08/news-longest-red-tide-wildlife-deaths-marine-life-toxins/.

Weiser, M. (2008). Ammonia from Sacramento waste could hurt delta ecosystem. *Sacramento Bee*. http://www.klamathbasincrisis.org/delta/ammoniaSacWaste060108.htm.

Whetton, P., Fowler, A., Haylock, M. & Pittock, A. (1993). Implications of climate change due to the enhanced greenhouse effect on floods and droughts in Australia. *Climatic Change* 25(3–4): 289–317.

Wiese, M., D'Agostino, P., Mihali. T., Moffitt, M. & Neilan, B. (2010). Neurotoxic alkaloids: saxitoxin and its analogs. *Marine Drugs* 8(7): 2185–2211.

Williams, C., Dittman, A., McElhany, P., Busch, D., Maher, M., Bammle, T. *et al.* (2018). Elevated CO2 impairs olfactory-mediated neural and behavioral responses and gene expression in ocean-phase coho salmon (*Oncorhynchus kisutch*). *Global Change Biology* 25(3): 963–977.

Wilson, C., Sastre, A., Hoffmeyer, M., Rowntree, V., Fire, S., Santinelli, N. *et al.* (2015). Southern right whale (*Eubalaena australis*) calf mortality at Península Valdés, Argentina: are harmful algal blooms to blame? *Marine Mammal Science* 32(2): 423–451.

Wingfield-Hayes, R. (2019). Japan sushi tycoon pays record tuna price. *BBC News*. https://www.bbc.com/news/world-asia-46767370.

Worland, J. (2016). A most beautiful death: an underwater investigation of coral bleaching in the South Pacific. *Time*. https://time.com/coral/.

Wright, S., Thomas, R. & Galloway, T. (2013). The physical impacts of microplastics on marine organisms: a review. *Environmental Pollution* 178: 483–492.

WWF. (2011a). Living Forests Report. World Wildlife Fund. https://wwf.panda.org/our_work/our_focus/forests_practice/forest_publications_news_and_reports/living_forests_report/.

WWF. (2011b). Race for the last bluefin. https://wwfeu.awsassets.panda.org/downloads/med_tuna_overcapacity.pdf.

Yasumoto, T., Oshima, Y., Sugawara, W., Fukuyo, Y., Oguri, H., Igaashi, T. *et al.* (1980). Identification of *Dinophysis fortii* as the causative organism of diarrhetic shellfish poisoning. *J-Stage* 46(11): 1405–1411.

Zhao, B., Wang, S., Liu, H., Xu, J., Kilmont, Z., Hao, J. *et al.* (2010). NOx emissions in China: historical trends and future perspectives. *Atmospheric Chememistry and Physics* 13: 9869–9897.

6

The threat of ecocrises interaction

As the world's population rises toward nine billion by mid-century, the inevitable logic of exponential growth in consumption is now hitting the real limits of global ecosystem and resource availability. The immediate pressure points... are converging rapidly in a manner not previously experienced. But those pressure points constitute only the tip of the broader global-sustainability iceberg

Ian Dunlap (2008, p. xi)

6.1 Compounded perturbations and ecological surprises

Ecocrises are of unique importance for three central reasons: 1) they are harm amplifiers; 2) they affect the health of the environment and the environment of health; and 3) their occurrence is often unexpected and not planned for. Each of these points is elaborated upon in this chapter through a discussion of a number of ecocrises interactions involving hazardous pollutants and polluted places, and the threats they pose to planetary and human health in a changing world.

From an ecological standpoint, the release or dumping of a hazardous pollutant constitutes a disturbance or perturbation of a natural system. White & Pickett (1985, p. 7) define a disturbance as "any relatively discrete event in time that disrupts ecosystem, community, or population structure and changes resources, substrate availability, or the physical environment." These events produce post-disturbance changes of various levels of magnitude of severity up through the occurrence of significant ecocrises, such as climate change at a global scale or collapse of a coral reef at a more local or regional scale. Multiple or compounded perturbations, be they simultaneous or sequential, yield ecological surprises. The concept of ecological surprise refers to unanticipated behavior in ecological systems. Filbee-Dexter *et al.* (2017) define it as a situation in which human expectations about natural system behavior deviate from past observed ecosystem behavior. Examples include the collapse of coastal fisheries (Hilborn *et al.* 2003) and outbreaks of insects that cause widespread mortality of old-growth forests (Safranyik & Wilson 2006). Disturbance interactions are of pronounced concern in ecology because of their potential to cause unexpected results (Buma & Wessman 2012). Paine *et al.* (1998, p. 542) note that "[a] world of ever-more-pervasive anthropogenic impacts on natural communities coupled with the increasing certainty of climatic response to human activities suggests that compounded perturbations and ecological surprises will become more common."

Ecosystem Crises Interactions: Human Health and the Changing Environment, First Edition. Merrill Singer.
© 2021 John Wiley & Sons, Inc. Published 2021 by John Wiley & Sons, Inc.

6.2 Climate change and polluted Superfund sites

Superfund sites are areas within the United States that have been contaminated by hazardous waste or pollutants, placed on the National Priorities List (NPL) by the Environmental Protection Agency (EPA), and designated as candidates for cleanup because of the long-term risks they pose to human health, the surrounding environment, or both.

The Superfund program, as it is informally known, was created by Congress in 1980 with the enactment of the Comprehensive Environmental Response, Compensation, and Liability Act, designed to address the most intensely contaminated places in the country. In responding to possible Superfund sites, the EPA is supposed to collaborate with affected local communities, scientists, officials from state and tribal governments, other federal agencies, and the potentially responsible parties (PRPs) who caused the problem in the first place. It is tasked with identifying sites, testing for levels and kinds of contamination, developing plans for amelioration, monitoring progress, and completing decontamination. PRPs are the current or former owners or operators of a facility at a site, transporters of a hazardous substance who selected the site for disposal, or agents who coordinate treatment or disposal of hazardous substances at the site. Sometimes, they are unwilling or unable to assume responsibility for cleanup, they have gone out of business, or they cannot be located. In such cases, the EPA or the relevant state or tribal government will undertake the cleanup with financial support from the Superfund program. The time between the identification of a hazardous site and actual cleanup can be 10 or more years. Over 800 substances (excluding petroleum and natural gas) that may be found at potential sites have been identified as hazardous. Analysis of flood zone maps, census data, and EPA records by the Associated Press (Dearen *et al.* 2017) concluded that about 2 million people in the United States live within a mile of a Superfund site. There are more than 300 sites across the country that are located in areas prone to flooding storms or vulnerable to sea-level rise caused by climate change.

One of the first actions taken under the Superfund Act was the cleanup of dioxin-contaminated waste oil that had been dumped as part of efforts to control dust on dirt roads in the rural community of Times Beach, Missouri, on the flood plain of the Meramec River. During the late 1960s, the Northeastern Pharmaceutical and Chemical Company (NEPACCO) had begun operating out of a facility located near Verona in southwestern Missouri. This facility was owned by Hoffman-Taff, a company that produced the Agent Orange herbicide used during the Vietnam War. In 1979, a former NEPACCO employee reported the burial of toxic waste on a farm located about 7 miles from Verona. On investigation, 90 drums, all of which were corroded and leaky, were unearthed at the site, 11 of which had dioxin concentrations as high as 2000 ppm. Russell Martin Bliss, the owner of a small local waste oil business in Missouri, was contracted by NEPACCO to remove heavily concentrated dioxin still bottoms (thick, oily residues) from its production facility. He collected six truckloads (nearly 18 500 gallons) of the waste and took it to his storage facility, where he mixed it into tanks containing used crankcase oils. This was the material he used when he was hired to control dust on the dirt roads of Times Beach.

Because of the extent of the contamination, the EPA purchased the entire town for $35 million and bulldozed it to the ground. Huge incinerators were erected to burn all of the homes and the belongings left behind by the 2000 people who had lived there, as well as the topsoil from the dirt roads (Hernan 2010). Also burned were over 265 000 tons of dioxin-contaminated materials collected from across Missouri. In 1985, the State of Missouri officially dis-incorporated the city of Times Beach. It is now covered by Route 66 State Park.

Times Beach is not alone. The EPA (2018a) has found that "[t]housands of contaminated sites exist nationally due to hazardous waste being dumped, left out in the open, or otherwise improperly managed. These sites, which have polluted the surrounding soil, water and air include manufacturing facilities, processing plants, landfills and mining sites." Hazardous sites are scattered across the United States in every state except North Dakota. New Jersey, which is the densest by population, has 114 Superfund sites, more than any other state. Research by Kirpich & Leary (2017) found that counties in Florida—the state with the sixth highest number of Superfund sites—which contain a Superfund site have a 6 percent higher cancer incidence rate than those without one. Similarly, based on an assessment of available research, the Agency for Toxic Substances and Disease Registry (ATSDR 2017) found elevated cancer risks at Marine Base Camp Lejeune, a Superfund site in North Carolina, among people exposed to drinking water from wells where an array of hazardous chemical had been dumped, compared to those who were not unexposed.

While leakage from Superfund sites can occur at any time, the risks are considerably enhanced under ecocrisis conditions, such as climate-enhanced hurricanes or other intense storms. When Hurricane Katrina barreled into the U.S. Gulf coast (Fig. 6.1), three Superfund sites in and around New Orleans were flooded, including the Agriculture Street Landfill. This site, where city residents had long dumped their trash, the city had discarded ash from waste incinerators, and debris from Hurricane Betsy was deposited in 1965, was developed for low-income residential use and promoted by city officials for first-time black home buyers in the late 1970s. When contamination became evident, the EPA excavated peoples' yards, installed a layer of protective geotextile sheeting, and replaced the top 2 feet of soil. During the events of Katrina, however, storm surge, funneled by the Mississippi River–Gulf Outlet Canal, flooded the neighborhood. The ability of standing floodwater to leach into the soil and release buried toxins became an issue of considerable concern (Eilperin 2009). In part of the neighborhood occupied by a townhouse complex, the EPA found dangerously high levels of benzopyrene, a known source of carcinogenic and mutagenic metabolites (Lyttle 2003).

In 2014, under the Obama administration, the EPA (2014, p. 4) affirmed concern about the climate change–Superfund issue in reporting that "Scientific evidence demonstrates that the climate is changing at an increasingly rapid rate, outside the range to which society has adapted in the past. These changes can pose significant challenges to the EPA's ability to fulfill its mission." A national assessment of the increased threat to Superfund sites posed by the contribution of climate change to rising sea levels, flooding, and stronger hurricanes found that of the more than 1600 sites reviewed, 521 were in 1-in-100-year or 1-in-500-year flood zones (Dearen *et al.* 2017). This terminology means that in any given year there is a 1-in-100 or 1-in-500 probability of a flood reaching a certain level. These calculations, based on a statistical technique called frequency analysis, can be used to estimate the probability of the occurrence of a specified precipitation event. However, because of climate change, the Memorial City section of West Houston, for example, which old models predict would flood once every 500 years, flooded in 2009, 2015, 2016, and 2017. Moreover, the assessment concluded that almost 50 Superfund sites in coastal areas were vulnerable to rising sea levels. The threats to the health of humans and wildlife from rising waters that inundate Superfund sites vary based on the specific contaminants and the concentrations involved. The danger is that floodwaters will carry toxins away from the site and distribute them over a wide area.

In 2017, when Hurricane Harvey swept up the Gulf of Mexico and inundated Houston, 7 of the 12 dangerously polluted Superfund sites in and around the city were flooded. Throughout Texas, the EPA reported 13 Superfund sites flooded by Harvey were put at risk for leakage. One of

Fig. 6.1 Hurricane Katrina flooding. *Source:* cgcolman/Pixabay.

these, the Patrick Bayou site in Deer Park, is located about 15 miles from downtown Houston, lying between the Houston Ship Channel and a thicket of petrochemical facilities. A highway separates it from residential dwellings. The site is susceptible to flooding because of its proximity to the channel. Sediments in the bayou contain various unsafe heavy metals, including arsenic, cadmium, chromium, copper, lead, manganese, mercury, nickel, selenium, and zinc. Pesticides and petrochemicals like polynuclear aromatic hydrocarbons and polychlorinated biphenyls (PCBs) add to the toxic load, according to the findings of EPA (2018b) testing. Three companies, Shell, Lubrizol Corp., and Occidental Chemical Corp., are the site's identified PRPs.

In the aftermath of Hurricane Harvey, the EPA reported that no toxins had leaked from the Patrick Bayou site while it was flooded. The same was not true in the case of the San Jacinto Waste Pits Superfund site, which juts into the San Jacinto River and is situated about 5 miles from Patrick Bayou. This site, which contains cancer-causing dioxins and has long been a source of complaint among local residents, was submerged by Harvey floodwaters. In September 2017, the EPA announced it had been damaged by flooding and that toxic waste had leaked from the pits. Testing revealed that surrounding areas contained significantly higher-than-normal dioxin levels (EPA 2017). International Paper and Industrial Maintenance Corporation are the identified PRPs: the former operated a paper mill that produced large quantities of sludge-like waste, which it contracted the latter to deposit in pits constructed alongside the river. Many residents from nearby communities had already testified during an EPA hearing about the extent of the harm being caused by pollution leaking to the river from the pits, including rare cancers, tumors, and fertility issues. They had also expressed their concerns, subsequently validated by Harvey, about what would happen at the pits during a hurricane (DeBenedetto 2016).

Just over 1000 miles southeast of Houston is Miami-Dade County in Florida (Fig. 6.2), an area also facing a climate change/Superfund ecocrisis. That, of course, is not the image that city and county officials want to project. Rather, they describe Miami-Dade, which in addition to Miami includes Hialeah, Coral Gables, Miami Beach, and 30 other cities, towns, and villages, as an international tourist destination that attracted almost 9 million visitors between January 1 and June 30, 2018, about equally divided between those coming from overseas and those from other parts of the country (Turner 2018). The leisure, hotel, tourism, and hospitality industries are big business in Miami-Dade as in other areas in Florida, and they want to keep growing. Warm weather, celebrated beaches, vibrant nightlife, and the availability of year-round outdoor activities are all touted by the tourism trade. Promoters aggressively market Miami-Dade as an appealing site for business meetings, sales events, conferences, expositions, and trade shows. The county also attracts many new residents. Population growth is at around 8 percent, about twice the U.S. national average. Moreover, in the second quarter of 2016 alone, 1.6 million square feet of office space and 1.8 million square feet of retail space were under construction (Rugget 2017).

Fig. 6.2 Miami-Dade County, Florida. *Source:* kendallpools/Pixabay.

But, in a time of climate change, there is another, less appealing side to the Miami-Dade story, some aspects of which have already been discussed. In 2018, red tide began appearing along the shore areas of the county, Florida's busiest coast. It remained at moderate levels, though high enough to trigger human respiratory symptoms and fish kills. Moreover, many people have heard about the risk southern Florida faces from ocean rise. Miami-Dade is considered a national hot spot for the health, social, and economic risks of a rising ocean. As Jeff Goodell, author of *The Water Will Come: Rising Seas, Sinking Cities, and the Remaking of the Civilized World* (2017), remarks, Miami, a mostly flat, low-lying place surrounded by water, "is kind of the poster child for a major city in big trouble . . . Miami as we know it today—there's virtually no scenario under which you can imagine it existing at the end of the century . . . It may be some smaller version of Miami that incorporates platform houses and floating structures" (quoted in Loria 2018b).

Global sea-level rise during the 20th century was faster than in any previous century for at least 2800 years (Kemp *et al.* 2011; Kopp *et al.* 2016). Reflecting regional differences in the level of acceleration in this process due to both natural and anthropogenic factors, it is estimated that the global average increase in ocean levels since 1870 is 8 inches, while in South Florida the ocean has risen by 12 inches (Church & White 2006; Tompkins & Deconcini 2014). Since 2006, sea level around Miami-Dade has risen more rapidly than in previous years—and considerably more rapidly than the global rate—going from 0.012 inches per year in 1998–2005 to 0.35 inches per year in

2006–2015 (Wdowinski *et al.* 2016). The numbers seem small, but they add up over the decades. Moreover, as the research of Zhang (2011, p. 537) shows, flooding "[i]nundation processes are non-linear: inundation is gradual before reaching a threshold, and speeds up rapidly afterwards due to the regional topography."

Natural factors that increase vulnerability to ocean changes in South Florida include the region's peninsular geomorphology, limestone geology, and low elevation. In everyday usage, the term "vulnerability" refers to the degree to which an organism, a community, an ecosystem, or another entity is likely to experience harm under a given set of conditions. With reference to the threat potential of climate change, the Fifth Assessment Report of the Intergovernmental Panel on Climate Change (IPCC 2014) defines it in terms of three factors: exposure to climate change effects; degree of sensitivity to changes in climate; and adaptive capacity. In the case of a peninsular area like South Florida, exposure is enhanced because it is surrounded by water on three sides. Climate change and the flooding of populated peninsular areas has become an issue of growing concern internationally (Kim & Choi 2011; Shahmohammadi-Kalalagh 2013). Limestone, the sedimentary rock on which Miami-Dade sits, is also a factor, because of its porousness, which is a product of its composition. The primary component of limestone is calcium carbonate, as in seashells and the skeletal structure of hard corals. This is because limestone forms through the accumulation and lithification of shell, coral, algal, and fecal debris, a process in which sediments are compacted under pressure, expel any entrapped fluids, and gradually become solid rock. Important but variable features of limestone are porosity (the proportion of the rock composed of the pores, voids, and cracks through which water can flow) and permeability (the rate of flow of water through the rock), as these control how water moves when in contact with a limestone area. Some limestone is so porous and permeable that a river flowing over it disappears as the water sinks into and through the limestone bed. This is a feature of both the River Manifold in the northern United Kingdom and the Popo Agie River in Wyoming.

With reference to elevation, even at the current sea level, over 10 percent of the land area of Miami-Dade sits at less than 1 foot above the ocean surface, while almost 20 percent is at less than 2 feet and another 25 percent is at less than 3 feet (Zhang 2011). This is why even very high tides, known as king tides, can cause flooding on storm-free days. In Miami Beach, the frequency of tide-induced flooding jumped by more than 400 percent between 2006 and 2013 (Wdowinski *et al.* 2016), a reflection of the effects of climate change. Though the timing of king tides is determined by the relative positions of the sun, the moon, and Earth, rising seas intensify their effects. As a result, in cities like Miami, "neighborhoods that flood on sunny days now didn't flood like this decades ago, even during high tides" (Loria 2018a). According to the National Centers for Environmental Information (NCEI 2018), king tide "flooding is increasing in frequency, depth and extent in many areas of the U.S. due to ongoing increases in local relative sea level. For example . . . the Southeast Atlantic coast is currently experiencing the fastest rate of increase in annual high tide flood days, with more than a 150 percent increase since 2000 predicted in 2018 at most locations."

In addition to climate change, an anthropogenic factor of importance is the population density in Miami-Dade, a county that has more people living less than 4 feet above sea level than any other place in the nation except for parts of Louisiana. The same tourist and land development industries that are attracting people to South Florida are contributing to the climate change that is putting cities like Miami at risk. This is the case for a number of reason, but one that has been less of a focus until recent years is the role of air traffic. As Baer (2019) points out, airplane flights, which emit nitrous oxide and other exhaust fumes, are one of the fastest growing sources of greenhouse gas emissions. In

Cancel the Apocalypse, Andrew Simms (2013) notes that research conducted at the Massachusetts Institute of Technology about the global airline industry found that 2000 airlines were operating around the world, serving 3700 airports, offering 28 million scheduled flights per year to 2 billion passengers, one-third of which involved flights to, from, or in the United States. Miami International Airport alone, the primary airport serving the Miami area, has over 1000 daily flights connecting it to 167 other airports in the United States and abroad. In 2018, over 40 million passengers traveled through Miami International, making it the 13th busiest U.S. airport and the 40th busiest worldwide by total passenger traffic. Moreover, in terms of international passenger levels, it is the 3rd busiest airport in the United States (Miami International Airport 2018).

King tide risk is always a social issue, as some people are more vulnerable to flooding than others. While it is the wealthier U.S. coastal areas that have been the hardest hit by flooding, floodwaters also push inland into less affluent communities. In these areas, the consequences of climate change and sea-level rise may be particularly damaging because low-income households tend to be less resilient to environmental disruptions, and they are at gravest risk of their needs being overlooked by government, which is more interested in saving the millions of dollars invested in developing high-priced waterfront real-estate.

There are a range of projections concerning the level of ocean rise likely to occur in South Florida in the coming decades. Calculations by the U.S. Army Corps of Engineers (Union of Concerned Scientists 2016) suggest a rise above current levels of 15 inches by the year 2045. As already noted, geology is a factor in the climate change/Superfund ecocrisis in Miami-Dade. At issue, as Flavelle (2018) indicates, is the rock structure beneath the metropolitan area:

> Miami-Dade is built on the Biscayne Aquifer, 4000 square miles of unusually shallow and porous limestone whose tiny air pockets are filled with rainwater and rivers running from the [Everglades] to the ocean. The aquifer and the infrastructure that draws from it . . . combine to form a giant but fragile machine. Without this abundant source of fresh water, made cheap by its proximity to the surface, this hot, remote city could become uninhabitable . . . Climate change [however] is slowly pulling that machine apart.

Drinking water supplied to homes and other customers by county authorities is drawn from specially designed production wells. The water that feeds these wells is primarily pumped from the Biscayne Aquifer, supplemented by the deeper Floridan Aquifer. In that it is so close to the land surface, as well as being one of the most permeable aquifers in the world, the Biscayne Aquifer directly interacts with natural and artificial bodies of surface water, including the Atlantic Ocean, streams, lakes, canals, and reservoirs. This hydraulic interconnection puts it at risk of contamination. Saltwater intrusion, an ever-present threat, can occur by way of artificial coastal and tidal canals if the water table is lowered through pumping for drinking and hygienic use by an expanding population and for farm irrigation, making untreated water drawn from the aquifer undrinkable. Another issue, as Flavelle (2018) points out, is that:

> As the ocean rises, salt water is being pushed into the limestone, forming a wall of brine that's creeping inland along the aquifer's floor. The county's wells are essentially giant straws drawing water from 60 feet to 80 feet beneath the ground. As the saltwater front advances westward across the aquifer, reaching each of those intake valves and enveloping them in saline water, it risks rendering them useless in succession.

Contamination through runoff from ground surfaces is an additional risk faced by the aquifer. It is threatened by contaminants from multiple sources, including: drainage from a growing number of local septic tanks; runoff from landfills, waste dumps, and sewage; pesticide and fertilizer runoff from farms and lawns; gasoline and toxic chemicals like benzene used in limestone mining; and hazardous materials from leaking or flooded Superfund sites.

Miami's first significant water treatment facility was built in the town of Hialeah in 1924. This critical facility, built of coral rock, cleanses water before it enters the municipal water system. Several blocks from the Hialeah plant and about 2 miles north of Miami International Airport, beneath what is now a maintenance yard for the county's Metrorail system, is a 1.2 acre Superfund site that the EPA ranks as the second most hazardous such location in Miami-Dade. From 1966 until 1981, the land was used by a company called Miami Drum Services, whose operations included cleaning and recycling containers used to hold an assortment of toxic chemicals. Residues from the cleaning process were disposed of on-site. Miami Drum was added to the NPL in 1983 after arsenic, cyanide, mercury, nickel, lead, cadmium, chromium, chloroform, and oil were detected contaminating the soil and groundwater. The county removed the contaminated soil and debris, and the EPA continues to monitor the site.

The Continental Cleaners site, meanwhile, is located in the residential Liberty City neighborhood, about 1000 feet from an elementary school. Prior to 1969, the property was used as a gas station, and three underground gasoline storage tanks were reportedly abandoned in place when it went out of business. It was then used for dry cleaning and laundry operations until 2005. Since 2011, the building and property have remained relatively vacant. The EPA has not yet initiated cleanup efforts at the site.

The dangers of the climate change–Superfund site ecocrisis will continue to mount as fiercer storms that bring massive amounts of flooding precipitation and rising ocean levels become the new "normal." This adverse interaction, recognized by some and dismissed by others (e.g., Scott Pruitt, the first head of the EPA under the Donald Trump administration), tends to fall through the cracks of urban and environmental planning. Initially, petroleum, chemical, and corporate taxes imposed by Congress funded much of the budget of the Superfund program, including expensive cleanups in cases where the polluters involved could not or would not pay to ameliorate the environmental problems they caused. With expiration of the Superfund taxes in 1995, the burden of payment shifted dramatically from corporations to individual taxpayers and local communities. As Congressional funding decreased, it became more difficult to oversee and enforce the Superfund program (Anderson 2017). Hostility in the Trump administration toward the EPA suggests that long-term planning to avert the ill effects of climate change–Superfund site interaction is likely to diminish at the very moment it is most critically needed (Sellers *et al.* 2018). Trump's FY19 budget, for example, proposed slashing the Superfund budget to below the level of the last year of the previous administration, and his policies concerning waste sites are seen by critics as providing cover for polluting companies (Stern 2017).

One kind of toxic site at particularly high risk for flooding because of the heavier rains triggered by climate change is that used for the storage of large quantities of coal ash from coal-burning power plants (Colman 2019). The residue ash contains arsenic, selenium, lead, mercury, boron, and other contaminants known to cause cancer, neurological damage, and heart ailments. Electric utilities plants commonly store it in massive landfills or unlined ponds. In 2008, spillage from such sites damaged dozens of homes. Also of concern are rain-caused rising water tables, which seep into the stored ash, contaminating groundwater used for agriculture and drinking.

6.3 Global toxic sites and climate change

Common practices for the disposal and management of hazardous waste have become a global threat that is exacerbated by the disruptive environmental effects of climate change. Poor and illegal waste management have been identified as the most important worldwide causes of contamination of soil and groundwater and, by extension, the human and environmental exposure to the poisonous byproducts of manufacturing, farming, construction, hospitals, and other industries (Nweke & Sanders 2009; Caravanos *et al.* 2013; van Liedekerke *et al.* 2014) Hazardous waste is deposited in the environment as liquids, solids, or sludges that contain noxious chemicals, heavy metals, radiation, or other damaging toxins. While, as the discussion of Superfund sites indicates, wealthy industrialized countries face major hazardous waste threats, "[t]he burden of diseases of waste-related exposures in middle-low income countries is increasing and not sufficiently recognized" (Fazzo *et al.* 2017). Diseases caused by exposure to pollutants and hazardous contaminants "undermine the development of poor countries by reducing the health, intelligence and economic productivity of entire generations" (Landrigan & Fuller 2014, p. 35). The risks of hazardous waste in developing countries "are compounded by a number of unique situations, viz., economic priorities, high burden of infectious diseases, impoverishment, and absence of a regulatory framework for the disposal of toxic chemicals" and other contaminants (Jamail & Davis 1991, p. 365). Inadequate preparedness for climate change, a condition that is linked directly to the global economic causes and historic political problems of underdevelopment, further complicates the risks posed by hazardous waste sites. Dowling *et al.* (2016, p. 701) note that:

> These countries often have limited governmental capacity and few incentives to formally regulate environmentally damaging industries or address contaminated sites. Even in cases where regulations exist, the capacity to manage or enforce laws can be limited.

A leader in the global effort to identify highly contaminated sites is Pure Earth (formerly the Blacksmith Institute), a New York-based nongovernment organization (NGO). Pure Earth (2018) spotlights what it refers to as "one of the biggest, yet most underreported, global problems—toxic pollution." In its assessment, over 100 million people, especially women and children, are at risk in some of the world's worst polluted places (Fuller 2015). As an exemplar of its identification, it initiated a systematic assessment to determine the total number of acutely polluted sites in Ghana. Team members focused on randomly selected geographic quadrats for comprehensive assessment, physically walking all accessible streets to visually identify every existing site. This visual identification was followed by field-based confirmation using portable x-ray fluorescence instruments to test soils for metals. Team members found 72 sites in the sampled quadrats. Extrapolating from this number to the entire country, it was estimated that there were between 1500 and 1900 significantly contaminated sites in Ghana. On a population basis, that is between 31 and 115 contaminated sites per million inhabitants.

E-waste containing dangerous levels of highly damaging dioxins and PCBs is a major component of hazardous waste in Ghana. Researchers from two groups, Ipen and the Basel Action Network, verified this assertion by analyzing eggs laid by free-range chickens that forage in the capital city's Agbogbloshie slum area, home to an estimated 80 000 people, who subsist primarily by recovering and selling copper cable and other metals from smoldering waste piles (Beaumont 2019). The analysis found that an adult eating just a single one of these eggs would consume enough toxins to exceed European Food Safety Authority (EFSA) limits on chlorinated dioxins 220 times over. Other toxic chemicals found in the eggs included PCBs and fire-retardant compounds.

Far away geographically, climatically, and culturally from Agbogbloshie, but sharing its pollution plight, is Norilsk in Central Siberia. Located 1800 miles northeast of Moscow, this is the world's northernmost city, as well as Russia's most polluted urban area. It is somewhat isolated in that there are no roads or railway lines that link it to parts of Russia outside the Arctic. Living above the Arctic Circle, residents experience polar day (or midnight sun) for 68 days each year and the perpetual darkness of polar night for 45 (Shiklomanv & Larulle 2017). In winter, temperatures frequently drop to −58 °F or even lower, forcing commuters to stand in shops instead of at bus stops for fear of freezing to death. An industrial nickel- and copper-producing hub with a population of about 175 000 people, Norilsk is home to the largest heavy metal smelting complex in the world. Named Norilsk Nichol (or Nornickel), it is a previously government-owned but now privatized business—one of the wealthiest in Russia—that has turned vast natural reserves of nickel, copper, and palladium (used in cell phones) into a multibillion-dollar enterprise. It also has produced an enormous ecocrisis. The mining and smelting company ranks first among Russian industrial polluters (Kotov & Nikitina 2010). Smokestacks from its plants annually release tons of carbon, cadmium, copper, lead, nickel, arsenic, selenium, and zinc into the city's smoggy air (Revich 1995). Nyland *et al.* (2017) note that "areas modified by human activities within the city and downwind from local pollution sources [have] experienced an expansion of barren ground due primarily to sulfur-dioxide-laden pollution." Smelters from two of its subsidiary units on the Kola Peninsula contribute additional high levels of sulfur dioxide into the atmosphere, causing acid rain throughout the region. Despite efforts in recent years to curb the environmental release of toxins, the area remains heavily polluted.

What are the health effects of living in a toxic hot spot like Norilsk? Public health research in the region is limited, and because many residents are exposed to both indoor (at work) and ambient contaminant sources, it is difficult to establish the precise impact of living in a place drenched in outdoor air pollutants. Further, the intensely cold weather adds health challenges (e.g., chronic polar tension caused by lowered immune system resistance, polar night syndrome characterized by anxiety, nervousness, drowsiness, and insomnia) that complicate comparisons with the health statics of other regions and national averages (Revich 2010). Summarizing available information, in its 2007 report on the 10 most polluted places in the world, the Blacksmith Institute (Pure Earth 2007) stated that in Norilsk:

> Children suffer from numerous respiratory diseases. Investigations evaluating the presence of ear, nose and throat diseases among schoolchildren revealed that children living near the copper plant were twice as likely to become ill than those living in further districts. Similarly, children living near the nickel plant were shown to become ill at a rate 1.5 times higher than children from further districts. Mortality from respiratory diseases is considerably higher than the average in Russia, accounting for 15.8% of all deaths among children. Premature births and late-term pregnancy complications are also frequent. Sulfur dioxide emissions contribute to chronic diseases of the lungs, respiratory tracts, and digestive systems—and can result in lung cancer.

Norilsk, along with the city of Yakutsk to the east, is the only large city located within the continuous permafrost zone that covers much of the Northern Hemisphere's upper latitudes. Permafrost is defined as ground that stays frozen for at least two consecutive years. With global warming, the permafrost pattern is changing, and rapidly so. Part of the change, at the local level, involves microclimate ground warming and degradation of permafrost soils produced by heated aboveground buildings and structures and underground utility lines. Moreover, Norisk is paved over with asphalt, which "promotes increase in the heat flow into the ground due to summer heating and to winter cooling with snow removal" (Grebenets *et al.* 2012, p. 105). Borehole analysis of samples taken from the center of the city show that the temperature of the permafrost at a depth between 65 and 195 feet has increased over a 30-year period (1955–1985) (Grebenets & Sadowski 1993). In an urban area like Norilsk, the consequence has been the development of significant structural damage. It has been estimated that 60 percent of all buildings in the city have been deformed as a result of climate change shrinking the permafrost zone (Luhn 2016). Studies in Norilsk during the first decade of the 21st century found approximately 250 buildings in the industrial district were badly damaged with cracking walls and sloping floors and 40 apartment buildings had been torn down or were scheduled for demolition because of the risk of collapse (Revich 2010).

At the macroclimate level, atmospheric warming is also having significant effects on permafrost by "deepening the thaw and speeding up the destruction" (Luhn 2016). While climate change has been driving up temperatures worldwide, in the Arctic the rise is twice the global average. Across the region, permafrost temperatures have been rising since at least the 1970s. Geostatistical analysis of shallowground heating due to climate change in Russian permafrost regions shows a substantially higher rate of change during the years 1999–2013 compared to the previous 50 years (Streletsky *et al.* 2015). Thawing and degradation of permafrost have been observed in many locations in the Arctic in recent decades, and these environmental alterations are expected to accelerate in the future as a result of climatic change. Mathematical modeling studies project that "by the mid-21st century, near-surface permafrost in the Northern Hemisphere may shrink by 15%–30%, leading to complete thawing of the frozen ground in the upper few meters, while elsewhere the depth of seasonal thawing may increase on average by 15%–25%, and by 50% or more in the northernmost locations" (Anisimov & Reneva 2006). Permafrost melting allows the greenhouse gases carbon dioxide and methane—produced from decomposed organic matter—to escape to the atmosphere, further heating Earth's climate. Additionally, climate change increases precipitation in the region, adding moisture to the soil, which freezes and expands in winter, slowly destroying the reinforced concrete piles intended to allow buildings to be constructed on permafrost.

In Norilsk, industrial pollution has contributed to a significant environmental and ecohealth crisis, in which: 1) industrial carbon release contributes to global warming, which drives permafrost melting, and which, in the Arctic, further pushes the release of greenhouse gases and more warming in a dangerous positive feedback loop; 2) global warming in turn contributes to arboreal forest fires in the Arctic, which release additional carbon and health-damaging particulate matter into the air; and 3) industrial heavy metal and other pollutant production harms human health while devastating local faunal ecosystems systems. As noted in the opening paragraph of this book, this kind of dynamic linkage of diverse factors exemplifies the extreme dangers of ecocrises interaction.

6.3.1 Camp Century, Greenland

A critical ecocrises interaction of a somewhat different sort may be unfolding in northwest Greenland, still in the (less and less) frozen north, but about 3000 miles east of Norilsk. The site, buried under the Greenland ice sheet, is called Camp Century and dates to 1960 and the Cold War nuclear standoff between the United States and the U.S.S.R. It was built by the U.S. Army Corps of Engineers as part of a secret program, called Project Iceworm, and was envisioned to eventually contain a network of 600 hidden medium-range Minuteman nuclear missiles targeted at the Soviet Union. Under the auspices of the U.S. Army's Polar Research and Development Center, the full plan was to build a vast and fully operational city buried about 26 feet under the ice, as well as a set of launch sites, with the missiles constantly rotating among them (Weiss 2011).

Included in the city were living quarters for about 200 Army personnel, an infirmary and operating room, a pharmacy, a recreation hall, a cold storage warehouse, offices, scientific laboratories, a communication center, machine and maintenance shops, a chapel, a barbershop, a laundry, a post office, a library, and other facilities. This subsurface urban complex was supplied by Thule Air Base, located 150 miles away on the coast over ice roads to the west. The enormously challenging construction project was presented to the public as a year-round Arctic research center designed "to probe deeper and deeper into the secrets of the universe" (U.S. Army Corps of Engineers 1962).

The name Camp Century was actually part of a cover story designed to shield the real mission from the government of Denmark, which at the time had full political, economic, and social jurisdiction over Greenland. The Camp was described to Denmark, a NATO ally, as an effort to test the feasibility of several different approaches for construction under Arctic conditions, installing and exploring the practical challenges encountered in operating a semi-mobile nuclear reactor and a backup diesel power plant (which, in fact, supplied power to the real military mission of the program), and to conduct various supporting scientific experiments (e.g., measuring the speed of glacial movement, assessing melting glacial ice to test for the presence of dormant infectious disease pathogens). The glacial tests sealed the Camp's fate, as they showed that the movement of the ice would crush the underground facilities within a few years. Of particular concern was the major deformation caused to the room carved out to hold the nuclear reactor. Consequently, the unsuccessful project was shut down and the site was abandoned in 1965. An Army team that revisited Camp Century in 1969 found that severe damage had occurred in the underground city, with all structures and equipment being systematically crushed under the extreme pressure of the moving snow; when the program had ended, the nuclear unit was removed but the underground city and various biological, chemical, and radioactive debris were left behind, on the false assumption that they would remain safely encased in ice "for eternity" (Clark *et al.* 1962). During the nuclear reactor's operational life, "a total of 47 078 gallons of radioactive liquid waste was discharged into the icecap" (U.S. Army 1963). Also present were carcinogenic industrial chemicals like PCBs.

The certainty of Camp Century's perpetual burial has significantly weakened since 1965 because of growing understanding of the effects of global warming. The Greenland Ice Sheet is melting, including in places where it previously did not regularly do so. Further, the melting area, where annual meltwater runoff exceeds snowfall, is expanding. In July 2012, 98 percent of the surface area, including the particularly cold polar areas at high altitudes, suffered from melting when a ridge of warm air became stagnant over Greenland (Nghiem *et al.* 2012). Over the coming decades, increased surface melting will affect ever higher ice sheet elevations. Based on an analysis of aerial imagery, Kjeldsen *et al.* (2015) have estimated the total ice mass loss from the Greenland Ice Sheet for three periods: ~75 billion tons per year for 1900–1983; ~74 billion tons per year for 1983–2003; and ~186 billion tons per year for 2003–2010. This pattern of change could expose hazardous waste left behind at the military site. To assess the risk, Colgan *et al.* (2016) developed computer simulations using recent trends in anthropogenic greenhouse gas emissions and found that what glaciologists call the equilibrium line altitude (ELA)—the point at which snow accumulation equals melting—will migrate inland across Camp Century by the end of the 21st century. This will lead to surface exposure and the flow of buried waste toward the Greenland coast.

While Camp Century is somewhat unique at the global level—though in Greenland it is only one of several buried U.S. bases—it signifies the kinds of unexpected ecocrises interaction produced by global climatic and environmental change.

6.4 The ecocrises of unfettered mining

From potash for fertilizers, to zinc for dry cell batteries, to lithium for laptops, to copper for cellphones, mining is the source of many of the essential raw materials needed by industrialized societies. However, it is also the cause of serious levels of health- and ecosystem-damaging pollution of land, air, and water. The environmental contaminants of mining include: toxic inorganic metals like mercury (which some bacteria can convert into the even more toxic chemical methylmercury); cyanide from gold mining; arsenic, which is soluble and winds up in water used for drinking, food preparation, and food crop irrigation; acids from substances like iron sulfide, which occur in the ground in association with metals like gold, copper, and silver; cadmium, a threat to the respiratory tract, liver, and kidneys; and coal dust, a cause of lung cancer, respiratory diseases, and low birth weight in newborns. Examples of unregulated or minimally regulated mining and pollution worst-case scenarios (see later) include the former mining towns of Picher, Wittenoom, and Centralia and the Navajo reservation.

Located in northeast Oklahoma, Picher, which the EPA at one time labeled among the most toxic places in the United States, was once home to 20 000 people and the world's richest lead and zinc mining fields. Vast deposits of these metals were found in a 1188 square mile area running along the southwestern flank of the Ozark Mountains known as the Tri-State District. During the Civil War, surface mining there was used to collect ore for the production of lead shot, which contributed to the more than 250 000 battlefield and post-battle wound-deaths in the deadliest war in U.S. history (Everett 2007). More than half of the lead and zinc consumed during World War I also came from the area. The largest ore strike occurred in 1913 at the site that developed into downtown Picher, a town with humble beginnings as a somewhat rowdy mining camp that practically sprang up overnight. It was a place of evening "fistfights just for the heck of it, plenty of bravado and wasted paychecks" spent at local saloons (Associated Press 2008). Between 1917 and 1947, over $20 billion worth of ore was extracted from Picher Field. U.S. Department of Interior, Bureau of Mines production records indicate that over 180 million tons of crude ore were produced from the Oklahoma portion of the district (EPA 2019). Concentrating mills separated excavated crude ore into lead, zinc, and discard material called tailings. By 1927, there were almost 250 mills operating in Picher Field, and these continued in operation until the late 1930s when centralized milling led to small mill buyout and consolidation. In that decade, the Eagle-Picher Company operated the largest lead-zinc concentrating mill in the world. Other big companies working in the Tri-State area included the Federal Mining and Smelting Company, Childers Mining Company, LaClede Lead and Zinc Company, and American Lead and Zinc Company. The largest smelting company turning ore into a marketable product was the Picher Lead Company (which, like the town, was named for O.S. Picher, its owner; it was later renamed the Eagle Picher Company). Mining continued in Picher until 1967 when Eagle-Picher shut down. In the last few years, the mines were only able to stay open because they were receiving federal subsidies.

One of the threats faced by mines is flooding from groundwater. In 2002, for example, Scotland's last deep coal mine completely flooded when millions of gallons poured in after a dam separating old works from new seams under the River Forth collapsed. The water poured into the mineshafts, located almost 2000 feet belowground, within a few minutes. Had the breakdown occurred 24 hours earlier, hundreds of miners would have drowned. In the case of the Picher area, water was continually seeping in from the Boone aquifer. Because of the presence of groundwater, many mines, including those at Picher, must use dewatering pumps, which can create an environment hazard through acid mine drainage and the dispersal of contaminated water into clean water sources and the general environment. Another danger arises when mines shut down. When the Picher mines closed, water pumping ended and they began to flood. It is estimated that there are over 100000 acre-feet of contaminated mine water under Picher. As the mines filled, oxidized minerals began to dissolve, creating an acidic solution. This acidic water further dissolved minerals within the mines, leading to increases in water concentrations of sulfate and heavy metals, especially cadmium, lead, nickel, and zinc (EPA 2019). Within a few years, this witch's brew of toxins began to seep into the surrounding environment. Acid mine water leaked into Tar Creek—a small stream that runs from Kansas through the Oklahoma mining area—staining its bottom a burned red color. Red stains also began to show up on downstream bridge abutments and surrounding rock in the Neosho River below its convergence with Tar Creek. This contamination threatened the Roubidoux Aquifer, the source of the area's public water supply. Moreover, honeycombed with at least 1400 mineshafts, the ground under Picher contains 70 million tons of waste trailing and 36 million tons of contaminated mill sand and sludge. Other trailing piles, some 13 stories high, have been left in the open air and subject to the wind, which has blown polluted dust into nearby ponds and lakes (Roosevelt 2004). Other wind-blown contaminated dust comes from evaporated flotation ponds dug at milling sites.

Exposure to this level of harmful contamination has taken its toll on the health of residents. As summarized by Roosevelt (2004), the grimmest legacy of the unfettered mining in Picher is the harm done to the town's children:

> As toddlers, they played in sandboxes of chat—the powdery output of mills after ore is extracted from rock. As preteens, they rode their bikes across the gravel mounds and swam in lime-green sinkholes. Their parents used mine tailings [which was sold as a construction product] to make driveways and foundations, never thinking that contaminated dust might blow through the heating ducts of their ranch houses . . . [S]tudies have shown that up to 38% of local children have had high levels of lead in their blood—an exposure that can cause permanent neurological damage and learning disabilities.

The blood lead studies Roosevelt mentions include the 1995–2002 Oklahoma State Department of Health Childhood Lead Poisoning Prevention Program assessment, the 1997 and 2000 Tribal Efforts Against Lead venous blood surveys, and the 1999–2004 Ottawa County Lead Poisoning Prevention Program blood-lead screening (CH2MHill 2006). Since they were conducted, public health officials have come to recognize that far lower levels of blood lead than they had earlier believed are harmful to all body organs, especially in infants and children, and that they are a significant source of learning disabilities. As the Centers for Disease Control and Prevention (CDC 2017) reports, "[n]o safe blood lead level in children has been identified. Even low levels of lead in blood have been shown to affect IQ, ability to pay attention, and academic achievement. And effects of lead exposure cannot be corrected."

In 1983, the EPA added the Picher area to its list of dangerously polluted areas as part of the Tar Creek Superfund Site. Testing for water and ground contamination led to various remedial actions, including plugging wells to stem leakage into the Roubidoux Aquifer and excavation of lead-contaminated soil at more than 2000 homes. Prior to the soil removal, door-to-door interviews were conducted by the EPA to identify community concerns. They revealed that many people were unaware that the area had become a Superfund site or that excavation was planned. People expressed eagerness to see contamination problems at the site resolved because of the potential risk to public health and the environment. The primary health concern voiced by residents was with the safety of the public water supply. Additionally, many parents wanted to know if their children were being exposed to lead and what they could do to prevent such exposure (EPA 2008).

The 2000 U.S. Census registered there were 1640 people living in 621 households in Picher. Thirty percent of those households included children under the age of 18. In June 2006, the Army Corps of Engineers found that most churches and homes and the high school were in serious danger of collapsing due to unrestricted subsurface excavation. A federal buyout was thus offered to the remaining Picher residents. Most accepted the offer and left. In 2007, it was decided that the local high school would only stay open for one more year as the senior class enrollment was down to eight students, compared to 26 the previous year. In 2008, six people were killed and 150 homes were destroyed when a tornado struck Picher, accelerating exodus from the town. The post office and city hall closed in 2009, and on September 1 of that year the state of Oklahoma dis-incorporated Picher, sealing its fate as a ghost town. By the 2010 census, the population was down to 20 people (U.S Bureau of the Census 2012). But the ecocrisis threat is not over.

One of the primary factors still driving the health and environmental risks to the areas around Picher is climate change. The climate future of Oklahoma is expected to be increasingly erratic and unpredictable, with generally warmer temperatures, periods of severe summer drought, and increased flooding due to intense storms. Overall annual rainfall in the state has been increasing, with more rain arriving as heavy downpours. As the EPA (2016) reports, "During the last 50 years, the amount of rain falling during the wettest four days of the year has increased about 15 percent . . . Over the next several decades, the amount of rainfall during the wettest days of the year is likely to continue to increase, which would increase flooding." The total annual rainfall on very wet days in Oklahoma has grown by more per year since 1948 than in any other state in the southern Great Plains area.

In May 2015, the National Weather Service issued a flash flood emergency for downtown Oklahoma City during heavy rain in the broader area, the first such warning in city history. The city airport received 6.9 inches of rain in one day (Scientific American 2015). The 6-month period from March to August 2015 set precipitation records in parts of Oklahoma. This was followed in 2016 by a sudden transition from severe drought to damaging flooding (Allen 2016). Intense flooding hit Oklahoma again in 2018 as a cold front stalled out over the state (Prickett 2018). In May of 2019, a storm system swept through southern Oklahoma, causing rivers to rise dramatically above flood stage and forcing evacuations. One of the worst hit areas was the town of Braggs in Muskogee County, with a population around 300, which was completely surrounded by flood water from the overflowing Arkansas River (FloodList 2019). These weather patterns, suggestive of new climate trends, raise the potential for the deserted mines of the Picher area to be a continued site of harmful ecocrises interaction far into the future.

6.5 Cement, asbestos, and climate change

Some ecocrises entail compound entwinements involving multiple cross-linked components and interactions. One such complex begins with the mining of the raw materials needed for the production of cement, a compound used to make the concrete employed in the construction of buildings, roads, infrastructure, and much more. While the terms "cement" and "concrete" commonly are used interchangeably, they actually denote different entities.

Several minerals must be mined in order to make cement. Limestone, one of the key ingredients, because it is primarily composed of calcium carbonate, typically is extracted through surface mining, though underground mining, complete with all of the water seepage and contamination problems discussed earlier, is also utilized (e.g., in Iowa). Surface mining, however, can be environmentally risky. Surface mining of limestone for cement production in Meghalaya, a small state in northeastern India, has been linked to the loss of biodiversity, degradation of soil and water quality, drops in the availability of potable and irrigation water, and production of air pollution during the blasting of explosives (Lamere & Singh 2017). Nartey *et al.* (2012) report similar environmental problems caused by limestone quarrying in Manya Krobo District in Ghana. In Laegerdorf, Northern Germany, surface limestone mines significantly affected the local environment through saline water infiltration of the mines and aquifer contamination during dewatering (Iwanoff 1998). Apart from limestone, other ingredients required for cement production include clay, silica, iron minerals, and gypsum. Mining for these materials has also been linked to water quality and other environmental problems (Mishra 2015; Asente-Kyei & Addae 2016; Omoti *et al.* 2016).

Once the raw materials are mined, they are ground into a fine powder, which is heated to extreme temperatures (as high as 2642 °F) in a cement kiln. During this process, the chemical bonds of the materials are broken down and they are recombined into new compounds. The result is called clinker. This is next ground to a fine powder in a cement mill and mixed with gypsum to create cement. The powdered cement is then mixed with water and aggregates to form concrete (Leetham 2018). Other ingredients like asbestos can be added to create desired properties such as fire resistance.

In the United States, standard cement production increased from about 65 million tons in 2010 to about 88 million tons in 2018 (U.S. Geological Survey 2019). The overall value of sales of cement was about $11 billion in 2016. While there are more than 20 types of cement (e.g., masonry cement, rapid-hardening cement, colored cement), the most common is Portland cement, named for its similarity to the stone quarried on the Isle of Portland in the United Kingdom. Because of global demand, it estimated that total cement production will increase between 12 and 20 percent by 2050 (IEA 2018).

Cement manufacturing is a highly energy- and emissions-intensive industry. As Rosenthal (2007) writes, "making cement means making pollution, in the form of carbon dioxide emissions." Globally, cement production is the third-largest industrial consumer of energy (using up about 7 percent of total global industrial energy production) and the second-largest industrial emitter of climate change-causing greenhouse gases, accounting for around 5 percent of global carbon dioxide release. Emissions are caused both by the decomposition of limestone, a necessary chemical reaction in the cement-making process, and the combustion of the fuels used in heating. Producing 1 ton of cement requires 4.7 million British thermal units (BTU) of energy, where 1 BTU is the amount of heat required to raise the temperature of 1 pound of water by 1 °F. Manufacturing a single ton of cement uses the BTU generated by about 400 pounds of coal, suggesting why cement production has such high energy demands and causes such high levels of greenhouse gas release. It is predicted that the cement sector will increase its direct CO_2 emissions by another 4 percent by 2050 in spite of industry efforts to diminish CO_2 levels by reducing waste, encouraging recycling, maximizing the design life of cement, and refitting production plants (IEA 2018). An additional risk generated by the cement industry is the production of serious health problems like cancer and chronic respiratory conditions caused by exposure to asbestos cement.

As the name suggests, asbestos cement is a mixture of cement and asbestos, a naturally occurring mineral composed of long thin fibrous crystals. Once profitably and widely marketed as a "magic mineral" that was resistant to heat, electricity, and chemical corrosion and could be used to enhance these capacities in cloth, paper, plastic, and cement, asbestos is now recognized as a notorious public health menace. In the construction industry, asbestos cement was once widely used to make heat-resistant corrugated roofing on houses, sheds, garages, and farms and to produce prefabricated walls, drainpipes, flues, watertanks, and other products. In the years between 1900 and 2003, the United States used over 3 million tons of asbestos cement, primarily for flooring and roofing (Virta 2006). This amounted to only about 1.5 percent of world use for those years. Asbestos cement usually contained white asbestos (chrysotile), but blue (crocidolite) and brown (amosite) forms were also used. All three share the same physical properties and health hazards.

As early as 1911, the Prudential Insurance Company completed an actuarial study that showed premature death in the asbestos industry. This was followed in 1929 by a Metropolitan Life Insurance study that showed high rates of asbestos-related illness (Asbestos.com 2018). Fifty years later, the U.S. Occupational Safety and Health Administration Asbestos Work Group (OSHA 1980) concluded that there was "no evidence for a threshold or for a 'safe' level of asbestos exposure."

The disease most linked with asbestos exposure is malignant mesothelioma, a slowly developing cancer that affects the thin layer of tissue that covers most internal organs, though it is most commonly found in the tissue that surrounds the lungs (Fig. 6.3). Mesothelioma begins as tiny, asymptomatic nodules that develop on the lining of the lungs or abdomen. It usually does not cause symptoms until it has grown or spread. The latency period for mesothelioma after initial exposure to asbestos is at least 30 years, but the median survival time after diagnosis is only 9–12 months (Robinson *et al.* 2005). Initial symptoms, such as chest pain, weight loss, and fatigue, usually do not appear until the tumors begin to press against nerves, organs, bones, or other body parts. In the United States, the annual incidence rate for mesothelioma is one new case per 100 000 people; approximately 3000 new cases were diagnosed annually in the years 1980–2015 (National Cancer Institute 2018). There were over 92 000 global mesothelioma deaths reported to the World Health Organization (WHO) between 1994 and 2008 (Delgermaa *et al.* 2011).

Another disease directly linked to asbestos is asbestosis, a noncurable, noncancer respiratory disease. It is caused by the inhalation of asbestos fibers, which become lodged deep within the lungs, producing scarring and prompting the growth of excess tissue. Sufferers of asbestosis have great difficulty breathing, finding it

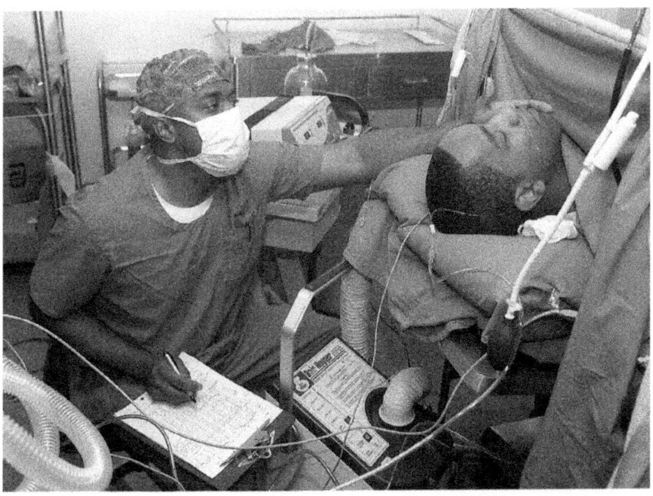

Fig. 6.3 Hospital patient. *Source:* PJF Military Collection / Alamy Stock Photo.

agonizingly painful. As the disease advances, it can cause suffocation. Lunder (2018) searched the WONDER database for deaths from asbestosis for the years 1999 to 2013. This database contains Multiple Causes of Death files reported to the CDC by public health departments around the country. She found that for this 14-year period there were over 20000 deaths, or about 1450 per year. Because its symptoms are similar to those of other respiratory conditions, asbestosis may not always be recorded or reported to the CDC as the cause of an individual's death.

Research by Korda *et al.* (2017) in the Australian Capital Territory on the health risks associated with living in houses insulated with asbestos identified additional asbestos–disease linkages. They found that in addition to comparatively high rates of mesothelioma among men living in asbestos-insulated homes, there was an elevated incidence of colorectal cancer in women and prostate cancer in men.

In response to the growing recognition of the dangers of asbestos exposure, over 50 countries, including the United Kingdom, Australia, Canada, Saudi Arabia, the United Arab Emirates, and the member countries of the European Union, have banned its use. While commercial use of the mineral is highly regulated in the United States, it continues to be imported and used in hundreds of consumer products, including insulation, car parts such as brake pads, spray-on fireproof fabric coatings, talcum powder, and vermiculite-potting soil. Its use is perfectly legal so long as it makes up less than 1 percent of a product's content.

There have been multiple attempts to ban asbestos in the United States over the years. These have been aggressively resisted by companies that profit from its continued use. This resistance began during the 1930s with asbestos industry efforts to cover up or manipulate the evidence that was emerging on the significant health toll among asbestos workers. Ultimately, it included the kinds of direct attacks on asbestos and health researchers that are today pepetrated against climate scientists by the energy industry. Exemplary is the case of Irving Selikoff, a researcher based at Mount Sinai Hospital in New York, who was the leading American medical expert on asbestos-related diseases between the 1960s and early 1990s. In the 1960s, he established the link between the inhalation of asbestos particles and lung-related ailments. He became well known to the public through his multiple media appearances on the growing asbestos problem. He also was well known by the asbestos industry, which, as McCulloch & Tweedale (2007, p. 619) indicate, mounted "a sustained and orchestrated campaign to discredit him." Despite this campaign, in 1989, the EPA issued the Asbestos Ban and Phase-Out Rule (ABPR), which was intended to impose a total ban on the manufacturing, importation, processing, and sale of asbestos-containing products. The ABPR triggered an angry response from the asbestos industry, which labeled it as "death by regulation," while emphasizing that it would lead to significant job losses and economic consequences. Asbestos product manufacturers filed a lawsuit against the ban, and on October 18, 1991, the Fifth Circuit Court of Appeals overturned it. The EPA, under the George H.W. Bush Administration from 1989 to 1993, did not appeal the ruling (Povtak 2018). Several subsequent asbestos-banning bills have failed to pass in Congress.

One of the places asbestos has been both extensively mined and used in construction is Australia. Mining of blue and white asbestos has a history going back more than 100 years. The center of blue asbestos was in Wittenoom in Western Australia, where mining began in the 1940s. A company town was soon built to house the workers, and Wittenoon grew into the largest town in the region, an arid area known as Pilbara that is renowned among ecologists for its subterranean invertebrate and cave-dwelling vertebrate biodiversity. In 1943, the Colonial Sugar Company, through its subsidiary Australian Blue Asbestos (later renamed Midalco), took over both the Wittenoom and nearby Yampire asbestos mines. During World War II, the demand for asbestos was high as governments wanted it for use in the production of tanks, planes, battleships, helmets, and gasmasks. During the 1950s and early 1960s, Wittenoom was Australia's sole supplier of asbestos. Over 160 000 tons of the fibrous mineral were extracted from the mine between 1943 and 1966 (Hills 1989). During this period, the drive to get at the asbestos was so great that consideration was briefly given to using nuclear explosives to speed up production (Zaunmayr 2018). Working conditions in the mines and mill at Wittenoom were horrendous. Asbestos dust fibers filled the air, ventilation was poor, and miners were not provided with protective gear. The town used asbestos tailings as road surfacing, with cars constantly stirring asbestos as they drove by, exposing the families of workers and others who never entered the mine. Jim McNulty (2011), a physician who worked as the medical officer for the Mines Department, occupational physician in the occupational health branch of the Health Department, and Commissioner for Health for the State of Western Australia, visited Wittenoom multiple times between 1957 and 1962. He repeatedly warned the Colonial Sugar Refinery's mine manager about the dangers to the miners and the people living in the town, but no corrective action was taken and his efforts were rebuffed. However, the mine was closed in 1966.

Turning a blind eye to the health problem created in Wittenoom did not make it go away. People began to get sick with asbestos-related diseases. By the 1970s, they instigated law suits against Colonial Sugar Refinery. While the courts at first dismissed the cases, eventually, after one dismissed suit was overturned on appeal, the company began to settle out of court. Then a mill worker was awarded substantial damages by Victorian Supreme Court in 1988 (Australian Asbestos Network 2018). In the wake of these developments, Wittenoom was declared a contaminated site and subsequently officially eliminated as a recognized town and removed from government maps. The local population had dropped to three by 2018—though it has become an occasional destination for extreme tourism in recent years—and Wittenoom, like Picher, is now a ghost town.

Not gone, however, is all of the asbestos extracted from Wittenoom and elsewhere that was used in constructing buildings in the country. As Gray *et al.* (2016, p. 292) affirm, "[t]here is a considerable amount of asbestos remaining *in situ* in the Australian built environment." The source of much of it was the industrial building materials company started by James Hardie, a late-19th-century Scottish immigrant to Melbourne. Hardie quickly launched an import business trading in oils and animal hide tanning products based on experience gained in his family's tannery and currier business in Linlithgow. He was joined by a fellow Scot, Andrew Reid, who became a full partner in James Hardie & Co. in 1895. While travelling overseas, Hardie encountered an opportunity to market a new type of roofing and lining slate called "fibro-cement," which during World War I became James Hardie & Co.'s core business. When Hardie retired in 1911, he sold his half of the business to Reid, whose family continued to run the expanding company until 1995.

During a housing boom that occurred after World War II, demand developed for building materials that were cheap, easy to manufacture, and quick to install. The Hardie company's "fibro house," consisting of an asbestos cement roof, walls, and fencing, fit the bill and rapidly became a dominant structure populating many of Australia's emergent suburbs, while other company products, like asbestos cement pipes and asbestos insulation, were installed throughout the county's industrial landscape. From the end of the war until 1954, in the Australian state of New South Wales alone, where there was a Hardie factory, at least 70 000 houses were constructed using asbestos cement—about half of the homes erected there during the post-war period. In Victoria, the site of another company factory, it is estimated that almost all homes constructed before 1976 contained asbestos products, as well as 20 percent of all domestic roofs.

Similarly, many buildings that went up in Queensland during the same period contained asbestos, and it is now found in one third of all Queensland homes (AHI 2018). More broadly, across Australia, until the 1960s, a quarter of all new houses were sided with asbestos cement. A survey conducted by the Australian Bureau of Statistics (1995) estimated that of the approximately 5.5 million occupied dwellings in Australia built before 1987, over 15 percent had asbestos cement sheeting as a main construction material. The true percentage may be higher, as the survey relied on the occupant having awareness of what kinds of materials contain asbestos.

In his book, *Killer Company: James Hardie Exposed*, investigative journalist Matt Peacock (2009) cites company and government documents, as well as his interviews with over 100 former company employees, to detail the stringent efforts of James Hardie Industries to cover up the known health hazards of its asbestos-containing products. Former workers afflicted with asbestos-related diseases ultimately sued the company for millions of dollars. In response, and in an effort to protect its reputation from the stigma of its considerable asbestos liabilities, the company restructured in 2001 and moved to the Netherlands (Haigh 2006). Greg Combet, the Secretary of the Australian Council of Trade Unions, labeled the movement of the Hardie assets overseas, and hence out of reach of asbestos disease sufferers in Australia, as "one of the most morally and legally repugnant acts in Australian corporate history" (Prince *et al.* 2004).

Asbestos has not been used in domestic building materials in Australia since the 1980s, but it was not banned from use until 2003 and removal of that installed before 2003 was not required (AHI 2018). The current popularity of do-it-yourself renovation, in combination with limited public asbestos awareness and lack of regulatory oversight, has raised public health concerns concerning a new wave of harmful asbestos exposures in Australia. With the second-highest mesothelioma death rate in the world, asbestos-related diseases cost the country over AUD500 million a year in direct health care costs, combined with loss of productivity in the workplace (CIE 2018). Three waves of mesothelioma have been identified: in people who were exposed to asbestos in the mining and milling of ore and the manufacturing of asbestos products; in people who used asbestos products; and in people involved in the repair, renovation, and demolition of buildings (Landrigan 1991). Another cause for concern is the potential for a fourth wave in people who suffer the effects of climate change on existing structures containing asbestos.

During Christmas 1974, Cyclone Tracy, a comparatively small but intense storm, demolished the rapidly expanding coastal city of Darwin in the Northern Territory. Prior to Tracy, the city's 45 000 residents had been enjoying a "record-breaking building spree" (Murphy 1984, p. 55). The cyclone wiped out more than 80 percent of Darwin's buildings, including most houses. Some journalists likened the level of destruction to Hiroshima after it was hit by the atomic bomb in 1945. The cleanup of the city began relatively quickly, with the approximately 10 000 people who had not been evacuated participating in the removal of the wreckage of fallen buildings and stores of not-yet used building materials that were scattered about. On Darwin's beaches there are still signs that read, "Attention: Asbestos may be seen on this beach. Do not pick up any material in this area," accompanied by a photograph of a piece of asbestos lying on the sand. Nearly 40 years after the cyclone, cases of asbestos-related diseases stemming from exposures during the cleanup effort began to emerge (Marriner 2011).

In 2011, Yasi, a Category 5 Severe Tropical Cyclone, with wind gusts reaching 175 miles per hour, devastated Northern Queensland. Numerous homes were completely destroyed, and many others sustained extensive damage. Roofs were ripped off buildings and sent flying down rain-soaked streets. Flooding was widespread. The cyclone created a hazard for thousands of workers who cleared the piles of debris from destroyed and damaged fibro homes. Three years later, Cyclone Christine passed by the Pilbara region of Western Australia, damaging the asbestos in buildings and creating a "health risk to Pilbara residents . . . exposed to the cancer-causing material" (Nicholson 2014). The same thing occurred again in 2015 when Cyclone Olwyn hit the town of Carnarvon, also in Western Australia (Collins 2015).

The lesson of these violent storms for a country that still has a considerable amount of asbestos in its buildings is evident. As the Australian Department of the Environment and Energy (2018) affirms, "[c]limate change exacerbates existing climate risks for our coasts and creates new risks . . . In the coastal zone, more intense storms and cyclones and rising sea levels could worsen storm surge, coastal inundation and erosion with impacts on the built environment." Research by the Australian Bureau of Meteorology (2015) suggests that future climate change will hit Australia harder than much of the rest of world, especially in terms of temperature rise and intense heat waves. The seas that surround the country are projected to warm further, providing additional energy for more extreme coastal storms. Moreover, there will be an increase in the number of days with severe wildfire danger.

Projected climate changes create the conditions for the corrosion of asbestos cement and the release of asbestos fibers into the environment. Rain causes the dissolution of soluble salts in the cement, leading to an increase in porosity and a surge in the speed of disintegration (Faucon *et al.* 1996). This produces surface corrosion and the loss of underlying layers of dissolved cement (Beddoe & Dorner 2005). Additionally, exposure to CO_2 can cause micro-fissures to form (Dias *et al.* 2008). Deterioration is also promoted by sudden thermal changes. Collectively, these physical changes under conditions projected to become more intense due to climate change suggest the ongoing dangers of this ecocrises interaction.

6.6 The climate change–nuclear ecocrisis nexus

Climate change reframes nuclear power as a potential solution to the problem of continued use of carbon-based, greenhouse gas-producing energy sources like oil, coal, and natural gas. As *Time Magazine* "Hero of the Environment" writer, policy analyst, and eco-pragmatist Michael Shellenberger (2017, p. 3) asserts, "state-led efforts to deploy nuclear power plants are the only proven way for governments to deliberately and rapidly decarbonize economies." As part of this argument, nuclear power is presented as a powerful, cost-efficient, and safe technology that produces "no greenhouse gases" (Vine & Juliani 2014, p. 3). Research in the United Kingdom suggests that many people who are concerned about the problem of nuclear waste buildup express a reluctant willingness to accept nuclear power to curtail global warming, seeing it as the lesser of two evils (Bickerstaff *et al.* 2008). This willingness requires overlooking the fact that "nuclear policies and postures are highly accident-prone and have brought the world very close to an accidental nuclear detonation on several occasions" (Scheffran *et al.* 2015, p. 10). What is more, the claims made for nuclear energy can be disputed on a number of grounds (Caldicott 2006), including the incalculable risks they introduce and the notion that nuclear power plants do not produce greenhouse gases (while technically true, this ignores the fact that at every step in the nuclear process, including the "mining and enrichment of uranium; the manufacturing, transport and reprocessing of nuclear fuel rods and waste; and the building and dismantling of the reactors," greenhouse gases are generated; Sortir du Nucléaire 2015). Regarding the latter point, the U.S. Energy Information Administration (EIA 2019) affirms that "[n]uclear power plants . . . have large amounts of metal and concrete, which require large amounts of energy to manufacture. If fossil fuels are used for mining and refining uranium ore, or if fossil fuels are used when constructing the nuclear power plant, then the emissions from burning those fuels could be associated with the electricity that nuclear power plants generate." Greenhouses gas emissions during these upstream activities include not only CO_2 but also a long list of other drivers of climate change, including carbon monoxide, methane, nitrous oxide, trichloroethane, carbon tetrachloride, nitrogen trifluoride, halons, chlorofluorocarbons, hydrochlorofluorocarbons, and hydrofluorocarbons (Scheffran *et al.* 2015). In other words, those who advocate for nuclear power as clean power that does not contribute to greenhouse gas production fail to consider the whole life cycle of nuclear technology, especially the front end of the nuclear fuel cycle. Moreover, far less attention has been given to a different climate change–nuclear nexus, namely the adverse "interplay between climate change and risks associated with nuclear weapons, facilities and materials" (Van Riet 2015, p. 4); in other words, to the pathways of their dangerous ecocrises interaction.

There are two pathways of interaction between climate change and nuclear weapons, facilities, and materials (Fig. 6.4). First, climate has been demonstrated to amplify social tensions and conflict and heighten the risk of political destabilization. Second, the sea-level rise and extreme storms pushed by climate change pose a rising threat to the security of nuclear facilities and radioactive disposal sites. A third possible pathway, not discussed here, involves the climate effects of a nuclear war—referred to as Nuclear Winter—including the severe and prolonged global climatic cooling caused by a potential layer of soot and dust blasted into the atmosphere to block the sun's warming rays (nuclear → climate change).

6.6.1 Radiation and health

On its own, independent of climate change, the nuclear industry has impacted human health by way of the release of radiation. Radiation is the movement of energy in either wave (e.g., gamma waves) or subatomic alpha or beta particle form through space or through a material entity, including a living organism. There are various kinds of radiation with differing properties, including visible light, microwaves, x-rays, ultrasound, and infrared. Based on the quantity of energy units involved, radiation is said to be either ionizing (more than 10 electron volts, a common unit of energy in nuclear physics) or nonionizing (less than 10 electron volts). The word "ionize" refers to the breaking away of one or more electrons from an atom. While, upon contact with a material entity, nonironizing radiation can shake or move molecules, ionizing radiation can break molecular bonds and cause chemical reactions. Though we can perceive some types of nonionizing radiation (e.g., light), we cannot see, feel, taste, smell, or hear ionizing radiation.

Living on Earth, we are routinely exposed to small amounts of nonionizing radiation (e.g.,

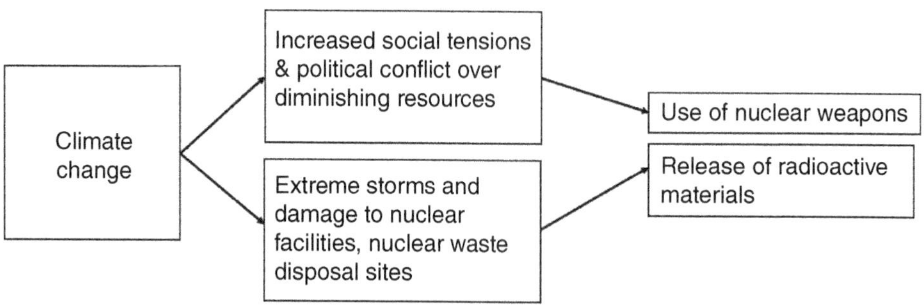

Fig. 6.4 Adverse climate change–nuclear nexus.

cosmic rays from outside of our solar system, radon in the ground). This exposure has health and mutational effects. Radon, a colorless, odorless, and tasteless gas, for example, is the second leading cause of lung cancer after smoking. Through activities like the mining and processing of uranium by the nuclear industry, radioactivity is released to flow through Earth's natural systems. This has resulted in biosphere and organism contamination. While human skin can stop alpha particles, if they enter the body through inhalation, ingestion, or a wound, they can do a great deal of damage and deliver a large dose of radiation to internal bodily cells. A fraction of the size of alpha cells, beta cells are more penetrating. They can breach and cause harm to the most sensitive layers of skin. Inhalation and ingestion, however, pose the greatest risks. Strontium-90 and tritium, two beta-emitting radionuclides, are regularly released from nuclear power plants during normal operation. Our bodies can misidentify strontium-90 as calcium, resulting in its storage in our bones, where it increases our risk of bone and blood cancers like leukemia. All of us have strontium-90 in our bodies because of the testing in the environment of nuclear bombs. Tritium is radioactive hydrogen, and can irradiate our DNA.

The effects of radiation exposure can be seen from examination of people living in the vicinity of the atmospheric testing of nuclear weapons. Between 1946 and 1958, for example, the United States detonated a series of 23 nuclear devices on or near Bikini Atoll, part of a mid-Pacific island chain called the Marshall Islands. On March 1, 1954, a thermonuclear hydrogen bomb called Castle Bravo was tested there, but significant miscalculations about the actual size and effect of the explosion led to fallout landing on the Rongelap and Utirik atolls, also of the Marshall Island group, and then spreading around the rest of the world. The inhabitants of the islands were not evacuated until 3 days after the bomb went off, by which time they were experiencing itchiness, vomiting, and fatigue, all common symptoms of radiation sickness. By 1963, the people of the Marshall Islands began to develop thyroid tumors, and many birth defects were reported (Simon *et al.* 2010). On the Marshall Islands, as elsewhere, the elderly, children, fetuses, and people with immune disorders were the most susceptible to ionizing radiation. Cancers linked to ionizing radiation exposure include blood cancers, lung cancer, and solid tumors of various body organs. Birth defects can include Down syndrome, cleft palate or lip, congenital malformations, spinal defects, and kidney and liver damage (Folkers 1999). Barbara Rose Johnston (2007, p. 2) stresses based on her research on the nuclear industry that "the communities that hosted the nuclear war machine—uranium mines, mills, and enrichment plants; weapons production facilities; military "proving" grounds; battlefields; and nuclear waste dumps—[have suffered] acute and lasting radioactive assaults on the environment and on human health."

The term "Chernobyl" has come to signify the radioactive risks of nuclear power. In her book, *Manual for Survival: A Chernobyl Guide to the Future*, Kate Brown (2019), a professor of the history of science, technology, and society at the Massachusetts Institute of Technology, reports on her investigation of the 1986 explosion at the Chernobyl nuclear power plant in Ukraine, which at the time of the catastrophe was a republic in the Soviet Union. On April 26, 1986, a team of 17 trained employees of the Chernobyl nuclear power plant turned off the emergency system at the plant's Reactor No. 4 to conduct a routine test. A chain reaction in the reactor core went "critical," meaning that the operators lost control. The reactor's power surged, causing two explosions that blew open the concrete lid of the reactor and sent a blast of radioactive gases into the atmosphere. As part of her study of what happened next, Brown pored through 27 archives from the event and found that much of what the world had been told about the Chernobyl accident was incomplete or false. Her basic finding was that "[p] eople were far sicker and far more people died than we are led to believe. Chernobyl contaminants were not safely enclosed within the

Chernobyl Zone . . . We are still ingesting Chernobyl fallout from 33 years ago" (quoted in MacPherson 2019). While the official record maintains that 300 people, mostly firemen and plant operators, were hospitalized after the accident, Brown found that at least 40 000 were hospitalized for Chernobyl exposures in the summer after the accident, many of them children. In two regions of the province that were closely monitored, she found that the rate of congenital malformations grew from 10 to 23 percent between 1986 and 1988. Rates of infant mortality were also high. Brown's (2013) earlier book, *Plutopia*, provides a history of the plutonium production and enduring plutonium disasters of both the United States (in the town of Hanford in east Washington State) and the Soviet Union (in the city of Ozersk in the southern Russian Ural Mountains). Together, these two volumes underline important reasons why nuclear power is not a solution to climate change.

6.6.2 Climate change and nuclear facilities

Paul Virilio, a French cultural theorist, urbanist, and technology analyst, argued that technology cannot exist without the potential for accidents. This argument is embodied in his most often repeated quote:

> When you invent the ship, you also invent the shipwreck; when you invent the plane you also invent the plane crash; and when you invent electricity, you invent electrocution . . . Every technology carries its own negativity, which is invented at the same time as technical progress. (Virilio 1999, p. 89)

Since World War II, there have been almost 450 major radiation accidents worldwide (Hasegawa *et al.* 2015). Fortunately, most involved radiation devices and radioisotopes and had only small, localized effects. There are, however, over 400 nuclear power plants in operation around the world today. Because of their potential to cause mass destruction, loss of life, disease, and environmental destruction, there is widespread concern about their safety in a volatile and changing world.

The International Nuclear and Radiological Event Scale (INES) is a system used to categorize the comparative magnitude of potentially catastrophic nuclear accidents. Much like Richter's magnitude scale for measuring the strength of earthquakes, the INES consists of seven levels, each eight times more severe than the one beneath it. An event labeled as a 7 is considered a major accident, while a Level 0 event is assessed to be a problematic deviation from routine operation. There have been six major nuclear facility accidents since the first nuclear reactor became operational in Chicago in 1942 (as well as numerous smaller accidents and breakdowns over the years, some of which have caused fatalities): in 1957 at Windscale in the United Kingdom (level 5); in 1957 at Kyshtym in the former Soviet Union (level 6); in 1979 at Three Mile Island in Pennsylvania (level 5); in 1986 at Chernobyl (level 7); in 2011 in Fukushima in Japan (level 7); and in 1952 at Chalk River in Ontario (level 5). Each caused fatalities or toxic human exposures (except perhaps Chalk River), environmental contamination, and social disruption in surrounding communities. While a combination of design and operator errors were the primary causes, one, Fukushima, was likely influenced by climate change factors, and climate change has significantly impacted or threatened other nuclear facilities.

6.6.2.1 Tsunamis

On March 11, 2011, the undersea Tōhoku earthquake, the biggest earthquake ever recorded in Japan, rattled the ocean bottom and triggered a powerful tsunami, with waves that reached heights of over 130 feet. Seawater washed over the top of the sizeable walls built to hold it back and protect nearby nuclear facilities and local residents. It flooded the Fukushima Daiichi and Fukushima Daini nuclear plants in the towns of Ōkuma and Futaba, which have seen repeated tsunamis over the decades (Nateghi *et al.* 2016). The plants suffered major damage, including cooling system malfunctions, inundation of backup diesel generators and power lines, overheating, and partial meltdown of nuclear fuel cells. This chain of events caused radiation leaks and permanently damaged several reactors, making them impossible to restart. Several studies and reports suggesting the need to increase protection from tsunami flooding had been rejected by Tokyo Electric Power Company Holdings (TEPCO), the operator of the plants. Subsequently, TEPCO admitted that it had not initiated stronger actions to prevent disaster because it feared they would signal that the nuclear facilities were not safe, triggering lawsuits or citizen outcry (McCurry 2012).

While multiple human and nonhuman factors were involved, one identifiable enhancer of the Fukushima nuclear disaster was the failure of TEPCO to anticipate the height of the tsunami wave. Sea-level rise challenges existing seawalls worldwide, as it raises both the mean ocean level and the height of waves during extreme weather events. It also increases both the frequency and the intensity of tsunami-induced flooding (Li *et al.* 2018). Research "shows that sea-level rise can significantly increase the tsunami hazard, which means that smaller tsunamis in the future can have the same adverse impacts as big tsunamis would today" (Robert Weiss, quoted in Virginia Tech Daily 2018).

There were almost 16 000 disaster-related deaths caused by the combined effects of the quake/tsunami in Japan and the massive evacuation it required (Hasegawa *et al.* 2015). Long-term displacement to unfamiliar locations and separation from established social support networks took a heavy toll on the physical and mental health of the 170 000 evacuees. Mortality among evacuated elderly people needing nursing care increased by about three times in the first 3 months after the evacuation. Stigma was also a factor. In Japan, the label "Hibakusha" ("explosion-affected people") is a derogatory term that historically was used to label women and children who survived the nuclear bombing of Nagasaki. It experienced a resurgence after the Fukushima event. Self-stigma and loss of self-esteem, based on the fear of being viewed negatively because of assumed radiation exposure, was another issue, and an increased rate of suicide was seen among Fukushima evacuees. Moreover, there was a rise in childhood obesity as protective parents kept their children indoors for fear of contamination outside. Exposure to radiation is not the only health risk of nuclear disaster. Nor is such a disaster merely a local or regional event. Radioactive fallout from Fukushima has been detected in environmental samples collected in the United States, Greece, and Russia, and researchers have projected long-term global health effects (Bolsunovsky & Dementyev 2011; Leon *et al.* 2011; Ten Hoevea & Jacobson 2012).

6.6.2.2 Hurricanes

The fission that occurs in a nuclear power plant produces an enormous amount of heat. This heat is used to generate steam, which drives the turbines that produce electricity. For a reactor to operate and not melt down or suffer containment failure, the heat must be removed as fast as it is produced. This transfer is achieved by passing a coolant through the areas where it is created, including the nuclear core where the fuel rods are located. The level of heat that is generated in a reactor is about 572 °F (for comparison, the sun's surface is about 9940 °F). Nuclear power stations are always built near a body of water to ensure that there is a reliable source for cooling purposes.

Because of their location in coastal areas, many nuclear plants are subject to being struck by hurricanes (also called cyclones or typhoons depending on location). Examples include:

- **2018.** Flooding caused by Hurricane Florence left a twin-reactor nuclear plant in Brunswick south of Wilmington, North Carolina with limited access due to impassable roads, though floodwaters did not enter the facility.

- **2018.** Typhoon Mangkhut, with 120 mph winds, threatened the Taishan and Yangjiang nuclear power stations in Guangdong, China, but did not damage the facilities.
- **2017.** Hurricane Harvey impacted the South Texas nuclear power plant, but the plant did not receive the full brunt of the storm that hit Houston 88 miles away. The plant continued to operate during the storm. Had it been necessary to shut down, the process would have taken 4–5 hours.
- **2012.** Flooding from Hurricane Sandy caused the Salem and Oyster Creek nuclear plants in New Jersey to shut down when high water levels threatened their water intake and circulation systems.
- **2005.** Hurricane Katrina hit the Waterford nuclear power plant in Louisiana. The plant was shut down after its electricity supply was lost.
- **2004.** Hurricane Jeanne caused the St. Lucie nuclear power plant in Florida to be shut down after it triggered a loss of the external energy supply. Flooding caused the loss of the plant's off-site power supply and knocked out several safety systems, resulting in a Level 2 event on the International Nuclear Event Scale.
- **1999.** At high tide, the winds from extratropical storm Martin pushed water over the seawall protecting the Blayais nuclear power plant in Southwestern France. Flooding led to a loss of the plant's off-site power supply and shut down several safety-related backup systems, resulting in a Level 2 event.
- **1992.** Hurricane Andrew caused the Turkey Point nuclear power plant located 25 miles south of Miami, Florida to shut off two units after it cut off exterior electric energy supply. While the reactors were not damaged, the wider plant site sustained $90 million worth of structural damage.

It is evident from this list that because of their (varying) hardened cement and steel outer shell containment structures, built to withstand major external events like airplane crashes, nuclear facilities have generally held up well during hurricanes. However, powerful storms have cut their external power, creating situations that, in combination with operator error (of the kinds that have occurred at several facilities over the years), could cascade into highly dangerous scenarios. Moreover, with climate change, the intensity of hurricanes, their rainfall rates, and the height of their storm surges have been increasing. These changes suggest a rise in the average destructive potential per storm. Unfortunately, as seen in Fukushima, only a single failure of a nuclear facility can create an enduring catastrophic outcome with both regional and global effects.

6.6.2.3 Sinking below rising seas

Included in the Enewetak Atoll, a group of 40 coral islands that is part of the Republic of the Marshall Islands, is Runit Island. Less than 2.26 square miles in size, with its highest point standing only 16.4 feet above the ocean surface, Runit was contaminated by nuclear explosions and radioactive waste produced by U.S. nuclear warhead tests between 1946 and 1958. In the final year of this program, an 18 kiloton nuclear device, roughly equivalent in size to the nuclear bomb dropped on Hiroshima, was detonated on the northern tip of the island. The blast produced a crater, which, from 1977 to 1980, the U.S. military used to bury loose waste and topsoil debris scraped off of six other islands in the Enewetak Atoll that were hit by radioactive fallout. Over the crater, a 14-inch-thick concrete dome about twice the size of a football field was constructed by military personnel (Fig. 6.5). It is called The Tomb by Enewetak islanders. Under the dome are deposited almost 100 000 cubic yards of highly radioactive debris, including plutonium-239, which has a half-life of 24 000 years. As a result, Runit will not be habitable again until the year 25958 (Wargo 2009).

Originally, the Defense Department design for the Runit radioactive waste storage facility included plans for the 30-foot-deep, 350-foot-wide crater to be lined with concrete, but these were scrapped because they were deemed to be

Fig. 6.5 Runit Dome, Enewetak Atoll, Republic of the Marshall Islands, aerial view. *Source:* US Defense Special Weapons Agency.

too expensive. Consequently, there is permeable coral rock at the bottom of the crater—rock that has been fractured by nuclear explosions. Already, the crater is below sea level at high tide (Committee on Evaluation of Enewetak Radioactivity Containment 1982). This makes it highly vulnerable to seawater intrusion. Concern has arisen that the depository is at increasing risk over time because of rising sea levels. Moreover, the dome, which was not constructed with much consideration for its long-term durability, has begun deteriorating, creating the possibility that it could be breached by a hurricane. In 2013, the Department of Energy issued a report that stated: "Catastrophic failure of the concrete dome facade covering the debris mount and instantaneous release of all its contents into the lagoon will not necessarily lead to any significant change in the radiation dose delivered to the local population" (Hamilton 2013, p. 1). The reason is not because there would be insignificant radiation leakage into marine water, but rather because "[t]he radiologic inventory buried beneath Runit Dome is dwarfed by the current inventory of fallout radionuclides in atoll lagoon sediments" (Hamilton 2013, p. 1). Further, the report indicates that "the potential does exist for contaminated groundwater from Runit Dome to flow into the nearby, subsurface marine environment" (Hamilton 2013, p. 1).

Michael Gerrard, director of the Sabin Center for Climate Change Law at Columbia Law School and associate chair of the Columbia University's Earth Institute, visited Runit in 2010 while working with Marshall Islands' government on a study of the legal consequences of the threat that sea-level rise poses to the Republic's existence as a nation. He followed a guide to the top of the unmarked, unfenced, unguarded, and (had he not already known what it was) mysterious-seeming dome. He observed that "[f]rom this low perch a few feet above sea level, the most prominent features are the twenty-three-mile-wide lagoon to the west . . . the Pacific Ocean to the east . . . and just a few hundred feet to the northeast, the almost-submerged crater from one of the other nuclear tests. In other words, as in most places in the Marshall Islands, what you see is mostly water" (Gerrard 2015, p. 90). With rising sea level, there will be even more water to see, but perhaps not while standing on an abovewater concrete perch. Already, plutonium isotopes discovered in the South China Sea have been traced to the Marshall Islands, some 2800 miles away (Wu *et al.* 2014; Graham 2017).

6.7 Concluding remarks

This chapter has reviewed six perilous ecocrises interactions around the world, including their environmental and human health implications. Notably, these threats do not come in isolation; they are overlapping and potentially interactive. People living near a hurricane-flooded Superfund site, for example, could be exposed to radioactive fallout from a distant nuclear meltdown, as could those living in storm-damaged homes containing asbestos or an apartment building whose foundation is crumbling due to permafrost melt. Climate change links all of the badly polluted places around the globe, uniting one anthropogenic threat with multiple others in what at times seems, from the human perspective, like a race toward planetary unsustainability.

References

AHI. (2018). Asbestos cement products. https://www.ahiasbestos.com.au/asbestos-cement-products/.

Allen, S. (2016). Heavy rains wipe away drought, but bring floods in Oklahoma. *The Oklahoman*. https://newsok.com/article/5492598/heavy-rains-wipe-away-drought-but-bring-floods-in-oklahoma.

Anderson, B. (2017). *Troubled water: EPA's lack of funds slows Superfund site cleanup*; Trump wants more cuts. TCPalm. https://www.tcpalm.com/story/news/investigations/2017/09/07/news-21-superfund/553262001/.

Anisimov, O. & Reneva, S. (2006). Permafrost and changing climate: the Russian perspective. *Ambio* 35(4): 169–175.

Asbestos.com (2018). Asbestos cover-up. https://www.asbestos.com/featured-stories/cover-up/.

Asente-Kyei, K. & Addae, A. (2016). The economic and environmental impacts on clay harvesting at Abonko in the Mfantsiman West District of Central Region, Ghana. *American Scientific Research Journal for Engineering, Technology, and Sciences* 18(1): 120–132.

Associated Press. (2008). Pollution brings end to Oklahoma mining town. http://www.nbcnews.com/id/24555711/ns/us_news-environment/t/pollution-brings-end-oklahoma-mining-town/#.W8zz4VVKjm4.

ATSDR. (2017). ATSDR assessment of the evidence for the drinking water contaminants at Camp Lejeune and specific cancers and other diseases. Agency for Toxic Substances and Disease Registry. https://www.atsdr.cdc.gov/sites/lejeune/docs/atsdr_summary_of_the_evidence_for_causality_tce_pce_508.pdf.

Australian Asbestos Network. (2018). Battling CSR. https://www.australianasbestosnetwork.org.au/asbestos-history/battles-2/battling-csr/.

Australian Bureau of Meteorology. (2015). Climate Change in Australia. https://www.climatechangeinaustralia.gov.au/media/ccia/2.1.6/cms:page_media/168/CCIA_2015_NRM_TechnicalReport_WEB.pdf.

Australian Bureau of Statistics. (1995). Australian social trends, 1995. http://www.abs.gov.au/AUSSTATS/abs@.nsf/2f762f95845417aeca25706c00834efa/cd6c2ea981520095ca2570ec007530a4!OpenDocument.

Australian Department of the Environment and Energy. (2018). Australia's climate change strategies. http://www.environment.gov.au/climate-change/climate-science-data/climate-science/impacts.

Baer, H. (2019). *Airplanes, the Environment, and the Human Condition*. London: Routledge.

Beaumont, P. (2019). Rotten eggs: e-waste from Europe poisons Ghana's food chain. *Guardian*. https://www.theguardian.com/global-development/2019/apr/24/rotten-chicken-eggs-e-waste-from-europe-poisons-ghana-food-chain-agbogbloshie-accra.

Beddoe, R. & Dorner, H. (2005). Modelling acid attack on concrete: part I. The essential mechanisms. *Cement and Concrete Research* 35(12): 2333–2339.

Bickerstaff, K., Lorenzoni, I., Pidgeon, N., Poortinga, W. & Simmons. P. (2008). Reframing nuclear power in the U.K. energy debate: nuclear power, climate change mitigation and radioactive waste. *Public Understanding of Science* 17(2): 145–169.

Bolsunovsky, A. & Dementyev, D. (2011). Evidence of the radioactive fallout in the center of Asia (Russia) following the Fukushima Nuclear Accident. *Journal of Environmental Radioactivity* 102(11): 1062–1064.

Brown, K. (2013). *Plutopia: Nuclear Families, Atomic Cities, and the Great Soviet and American Plutonium Disasters*. Oxford: Oxford University Press.

Brown, K. (2019). *Manual for Survival: A Chernobyl Guide to the Future*. New York: Penguin.

Buma, B. & Wessman, C. (2012). Differential species responses to compounded perturbations and implications for landscape heterogeneity and resilience. *Forest Ecology and Management* 266: 25–33.

Caldicott, H. (2006). *Nuclear Power Is Not the Answer*. London: The New Press.

Caravanos, J., Chatham-Stephens, K., Ericson, B., Landrigan, P. & Fuller, R. (2013). The burden of disease from pediatric lead exposure at hazardous waste sites in 7 Asian countries. *Environmental Research* 120: 119–125.

CDC. (2017). *Blood lead levels in children*. Centers for Disease Control and Prevention. https://www.cdc.gov/nceh/lead/acclpp/blood_lead_levels.htm.

CH2MHill. (2006). Human Health Risk Assessment: Tar Creek Superfund Site Operable Unit No. 4 Ottawa County, Oklahoma. Prepared for U.S. Environmental Protection Agency. https://semspub.epa.gov/work/06/9223551.pdf.

Church, J. & White, N. (2006). A 20th century acceleration in global sea-level rise. *Geophysical Research Letters* 33(1): L01602.

CIE. (2018). *The Economic Burden of Asbestos-Related Disease*. Canberra: Centre for International Economics.

Clark, L., Alter, A. & Blake, L. (1962). Sanitary waste disposal for navy camps in polar regions. *Journal of the Water Pollution Control Federation* 34: 1219–1234.

Colgan, W., Machguth, H., MacFerrin, M., Colgan, J., van As, D. & MacGregor, J. (2016). The abandoned ice sheet base at Camp Century, Greenland, in a warming climate. *Geophysical Research Letters* 43(16): 8091–8096.

Collins, B. (2015). Cyclone Olwyn: asbestos risk hampers Carnarvon's cleanup. ABC Northwest WA. http://www.abc.net.au/local/photos/2015/03/16/4198600.htm.

Colman, Z. (2019). The toxic waste threat that climate change is making worse. *Politico*. https://www.politico.com/story/2019/08/26/toxic-waste-climate-change-worse-1672998.

Committee on Evaluation of Enewetak Radioactivity Containment. (1982). *Evaluation of Enewetak Radioactivity Containment 019026391*. Washington, D.C.: National Research Council.

Dearen, J., Biesecker, M. & Kastanis, A. (2017). *AP finds climate change risk for 327 toxic Superfund sites. Associated Press*. https://apnews.com/31765cc6d10244588805ee738edcb36b.

DeBenedetto, P. (2016). Along the San Jacinto River, cleaning up a decades-old disaster. *Houstonia*. https://www.houstoniamag.com/articles/2016/12/30/san-jacinto-river-plan-toxic-waste.

Delgermaa, V. Takahashi, K., Park, E., Le, G., Hara, T. & Sorahan, T. (2011). Global mesothelioma deaths reported to the World Health Organization between 1994 and 2008. *Bulletin of the World Health Organization* 89: 716–724.

Dias, C., Cincotto, M., Savastano, H. & John, V. (2008). Long-term aging of fiber-cement sheets—the effect of carbonation, leaching and acid rain. *Cement and Concrete Composites* 30(4): 255–565.

Dowling, R., Caravanos, J., Grigsby, P., Rivera, A., Ericson, B. Amoyaw-Osei, A. *et al.* (2016). Estimating the prevalence of toxic waste sites in low- and middle-income countries. *Annals of Global Health* 82(5): 700–710.

Dunlap, I. (2008). Forwards. In: D. Spratt & P. Sutton, *Climate Code Red: The Case for Emergency Action*, pp. xi–xiii. Melbourne, VIC: Scribe.

EIA. (2019) Nuclear explained: nuclear power and the environment. U.S. Energy Information Administration. https://www.eia.gov/energyexplained/index.php?page=nuclear_environment.

Eilperin, J. (2009). Flooded toxic waste sites are potential health threat. *Washington Post*. http://www.washingtonpost.com/wp-dyn/content/article/2005/09/09/AR2005090902074.html.

EPA. (2008). *Community Involvement Plan*, Tar Creek Superfund Site. U.S. Environmetnal Protection Agency. https://semspub.epa.gov/work/06/825646.pdf.

EPA. (2014). Climate Change Adaptation Plan. U.S. Environmental Protection Agency. https://www.epa.gov/sites/production/files/2015-08/documents/adaptationplans2014_508.pdf.

EPA. (2016). What climate change means for Oklahoma. U.S. Environmental Protection Agency. https://19january2017snapshot.epa.gov/sites/production/files/2016-09/documents/climate-change-ok.pdf.

EPA. (2017). *San Jacinto River Waste Pits Superfund Site Harris County*, Texas. U.S. Environmental Protection Agency. https://www.epa.gov/tx/sjrwp.

EPA. (2018a). *What is Superfund?* U.S. Environmental Protection Agency. https://www.epa.gov/superfund/what-superfund.

EPA. (2018b). *Patrick Bayou, Deerfield Park*, TX. U.S. Environmental Protection Agency. https://cumulis.epa.gov/supercpad/SiteProfiles/index.cfm?fuseaction=second.Cleanup&id=0605329#bkground.

EPA. (2019). *Tar Creek (Ottawa County) Ottawa County*, OK. U.S. Environmental Protection Agency. https://cumulis.epa.gov/supercpad/cursites/csitinfo.cfm?id=0601269.

Everett, D. (2007). Tri-state lead and zinc district. *The Encyclopedia of Oklahoma History and Culture*. Oklahoma City, OK: Oklahoma Historical Society.

Faucon, P., Le Bescop, P., Adenot, F., Bonville, P., Jacquinot, J., Pineau, F. *et al.* (1996) Leaching of cement: study of the surface layer. *Cement and Concrete Research* 26(11): 1707–1715.

Fazzo, L., Minichilli, F., Santoro, M., Ceccarini, A., Della Seta, M., Bianchi, F. *et al.* (2017). Hazardous waste and health impact: a systematic review of the scientific literature. *Environmental Health* 16: 107.

Filbee-Dexter, K., Pittman, J., Haig, H., Steven, M., Alexander, S., Symons, C., *et al.* (2017). Ecological surprise: concept, synthesis, and social dimensions. *Ecosphere* 8(12): ecs2.2005.

Flavelle, C. (2018). *Miami will be underwater soon*. Its drinking water could go first. *Bloomberg News*. https://www.bloomberg.com/news/features/2018-08-29/miami-s-other-water-problem.

FloodList. (2019). Oklahoma and Arkansas—evacuations as rivers approach record levels. http://floodlist.com/america/usa/oklahoma-arkansas-river-flood-record-levels-may-2019.

Folkers, C. (1999). *Radiation Basics*. Washington, D.C.: Nuclear Information and Resource Service.

Fuller, R. (2015). *The Brown Agenda: My Mission to Clean Up the World's Most Life-Threatening Pollution*. Solana Beach, CA: Santa Monica Press.

Gerrard, M. (2015). America's forgotten nuclear waste dump in the Pacific. *SAIS Review* 35(1): 87–97.

Goodell, J. (2017). *The Water Will Come: Rising Seas, Sinking Cities, and the Remaking of the Civilized World*. Boston, MA: Little, Brown.

Graham, K. (2017). Rising sea levels threatens U.S. nuclear waste dump in Pacific. *Digital Journal*. http://www.digitaljournal.com/news/environment/usa-nuclear-waste-dump-threatens-to-contaminate-pacific-ocean/article/508854#ixzz5W1OT9esn.

Gray, C., Carey, R. & Reid, A. (2016). Current and future risks of asbestos exposure in the Australian community. *International Journal of Occupational and Environmental Health* 22(4): 292–299.

Grebenets, V. & Sadowski, A. (1993) Climate warming and thermal regime of foundations of a northern city. *Sub-bases, Foundations, and Soil Mechanics* 5: 27–30.

Grebenets, V., Streletskiy, D. & Shiklomanov, N. (2012). Geotechnical safety issues in the cities of polar regions *Geography Environment Sustainability* 5: 104–119.

Haigh, G. (2006). *Asbestos House: The Secret History of James Hardie Industries*. Brunswick, VIC: Scribe.

Hamilton, T. (2013). A Visual Description of the Concrete Exterior of the Cactus Crater

Containment Structure. Prepared for the U.S. Department of Energy. Berkeley, CA: Lawrence Livermore Laboratory, LLNLTR-648143.

Hasegawa, A., Tanigawa, K., Ohtsuru, A., Yabe, H., Maeda, M., Shigemura, J. *et al.* (2015). Health effects of radiation and other health problems in the aftermath of nuclear accidents, with an emphasis on Fukushima. *Lancet* 386: 479–488.

Hernan, R. (2010) Time Beach, Missouri, 1882. In: *This Borrowed Earth: Lessons from the Fifteen Worst Environmental Disasters Around the World*, pp. 91–100. New York: Palgrave Macmillan.

Hilborn, R., Branch, T., Ernst, B., Magnusson, A., Minte-Vera, C., Scheuerell, M. & Valero, J. (2003). State of the world's fisheries. *Annual Review of Environment and Resources* 28: 359–399.

Hills, B. (1989). *Blue Murder: Two Thousand Doomed to Die, the Shocking Truth About Wittenoom's Deadly Dust*. Melbourne, VIC: Sun Books.

IEA. (2018). Technology Roadmap: Low-Carbon Transition in the Cement Industry. International Energy Agency. https://www.iea.org/reports/technology-roadmap-low-carbon-transition-in-the-cement-industry.

IPCC. (2014). *Climate Change 2014: Synthesis Report. Contribution of Working Groups I, II and III to the Fifth Assessment Report of the Intergovernmental Panel on Climate Change*. Geneva: Intergovernmental Panel on Climate Change.

Iwanoff, A. (1998). Environmental impacts of deep opencast limestone mines in Laegerdorf, Northern Germany. *Mine Water and the Environment* 17(1): 52–61.

Jamail, I. & Davis, B. (1991). Chemicals and environmentally caused diseases in developing countries. *Infectious Disease Clinics of North America* 5(2): 365–375.

Johnston, B. (2007). Half-lives and half-truths, and other radioactive legacies of the Cold War. In: B. Johnston (Ed.), *Half-Lives And Half-Truths: Confronting the Radioactive Legacies of the Cold War*, pp. 1–24. Santa Fe, NM: School for Advanced Research Press.

Kemp, A., Horton, B., Donnelly, J., Mann, M., Vermeer, M. & Rahmstorf, S. (2011). Climate related sea-level variations over the past two millennia. *Proceedings of the National Academy of Sciences of the U.S.A.* 108(27): 11 017–11 022.

Kim, E. & Choi, H. (2011). Assessment of vulnerability to extreme flash floods in design storms. *International Journal of Environmental Resesearch and Public Health* 8(7): 2907–2922.

Kirpich, A. & Leary, E. (2017). Superfund locations and potential associations with cancer incidence in Florida. *Statistics and Public Policy* 4(1).

Kjeldsen, K., Korsgaard, N., Bjørk, A., Khan, S., Box, J., Funder, S. *et al.* (2015). Spatial and temporal distribution of mass loss from the Greenland Ice Sheet since AD 1900. *Nature* 528: 396–400.

Kopp, R., Kemp, A., Bittermann, K., Horton, B., Donnelly, J., Gehrels, W., *et al.* (2016). Temperature-driven global sea-level variability in the Common Era. *Proceedings of the National Academy of Sciences of the U.S.A.* 113(11): E1434–E1441

Korda, R., Clements, M., Armstrong, B., Law, H., Guiver, T., Anderson, P. *et al.* (2017). Risk of cancer associated with residential exposure to asbestos insulation: a whole-population cohort study. *Lancet Public Health* 2(11): 522–528.

Kotov, V. & Nikitina, E. (2010). Norilsk Nickel: Russia wrestles with an old polluter. *Environment: Science and Policy for Sustainable Development* 38(9): 6–37.

Lamere, R. & Singh, O. (2017). Limestone mining and it's environmental implications in Meghalaya, India. *ENVIS Bulletin Himalayan Ecology* 24: 87–100.

Landrigan, P. (1991). The third wave of asbestos disease: exposure to asbestos in place. Public health control. Introduction. *Annals of the New York Academy of Sciences* 643: xv–xvi.

Landrigan, P. & Fuller, R. (2014). Environmental pollution: an enormous and invisible burden on health systems in low- and middle-income counties. *World Hospitals and Health Services* 50(4): 35–40.

Leetham, D. (2018). The cement manufacturing process. ThermoFisher Scientific. https://www.thermofisher.com/blog/mining/the-cement-manufacturing-process/.

Leon, J., Jaffe, D., Kaspar, J., Knecht, A., Miller, M., Robertson, R. & Schubert, A. (2011). Arrival time and magnitude of airborne fission products from the Fukushima, Japan, reactor incident as measured in Seattle, WA, U.S.A. *Journal of Environmental Radioactivity* 102: 1032–1038.

Li, L., Switzer, A., Wang, Y., Chan, C.-H., Qiu, Q. & Weiss, R, (2018). A modest 0.5-m rise in sea level will double the tsunami hazard in Macau. *Science Advances* 4(8): eaat1180.

Loria, K. (2018a). Cities around the U.S. are flooding at high tide and on sunny days at record rates—here's what it's like. *Business Insider*. https://www.businessinsider.com/sea-level-rise-high-tides-sunny-day-flooding-coastal-cities-2018-4.

Loria, K. (2018b). Miami is racing against time to keep up with sea-level rise. *Business Insider*. https://www.businessinsider.com/miami-floods-sea-level-rise-solutions-2018-4.

Luhn, A. (2016). Slow-motion wrecks: how thawing permafrost is destroying Arctic cities. *Guardian*. https://www.theguardian.com/cities/2016/oct/14/thawing-permafrost-destroying-arctic-cities-norilsk-russia.

Lunder, S. (2018). Asbestos kills 12 000–15 000 people per year in the U.S. *Asbestos Nation*. http://www.asbestosnation.org/facts/asbestos-kills-12000-15000-people-per-year-in-the-u-s/.

Lyttle, A. (2003). *Agriculture Street Landfill: environmental justice case study*. University of Michigan. http://www.umich.edu/~snre492/Jones/agstreet.htm.

MacPherson, C. (2019). U.N. and Western countries covered up the facts on the huge health toll of Chernobyl radiation. Fairewinds Energy Education. https://www.fairewinds.org/nuclear-energy-education/un-and-western-countries-covered-up-the-facts-on-the-huge-health-toll-of-chernobyl-radiation.

Marriner, C. (2011). Asbestos fears as crews assess Cyclone Yasi clean-up. *Sydney Morning Herald*. https://www.smh.com.au/environment/weather/asbestos-fears-as-crews-assess-cyclone-yasi-clean-up-20110219-1b0a7.html.

McCulloch, J. & Tweedale, G. (2007). Shooting the messenger: the vilification of Irving J. Selikoff. *International Journal of Health Services* 37(4): 619–634.

McCurry, J. (2012). Fukushima disaster could have been avoided, nuclear plant operator admits. *Guardian*. https://www.theguardian.com/environment/2012/oct/15/fukushima-disaster-avoided-nuclear-plant.

McNulty, J. (2011). Asbestos: the Wittenoom tragedy. WorkSafe. https://www.commerce.wa.gov.au/sites/default/files/atoms/files/wittenoom_transcript.pdf.

Miami International Airport. (2018). Miami International Airport—traffic report. *Ttwelve months ended: December* 31, 2018. http://www.miami-airport.com/library/pdfdoc/Monthly%20Traffic%20Reports/December%202018%20Moving%20Twelve%20Months%20Traffic%20Report.pdf.

Mishra, A. (2015). Impact of silica mining on environment. *Journal of Geography and Regional Planning* 8(6): 150–156.

Murphy, K. (1984). *Big Blow Up North: A History of Tropical Cyclones in Australia's Northern Territory*. Darwin, NT: University Planning Authority.

Nartey, V., Nanor, J. & Klake R. (2012). Effects of quarry activities on some selected communities in the Lower Manya Krobo District of the Eastern Region of Ghana. *Atmospheric and Climate Sciences* 2: 362–372.

Nateghi, R., Bricker, J., Guikema, S. & Bessho, A. (2016). Statistical analysis of the effectiveness of seawalls and coastal forests in mitigating tsunami impacts in Iwate and Miyagi Prefectures. *PLoS One* 11(8): e0158375.

National Cancer Institute. (2018). Mesothelioma. Age-adjusted SEER cancer incidence rates, 1980–2015 by race, sex and year of diagnosis.

https://seer.cancer.gov/csr/1975_2015/results_merged/sect_17_mesothelioma.pdf.

NCEI. (2018). National Climate Report—May 2018: 2017 state of U.S. high tide flooding and a 2018 outlook. National Centers for Environmental Information. https://www.ncdc.noaa.gov/sotc/national/2018/05/supplemental/page-1.

Nghiem, S., Hall, D., Mote, T., Tedesco. M., Albert, M., Keegan, K. *et al.* (2012). The extreme melt across the Greenland ice sheet in 2012. *Geophysical Research Letters* 39(20).

Nicholson, L. (2014). Pilbara asbestos fears follow in the wake of cyclone Christine. *WAToday*. https://www.watoday.com.au/national/western-australia/pilbara-asbestos-fears-follow-in-the-wake-of-cyclone-christine-20140102-3075i.html.

Nweke, O. & Sanders, W. (2009). Modern environmental health hazards: a public health issue of increasing significance in Africa. *Environmental Health Perspectives* 117(6): 863–870.

Nyland, K., Streletsky, D. & Shiklomanov, N. (2017). Climatic- and anthropogenic-induced land cover change around Norilsk, Russia. *Polar Geography* 40(4): 257–272.

Omoti, K., Kitetu J. & Keriko, J. (2016). An assessment of the impacts of gypsum mining on water quality in Kajiado County, Kenya. *Kabarak Journal of Research & Innovation* 4(1): 91–104.

OSHA. (1980). Workplace Exposure to Asbestos. Review and Recommendations. Publication No. 81-103. U.S. Occupational Safety and Health Administration Asbestos Work Group. https://www.cdc.gov/niosh/nioshtic-2/00186632.html.

Peacockm M. (2009). *Killer Company: James Hardie Exposed*. Sydney, NSW: ABC Books.

Povtak, T. (2018). No asbestos ban in the U.S. Asbestos.com. https://www.asbestos.com/legislation/ban/.

Prickett, A. (2018). Intense flooding hits Oklahoma with some areas getting over 8 inches of rain. *Fox News Watch*. https://okcfox.com/news/local/roads-begin-to-flood-as-heavy-rains-hit-oklahoma.

Prince, P., Davidson, J. & Dudley, S. (2004). In the shadow of the corporate veil: James Hardie and asbestos compensation. Department of Parliamentary Services. https://www.aph.gov.au/binaries/library/pubs/rn/2004-05/05rn12.pdf.

Pure Earth. (2007) Top 10 most polluted places: Norilsk, Russia. http://www.worstpolluted.org/projects_reports/display/43.

Pure Earth. (2018). About: a global killer, a solvable problem. https://www.pureearth.org/blog/about/.

Revich, B. (1995). Public health and ambient air pollution in arctic and subarctic cities of Russia. *Science of the Total Environment* 160–161: 585–592.

Revich, B. (2010). *Climate Change Impact on Public Health in the Russian Arctic.* Moscow: United Nations in the Russian Federation.

Robinson, B., Musk, A. & Lake, R. (2005). Malignant mesothelioma. *Lancet* 366: 397–408.

Roosevelt, M. (2004). The tragedy of Tar Creek. *Time.* http://content.time.com/time/magazine/article/0,9171,612395,00.html.

Rosenthal, E. (2007). Cement industry is at center of climate change debate. *New York Times.* https://www.nytimes.com/2007/10/26/business/worldbusiness/26cement.html.

Rugget, A. (2017). Miami's fight against rising seas. *BBC Future Now*. http://www.bbc.com/future/story/20170403-miamis-fight-against-sea-level-rise.

Safranyik, L. & Wilson, B. (2006). *The mountain pine beetle: a synthesis of biology, management and impacts on lodgepole pine. Mountain Pine Beetle Initiative*. Victoria, BC: Canadian Forest Service.

Scheffran. J., Borroughs, J., Leidreiter, A. & Ware, A. (2015). *The Climate–Nuclear Nexus*. London: World Futures Council.

Scientific American. (2015). Tornadoes, severe flooding hit Oklahoma City area. https://www.scientificamerican.com/article/

tornadoes-severe-flooding-hit-oklahoma-city-area/.

Sellers, C., Dillon, L. & Brown, P. (2018). EPA staff say the Trump administration is changing their mission from protecting human health and the environment to protecting industry. *The Conversation*. https://theconversation.com/epa-staff-say-the-trump-administration-is-changing-their-mission-from-protecting-human-health-and-the-environment-to-protecting-industry-96256.

Shahmohammadi-Kalalagh, S. (2013). Flood vulnerability index as a knowledge base for flood risk assessment in urban area. *Journal of Novel Applied Sciences* 2: 266–269.

Shellenberger, M. (2017). Introduction. In: M. Nelson, A. Ramamirthy, M. Czerwinski, M. Light & M. Shellenberger (Eds.), *The Power to Decarbonize*, pp. 3–7. Berkeley, CA: Environmental Progress.

Shiklomanv, N. & Larulle, M. (2017). A truly Arctic city: an introduction to the special issue on the city of Norilsk, Russia. *Polar Geography* 40(4): 251–256.

Simms, A. (2013). *Cancel the Apocalypse: The New Path to Prosperity*. London: Little, Brown.

Simon, S. Land, B. & Beck, H. (2010). Radiation doses and cancer risks in the Marshall Islands associated with exposure to radioactive fallout from Bikini and Enewetak nuclear weapons tests: summary. *Health Physics* 99(2): 105–123.

Sortir du Nucléaire. (2015). Nuclear power: a false solution to climate change. http://www.sortirdunucleaire.org/Nuclear-power-a-false-solution-to-climate-change-44206.

Stern, M. (2017). EPA using controversial process to push cleanup of America's most toxic sites. *The Weather Channel*. https://www.environmentalprotectionnetwork.org/news/epa-using-controversial-process-to-push-cleanup-of-americas-most-toxic-sites/.

Streletsky, D., Sherstiukov, A., Frauenfeld, O. & Nelson, F. (2015). Changes in the 1963–2013 shallow ground thermal regime in Russian permafrost regions. *Environmental Research Letters* 10(12): 125005.

Ten Hoevea, J. & Jacobson, M. (2012). Worldwide health effects of the Fukushima Daiichi nuclear accident. *Energy & Environmental Science* 5: 8743–8757.

Tompkins, F. & Deconcini, C. (2014). *Sea-level rise and its impact on Miami-Dade County*. World Resources Institute. https://www.wri.org/publication/sea-level-rise-and-its-impact-miami-dade-county.

Turner, J. (2018). Florida is on pace to set a new tourist record. *Miami's numbers are up, too*. Miami *Herald*. https://www.miamiherald.com/news/business/tourism-cruises/article218351155.html.

U.S. Bureau of the Census. (2012). Picher, Oklahoma Population: Census 2010 and 2000 interactive map, demographics, statistics, quick facts. http://censusviewer.com/city/OK/Picher.

U.S. Army. (1963). Camp Century, Greenland. http://gombessa.tripod.com/scienceleadstheway/id9.html.

U.S. Army Corps of Engineers. (1962). The story of Camp Century: the city under ice; research and development, progress report no. 6. https://usace.contentdm.oclc.org/digital/collection/p16021coll1/id/34/.

U.S. Geological Survey. (2019). *Mineral commodity summaries* 2019. https://prd-wret.s3-us-west-2.amazonaws.com/assets/palladium/production/atoms/files/mcs2019_all.pdf.

Union of Concerned Scientists. (2016). Tidal flooding and sea level rise in Miami-Dade County, Florida. https://www.ucsusa.org/global-warming/science-and-impacts/impacts/tidal-flooding-sea-level-rise-miami-dade-county-florida#.W8VVvHtKjm5.

van Liedekerke, M., Prokop, G., Rabl-Berger, S., Kibblewhite, M. & Louwagie. G. (2014). Progress in the Management of Contaminated Sites in Europe. *JRC Reference Reports*. Luxembourg City: Publications Office of the European Union Joint Research Centre.

Van Riet, R. (2015). Foreward. In: J. Schefran, J. Borroughs, A. Leidreiter & A. Ware (Eds.), *The Climate–Nuclear Nexus*, pp. 4–5. London: World Futures Council.

Vine, D. & Juliani, T. (2014). Climate solutions: the role of nuclear power. Center for Climate and Energy Solutions. https://www.c2es.org/site/assets/uploads/2014/04/climate-solutions-role-nuclear-power.pdf.

Virginia Tech Daily. (2018). Climate change sea-level rises could increase risk for more devastating tsunamis worldwide. https://vtnews.vt.edu/articles/2018/08/Science-tsunamis_increase_climate_change.html.

Virilio, P. (1999). *Politics of the Very Worst.* New York: Semiotext(e).

Virta, R. (2006). *Worldwide asbestos supply and consumption trends from 1900 through 2003. U.S. Geological Survey Circular 1298: 80.* Reston, VA: U.S. Geological Survey.

Wargo, J. (2009). *Green Intelligence: Creating Environments that Protect Human Health.* New Haven, CT: Yale University Press.

Wdowinski, S., Bray, R., Kirtman, B. & Wu, Z. (2016). Increasing flooding hazard in coastal communities due to rising sea level: case study of Miami Beach, Florida. *Ocean & Coastal Management* 126: 1–8.

Weiss, E. (2011). Cold War under the ice: the army's bid for a long-range nuclear role, 1959–1963. *Journal of Cold War Studies* 3: 31–58.

White, P. & Pickett, S. (1985). Natural disturbance and patch dynamics: an introduction. In: S. Pickett & P. White (Eds.), *The Ecology of Natural Disturbance and Patch Dynamics*, pp. 3–13. New York: Academic.

Wu, J., Zheng, J., Dai, M., Huh. C., Chen, W., Tagami, K. *et al.* (2014). Isotopic composition and distribution of plutonium in northern South China Sea sediments revealed continuous release and transport of Pu from the Marshall Islands. *Environmental Science & Technology* 48(6): 3136–3144.

Zaunmayr, T. (2018). Group gives rare glimpse into Wittenoom's heyday. *Pilbara News.* https://www.pilbaranews.com.au/news/pilbara-news/group-gives-rare-glimpse-into-wittenooms-heyday-ng-b88728531z.

Zhang, K. (2011). Analysis of non-linear inundation from sea-level rise using LIDAR data: a case study for South Florida. *Climatic Change* 106(4): 537–565.

Part 2

Environmental crisis

7

Encountering degrading environments

[We] have been mortgaging the health of future generations to realise economic and development gains in the present. By unsustainably exploiting nature's resources, human civilisation has flourished but now risks substantial health effects from the degradation of nature's life support systems in the future. Health effects from changes to the environment including climatic change, ocean acidification, land degradation, water scarcity, overexploitation of fisheries, and biodiversity loss pose serious challenges to the global health gains of the past several decades and are likely to become increasingly dominant during the second half of this century and beyond.

Lancet Commission on Planetary Health (2015, p. 1973)

7.1 Complexities of the environment–health nexus

From the perspective of planetary health, introduced in Chapter 1, the health and well-being of human communities and individuals are multiply linked to the environment and to other species. Environmental health researchers have developed ways to measure the correlation between a healthy environment and a healthy human body (Institute of Medicine 2001). This work is challenging because the causal pathway between environmental factors and disease outcomes is complex; the connections involved include many different components related to one another in complicated ways. The extent of this complexity is seen by considering the case of malaria. In a review of eco-epidemiological efforts to reduce the disease burden of malaria, Keiser *et al.* (2005) found 16 studies that applied environmental modification (e.g., draining swamps where mosquitoes breed, increasing drainage to speed the drying up of temporary standing pools of water) in which the risk ratio of malaria was reduced by 88 percent. This research, in short, suggests the existence of significant health benefits flowing from planned environmental alteration. Similarly, research conducted in the Brazilian Amazon (Valle & Clark 2013) found that the greater the extent of forested areas (which was used as a proxy for living near forest fringe areas), the higher the incidence of malaria, and that this "forest cover effect" was 25 times greater than the "land clearing effect," which is the oft cited culprit in research on the causes of malaria in the region (e.g., Vittor *et al.* 2006, 2009). In other words, there is research supporting both sides of the question of environmental modification and the frequency of malaria, suggesting the complexity of factors involved in the relationship between health and the environment.

An additional challenge arises because the causes of many health problems are multifactorial. As Myers *et al.* (2013) note, "We must ask: how much of the malaria in the Brazilian

Ecosystem Crises Interactions: Human Health and the Changing Environment, First Edition. Merrill Singer.
© 2021 John Wiley & Sons, Inc. Published 2021 by John Wiley & Sons, Inc.

Amazon is a result of deforestation, how much of the iron deficiency in Madagascar is due to wildlife population declines, and how much of the cardiopulmonary disease in Singapore is due to forest fires in Indonesia?" While determining causal influences among coexistent and potentially linked factors is trying (McMichael & Woodruff 2005), these questions draw attention to the often somewhat concealed nature of environmental causes of health and why there is a need to carefully and vigorously pursue this line of inquiry.

The benefits of this work are evident in the accomplishments of the Superfund Research Program (SRP), an academically based, multidisciplinary effort that funds grants for the development of scientific solutions to health and environmental problems associated with hazardous waste sites. Research funded by the SRP led to the "discovery of arsenic's toxicity to the developing human central nervous system; documentation of benzene toxicity to hematologic progenitor cells in human bone marrow; development of novel analytic techniques such as the luciferase expression assay and laser fragmentation fluorescence spectroscopy [tools for studying gene expression and monitoring biochemicals]; demonstration that PCBs [polychlorinated biphenyls] can cause developmental neurotoxicity at low levels and alter the genomic characteristics of sentinel animals; elucidation of the neurodevelopmental toxicity of organophosphate insecticides; documentation of links between antimicrobial agents and alterations in hormone response; discovery of biological mechanisms through which environmental chemicals may contribute to obesity, atherosclerosis, diabetes, and cancer; tracking the health and environmental effects of the attacks on the World Trade Center and Hurricane Katrina; and development of novel biological and engineering techniques to facilitate more efficient and lower-cost remediation of hazardous waste sites" (Landrigan *et al.* 2015). The SRP has supported research that informs public policy. For example, the U.S. Food and Drug Administration (FDA) issued a rule banning 19 antibacterial chemicals as ingredients in over-the-counter (OTC) antibacterial hand and body washes based on its findings and the inability of manufacturers to demonstrate that these ingredients are both safe for long-term daily use and more effective than plain soap and water in preventing the spread of certain infections. Similarly, SRP-supported research informed efforts by the Environmental Protection Agency (EPA) to remove the organophosphate pesticides chlorpyrifos and diazinon from retail sale and home use.

In one area of environmental-caused disease that is bound up in several ecocrises interactions, progress has followed the introduction of the concept of the "exposome," the totality of adverse environmental exposures (e.g., radiation, infectious agents, chemical contaminants, environmental pollutants, diet) from gestation onward throughout one's life (Wild 2005, 2012) and the ways this accumulation impacts health. Developed initially in cancer research, where an interaction of genetic and environmental exposures to carcinogenic factors is known to occur (Vineis 2004), the exposome concept is intended to match that of the "genome" half of the genetic/environment equation. For example, primary liver cancer is one of the most common cancers globally, accounting for half of all cases in some lower-income countries (e.g., Thailand). On the genetic side of the equation, individual variation in genome influences susceptibility to hemochromatosis (deleterious overabsorption of iron). On the environmental side, epidemiological studies demonstrate a strong association between exposure to aflatoxins (a group of toxins produced by Aspergillus fungi, found on some agricultural crops such as nuts and corn) and an increased incidence of human liver cancer (Su *et al.* 2013). The role of aflatoxins in liver cancer has promoted the development of accurate techniques (e.g., molecular biomarkers) for assessing accumulated aflatoxin exposure to disease development (Makarananda *et al.* 1998; Wild & Turner 2002).

A primary challenge in developing the exposome concept has been the identification of accurate and reliable measurements for many environmental exposures in the external environment. There are statistical challenges as well, including in differentiating correlational and causal factors in disease development and identifying synergistic effects among various exposures. It is also necessary to address the dynamic, life-course, and changing life-stage nature of the exposome. Additionally, some exposures are transient, and the chemicals underlying an exposure that occurred months or years ago may be hard to detect in the body even if they produced an adverse effect. Advances in biomarker technology, the development of personal monitors, improved imaging, the fielding of large-scale prospective cohorts and intervention studies, the creation of data-mining techniques to find statistical associations between exposures and their effects, and the emergence of big data computational capacity have all contributed to addressing these issues, but major hurdles remain.

Wild's (2012) discussion of the incredible advances that have occurred in genome description in recent years suggests what is possible for work on the exposome, but there is a fundamental difference in the political economies of these two areas of health research. Work on sequencing the human and other genomes has had broad support and public financing. By contrast, there is active, aggressive, and well-funded opposition to exposure research. As the Union of Concerned Scientists (2015) reports, following a pattern modeled by the tobacco industry, and before that the asbestos industry, the Chemical Industry's Trade Association's game plan for handling the health risks of numerous chemicals introduced on to the market is to "deny the science, bring in its own experts to counter the evidence, launch misleading advertising campaigns, and pressure decision makers to abandon restrictions on the chemical's use." The goal is to obscure scientific findings and block policies designed to protect public health and safety. For example, the Chemical Industry's Trade Association has worked to fight regulation of formaldehyde, a known carcinogen (National Academy of Sciences 2014). This chemical is widely used in building materials such as particleboard, plywood, and fiberboard, as well in glues and adhesives, medicinal and personal-care products, and furnishings. The reason this continues is that the chemical industry, through lobbying and political campaign contributions, has been able to block or water down science-based legislation designed to reduce public exposure to formaldehyde (Goldman *et al.* 2015).

Well-funded opposition notwithstanding, health researchers continue to explore lifetime environmental exposures. One area of ongoing concern involves prenatal and early childhood exposures as they are recognized as affecting later health outcomes. It is widely understood that organ development during prenatal life and infancy is particularly vulnerable to the effects of environmental risk factors (Robinson & Vrijheid 2015). A body of existing studies now shows that there are health impacts of prenatal exposure to various environmental contaminants, including adverse effects of air pollution, metals, and industrial agents on fetal growth, neurological development, the immune system, and the respiratory system (Wigle *et al.* 2008; Gascon *et al.* 2013; Pedersen *et al.* 2013; Cao *et al.* 2016).

The Human Early-Life Exposome (HELIX) study, for example, was developed to measure and describe multiple environmental exposures during early life (pregnancy and childhood) in a six-country European prospective cohort and associate these exposures with child health outcomes at age 6–11 years. Exposure variables include air pollution, cotinine (an alkaloid found in tobacco), various heavy metals, persistent organic pollutants (degradation-resistant compounds found in industrial chemicals, pesticides, and solvents), perfluorooctanesulfonic acids, phthalates, phenols, and organophosphates. Members of the HELIX research team are examining a range of outcomes, such as hypertension, obesity, lung function, and neurodevelopment (Warembourg *et al.* 2018). Thus far, this ongoing study has found exposure biomarkers with high detection frequencies in all individuals in the multicountry sample. The concentrations of chemical compounds found in the maternal samples are in general higher than those measured in the children's ones. For most compounds that tend to persist in the environment, the correlation between maternal and children's levels is high, while considerably lower correlations are observed for most nonpersistent compounds. Additionally, compound concentrations are significantly different between country cohorts for almost all compounds. As this suggests, even in a heavily polluted world, where you live matters, and communities can suffer what Taylor (2019) calls "spatial injustice."

Today, because the scientific literature exploring the relationship of the environment and health is extensive (Myers & Patz 2009), many of the negative impacts of human alterations have become clearer. There are over 600 chemicals used during natural gas "fracking" (unconventional natural gas extraction) operations (Fig. 7.1), more than 350 of which have been determined to have potential negative health effects (McDermott-Levy *et al.* 2013). A review of the scientific literature on human health and fracking from 2009 to 2015 (Hays & Shonkoff 2016) found 685 papers published in peer-reviewed scientific journals. The majority of the public health studies (84 percent) reveal public health hazards, elevated risks, or adverse health outcomes associated with fracking. Similarly, 69 percent of water quality studies contain findings that indicate potential, positive association, or actual incidence of water contamination. In the same vein, 87 percent of air quality studies indicate elevated air pollutant emissions and/or atmospheric concentrations associated with fracking. In short, the weight of the findings in the scientific literature shows that there are specific health hazards and elevated risks produced by hydraulic fracking. Among these, based on studies in Pennsylvania, are kidney infections, calculus of ureter, and urinary tract infections in women with long-term exposure (Denham *et al.* 2019), hospitalizations for pneumonia among the elderly (Peng *et al.* 2018), lower birth weight and a higher incidence of smaller size for gestational age

Fig. 7.1 Fracking drilling site. *Source:* sobrevolandopatagonia/123RF.

among newborns (Stacy *et al.* 2015), and preterm birth (Casey *et al.* 2016). While there remains a paucity of research on the health effects of fracking, these studies strongly suggest the need for more research given the increasing numbers of people impacted by this fossil fuel industry. Additionally, industry officials and some governments promote natural gas as a climate- and health-friendly "clean, green" fuel, but studies show that fracked natural gas can produce as many greenhouse gas emissions as burning coal. Fracking releases large amounts of natural gas—which consists of both CO_2 and methane—directly into the atmosphere. In fact, there is indication that fracking wells leak 40–60 percent more methane than conventional natural gas wells (Fischetti 2012), leaking "stealthily at every point in the gas supply chain" (Tomasula Y Garcia 2019), and therefore are a source of the adverse health impacts of climate change.

There is, moreover, a mounting body of research that describes the large number of synthetic micropollutants in freshwater and their effects on aquatic life and human health (Schwarzenbach *et al.* 2010). The term "micropollutants" refers to organic or mineral substances whose toxic, persistent, and bioaccumulative properties may have adverse effects. Hazardous micropollutants include heavy metals such as chromium, nickel, copper, zinc, cadmium, lead, and mercury and metalloids such as selenium and arsenic. These elements are not subject to degradation like many of the organic pollutants. Organic micropollutants include solvents, detergents, and cosmetics. An increasingly complex cocktail of micropollutants is finding its way into drinking water supplies. Notably, some are endocrine disruptors; these interfere with normal hormonal processes by mimicking natural hormones, blocking the effects of natural hormones, or directly stimulating or inhibiting hormone production in the body. As these diverse bodies of research suggest, understanding the environment–health interface is critical to assuring the sustainability of human ecosystems and health.

Virulent and often lethal new infectious diseases such as COVID-19, SARS, Ebola, Nipah, and Hantavirus Pulmonary Syndrome have spread to humans following anthropogenic ecosystem changes (Atlas & Maloy 2014; Khan & Patrick 2016). Based on a review of the literature, Nava *et al.* (2017) found that most emergent infectious disease outbreaks are linked to anthropogenic drivers, such as changes in land-use patterns (e.g., deforestation, mining, oil extraction), changes in food production (e.g., livestock intensification), or global trade and travel. The simultaneous expansion of agriculture and urbanization, they found, has significantly modified the structure and functioning of ecosystems and patterns of species distribution and biodiversity. Fragmentation and habitat loss can change the community structure of species and may result in areas with greater risk of zoonotic disease outbreaks (Rubio *et al.* 2014).

More broadly, based on a global assessment covering 102 major diseases, the World Health Organization (WHO) estimates that 24 percent of the disease burden suffered by humans (i.e., the total number of healthy life years lost) and 23 percent of all deaths (i.e., premature mortality) are attributable to environmental factors (Prüss-Üstün & Corvalán 2006). Among children 14 years of age and younger, the rate is particularly high, at an estimated 36 percent. Regional and national wealth differences are significant in the distribution of environmental health threats and their contribution to disease burden. Low-income countries, for example, suffer the most from environmental health problems, losing up to 20 times more healthy years of life per person per year than high-income countries (Prüss-Üstün *et al.* 2008).

7.2 Ecosystem distress syndrome

In its epidemiological assessment, the WHO followed a generally broad definition of the environment that included all of the physical, chemical, and biological factors external to the human host, as well as all related behaviors, though it excluded natural environmental factors that could not reasonably be modified through environmental management. The field of ecohealth, by contrast, focuses specifically on the ways that alterations in identifiable ecosystems affect human health. This requires detailed understanding of ecosystem characteristics, which helps in resolving some of the complexities of the environment–health relationship. As emphasized in Chapter 2, mature ecosystems tend to be complex and intricately entwined networks composed of biotic and abiotic components. Central to an ecosystem's complexity is its tightly interconnected trophic (i.e., food) chain, beginning with plants and extending to herbivores, first-level carnivores, and second-level carnivores, which, like us, prey on all other trophic levels. Also at play are microorganisms and arthropods that decompose waste, creating nutrients for the herbivores.

While there is no agreement among ecologists about what it means to have a healthy ecosystem (Palmer & Febria 2012), a commonly used approach is the comparison of current ecosystem structure to a least-impacted, reference, or historical state. This, of course, requires the availability of baseline data, which for many ecosystems are limited. Moreover, in addition to structure, there are dynamic biophysical processes at work. Alternately, Mageau *et al.* (1995) propose that a healthy ecosystem is one that is sustainable because it has the ability to maintain its structure and function over time despite external stress.

A number of critics have argued against the adoption of the ecosystem health concept. A primary criticism is that ecosystems are neither structurally nor functionally like organisms and do not have the features of organisms, such as health (Ehrenfeld 1992; Sutter 1993). Where such criticisms fail is in overessentializing health and disease in humans. Health does not exist per se in nature, it is a conception developed by and useful to humans. The same is true of disease. Consider malaria again. *Plasmodium falciparum* is a unicellular protozoan organism with a complex developmental process consisting of several stages. It is noteworthy because it is linked to at least half of all malaria cases in humans. The spread of malaria is dependent upon mosquitoes "biting" humans at least twice: once to draw in the gametocyte stage of the *P. falciparum* organism present in the blood of an infected person, and the second time when the sporozoite stage of the organism is present in the mosquito's salivary glands. While the mosquito is biting the second person, *P. falciparum* sporozoites enter their body and travel into the liver, where they mature and multiply. After several days, the mature organisms enter the bloodstream and begin to pass into red blood cells in an attempt to evade the immune system (not that pathogens have intentions) and to remodel the blood cells as a useful place to multiply asexually using the cells' resources. This process ultimately destroys large numbers of the red blood cells, creating the symptoms associated with malaria. But for *P. falciparum*, this whole "disease" process is a food-getting and reproduction process. In other words, "health" is a label (English-speaking) humans use to talk about a desired state of being that entails projecting human desires on to natural processes. Health, in this sense, is no less an artificial feature of ecosystems than it is of human bodies, and probably as useful and hence as likely to be retained.

While ecosystems in principle tend to move toward a stabilization of their component parts, they are subject to stress. The term "ecosystem distress syndrome" was coined to label the situation "that arises when the balanced equilibrium of an ecosystem is disrupted by a failure or disruption of an essential link or links in its chain of being" (Oxford Reference 2018). Ecosystem breakdown can occur naturally over time (or even suddenly, as with the eruption of a volcano) or as a result of human-induced agents of change, such as pollution, alien species, modification of the prevailing temperature range or other environmental conditions, physical restructuring, overharvesting, or toxic dumping (or any combination of these variables) (Rapport *et al.* 1985, 1998b). Breakdown is a consequence of these stressors cascading through the trophic layers and processes of an ecosystem. Commonly, as stressed ecosystems degrade, their role in contributing to the health of humans and other species is diminished, while their contribution to disease is increased (Rapport *et al.* 1998a). Importantly, ecosystem alterations may produce health benefits for one segment of a population while generating health costs for another. These issues present further and ongoing challenges to the effort to fully link the health of the environment to the environment of human health.

7.3 Case studies of degraded environments

The concept of environmental degradation refers to the deterioration of an ecosystem through the depletion or damaging of vital components. As discussed in Chapter 6, human overexploitation of groundwater, for example, can severely degrade ancient aquifer systems. Three semi-independent facets of aquifer degradation have been identified: 1) depletion of aquifer storage and its effects on groundwater availability and both terrestrial and aquatic ecosystems; 2) groundwater salinization arising from various anthropogenic hydraulic disturbances; and 3) vulnerability of aquifers to pollution from land-use and waste-discharge practices like fracking (Foster & Chilton 2003). Degradation occurs in multiple environments, as examined in the following case studies.

7.3.1 Degrading Arctic permafrost

One quarter of the Northern Hemisphere, and 17 percent of the Earth's total land surface, is covered by permafrost (Gruber 2012). Using ice borehole data from a global network of permafrost test sites in the Arctic, Antarctic, and high mountains, Biskaborn *et al.* (2019) found that on average, permafrost regions in the Arctic, Antarctic, and high mountains warmed by 0.5 °F between the years 2007 and 2016. The most dramatic warming occurred in the Siberian Arctic, where temperatures in the deep permafrost rose by 1.6 °F. The growing risks of climate change and permafrost warming, thawing, and degradation in Siberia were discussed in Chapter 6. Another perspective on the impact of climate change on permafrost ecology is provided in a study by Hjort *et al.* (2018), the first research project to explicitly show the amount of fundamental infrastructure potentially at risk across the Northern Hemisphere permafrost zone in a time of comparatively rapid climate change. Building on a tradition of geohazard mapping in permafrost regions that dates to the mid-20th century, this study uses unprecedentedly high-spatial-resolution aerial photography to map infrastructure hazard areas in the whole pan-Arctic permafrost region. It then quantifies the amount and proportion of human-engineered structures in areas where damaged ground subsidence and loss of structural-bearing capacity could harm critical infrastructure by the year 2050 under several computer-generated climatic change scenarios. Previous studies have predicted widespread collapse of existing buildings, roads, and other infrastructure due to the loss of structural integrity associated with permafrost thaw. In an analysis that went beyond the Siberian region around Norilsk discussed in Chapter 6, Shiklomanov *et al.* (2016), for example, projected a significant (i.e., at least 25 percent) climate-induced reduction in the urban infrastructure stability throughout the entire Russian permafrost region by 2050. To date, such studies generally have been limited to a regional focus, but Hjort *et al.* (2018) recognized the need for the kind of pan-Arctic geohazard exploration and infrastructure risk assessment called for by the Arctic Monitoring and Assessment Program working group of the intergovernmental Arctic Council (Hassol 2004).

The focus of the Hjort study is on the thaw occurring in near-surface permafrost (less than 50 feet), because of its central role in sustaining infrastructure. Near-surface permafrost is considered to thaw when mean annual ground temperature is projected to change from below to above the freezing point of water (32 °F) between the baseline (2000–2014) period and the future target period (2050). This target period was selected because of the relatively short lifespan (often 20–50 years) of infrastructure in Arctic regions. Assessment concentrates on current infrastructure considered essential to Arctic communities and economic activity, such as residential settlements and building, transportation systems including roads, railways, and airports, and industrial facilities including pipelines. Hjort *et al.* (2018) reveal that the level of threat to artificial structures is significant. About 70 percent of existing infrastructure in the permafrost domain is located in areas with a high potential for thaw by 2050. Specifically, one third of pan-Arctic infrastructure and 45 percent of the fossil fuel extraction fields in the Russian Arctic are located in high-hazard regions. The results show that the most fundamental Arctic infrastructure will be at risk, even if the Paris Agreement targeted fossil fuel emission levels are achieved, which now looks highly unlikely. This means that by 2050, 3.6 million people, or about three quarters of the current population in the Northern Hemisphere permafrost area, could be impacted by damage to infrastructure associated with permafrost thaw. The potential risk to railways appears to be especially great, with almost 300 miles of the Qinghai–Tibet railway and about 175 miles of the Obskaya–Bovanenkovo railway, the world's northernmost railway, lying in areas of thawing permafrost. Risk to residential infrastructure and pipelines is are high. There are currently more than 1200 human settlements (about 40 of which have populations of more than 5000 people) in the at-risk permafrost thaw zone. The area with the highest risk (greatest projected thaw) holds a human population of nearly 1 million people and about one third of existing pan-Arctic infrastructure. This "hot zone" contains more than 36 000 buildings, 8000 miles of roads, and 100 airports.

For the indigenous peoples of Alaska, permafrost thaw presents a mounting threat to every aspect of their lives. Already, dozens of indigenous communities are directly threatened by climate change. One of these is the city of Utqiaġvik, a place known traditionally among local peoples as "where snowy owls are hunted." Located on the Chukchi and Beaufort Sea coasts north of the Arctic Circle, and previously known as Barrow (until 2016), Utqiaġvik is one of the northernmost cities in the United States. A place of months without sunlight during the day in winter and months of light during the night in summer, it had an estimated population of just under 4500 in 2017 (U.S. Bureau of the Census 2017), 61 percent of whom were Iñupiaq Eskimo. As described on its website, "Although Utqiaġvik is a modern community, subsistence hunting, fishing and whaling are still very important to the local economy. Many residents who work full- or part-time continue to hunt and fish for much of their food" (City of Utqiaġvik 2014). There are no roads that lead to Utqiaġvik. Getting there requires flying into Wiley Post-Will Rodgers Memorial Airport or, if sea ice conditions permit, going by ship. In town, residents use cars or four-wheel-drive ATVs, but they turn to snowmobiles to hunt caribou, moose, and other game.

In December 2017, climate scientists at the National Oceanic and Atmospheric Administration (NOAA) noticed that the data for all of 2017 and the last few months of 2016 from the Utqiaġvik climate monitoring station were missing from the agency's monthly climate report. The cause of this glitch, the scientists realized, was that the temperature in Utqiaġvik had been warming so fast that data were automatically flagged by a programed quality-control algorithm as unreal and removed from the database. Upon checking, it was found that they were in fact accurate and the area around Utqiaġvik was reaching levels of temperature unanticipated when the algorithm was designed (Arndt 2017). This problem reflects the fact that the annual temperature at Utqiaġvik has changed dramatically over the last two decades. Temperatures in October–December 2017 were notably warmer than the average for these months during 1979–1999: October was 7.9 °F warmer, November 6.9 °F warmer, and December 4.7 °F warmer.

The change in climate at Utqiaġvik is not just measured by scientific equipment; it is ever-present and etched in people's experience. Craig George, the son of award-winning children's author Jean Craighead George and the Senior Wildlife Biologist in the Alaska Department of Wildlife Management, who studies bowhead whales and other animals from his home in Utqiaġvik, says that:

> The term is no longer "climate change" at Utqiaġvik. It is "climate changed." No doubt about it, based on my 40 years . . . This year, we had crashing waves onshore and 34 degrees F on winter solstice . . . It's almost like a different planet . . . I've been waiting, almost wishing, for the climate models to be wrong . . . No question now. They're right. Argue about the reason, but the fact is, our world has changed. (quoted in Rozell 2018)

The change was never more apparent than in September 2017 when residents, aided by local construction companies, fought to save Utqiaġvik from being washed away as gale-force winds in excess of 30 mph, combined with wind gusts of more than 45 mph, blew the open sea against the coast. Residents saw waves and surf of 8 feet or higher come crashing ashore, and the city's freshwater source was almost flooded. In the past, sea ice would have covered the water and served as a protective barrier for the shore, but those days are gone (Oliver 2018). Beginning on September 28, heavy equipment was used to build up a gravel embankment to hold back the pressing seawater. Craig George reports:

> Our Public Works people made an amazing save. They've worked around the clock since Thursday night (and are still at it) piling hundreds of tons of gravel on the beach against the big seas, and dozing up berms along the coast to the NARL (Naval Arctic Research Laboratory) hangar area . . . As a result we averted a major disaster . . . No question that a big fall storm poses the biggest threat to our town . . . this will be an expensive storm to recover from. Another 20 knots [of wind] and we would have been devastated. (quoted in Medred 2017)

Even without that extra wind speed, the storm caused millions of dollars in damage to roads, buildings, and archeological sites and led the governor to declare the area a disaster zone.

The change in sea ice at Utqiaġvik is detailed by Mark Johnson and Hajo Eicken (2016) in their study of local sea-ice freeze-up and break-up based on satellite records for the period from 1979 to 2013. Their results show that over the study period, sea-ice freeze-up was delayed by 2 weeks per decade for the Chukchi coast and 1 week per decade for the Beaufort coast. They note: "[f]or both regions, break-up start is arriving earlier by 5–7 days per decade and break-up end is arriving earlier by 10–12 days per decade. In the Chukchi Sea, 'early' break-up arrived earlier by one month over the 34-year period and alternated with a 'late' break-up" (Johnson & Eicken 2016). This suggests that by around 2030, the open-water season between the completion of ice break-up and the start of freeze-up will run from early June until early December, nearly twice the 1978–2013 mean range of 97 days. This means that the threat to Utqiaġvik from late-year storms will continue to increase. Quite literally, the loss of sea ice has put Utqiaġvik on the embattled frontline of climate change. Particularly vulnerable is the city's sewer lift station, at the lowest part of town. Craig George observes: "If that lift station goes out, nearly the entire sewer system for Barrow goes with it" (quoted in Medred 2017).

George's experience is echoed by Stacy Bowen, manager of the family-owned The Fur Shop, which sells an assorted stock of animal skins, inexpensive jewelry, leather goods, bolts of fabric, local crafts, and phone cards. She says: "[t]he weather is definitely not the same as it was 20 years ago. It's been a big concern not just because of erosion but because of subsistence and our way of life" (quoted in Dobbyn 2018). Coastal erosion due to the loss of sea ice is a significant threat to Utqiaġvik. A bluff overlooking the coast has retreated about 20 feet in recent years; community elders report that it has lost several hundred feet during their lifetimes. With each fall storm, the bluff retreats even further (Stone 2018). Micheal Judkins, a Utqiaġvik resident trained as a shoreline monitor, is well aware of the city's vulnerability. He laments, "[t]here's not much you can do to reverse the inevitable . . . I don't think the federal or state governments care about it. If they cared, something would be done about it" (quoted in Dobbyn 2018).

Intense storms are not the only climate change-related threat to Utqiaġvik. Virtually all of the land in the area is underlain by permafrost, and it is thawing. Herman Ahsoak, a local resident, whaling captain, and subsistence hunter, is one of those already feeling the impact of the permafrost degradation (Shankman 2017). One day in August 2017, as he went into his backyard, he found the permafrost covered with large puddles of water from the rainiest summer anyone could remember. In the yard, the traditional skin boat he uses to hunt bowhead whale was covered in plastic to keep it from getting moldy, a practice he has had to adopt as the weather has gotten warmer and more humid. Ahsoak was headed to his family's *siġluaq*, or ice cellar, a critical facility for the subsistence hunters in the community. Meat from whales and land animals is stored underground in the *siġluaq*, where it ferments before it is eaten. As permafrost thaws, however, meat is beginning to spoil. When Ahsoak pried open the heavy wooden top of his *siġluaq*, he saw slush at the bottom—something that had never happened before.

The permafrost melt signaled by the decay of ice cellars in Utqiaġvik is directly tied in two ways to the loss of sea ice. The first is simply that both are being driven by a warming climate (Jorgenson *et al.* 2006). The second involves the albedo effect (the absorption of solar energy by open sea that has lost its reflective ice cover). As Lawrence *et al.* (2008) note based on their research, "rapid sea ice loss forces a strong acceleration of Arctic land warming . . . (3.5-fold increase, peaking in autumn) which can trigger rapid degradation of currently warm permafrost and may increase the vulnerability of colder permafrost for subsequent degradation under continued warming."

Another Inupiaq community suffering the effects of climate change is Shishmaref, on the tiny island of Sarichef in the Chukchi Sea, which has been dubbed the "poster child for climate change" in the media (Sheppard 2014). The 600 people who make up its population are probably the first to be the subject of a children's book on climate refugees. Titled *My Wounded Island*, this engaging book by Jacques Pasquet and Marion Arbona (2018) tells the story of the threat to Shishmaref from the viewpoint of Imarvaluk ("The Song of the Waves" in Inupiaq), a young girl created by the author to relay her fears about the mysterious sea creature that is slowly devouring her island. Outsiders who have studied this creature, she explains, call it "climate change." Imarvaluk describes how her home will eventually disappear into the sea and her people may have to relocate to the city of Nome. She worries that the culture of her people and their traditional foodways—based on hunting bearded seal, walrus, elk, and caribou, catching fish, and picking berries—are at risk as they confront the demise of their island. Pasquet is not Inupiaq but lived with Eskimo people in Alaska for a decade. In this fictional account, he tells a very true and urgent story.

Because of warming temperatures, Shishmaref, like Utqiaġvik, is threatened during the closing months of each year by punishing storm surges that bring flooding and erosion of the village's shoreline. Without the buildup of protective shoreline pack ice, it has become ever more vulnerable and is being pulled into the sea at a rate of about 10 feet per year, in a process that has been going on for decades. Satellite images taken of the island over the past 20 years show how far the shoreline has retreated. As a result, a number of buildings have tumbled into the sea. According to John Sinnock, a carving and traditional crafts teacher at the Shishmaref School, "[i]n front of my mother's house we used to look out at the gentle slope . . . In my life time we've probably lost about 400 feet" (quoted in Brown & Morgan 2007). Many houses have had to be moved away from the area of greatest erosion. Those that are not moved in time tumble down the slope to the water's edge. Some people who have had to move their homes away from the sea realize they have not moved far enough and in time the hungry coastline puts their new homes at risk. Esau Sinnok, a 19-year-old community member, who, with the help of the Sierra Club, attended the international climate change negotiations in Paris in December 2015, posted a statement on his personal blog in 2016 saying that his community had "lost 2500 to 3000 feet of land to coastal erosion" over the past 35 years (quoted in Kennedy 2016). As a result, he wrote, his family had had to move 13 times in 15 years.

The dilemma facing the residents of Shishmaref is a product of both a changing environment and their subordinate relationship with American society. Traditionally, the Inupiaq did not live in permanent year-round communities. Their ancestors inhabited the coastal and river drainage areas around Sarichef Island for thousands of years, and had a rich cultural tradition that included environmental expertise and the social organization, skills, and pre-industrial technologies needed for life in this part of the Arctic. During that era, prior to the turn of the 20th century, the Inupiaq were mobile and followed a regular rotation, spending the summer months inland and winter and spring along the coast (Burch 1988, 2006). Barrier islands such as Sarichef Island were used traditionally only as settlements in winter and hunting grounds in spring—periods of local food abundance—and remained uninhabited during the fall. Worsening environmental conditions would normally trigger community decisions to relocate to more promising locales. Then things changed. The U.S. government began promoting sedentarization throughout Alaska. This strategy, touted as both efficient and part of a civilizing process intended for the good of the "backward" indigenous peoples, was, in fact, tied to the state project of population control and nation-state construction (Scott 1998). Even the community's name was imposed upon it. The original Inuit name for the island was Kigiktaq, but in 1816 Lt. Otto Von Kotzebue, a navigator in the Imperial Russian Navy on a voyage to find a passage across the Arctic Ocean, named the inlet Shishmarev after a member of his crew. The village was named after the inlet.

The changes wrought by government's directed centralization and population concentration efforts were facilitated through a state–church nexus. Infrastructure development (e.g., electricity, sanitation, public services, gravel roads, a barge landing facility, a paved airstrip) was accompanied by enforced consolidation of mobile family groups into sedentary villages. On Sarichef Island, the government constructed a post office in 1901 and a school in 1906. The Lutherans built a church in 1929, and missionaries pushed for converted indigenous people to live near them (Koutsky 1981). Western-introduced infectious epidemics also promoted residential consolidation. Of pivotal importance was the establishment of a school infrastructure by the U.S. Department of the Interior's Bureau of Education and associated policies that required attendance (Berardi 1999). Over time, everyday life on the island became deeply rooted in the imposed infrastructure of education, paid government employment, purchased commodities, and fossil fuel transportation, though a wide range of traditional subsistence, artistic, and social network practices survived. As Marino (2012b) points out, today Shishmaref "is significant both to residents and to far reaching networks of people [in places like Nome and Anchorage] who share in identity and practice through food that is widely distributed and praised." Along the way, meaningful social identity—under the threat of ever-present Western values, symbols, and artifacts—came to be invested in a place that in the past was but part of a wider seasonal subsistence rotation cycle. Going back fully to a pre-contact lifeway is impossible, in part because of the changing climate.

For years, people in the community have grappled with the issue of relocation. Some favor the idea, recognizing that it is a hard but unavoidable choice. Others do not want to move away from the island. As one resident told a reporter, she is disposed toward remaining in place because of the island's proximity to traditional subsistence foods that people rely on for much of their diet. According to community member Annie Wyiouanna: "You are taking away our identity . . . That's like a punch in the gut. That's who we are—we are tied to this land and the sea . . . It's a dividing decision; it's hard for me to say or even talk about it" (quoted in Hughes 2017).

The site where the community might relocate is about 20 miles away on the mainland. It is surrounded by the 2.5 million-acre Bering Land Bridge National Reserve. This means the federal government will have to grant permission to construct roads to the new site and to develop a quarry for foundation materials. In response to a Congressional request, the U.S. Army Corps of Engineers estimated in 2004 that the cost of moving (building by building) over a 5-year period would be $180 million, an exorbitant amount for a community in which the median household income was then under $30 000 annually. The Corps asserted that moving and assimilating the entire population of Shishmaref into the city of Nome, 120 miles away, would cost about half the price of rebuilding the town in a new location (TetraTech 2004). Still, residents have voted multiple times to relocate the community to the mainland, and several community members have testified about their plight before Congress. Not much has changed, however, and the clock continues to tick on the island's future. Martin (2018) notes that "[t]he U.S. government was very effective when it wanted to make people settle in this particular place, but now that it's clear they need to relocate, it's so far proven completely ineffective in helping them to get out."

To understand local, embedded practices of social and political relocation planning in Shishmaref, Marino (2012a) and colleagues spent a total of 6 months in the village collecting interview, survey, life history, oral tradition, and ethnographic data. Over 40 interviews and 30 household surveys were completed, along with extensive informal conversations with local residents. In the community, researchers participated in the activities of daily life to the degree possible. Potential interview and survey questions were discussed with local participants to confirm their cross-cultural and linguistic accuracy. Survey respondents were asked to indicate the extent to which they agreed with a set of attitude statements on a Likert-type scale ranging from 1 (strongly disagree) to 5 (strongly agree). Findings showed that respondents who reported understanding which government agencies were working toward funding relocation were more likely to believe that a disaster would occur before relocation could take place. Government-driven adaptation and relocation planning was met by profound suspicion in the community. Interview data corroborated and extended these findings. Exemplary comments included: "I don't believe the political structure/

process can do it [fund relocation]. It is too slow. It's always in the planning stages, but there's no funding for it. One day we are going to be evacuated"; "Nothing's being done. Look, we're still here"; and "Let the federal agencies come here and experience a whole storm, not come for the day and leave. Let them be here two weeks, so they could see it for themselves, 'cause it always seems like they don't believe us." Overall, the research findings indicated that global climate change is a pressing worry among community members, that there is significant distrust in the government, and that the community feels misunderstood by government representatives. The perceived severity of climate change combined with a high level of perceived vulnerability thrust community members into a quandary that seems irresolvable barring significant government or other outside assistance. While the effects of climate change whittle away at the island, residents and the government appear to be at a standstill—a troubling reflection on the wider inability of the nations of the world to take the kinds of bold actions needed to limit the growing environmental disruptions of climate change.

Meanwhile, life goes on in Shishmaref. In 2019, efforts began to construct an expansion for the community school so that the students would no longer be forced to attend class in temporary trailers during the harsh winter months. Planning is underway to reconstruct a road that was severely damaged by a storm in 2013 (Hofstaedter 2019). The village remains caught between current needs and an uncertain future. Complicating the challenges facing the community is COVID-19. To avert introduction of the disease, it "has sealed itself off from the rest of the state: No one may enter the village except medical workers, emergency workers and essential maintenance workers; anyone who has traveled must self-quarantine for two weeks" (Hilleary 2020). This same strategy was used effectively during the 1918 influenza epidemic.

In 2009, the U.S. Government Accountability Office issued a report titled "Alaska Native Villages: Limited Progress Has Been Made on Relocating Villages Threatened by Flooding and Erosion," referring to 31 Alaska villages facing imminent threat from climate change (out of a total of more than 200 indigenous villages identified as being affected to some degree) (GAO 2019). A decade later, the same report with the same title could be issued again (with limited edits), except that the adverse effects of flooding, coastal erosion, and permafrost thaw are now worse. The year 2009 is also of note in this regard because it was the date of the Climate Summit in Copenhagen, an event that Bruno Latour (2010, p. 473) described as "the greatest assembly of representatives of the human race [who] managed to sit on their hands for days doing nothing and making no decisions whatsoever," a portrayal that in its concluding phrase applies as well to the government response to the environmental crisis faced by many indigenous Alaskans.

Amidst the general pattern of denial about the descending fate of indigenous Alaskans buffeted about by climate change, there is the noteworthy baby step taken by the federal government in 2018 on the day President Trump signed the nation's annual budget. Included in this $1.3 trillion budget, on page 458, was $15 million of funding for the Denali Commission, an independent federal agency designed to provide critical utilities, infrastructure, and economic support in Alaska. The Commission had previously provided over $2 million in funding to the threatened indigenous community of Newtok to support its relocation. It announced in March 2018 that most of the money it received in the new federal budget would also go to advance the relocation of Newtok, though it is estimated that the total cost of moving the village will be well over $100 million (Waldholz 2018).

With a population of just over 300 people living in about 60 households, Newtok is located near Alaska's western coast. It has a school but no running water or sewage pipes. Community members have been watching houses slide into the Ninglick River, which wraps around the settlement on three sides, for a number of years. Coastal storms and thawing permafrost have further worn down the land on which Newtok sits. About 70 feet of land a year erode away between the river and the village, putting the community's buildings ever closer to the water's edge (Waldholz 2017). Newtok has already lost its barge landing, sewage lagoon, and landfill. As river water flows in and land sinks, the community expects to lose its drinking water source, school, and airstrip within the next few years.

Residents know they must relocate—a decision they made in 1994. A new village site called Mertarvik (Yup'ik for "place to get water"), located at the top of a high ridge of dark volcanic rock, has been selected. It is 9 miles away from the current community, a half-hour journey by snowmobile. The village plans to retrofit old military barracks it acquired from the Joint Base Elmendorf-Richardson, a U.S. military facility in Anchorage, and move them by barge to the new site. Engineers have estimated that this strategy is cheaper than building brand new houses. In June 2019, Air Force and Marine units began work constructing the new site. The Newtok community had invited the Department of Defense to conduct its Innovations Readiness Training in Mertarvik, allowing reserve forces to practice setting up infrastructure in unfamiliar environments. Military personnel are building a gravel road, a landfill, a heavy equipment shop, and four homes. Other construction workers at the new townsite include Newtok residents, staff from the Alaska Native Tribal Health Consortium, and employees from Native-owned contracting companies. Additionally, the Lower Kuskokwim School District Board has passed a resolution to establish a new school in Mertarvik (Kim 2019), while the Cold Climate Housing Research Center in Fairbanks has designed an energy-efficient model for new homes built there (Rozell 2019).

7.3.2 Drugged aquatic environments

Pollution of various kinds is a major source of environmental degradation. In recent years, there has emerged a growing recognition of the damage being caused by drug pollution of the aquatic environment involving pharmaceutical products and their metabolites. This form of pollution has four primary sources: 1) pharmaceutical drugs excreted in urine by individual users; 2) drugs flushed by individual users and hospitals; 3) agricultural runoff of antibiotics and other drugs in livestock; and 4) waste dumping by pharmaceutical manufacturers. These bioactive substances reach water systems because of inadequate sewer, septic, and water treatment systems or other systemic problems (e.g., corruption in the case of dumping by pharmaceutical manufacturers) (Vijay 2017).

In assessing freshwater systems, ecologists differentiate lotic systems (those such as rivers and streams with flowing waters) and lentic systems (self-contained bodies of water like ponds and lakes). Both prescription and OTC drugs have been detected for many decades in lotic systems across Europe and the United States (Garrison *et al.* 1976; Aherne *et al.* 1990). The most commonly found are antibiotics, analgesic and anti-inflammatory drugs, steroid hormones, antihypertensives, and antidepressants. More recently, researchers began testing lentic systems for pharmaceuticals. Surface water samples taken at various Great Lake sites in Canada by Metcalfe *et al.* (2009) in 2000 and 2002 found evidence of clofibric acid (a metabolite of the cholesterol-lowering pharmaceutical drug clofibrate), the anti-inflammatory drugs ketoprofen and fenoprofen, the antidepressant drug fluoxetinem, the antibiotic drug trimethoprim, and the antiepileptic drug carbamazepine. Caffeine and cotinine, a metabolite of nicotine, were also detected. Research in the United States at four southern Lake Michigan sites found the OTC pain relievers acetaminophen and ibuprofen, as well as the anticonvulsant drug carbamazepine, the antibiotics lincomycin and sulfamethoxazole, the anti-inflammatory drug naproxen, and the antimicrobial drug sulfadimethoxine, used in veterinary medicine (Ferguson *et al.* 2013).

In a review of the literature on the presence of antihypertensive drugs in different aquatic environments, Godoy *et al.* (2015) identified 43 articles presenting quantitative data on acute and chronic ecotoxicological effects of 34 different drugs and their metabolites. They found that the beta-blockers atenolol, metoprolol, and propranolol were most frequently detected in the aquatic environment; they also had the highest maximum concentrations in surface waters. The highest percentages of ecotoxicity data regarding antihypertensives were likewise related to these beta-blockers. In the researchers' assessment, all three may pose a long-term risk for organisms that inhabit fresh and marine waters, though data on saltwater species are particularly limited.

While the presence of pharmaceuticals in freshwater aquatic environments has been detected in multiple locations, assessment of their effects on aquatic organisms is less advanced (Fent *et al.* 2006; Puckowski *et al.* 2016). Several effects have been shown to occur due to continuous exposure to low drug concentrations during the life cycles of aquatic vertebrates and invertebrates. These accumulate slowly and manifest themselves as enhanced newborn length, which is frequently only detected after several generations and may affect the sustainability of aquatic organism populations (Santos *et al.* 2010). However, this branch of research on biologically active compounds is still in its infancy, and findings thus far are limited.

Pharmaceuticals are designed to be effective at very low concentrations, and they are continuously being discharged in natural ecosystems, exposing aquatic species to admixtures of drugs over their entire life course (Bebianno & Gonzalez-Rey 2015). Available findings suggest that the multiplicity of pharmaceuticals in aquatic environments may pose a growing threat to aquatic lifeforms and, possibly, even to humans (Hester & Harrison 2016). Based on a review of published studies, Mezzelani *et al.* (2018) concluded that complex ecotoxicological effects of pharmaceutical production and products are increasingly being documented, and that multidisciplinary, integrated approaches are needed to clarify the environmental hazard they present to aquatic environments. Given exposure of the lack of pharmaceutic vigilance for humans (McGreal & Walters 2019), it is not likely that there has been much ecopharmacovigilance to date.

7.4 Case studies of fragmented environments

Environmental fragmentation is a form of disruption that involves the division of large, contiguous areas with a shared ecosystem into smaller, separated segments. Common fragmenting features are roads, agricultural development, utility corridors, pipelines, logging, housing subdivisions, and other construction. As natural history writer Jason Daley (2018) observes, "[c]ountry musicians may sing about the highway as a symbol of freedom, but for animals it's a borderline where few cross and many never to return." Fragmentation often occurs incrementally over time, beginning with smaller cleared patches or corridors such as roads and spreading outward to encompass contiguous areas. Wilcox & Murphy (1985, p. 884) see it as "the most serious threat to biological diversity and . . . the primary cause of the present extinction crisis." Fragmentation, by creating a mosaic landscape of isolated patches, drives population declines, alters ecological relationships among species, and damages overall ecosystem structure.

Fragmented landscapes have important features and dynamics that impact biodiversity, species population densities, and species relationships. One key factor that influences diversity is the degree of habitat heterogeneity within isolated fragments. Fragments of the same approximate size that are more heterogeneous in their features can support a larger number and variety of species than more homogeneous ones. Fragment size is also critical as larger fragments have a greater likelihood of supporting a broader variety of soil types, a higher degree of topographic variation, more microclimatic variation, and a larger number of habitat subtypes. Fragment features are linked. In New Zealand, Ewers *et al.* (2007) found in a study of 900 beetle species that a synergistic interaction occurs between the size of a habitat and the distance to a fragment edge, and that this interaction is a more important determinant of patterns in beetle community composition than direct edge or area effects alone. Additionally, fragments that support wider floral diversity tend to have greater microclimate variation (Collinge 1996). Fragmentation threshold is species-specific, as spatial needs vary across different kinds of organisms and thus there is a wide range of responses to fragmentation across species (Debinski & Holt 2000). At one extreme are species at higher trophic levels such as large predators. A young female Arctic fox that had been fitted with a satellite transmitter amazed scientific observers by traveling over 2175 miles from Norway to Canada in just 76 days, one of the longest movements from birthplace to potential breeding location ever recorded for the species (Guy 2019). From 2002 to 2009, Field Museum's Curator of Mammals Bruce Patterson tracked three lions living in Tsavo National Park in Kenya using GPS collars (Field Museum 2018). He found that one traversed an area twice the size of Chicago in order to find adequate food and attract mates. Apex predators, like lions, tend to need large spaces. By contrast, some ground beetles of the family Carabidae (Fig. 7.2) are not significantly impacted by fragmentation (Davies & Margules 2002). Consequently, there is considerable interest among researchers in traits that might confer susceptibility or resilience in the face of expanding fragmentation. Based on a review of the literature, Henle *et al.* (2004) identified 12 traits or trait groups that have been suggested as predictors of species sensitivity to fragmentation: population size; population fluctuation (ability to maintain stable populations) and storage effect (in plants, the storage of reproductive potential across generations); dispersal power (ability to migrate); reproductive potential; annual survival; sociality; body size; trophic position; ecological specialization, microhabitat, and wider environmental use patterns; level of disturbance and competition sensitivity; rarity (low abundance within a habitat); and biogeographic range. Assessment of the empirical data found support for six of these traits: population size; population fluctuation and storage effects; competitive ability and disturbance sensitivity in plants; microhabitat specialization and wider environmental use; rarity within a habitat; and relative biogeographic position.

Fig. 7.2 Ground beetle. *Source:* Kathy2408/Pixabay.

Species with greater special needs and relatively narrow ecological tolerances are at greater risk of extinction under conditions of fragmentation, a trait referred to as "extinction proneness." But the pace of fragmentation is significant. Long evolutionary timescales allow for adaptation to fragmentation trends through selection for traits that strengthen dispersal ability and diminish narrow resource specialization. Fast-paced anthropogenic change, by exceeding the capacity of most species to adapt, generally does not allow this kind of response, though some species in sometimes uncanny ways do adapt well to living in human-shaped environments. The Eurasian blackcap, a songbird that is native to Germany and Austria, typically migrates to Spain for the winter. But in recent decades, some blackcaps have begun to fly to the United Kingdom instead. They are attracted by backyard bird feeders, which provide a dependable source of food. In a parallel adaption in Florida, American manatees have found an unusual place to keep warm during prolonged cold weather: the discharge pipes of coastal power plants. Making the best of a bad situation, even roadbuilding for logging has been taken advantage of by orangutans in the Wehea Forest on the Indonesian part of Borneo. Orangutans there have begun to show a preference for walking along roads built by the timber industry to get to other parts of the forest using less energy (Bittel 2015).

Of special concern here are synergistic interactions between fragmentation and other components of global environmental change (Didham 2010). In the past, for example, inadequate attention was paid to the synergetic effects of habitat fragmentation and climate change. As Opdam & Wascher (2004, p. 286) stress, "[i]n most studies it [was] assumed that the habitat is homogeneous, allowing a geographical response only limited by the shift of temperature, and not by the density and the configuration of habitat at the landscape level." Increasingly, it has come to be recognized that all known features of landscapes and climate must be considered in the assessment of synergy to predict the extent and nature of ecocrises interaction.

For example, the geographical distribution of butterflies has been found to track climate warming, but in fragmented landscapes the movement of habitat specialist species is limited by deteriorating conditions, causing climate change impacts to be most severe in those landscapes that are heavily fragmented (Thomas *et al.* 2006). This adverse interaction is seen in the case of monarch butterflies (*Danaus plexippus*), an iconic pollinator species. Most monarchs spend the winter in Mexico, clustering together in the high-altitude oyamel fir trees (*Abies religiosa*) of Michoacan and Mexico states, where the cool temperatures slow their metabolism and allow a conservation of lipid reserves (Anderson & Brower 1996). Most monarchs living west of the Rocky Mountains, however, spend the winter in and around Pacific Grove, California (nicknamed "Butterfly Town, USA"), congregating in pine, cypress, and eucalyptus trees (though some do migrate to Mexico). The remainder of the year, monarchs can be found throughout much of the continental United States and southern parts of Canada, as long as milkweed plants (of which there are several species) are available. Monarchs lay their eggs on these plants, as milkweed is the only food the black and yellow monarch caterpillars consume. Increasingly, however, human land development and agricultural practices, including the use of herbicides such as glyphosate, have eliminated or fragment milkweed habitat (Stenoien *et al.* 2018). This alteration is making it harder for monarchs to find stopover sites where they can lay their eggs during their summer migration. Global warming is putting additional pressure on a species that is already environmentally stressed. With the earlier arrival of spring, the monarchs that head north to Texas are beginning to arrive before milkweed has developed enough for to allow them to lay their eggs. Moreover, monarch eggs, caterpillars, and cocoons develop more quickly under milder weather conditions. Temperatures that rise above 95 °F can be lethal for caterpillars, while hotter, more arid conditions can cause eggs to dry out, significantly reducing hatch rate. Research by Lemoine (2015) examining the impact of warming temperatures on the monarch's breeding range indicates that over the next 50 years, milkweed will expand its north range further into Canada, and that the monarchs will follow this needed plant as they encounter reduced habitat suitability in the United States. While this will expand the breeding locations of the butterfly, it will also increase the distance it has to travel southward to reach its overwintering habitat in Mexico, resulting in greater migration mortality. In the words of monarch researcher Andy Davis (2015): "It would be like moving the finish line further back each year in the Boston Marathon and watching how the number of finishers goes down each year. Travelling farther means greater risks of storms, predation, road mortality [e.g., crashes into vehicle windshields], etc." The end result will be fewer monarchs reaching Mexico and fewer migrating northward again each spring.

7.4.1 Fragmenting sky islands

One unique place where anthropogenic fragmentation and anthropogenic climate change are interacting adversely is in the sky island region of southeastern Arizona, southwestern New Mexico, and northern Mexico (Gottfried *et al.* 2004). Comprising 40 distinct mountain ranges—which constitute biogeographical islands, in that their temperate and subtropical climates contrast sharply with the sea of arid land at their bases—the sky islands are characterized by a degree of biological diversity and ecological interactions not found anywhere else on Earth. In the words of Aldo Leopold (1995), considered by many to be the father of wildlife ecology, they "come near to being the cream of creation." Native species of this 47 million acre biodiversity hot spot include jaguars, thick-billed parrots, grizzly bears, black bears, coati, bighorn sheep, bison, Mexican gray wolves, black-tailed prairie dogs, Apache trout, eared quetzal, flame-colored tanager, and gray hawks. Bee, reptile, and ant diversity is unparalleled. Migratory birds that spend time in the region represent a species range that rivals the tropical forests of Central America. Similarly, numerous plants grow in the sky islands' diverse faunal communities. Some, such as the Chiltepine (*Capsicum annuum* var. *galbriusculum*), a wild pepper, only grow in this region. All together, it is estimated that there are more than 7000 plant and animal species living in the sky islands. Included in this bountiful menagerie are at least 41 endangered species and many others considered to be threatened—the highest number of endangered and threatened species in all U.S. National Forest lands.

In the U.S. sector of the sky island region, fragmentation is occurring because of human population increases and the building of habitat-isolating paved roads, four-lane freeways, spur roads created by recreational all-terrain vehicles, and rooftops that disrupt the natural movements and migratory patterns of native wildlife. In the Mexican portion, conditions are more stable, but illegal logging and the illicit drug trade near the border are beginning to damage and degrade environmental conditions. Kristian Beadle (2017) notes, "4 acres of forest are clear-cut on average to produce 1 acre of [illicit] crop. The result is habitat destruction, emission of greenhouse gases and erosion that fills creeks with silt. Fertilizers and herbicides are used in excessive quantities, which poison and deplete the land and waterways." Also a threat is the border wall promoted by President Donald Trump, which, if built, would block migratory corridors and wildlife pathways (Misztal 2018). Though the sky island region is enormous, ideal habitats are scarce and water availability shifts over time and place. Survival in such an environment requires mobility for many species, and mobility and "big, beautiful walls" are not compatible (Carswell 2017).

Climate change is adding to the woes of the sky islands, as they sit in one of the fastest-warming regions in North America. Annual average temperatures are projected to increase between 2.5 and 5.5 °F by the years 2041–2070 and between 5.5 and 9.5 °F by the end of the century. Summertime heat waves are projected to become longer and hotter, with less soil moisture and increased frequency of large wildfires (National Climate Assessment 2018). These changes are expected to produce enhanced habitat fragmentation because of their negative impacts on forest health, waterway morphology, watershed moisture retention, and water table levels (Misztal *et al.* 2013). As a result of the dual action of direct (e.g., road building) and indirect (climate change) human impacts, the species-rich sky island region is fragmenting, with dire consequences for native species.

7.4.2 Fragmenting forests

Fragmentation is especially common in forested areas, resulting, at its extreme, in dispersed and disconnected forest islands. Critical components of forests that are most affected by fragmentation are their edges, where the forest abuts nonforest areas (Fig. 7.3), and their core interior areas, which are at some distance from the nearest edge. Forest fragmentation is globally ubiquitous, with nearly 20 percent of the world's existing forest now found within 325 feet of an edge and 50 percent within 1650 feet (Pfeifer *et al.* 2017).

In ecology, the term "edge effects" refers to the changes that tend to occur in the border areas. Edge effects are dominant drivers of environmental change in many fragmented landscapes. At the edges, light, wind, heat, moisture, and invasive species (including human pets) can enter and change the forest. As Collinge (1996, p. 63) notes, "[i]ncreased wind velocities at the forest edge may result in increased incidence of mortality due to treefall, especially for shallow-rooted tropical trees." Research in Brazilian tropical forest shows far greater tree mortality on windswept edges compared to edges receiving less wind (Bierregaard *et al.* 1992). Small forest fragments have a high edge/area ratio, resulting in greater disruption. Edges vary depending on several factors, especially what lies beyond them (e.g., meadow, road, suburb, agricultural field, city) and their orientation (depending on the hemisphere, forest edges on the north or south receive less or more sunlight than those on the opposite side, producing differing

Fig. 7.3 Forest edge. *Source:* Free-Photos/Pixabay.

faunal assemblages). Forest fires are more common near edges because of increased light availability and resulting desiccation and understory plant growth, which fuels burning. Edges also have higher contact with human-initiated fires (e.g., to clear land for farming, cigarette discard along roads).

In the Amazon, for example, rainforest fragments have been found to experience a dramatic loss of aboveground tree biomass (Laurance *et al.* 1997), which is greatest within 325 feet of fragment edges. In this space, tree mortality sharply increases due to microclimatic changes and elevated wind turbulence. Losses are as high as 36 percent of tree biomass in the first 10–17 years after fragmentation in some areas. While forest fragmentation is considered a greater threat to vertebrates than to tree communities because of the long lifespan of individual trees and their stationary nature, in the Amazon it can produce abrupt shifts in floral composition, tree population declines, and local extinctions of large-seeded, slow-growing, old-growth tree taxa near forest edges (Laurance *et al.* 2006). Fragmentation has also been found to lead to a significant drop in tree seedling density in this region (Benetiz-Malvido 1998).

A very different environment, the core interior of forests—free of edge effects—is found about 200–300 feet from the nearest edge, depending on location and dominant tree species. In deciduous forest fragments in Pennsylvania and Delaware, temperature and light are known to decrease within relatively short distances, while edge-like humidity and leaf litter moisture continue to change 165 feet into the forest interior (Matlack 1993). In old-growth Douglas fir forests of the Pacific Northwest, edge effects have been shown to extend as far as 785 feet in (Chen *et al.* 1995). A forest needs to be more than 14 acres in size to encircle just 1 acre of interior forest. Extensive forest fragmentation can lead to the creation of "forest patches" that are too small to contain any core forest areas at all. Though some species prefer forest fringes, the core areas provide a more stable home for many, protecting biodiversity. Forest continuity allows inhabitants a wider range in which to live, search for food, and find shelter. Species that rely on forest core environmental conditions are especially put at risk by forest fragmentation.

In India, the seventh-largest country in the world, Reddy *et al.* (2013) carried out a national assessment of forest fragmentation based on the division of the country into biogeographic zones defined by topography, land use, geography, and climate for the 30-year period from 1975 to 2005. Their analysis revealed that in all of the biogeographic zones, more than 90 percent of identified forest fragments consisted of patches of less than 250 acres. Temporal analysis showed that the loss of Indian forested area was almost 6 percent during the study period. Decreasing mean patch size, increasing number of patches, and increasing extent of forest edges suggest a national pattern of ongoing fragmentation in all of India's biogeographic zones.

India has had a long history of deforestation dating to British colonial rule (1880–1947) and the early decades following independence. The Colonial British government passed several pieces of legislation to facilitate its control of Indian forests and increase the revenue it extracted from the region through the export of timber and the construction of infrastructure (e.g., railroads) (Tian *et al.* 2015). The Forest Act of 1865 empowered it to appropriate any forested area not covered by the traditional rights of individuals and communities (Mohapatra & Mohapatro 1997). But this law was superseded by the Indian Forest Act of 1878, which was designed to remove all land rights that were not explicitly granted by the colonial state to insure strict control over forest resources. This act constituted a large-scale annexation of Indian forests by the colonial state and began a process of forest commodification that continued following independence (Gadgil & Guha 1992).

Exemplary of forest fragmentation in a particular biogeographical zone in India are the changes in the Shopian district in the northernmost state of Jammu and Kashmir for the period 1995–2015, as reported by Amin & Fazal (2017). This region is characterized by temperate forests dominated by coniferous vegetation. Common tree species include the Deodar cedar (*Cedrus deodara*), Bhutan pine (*Pinus wallichian*), silver fir (*Abies pondrow*), spruce (*Piecia simithiana*), and yew (*Taxus baccata*). Amin & Fazal (2017) found that the total forest area in Shopian declined by 15 percent during the study period. The core forest areas declined by 17 percent, while core forests with areas of less than 615 acres increased by 34 percent. Overall, there was a widespread degradation of the tree cover and forest density, and an increase in the number of smaller forest patches.

Various anthropogenic factors have contributed to fragmentation in the Shopian forests, including the expansion of agriculture, road construction, increasing population and associated pressures (e.g., firewood, livestock fodder, food, medicinal plants), and logging. Illegal logging, in particular, has been a major driving force in the fragmentation and loss of these forests (Amin & Fazal 2017). Civil unrest in the area beginning in 1989 (against Indian rule of this Muslim-majority region) has also contributed to deforestation. Several thousand acres of woodland in the district have been converted into orchards or maize fields by local residents struggling to make a living during unsettled times. But climate change is a synergistic factor (Ishtiyak *et al.* 2016). The region has seen unpredictable weather, increasing rainfall, and increasing temperatures associated with climate change. These changes exacerbate other forms of human-caused forest fragmentation. As has been emphasized in prior chapters, deforestation in turn produces increasing emissions of CO_2 into the atmosphere, driving further climate change.

7.4.3 Fragmenting grasslands

Landscape fragmentation does not only occur in forests. It is also a damaging factor in grasslands, though historically this habitat has received less attention in fragmentation studies (Fadila *et al.* 2017). Grasslands are areas dominated by grasses—plants with narrow leaves that emanate from a bulb—rather than trees or larger shrubs. Wild grasses are an important food source for many animals, including both large grazing mammals like bison, deer, buffalo, wildebeest, plains zebra, rhinos, giraffes, and elephants, and smaller species such as mice, gophers, rabbits, prairie dogs, hyrax, grouse, and grasshoppers. Because grasses grow from the bottom, when animals eat the grassy leaves they usually do not destroy the base, from which the leaves can regrow. If grass bulbs are destroyed, however, rainstorms can wash away top soils and create erosion.

Grasslands in temperate zones represent one of the world's 14 major biomes, many of which have already been examined in this book (e.g., rainforests, tundra, coral reefs, estuaries, deserts, oceans). During Earth's long history, they have been home to some of the largest assemblages of wildlife the planet has ever known. They are categorized as the most altered terrestrial ecosystem on Earth and have become one of the most endangered on most continents. Their precarity is seen in the fact that:

> less than half remain . . . in an intact, natural condition. Intensive agriculture has replaced 41 percent of the world's temperate grasslands and another 13.5 percent have been converted to urban, industrial and other uses. Much of the remainder, although still under grassland vegetation, is degraded. (Heidenreich 2009, p. II)

There are several kinds of grasslands. Prairies are temperate grasslands with tall grasses, steppes are dry grasslands with short grasses, and savannas are tropical grasslands sporting scattered individual trees. Agricultural conversion, urban development, industrial development, and other human restructuring of the environment is cutting up grassland areas in ways that diminish wind-dispersed grassland plant seeds, a process that increases the probability for species extinction. As a grassland area is fragmented into unconnected spaces, the characteristics of individual patches—such as the number of remaining patches present in the area, the size of those patches, the degree of patches' isolation from one another, and the shapes of patches (which influences the ratio of interior to the length of edges)—are affected.

Regional research on grassland areas has clarified aspects of fragmentation in this type of ecosystem. In the Netherlands, grasslands were once widespread, though they grew in nutrient-poor soils. To assess changes over time, Soons *et al.* (2005) created a time series of maps for the years 1900, 1950, and 2000 to examine three selected study areas. They quantified habitat loss and fragmentation from these habitat maps using GIS and four easily interpreted quantitative habitat descriptors: total habitat area, number of grassland patches, average and range of patch size, and average and range in distance of patches to their nearest neighbors. They found that since 1900, habitat destruction has drastically reduced both the total habitat size and the number and size of habitat patches and increased the distance between patches, resulting in decreased habitat connectivity. The largest reductions in habitat area and connectivity occurred between 1900 and 1950. By 2000, only very limited grassland remained, comprising mainly small patches separated by long distances. Most of these patches have survived because they are now within established and protected nature reserves. Fragmentation of wild grassland in the study areas is thus complete; no large zones of continuous grassland remain.

A similar pattern has been described in South Africa, where grasslands cover approximately 28 percent of the land surface. Remaining grassland areas face growing risks because of the development of exotic timber plantations, agriculture, domestic herd overgrazing, mining, and urbanization. In 2006, the South African National Biodiversity Institute brought national attention to the Woodbush Granite Grassland as the only critically endangered ecosystem in Limpopo Province, the country's northernmost province. This grassland has been dated back to about 18 000 years ago using fossil pollen. In addition to many kinds of grasses (divided among 53 species), it supports almost 100 types of wildflowers. Niemandt & Greve (2016) used repeat aerial photography to assess anthropogenically driven habitat loss and fragmentation in this area over the 60-year period from 1948 to 2008. They found that only an estimated 6 percent of the original grassland remained in 2008. Increases in patch isolation were greatest in the years between 1948 and 1977, while patch sizes decreased and edge effects increased between 1977 and 2008.

These two examples tell the global story of grassland fragmentation and loss in the face of human activities. Moreover, being comparatively drier environments, grasslands are vulnerable to shifts in climatic regime, including increased risk of wildfires and the processes of desertification (Schlesinger *et al.* 1990; Pitman & McAneny 2007). As with forests, but with their own unique ecological features and dynamics, grasslands, traditionally rich in resources for humans and other species, are enduring the effects of ecocrises interactions—a process that appears likely to worsen over time.

7.5 The dilemma of simplified environments

A central contradiction in the human–environment relationship is that, on the one hand, "[e]cosystem simplification is the ecological hallmark of humanity and the reason for our evolutionary success," and, on the other, "the side effects of our profligacy and poor resource practices are now so pervasive as to threaten our future no less than that of biological diversity itself" (Western 2001, p. 548). Evolutionary success can be measured in several ways, including degree of dispersal around the planet, adaptation to diverse environments, overall population size, and persistence through time. While *Homo sapiens* is a comparatively young species, we have proven to be very effective in meeting most of these criteria. Yet, because of our unprecedented ability to reshape and simplify environments to meet our biological, social, and political economic needs, and in the process cause destructive unintended consequences, we have been forced to begin pondering our own demise and that of untold other species. When we think of dinosaurs, we tend to see them as a failed life-form because (except for birds) they are long gone and we only know them through fossilized remains. We even use the word "dinosaur" colloquially to refer to things that are old, throw-backs to a world left behind. But dinosaurs as a group lived on the planet far longer than humans have thus far. And dinosaurs did not cause their own extinction. In the long run, the same may not be said of humans.

Peipoch *et al.* (2015) proposed the concept of ecological simplification "as the general mechanism by which human impacts have modified cross-scale relationships among landscape complexity, integrity, and niche diversity in ecosystems." As stressed in Chapter 2, the science of ecology seeks to understand the complex interconnections that make up the environment. At the same time, it is also deeply concerned with the effects of the unraveling of that complexity and the consequences of movement—especially as driven by anthropogenic activity—toward simplified ecosystems. As Peipoch *et al.* (2015) indicate, simplification "transforms highly complex landscapes into more homogenous, less complex entities, reflecting exogenous control (i.e., human influences) over endogenous properties such as complex behavior." The consequences reverberate on us, the drivers of simplification. As Howarth *et al.* (2013) maintain with reference to fisheries:

> Many over-exploited marine ecosystems worldwide have lost their natural populations of large predatory finfish and have become dominated by crustaceans and other invertebrates . . . [T]he transition from multispecies fisheries to simplified invertebrate fisheries is causing a global decline in biodiversity and is threatening global food security

The rationale behind this assertion is the demonstrated fact that biodiversity confers ecosystem resilience and that ecosystems with lower levels of biodiversity are more susceptible to suffering the effects of human-driven ecocrises. In the case of marine environments, this could include the interaction of overfishing with climate change, eutrophication, and species invasions (Loreau *et al.* 2001). For example, overfishing in the Black Sea, the most severely degraded and exploited large marine ecosystem in the world, involving the demise of the top predator species (first pelagic fish such as tuna in the 1970s, and then planktivorous fish such as anchovy in the 1990s), has made the entire ecosystem more vulnerable to climate change, ocean acidification, nutrient loading from agricultural and urban runoff, and invasion by both native and non-native species (Llope *et al.* 2011). As a result, many shallow areas have become oxygen-depleted, and the biotic ecosystem has moved from a fish-dominated community to one dominated by jellyfish and ctenophores or comb jellies such as *Mnemiopsis leidyi,* a bioinvader from the North American East Coast. This alteration reprsents the most dramatic, large-scale gelatinous plankton invasion event recorded to date (Oguz *et al.* 2008). According to Oguz (2017), the multiple pressures on food web functioning introduced by anthropogenic activity continue to exacerbate the situation.

7.6 Fragmented environments, ticks, and human health

Lyme disease is an increasingly widely spread vector-borne infection, named after the town of Lyme, CT. Its immediate cause is a bacterium called *Borrelia burgdorferi* that is transmitted through the bite of the blacklegged tick (*Ixodes scapulris*), also known as the deer tick. Ticks are known to transmit a wide array of pathogens, and new disease-causing tick-borne microorganisms are being discovered regularly (Paddock & Telford 2011). Lyme disease, the best-known tick-borne disease, is now the most prevalent vector-borne disease in North America, and both its "annual incidence and geographic range are increasing" (Levia *et al.* 2012). It is characterized by flu-like symptoms, which, left untreated, can lead to serious chronic consequences such as arthritis, neurological problems, cardiac arrhythmia, and cognitive defects.

Though the significant upsurge in Lyme disease cases is relatively recent, ticks are not new to the area where the disease was first identified and named. The presence of ticks in the Northeast was established in the mid-18th century by the Swedish botanist Pehr Kalm (1773), who visited forested areas in the state of New York and reported ticks flourished there. Much of the original forests that once covered New England, however, were cut down and cleared as settlers developed farms, built towns, and created roads during the 18th and 19th centuries. It is estimated that 60–80 percent of the land was cleared for pasture, tillage, orchards, and buildings. Hunting and trapping diminished the larger mammals that lived there. Then, with the rise of the Industrial Revolution, the small New England farms were abandoned in large numbers as younger generations migrated to the growing cities and their rising demand for industrial labor. The result was a considerable drop in rural tree cutting, which allowed new forests to grow back over the abandoned farmlands. By the end of the 19th century, forests and thick woods once again covered broad areas throughout the region. With the new trees came a return of indigenous species like the white-tailed deer (*Odocoileus virginianus*) and the white-footed mouse (*Peromyscus leucopus*)—and aboard them came the blacklegged tick. All of these species thrived in the new forests. With wolves driven from the environment by settlers, deer populations exploded to levels probably never seen in pre-colonial times and forest rodents flourished.

These changes affected the 2-year life cycle of the blacklegged tick, which is divided into three main developmental stages: larva, nymph, and adult. In the spring and summer of the first year of this cycle, eggs hatch into larvae, which feed once on a host animal (often a white-footed mouse) and molt into nymphs (which look like tiny, less oblong versions of the adults). They then become dormant during the fall and winter. In the second year, the nymphs feed on hosts from May through July. During this time, they can transmit disease if they acquired pathogens from a host during a previous blood meal. In the fall, they molt into adults. The females feed on deer and other large mammals (including, if present, humans), seek mates, lay as many as 2000 eggs, and then die. Males attach to a host to wait for females to mate with, but they do not take blood meals. The abundant host populations following the Industrial Revolution allowed an expansion of tick populations, while reforestation facilitated an expansion of their range.

With the emergence of suburbia and a return to deforestation and forest fragmentation, people have taken residence in areas and engaged in activities (e.g., hiking) that thrust them into growing contact with the vectors of Lyme disease. According to the Harvard Forest report (Foster *et al.* 2017), New England is now losing 65 acres of forest to development each day. At the current pace, the region will lose another 1.2 million acres by the year 2060—an area nearly double the size of Rhode Island. As the report stresses, "[t]he New England landscape faces the increasing loss of forest and farmland to residential and commercial development; ongoing parcelization and fragmentation of land" (Foster *et al.* 2017, p. 2). Forest refragmentation has increased the risk of human exposure to Lyme disease. White-footed mice are more abundant in forest fragments, probably because fewer of their common predators (e.g., wolves and foxes) and competitors (e.g., rabbits) remain in the fragmented patches. The mice, by contrast, tend to be particularly abundant in forested land segments smaller than 5 acres.

Research by Allan *et al.* (2003) shows that forest patches under 3 acres have an average of three times as many ticks as do larger fragments, and seven times more infected ticks. In the smallest patches, as many as 80 percent of the ticks are infected—the highest rate these researchers have seen in their various tick studies.

Among those caught up in the Lyme epidemic was Pamela Weintraub, a senior editor at *Discover Magazine* who authored the popular book *Cure Unknown: Inside the Lyme Epidemic* (2009) in response to her own illness experience and those of her family members. Weintraub and her family only began to experience symptoms after moving in 1993 to a comfortable suburban home in the hamlet of Chappaqua, in Westchester County, New York, that abutted a spruce forest. Family members began to experience an array of perplexing symptoms, including intense headaches, memory loss, swollen knee joints, sharp mood swings, painful limbs, and withering fatigue. Eventually, they were all diagnosed with Lyme disease and other infections spread by tick bites. Weintraub (2011, p. 31) writes that many people with Lyme disease are co-infected "with babesiosis, anaplasmosis, ehrlichiosis, or some other tick-borne infection. Surveys around the country report that ticks can transmit these well-known human diseases, yet primary care physicians almost never consider or test for them, even if they seriously consider Lyme disease." Tick-borne disease, and poly-infected ticks, are of global health significance in light of: 1) the continued identification of new tick-borne pathogens; 2) the increasing geographic diffusion of ticks and tick species caused by ongoing anthropogenic environmental/climate changes; and 3) shifting human–environment interaction patterns (Singer & Bulled 2016).

7.7 Solastalgia: distress linked to environmental change

Environmental breakdown can have reverberations on human mental health. The term "solastalgia" is used by some researchers to label environmentally induced distress. By contrast with nostalgia, which is a homesickness experienced by individuals when separated from a beloved and fondly remembered place, solastalgia refers to the emotional distress produced by significant environmental change that impacts people's mental well-being while they are still directly connected to their home environment (Albrecht *et al.* 2007). Kahn (2002, p. 106) points out that "[p]eople take the natural environment they encounter during childhood as the norm against which they measure environmental degradation later in their lives. With each ensuing generation, the amount of environmental degradation increases, but each generation in its youth takes that degraded condition as the non-degraded condition—as the normal experience." As this suggests, how human populations perceive landscape fragmentation and degradation as they occur during their life course influences their emotional health, which in turn affects their physical health.

In Australia, a country experiencing escalating ecosystem degradation, Speldewinde *et al.* (2009) found that landscape degradation is associated with adverse mental health problems in the resident rural population. Specifically, they found an association between farmland salinification and hospitalization for depression. Salinization can dramatically alter the appearance and social meaning of landscape, producing denuded surfaces populated by dead trees and salt pans and, in worst-case scenarios, previously farmed land that is unusable for farming. Mental health impacts occur across a number of interrelated pathways, including significant economic loss and increased financial pressures that force some people to move away from established communities and dependable routines, fragmenting social networks and avenues of social support. In another rural Australian study, Obrien *et al.* (2014) quantified drought in terms of both duration and degree of dryness and then identified drought characteristics associated with poor mental health. Their findings show that during a 7-year period of major and widespread drought, one pattern of relative dryness, namely extreme prolonged dryness lasting a year or more, was associated with increased emotional distress.

The field of environmental health highlights the degree to which humans are part of the environment and how our well-being depends on a healthy environment, but at the same time it recognizes that we are the biggest threat to that which we need to sustain ourselves. This is the fundamental human environmental dilemma. While on an individual level we may feel powerless with regard to ecocrises, it is our behavior that produces them; collectively, our fate is in our own hands. One thing has become clear. The lifestyles, values, desires, political structures, and economic systems that have been central to creating our environmental dilemma are probably not the place to find a solution to it. Rather, solutions likely lie in the development of new lifestyles, values, desires, political structures, and economic systems.

References

Aherne, G., Hardcastle, A. & Nield, A. (1990). Cytotoxic drugs and the aquatic environment: estimation of bleomycin in river and water samples. *Journal of Pharmacy and Pharmacology* 42: 741–742.

Albrecht, G., Sartore, G., Connor, L., Nick Higginbotham, N. & Freeman, S. (2007). Solastalgia: the distress caused by environmental change. *Australasian Psychiatry* 15(Suppl. 1): S95–S98.

Allan, B., Keesing, F. & Ostfeld, R. (2003). Effect of forest fragmentation on Lyme disease risk. *Conservation Biology* 17(1): 267–272.

Amin, A. & Fazal, S. (2017). Assessment of forest fragmentation in district Shopian using multi-temporal land cover (a GIS approach). *Journal of Geosciences and Geomatics* 5(1): 12–23.

Anderson, J. & Brower, L. (1996). Freeze-protection of overwintering monarch butterflies in Mexico: critical role of the forest as a blanket and an umbrella. *Ecological Entomology* 21: 107–116.

Arndt, D. (2017). Alaskan North Slope climate change just outran one of our tools to measure it. Climate.gov. https://www.climate.gov/news-features/blogs/beyond-data/alaskan-north-slope-climate-change-just-outran-one-our-tools-measure.

Atlas, R. & Maloy, S. (2014). *One Health: People, Animals, and the Environment*. Washington, D.C.: ASM Press.

Beadle, K. (2017). The drug destruction of Mexico, part II. *Pacific Standard*. https://psmag.com/environment/the-drug-destruction-of-mexico-part-ii-19343.

Bebianno, M. & Gonzalez-Rey, M. (2015). Ecotoxicological risk of personal care products and pharmaceuticals. In: C. Amiard-Triquet, J.-C. Amiard & C. Mouneyrac (Eds.), *Aquatic Ecotoxicology: Advancing Tools for Dealing with Emerging Risks*, pp. 383–416. London: Academic Press.

Benetiz-Malvido, J. (1998). Impact of forest fragmentation on seedling abundance in a tropical rain forest. *Conservation Biology* 12: 380–389.

Berardi, G. (1999). Schools, settlement, and sanitation in Alaska Native villages. *Ethnohistory* 46(2): 329–359.

Bierregaard, R., Lovejoy, T., Kapos, V., Augusto dos Santos, A. & Hutchings, R. (1992). The biological dynamics of tropical rainforest fragments. *Bioscience* 42: 859–866.

Biskaborn, B., Smith, S., Moetzli, K., Mattes, H., Vieira, G., Streletsky, D., *et al.* (2019). Permafrost is warming at a global scale. *Nature Communication* 10: 264.

Bittel, J. (2015). 4 clever (and kind of sad) ways animals adapt to humans. *National Geographic*. https://news.nationalgeographic.com/news/2015/02/150211-climate-change-science-animals-adaptation-orangutans-manatees/.

Brown, E. & Morgan, T. (2007). Alaskan village threatened by warming. *ABC News*. https://abcnews.go.com/Nightline/story?id=2971039&page=1.

Burch, E. (1988). *The Inupiaq Eskimo Nations of Northwest Alaska*. Fairbanks, AK: University of Alaska Press.

Burch, E. (2006). *Social Life in Northwest Alaska: The Structure of Iñupiaq Eskimo Nations*. Fairbanks, AK: University of Alaska Press.

Cao, J., Xu, X., Hylkema, M., Zeng, E., Sly, P., Suk, W. *et al*. (2016). Early-life exposure to widespread environmental toxicants and health risk: a focus on the immune and respiratory systems. *Annals of Global Health* 82(1): 119–131.

Carswell, C. (2017). Trump's wall may threaten thousands of plant and animal species on the U.S.–Mexico border. *Scientific American*. https://www.scientificamerican.com/article/trump-rsquo-s-wall-may-threaten-thousands-of-plant-and-animal-species-on-the-u-s-mexico-border/.

Casey, J., Savitz, D., Rasmussen, S., Ogburn, E., Pollak, J., Mercer, D. *et al*. (2016). Unconventional natural gas development and birth outcomes in Pennsylvania, USA. *Epidemiology* 27(2): 163–172.

Chen, J., Franklin, J. & Spies, T. (1995). Growing-season microclimatic gradients from clearcut edges into old growth Douglas-fir forests. *Ecological Applications* 5(1): 74–86.

City of Utqiaġvik. (2014). About Utqiaġvik. https://www.utqiagvik.us/index.php/about-utqiagvik.

Collinge, S. (1996). Ecological consequences of habitat fragmentation: implications for landscape architecture and planning. *Landscape and Urban Planning* 36(1): 59–77.

Daley, J. (2018). New animal overpass is already protecting critters in Washington state. Smithsonian.com. https://www.smithsonianmag.com/smart-news/animal-overpass-already-protecting-animals-washington-state-180971124/.

Davies, K. & Margules, C. (2002). Effects of habitat fragmentation on carabid beetles: experimental evidence. *Journal of Animal Ecology* 67(3): 460–471.

Davis, A. (2015). The quietly-published paper on climate change and monarchs. *The Science of Monarch Butterflies*. http://akdavis6.wixsite.com/monarchscience/single-post/2015/08/10/The-quietlypublished-paper-on-climate-change-and-monarchs.

Debinski, D. & Holt, R. (2000). A survey and overview of habitat fragmentation experiments. *Conservation Biology* 14(2): 342–555.

Denham, A., Willis, M., Zavez, A. & Hill, E. (2019). Unconventional natural gas development and hospitalizations: evidence from Pennsylvania, United States, 2003–2014. *Public Health* 168: 17–25.

Didham, R. (2010). Ecological consequences of habitat fragmentation. *Landscape and Urban Planning* 36(1): 59–77.

Dobbyn, P. (2018). Study aims to help Alaska's North Slope officials adapt to climate change hazards. *Sea Grant Alaska*. https://alaskaseagrant.org/2018/08/06/study-aims-to-help-alaskas-north-slope-officials-adapt-to-climate-change-hazards/.

Ehrenfeld, D. (1992). Ecosystem health and ecological theories. In: R. Costanza (Ed.), *Ecosystem Health: New Goals for Environmental Management*, pp. 135–143. Washington, D.C.: Island Press.

Ewers, R., Thorpe, S. & Didham, R. (2007). Synergistic interactions between edge and area effects in a heavily fragmented landscape. *Ecology* 88(1): 96–106.

Fadila, D., Kelly, L., Moore, J. & McCarthy, M. (2017). A systematic review reveals changes in where and how we have studied habitat loss and fragmentation over 20 years. *Biological Conservation* 212(Part A): 130–138.

Fent, K., Weston, A. & Caminada, D. (2006). Ecotoxicology of human pharmaceuticals. *Aquatic Toxicity* 76(2): 122–159.

Ferguson, P., Bernot, M., Doll, J, & Lauer, T. (2013). Detection of pharmaceuticals and personal care products (PPCPs) in near-shore habitats of southern Lake Michigan. *Science of The Total Environment* 458–460: 187–196.

Field Museum. (2018). Giving mammals like Tsavo lions room to roam. https://www.fieldmuseum.org/blog/giving-mammals-tsavo-lions-room-roam.

Fischetti, M. (2012). Fracking would emit large quantities of greenhouse gases. *Scientific American*. https://www.scientificamerican.com/article/fracking-would-emit-methane/.

Foster, D., Donahue, B., Kittredge, D., Lambert, K., Hunter, M., Hall, B. *et al.* (2017). *Wildlife & Woodlands: A Vision for the New England Landscape*. Petersbam, MA: Harvard University.

Foster, S. & Chilton, P. (2003). Groundwater: the processes and global significance of aquifer degradation. *Philosophical Transactions of the Royal Society of London B* 358(1440): 1957–1972.

Gadgil, M. & Guha, R. (1992). *This Fissured Land: An Ecological History of India*. Oxford: Oxford University Press.

GAO. (2009). Alaska Native Villages: Limited Progress has Been Made on Relocating Villages Threatened by Flooding and Erosion. U.S. Government Accountability Office. https://www.gao.gov/new.items/d09551.pdf.

Garrison, A., Pope, J. & Allen, F. (1976). GC/MS analysis of organic compounds in domestic wastewaters. In: C. Keith (Ed.), *Identification and Analysis of Organic Pollutants in Water*, pp. 517–556. Ann Arbor, MI: Ann Arbor Science Publishers.

Gascon, M., Morales, E., Sunyer, J. & Vrijheid, M. (2013). Effects of persistent organic pollutants on the developing respiratory and immune systems: a systematic review. *Environment International* 52: 51–65.

Godoy, A., Kummrow, F. & Pamplin, P. (2015). Occurrence, ecotoxicological effects and risk assessment of antihypertensive pharmaceutical residues in the aquatic environment—a review. *Chemosphere* 138: 281–292.

Goldman, G., Carlson, C. & Zhang, Y. (2015). *Bad Chemistry: How the Chemical Industry's Trade Association Undermines the Policies that Protect Us*. Washington, D.C.: Union of Concerned Scientists.

Gottfried, G., Gebow, B., Eskew, L. & Edminster, C. (2004). *Connecting mountain islands and desert seas: biodiversity and management of the Madrean Archipelago II. Proceedings RMRS-P-36*. Fort Collins, CO: USDA Forest Service.

Gruber, S. (2012). Derivation and analysis of a high-resolution estimate of global permafrost zonation. *Cryosphere* 6: 221–233.

Guy, J. (2019). Arctic fox amazes scientists with 2000-mile trek in 76 days. *CNN*. https://www.cnn.com/2019/07/02/europe/arctic-fox-journey-scli-intl/index.html.

Hassol, S. (2004). *Impacts of a Warming Arctic: Arctic Climate Impact Assessment*. Cambridge, MA: Cambridge University Press.

Hays, J. & Shonkoff, S. (2016). Toward an understanding of the environmental and public health impacts of unconventional natural gas development: a categorical assessment of the peer-reviewed scientific literature, 2009–2015. *PLoS One* 11(4): e0154164.

Heidenreich, B. (2009). *What Are Global Temperate Grasslands Worth? A Case for their Protection*. Vancouver, BC: Temperate Grassland Conservation Initiative.

Henle, K., Davies, K., Kleyer, M., Margules, C. & Settele, J. (2004). Predictors of species sensitivity to fragmentation. *Biodiversity and Conservation* 13(1): 207–251.

Hester, R. & Harrison, R. (2016). *Pharmaceuticals in the Environment*. London: Royal Society of Chemistry.

Hilleary, C. (2020). Alaska Natives in Bering Strait region brace against COVID-19. *VOA*. https://www.voanews.com/covid-19-pandemic/alaska-natives-bering-strait-region-brace-against-covid-19.

Hjort, J., Karjalainen, O., Aalto, J., Westermann, S., Romanovsky, V., Nelson, F. *et al.* (2018). Degrading permafrost puts Arctic infrastructure at risk by mid-century. *Nature Communication* 9: 5147.

Hofstaedter, E. (2019). *Amid an erosion crisis,* Shishmaref takes small steps toward expansion. Knom Radio Mission. https://

www.knom.org/wp/blog/2019/06/28/amid-an-erosion-crisis-shishmaref-takes-small-steps-toward-expansion/.

Howarth, L., Roberts, C., Thurstan, R. & Stewart, B. (2013). The unintended consequences of simplifying the sea: making the case for complexity. *Fish and Fisheries* 15(4): 690–711.

Hughes, T. (2017). Residents on remote Alaska island fear climate change will doom way of life. *USA Today*. https://www.usatoday.com/story/news/nation/2017/06/02/alaska-island-fear-climate-change-dooms-town/102398636/.

Institute of Medicine. (2001). *Rebuilding the Unity of Health and the Environment: A New Vision of Environmental Health for the 21st Century*. Washington, D.C.: National Academies Press.

Ishtiyak, P., Reddy, M., Panse, S., Wani, I. & Peer, Q. (2016). Impact of climate change and anthropogenic interventions on natural vis-a-vis human resources in Kashmir, India: an overview. *Journal of Applied and Natural Science* 8(1): 489–493.

Johnson, M. & Eicken, H. (2016). Estimating Arctic sea-ice freeze-up and break-up from the satellite record: a comparison of different approaches in the Chukchi and Beaufort Seas. *Elementa: Science of the Anthropocene* 4: 000124.

Jorgenson, M., Shur, Y. & Pullman, E. (2006). Abrupt increase in permafrost degradation in Arctic Alaska, *Geophysical Research Letters* 25: L02503.

Kahn, P. (2002). Children's affiliations with nature: structure, development, and the problem of environmental generational amnesia. In: P. Kahn & S. Kellert (Eds.), *Children and Nature*, pp. 92–115. Cambridge, MA: MIT Press.

Kalm, P. (1773). *Travels into North America*, vol. 1 (trans. J. Forster). American Journeys: Eye Witnessed Accounts of Early American Explorations and Settlement. http://www.americanjourneys.org/aj-117a/summary/index.asp.

Keiser, J., Singer, B. & Utzinger, J. (2005). Reducing the burden of malaria in different eco-epidemiological settings with environmental management: a systematic review. *Lancet Infectious Diseases* 5(11): 695–708.

Kennedy, M. (2016). Alaska village decides to relocate. *NPR*. https://www.npr.org/sections/thetwo-way/2016/08/18/490519540/threatened-by-rising-seas-an-alaskan-village-decides-to-relocate.

Khan, A. & Patrick, W. (2016). *The Next Pandemic: On the Front Lines Against Humankind's Gravest Dangers*. New York: Public Affairs.

Kim, G. (2019). *Work begins on new site for village of Newtok*. *Alaska Public Media*. https://www.alaskapublic.org/2019/06/12/work-begins-on-new-site-for-village-of-newtok/.

Koutsky, K. (1981). *Early Days on Norton Sound and Bering Strait: An Overview of Historic Sites in the BSNC Region: Volume 1, The Shishmaref Area*. Fairbanks, AK: Anthropology and Historic Preservation Cooperative Park Studies Unit.

Lancet Commission on Planetary Health. (2015). Safeguarding human health in the Anthropocene epoch: report of the Rockefeller Foundation. *Lancet* 386(10007): 1973–2028.

Landrigan, P., Wright, R., Cordero, J., Eaton, D., Goldstein, B., Hennig, B. *et al.* (2015). The NIEHS superfund research program: 25 years of translational research for public health. *Environmental Health Perspectives* 123(10): 909–918.

Latour, B. (2010). An attempt at a "Compositionist Manifesto." *New Literary History* 41: 471–490.

Laurance, W., Laurance, S., Ferreira, L., Rankin-de Merona, J., Gascon, C. & Lovejoy, T. (1997). Biomass collapse in Amazonian forest fragments. *Science* 278(5340): 1117–1118.

Laurance, W., Nascimento, H., Laurance, S., Andrade, A., Ribeiro, J., Giraldo, J. *et al.* (2006). Rapid decay of tree-community composition in Amazonian forest fragments. *Proceedings of the National Academy of Science of the U.S.A.* 103(50): 19010–19014.

Lawrence, D., Slater, A., Tomas, R., Holland, M. & Deser, C. (2008). Accelerated arctic land warming and permafrost degradation during rapid sea ice loss. *Geophysical Research Letters* 35(11).

Lemoine, N. (2015). Climate change may alter breeding ground distributions of eastern migratory monarchs (*Danaus plexippus*) via range expansion of Asclepias host plants. *PLoS One* 10(2): e0118614.

Leopold, A. (1995). Conservationist in Mexico. In: D. Brown & N. Carmony (Ed.), *Aldo Leopold's Southwest*. Albuquerque, NM: University of New Mexico Press.

Levia, T., Kilpatrick, M., Mangel, M. & Wilmers, C. (2012). Deer, predators, and the emergence of Lyme disease. *Proceedings of the National Academy of Sciences of the U.S.A.* 109: 10942–10947.

Llope, M., Daskalov, G., Rouyer, T., Mihneva, V., Chan, K.-S., Grishin, A. *et al.* (2011) Overfishing of top predators eroded the resilience of the Black Sea system regardless of the climate and anthropogenic conditions. *Global Change Biology* 17: 1251–1265.

Loreau, M., Naeem, S., Inchausti, P., Bengtsson, J., Grime, J., Hector, A. *et al.* (2001). Biodiversity and ecosystem functioning: current knowledge and future challenges. *Science* 294: 804–808.

Mageau, M., Costanza, R. & Ulanowicz, R. (1995). The development and initial testing of a quantitative assessment of ecosystem health. *Health* 1: 201–213.

Makarananda, K., Pengpan, U., Srisakulthong, M., Yoovathaworn, K. & Sriwatanakul, K. (1998). Monitoring of aflatoxin exposure by biomarkers. *Journal of Toxicological Sciences* 23(Suppl. 2): 155–159.

Marino, E. (2012a). Losing ground: an ethnography of vulnerability and climate change in Shishmaref, Alaska. Ph.D. dissertation. Fairbanks, AK: University of Alaska Fairbanks. https://scholarworks.alaska.edu/bitstream/handle/11122/9151/Marino_E_2012.pdf?sequence=1.

Marino, E. (2012b). The long history of environmental migration: assessing vulnerability construction and obstacles to successful relocation in Shishmaref, Alaska. *Global Environmental Change* 22(2): 374–381.

Martin, A. (2018). *An Alaskan village is falling into the sea.* Washington is looking the other way. *Public Radio International.* https://www.pri.org/stories/2018-10-22/alaskan-village-falling-sea-washington-looking-other-way.

Matlack, G. (1993). Microenvironment variation within and among forest edge sites in the eastern United States. *Biological Conservation* 66: 185–194.

McDermott-Levy, R., Kaktins, N. & Sattler, B. (2013). Fracking, the environment, and health. *American Journal of Nursing* 113: 45–51.

McGreal, C. & Walters, J. (2019). Drug makers flooded U.S. with billions of opioid pills as epidemic surged, data shows. *Guardian.* https://www.theguardian.com/us-news/2019/jul/17/opioid-crisis-drug-makers-pills-federal-data-epidemic.

McMichael, A. & Woodruff, R. (2005). Detecting the health effects of environmental change: scientific and political challenge. *Ecohealth* 2(1): 1–3.

Medred, C. (2017). No whales now. https://craigmedred.news/2017/10/01/no-whales-now/.

Metcalfe, C., Miao, X.-S., Koenig, B. & Struger, J. (2009). Distribution of acidic and neutral drugs in surface waters near sewage treatment plants in the lower Great Lakes, Canada. *Environmental Toxicology and Chemistry* 22(12): 2881–2889.

Mezzelani, M., Gorbi, S. & Regoli, F. (2018). Pharmaceuticals in the aquatic environments: evidence of emerged threat and future challenges for marine organisms. *Marine Environment Research* 140: 41–60.

Misztal, L. (2018). Too much at stake—no border wall in our Sky Islands! Sky Island Alliance. https://www.skyislandalliance.

org/2018/03/much-stake-no-border-wall-sky-islands/.

Misztal, L., Garfin, G., & Hansen, L. (2013). USDA Forest Service Proceedings RMRS-P-67. https://www.fs.fed.us/rm/pubs/rmrs_p067/rmrs_p067_060_067.pdf.

Mohapatra, P. & Mohapatro, P. (1997). *Forest Management in Tribal Areas; Forest Policy and Peoples' Participation*. New Delhi: Concept Publishing Company.

Myers, S. & Patz, J. (2009). Emerging threats to human health from global environmental change. *Annual Review of Environmental Resources* 234: 223–252.

Myers, S., Gaffikin, L., Golden, C., Ostfeld, R., Redford, K., Ricketts, T. *et al.* (2013). Human health impacts of ecosystem alteration. *Proceedings of the National Academy of Sciences of the U.S.A.* 110(47): 18 753–18 760.

National Academy of Sciences. (2014). *Review of the Formaldehyde Assessment in the National Toxicology Program 12th Report on Carcinogens*. Washington, D.C.: National Research Council.

National Climate Assessment. (2018). Fourth National Climate Assessment. https://nca2018.globalchange.gov/.

Nava, A., Shimabukuro, J., Chmura, A. & Luz, S. (2017). The impact of global environmental changes on infectious disease emergence with a focus on risks for Brazil. *ILAR Journal* 58(3): 393–400.

Niemandt, C. & Greve, M. (2016). Fragmentation metric proxies provide insights into historical biodiversity loss in critically endangered grassland. *Agriculture, Ecosystems & Environment* 235: 172–181.

Obrien, L., Berry, H., Coleman, C. & Hanigan, I. (2014). Drought as a mental health exposure. *Environmental Research* 131: 181–187.

Oguz, T. (2017). Controls of multiple stressors on the Black Sea fishery. *Frontiers in Marine Science* 4: 110.

Oguz, T., Fach, B. & Salihoglu, B. (2008). Invasion dynamics of the alien ctenophore *Mnemiopsis leidyi* and its impact on anchovy collapse in the Black Sea. *Plankton Research* 30(12): 1385–1397.

Oliver, S. (2018). Autumn storm that battered Utqiaġvik coastline caused more than $10 million in damage. *Anchorage Daily News*. https://www.adn.com/alaska-news/rural-alaska/2017/10/08/autumn-storm-that-battered-utqiagvik-coastline-caused-more-than-10-million-in-damage/.

Opdam, P. & Wascher, D. (2004). Climate change meets habitat fragmentation: linking landscape and biogeographical scale levels in research and conservation. *Biological Conservation* 117(3): 285–297.

Oxford Reference. (2018). Ecosystem distress syndrome. http://www.oxfordreference.com/view/10.1093/oi/authority.201108030957 41562.

Paddock, C. & Telford, S. (2011). Through a glass darkly: the global incidence of tick-borne diseases. Commissioned paper for critical needs and gaps in understanding prevention, amelioration, and resolution of Lyme and other tick-borne diseases: the short-term and long-term outcomes. *Workshop report*. Washington, D.C.: National Academy of Medicine.

Palmer, M. & Febria, C. (2012). The heartbeat of ecosystems. *Science* 336: 1393–1394.

Pasquet, J. & Arbona, M. (2018). *My Wounded Island*. Olympia, WA: Orca Books.

Pedersen, M., Giorgis-Allemand, L., Bernard, C., Aguilera., I, Andersen, A., Ballester, F. *et al.* (2013). Ambient air pollution and low birthweight: a European cohort study (ESCAPE). *Lancet Respiratory Medicine* 1: 695–704.

Peipoch, M., Brauns, M., Hauer, R., Weitere, M. & Valett, H. (2015). Ecological simplification: human influences on riverscape complexity. *BioScience* 65(11): 1057–1065.

Peng, L., Meyerhoefer, C. & Chou, S.Y. (2018). The health implications of unconventional natural gas development in Pennsylvania. *Health Economics* 27(6): 956–983.

Pfeifer, M., Lefebvre, V., Peres, C., Banks-Leite, C., Wearn, O., Marsh, C. *et al.* (2017). Creation of forest edges has a global impact on forest vertebrates. *Nature* 551(7679): 187–191.

Pitman, A. & McAneny, G. (2007). The impact of climate change on the risk of forest and grassland fires in Australia. *Climate Change* 84(3–4): 383–401.

Prüss-Üstün, A. & Corvalán, C. (2006). *Preventing Disease through Healthy Environments: Towards an Estimate of the Environmental Burden of Disease*. Geneva: World Health Organization.

Prüss-Üstün, A., Bonjour, S. & Corvalán, C. (2008). The impact of the environment on health by country: a meta-synthesis. *Environmental Health* 7: 7.

Puckowski, A., Mioduszewska, K., Łukaszewicz, P., Borecka, M., Caban, M., Maszkowska, J. *et al.* (2016). Bioaccumulation and analytics of pharmaceutical residues in the environment: a review. *Journal of Pharmaceutical and Biomedical Analysis* 127: 232–255.

Rapport, D., Regier, H. & Hutchinson, T. (1985). Ecosystem behavior under stress. *American Naturalist* 125(5): 617–640.

Rapport, D., Constanza, R. & McMichael, A. (1998a). Assessing ecosystem health. *Trends in Ecology & Evolution* 10: 307–402.

Rapport, D., Whiteford, W. & Hildén, M. (1998b). Common patterns of ecosystem breakdown under stress. *Environmental Monitoring and Assessment* 51(1–2): 171–178.

Reddy, C., Sreelekshmi, S., Jha, C. & Dadhwal, V. (2013). National assessment of forest fragmentation in India: landscape indices as measures of the effects of fragmentation and forest cover change. *Ecological Engineering* 60: 453–464.

Robinson, O. & Vrijheid, M. (2015). The pregnancy exposome. *Current Environmental Health Reports* 2(2): 204–213.

Rozell, N. (2018). Utqiaġvik, where the climate has changed. *Daily News-Miner*. http://www.newsminer.com/features/sundays/alaska_science_forum/utqiagvik-where-the-climate-has-changed/article_f40cfbf6-02ed-11e8-81ea-c35dcaec3845.html.

Rozell, N. (2019). *Village move intensifying in summer 2019*. University of Alaska, Fairbanks News and Information. https://news.uaf.edu/village-move-intensifying-in-summer-2019/.

Rubio, A., Avila-Flores, R., & Suzán, G. (2014). Responses of small mammals to habitat fragmentation: epidemiological considerations for rodent-borne hantaviruses in the Americas. *EcoHealth* 11: 526–533.

Santosa, L., Araújoa, A., Fachinia, A., Pena, A., Delerue-Matos, C. & Montenegro, M. (2010). Ecotoxicological aspects related to the presence of pharmaceuticals in the aquatic environment. *Journal of Hazardous Materials* 175(1–3): 45–95.

Schlesinger, W., Reynolds, J., Cunningham, G., Huenneke, L., Jarrell, W., Virginia, R. *et al.* (1990). Biological feedbacks in global desertification. *Science* 247(4946): 1043–1048.

Schwarzenbach, R., Elgi, T., Hofstetter, T., von Gunten, U. & Wehrli, B. (2010). Global water pollution and human health. *Annual Review of Environment and Resources* 35: 109–136.

Scott, J. (1998). *Seeing Like a State: How Certain Schemes to Improve the Human Condition Have Failed*. New Haven, CT: Yale University Press.

Shankman, S. (2017). In Alaska's thawing permafrost, humanity's "library is on fire." *Inside Climate News*. https://insideclimatenews.org/news/30112017/alaska-global-warming-archaeology-permafrost-history-artifacts-sea-ice-hunting-whaling-traditions.

Sheppard, K. (2014). As the planet warms a remote Alaskan town shows just how unprepared we are. *Huffington Post*. https://www.huffingtonpost.com/2014/12/14/shishmaref-alaska-climate-changerelocation_n_6296516.html.

Shiklomanov, N., Streletskiy, D., Swales, T. & Kokorev, V. (2016). Climate change and stability of urban infrastructure in Russian permafrost regions: prognostic assessment based on GCM climate projections. *Geographical Review* 107(1): 125–142.

Singer, M. & Bulled, N. (2016). Ectoparasitic syndemics: polymicrobial tick-borne disease interactions in a changing anthropogenic landscape. *Medical Anthropology Quarterly* 30(4): 442–461.

Soons, M., Messelink, J., Jongejans, E. & Weil, G. (2005). Habitat fragmentation reduces grassland connectivity for both short-distance and long-distance wind-dispersed forbs. *Journal of Ecology* 93(6): 1212–1225.

Speldewinde, P., Cook, A., Davies, P. & Weinstein, P. (2009). A relationship between environmental degradation and mental health in rural Western Australia. *Health & Place* 15(3): 880–887.

Stacy, S., Brink, L., Larkin, J., Sadovsky, Y., Goldstein, B., Pitt, B. & Talbott, E. (2015). Perinatal outcomes and unconventional natural gas operations in Southwest Pennsylvania. *PLos One* 10(6): e0126425.

Stenoien, C., Nail, K., Zalucki, J., Parry, H., Oberhauser, K. & Zalucki, M. (2018). Monarchs in decline: a collateral landscape-level effect of modern agriculture. *Insect Science* 25(4): 528–541.

Stone, M. (2018). Climate change is erasing human history. *Gizmodo*. https://gizmodo.com/climate-change-is-erasing-human-history-1819960026.

Su, C., Lin, Y. & Cai, L. (2013). Genetic factors, viral infection, other factors and liver cancer: an update on current progress. *Asian Pacific Journal of Cancer Prevention* 14(9): 4953–4960.

Sutter, G. (1993). A critique of ecosystem health concepts and indexes. *Environmental Toxicology and Chemistry* 12: 1533–1539.

Taylor, M. (2019). *Rural Health Disparities: Public Health, Policy, and Planning Approaches*. Basel: Springer Nature Switzerland AG.

TetraTech. (2004) *Shishmaref Relocation and Collocation Study, Shishmaref*, Alaska. Report to the U.S. Army Corps of Engineers, Alaska District, Anchorage, AK.

Thomas, C., Franco, A. & Hill, J. (2006). Range contractions and extinction in the face of climate change. *Trends in Ecology and Evolution* 21: 415–416.

Tian, H., Banger, K., Bo, T. & Dadhwal, V. (2015). History of land use in India during 1880–2010: large-scale land transformations reconstructed from satellite data and historical archives. *Global and Planetary Change* 121: 78–88.

Tomasula Y Garcia, A. (2019). How fracking's methane leaks aggravate climate change. AIDA. https://aida-americas.org/en/blog/how-fracking-s-methane-leaks-aggravate-climate-change.

Union of Concerned Scientists. (2015). Bad chemistry: how the chemical industry's trade association undermines the policies that protect us. https://www.ucsusa.org/center-science-and-democracy/fighting-misinformation/american-chemistry-council-report#.XA7HLlVKjm4.

U.S. Bureau of the Census. (2017). Barrow City, Alaska. https://data.census.gov/cedsci/table?q=Barrow%20City,%20Alaska&tid=ACSDP5Y2017.DP05&hidePreview=false.

Valle, D. & Clark, J. (2013). Conservation efforts may increase malaria burden in the Brazilian Amazon. *PLoS One* 8: e57519.

Vijay, G. (2017). Systemic failure of regulation: the political economy of pharmaceutical and bulk drug manufacturing. In: H. Lofgren (Ed.), *The Politics of the Pharmaceutical Industry and Access to Medicines—World Pharmacy and India*. New Delhi: Social Science Press.

Vineis, P. (2004). A self-fulfilling prophecy: are we underestimating the role of the environment in gene-environment interaction research? *International Journal of Epidemiology* 33: 945–946.

Vittor, A., Gilman, R., Tielsch, J., Glass, G., Shields, T., Lozano, W. *et al*. (2006). The effect of deforestation on the human-biting rate of *Anopheles darlingi*, the primary vector of falciparum malaria in the Peruvian Amazon.

American Journal of Tropical Medicine and Hygiene 74: 3–11.

Vittor, A., Pan, W., Gilman, R., Tielsch, J., Glass, G., Shields, T. *et al.* (2009). Linking deforestation to malaria in the Amazon: characterization of the breeding habitat of the principal malaria vector, *Anopheles darlingi*. *American Journal of Tropical Medicine and Hygiene* 81(1): 5–12.

Waldholz, R. (2017). Alaskan village, citing climate change, seeks disaster relief in order to relocate. *New England Public Radio*. https://www.npr.org/2017/01/10/509176361/alaskan-village-citing-climate-change-seeks-disaster-relief-in-order-to-relocate.

Waldholz, R. (2018). Newtok to Congress: thank you for saving our village. *Alaska Public Media*. https://www.alaskapublic.org/2018/03/27/newtok-to-congress-thank-you-for-saving-our-village/.

Warembourg, C., Tamayo, I., Maitre, L., Chatzi, L., Fossati, S., Gonzalez, J. *et al.* (2018). Environmental exposures during early-life and child blood pressure: an exposome approach. *Environmental Health Perspectives* 74(10): 1317–1328.

Weintraub, P. (2009). *Cure Unknown: Inside the Lyme Epidemic*. New York: St. Martin's Griffin.

Weintraub, P. (2011). The human face of human tick-borne infections. Commissioned paper for critical needs and gaps in understanding prevention, amelioration, and resolution of Lyme and other tick-borne diseases: the short-term and long-term outcomes. *Workshop report*. Washington, D.C.: National Academy of Medicine.

Western, D. (2001). Human-modified ecosystems and future evolution. *Proceedings of the National Academy of Sciences of the U.S.A.* 98(10): 5458–5465.

Wigle, D., Arbuckle, T., Turner, M., Berube, A., Yang, Q., Liu. S. *et al.* (2008). Epidemiologic evidence of relationships between reproductive and child health outcomes and environmental chemical contaminants. *Journal of Toxicology and Environmental Health, Part B: Critical Reviews* 11: 373–517.

Wilcox, B. & Murphy, D. (1985). Conservation strategy: the effects of fragmentation on extinction. *American Naturalist* 125: 879–887.

Wild, C. (2005). Complementing the genome with an "exposome": the outstanding challenge of environmental exposure measurement in molecular epidemiology. *Cancer Epidemiology, Biomarkers & Prevention* 14(16): 1847–1850.

Wild, C. (2012). The exposome: from concept to utility. *International Journal of Epidemiology* 41: 24–32.

Wild, C. & Turner, P. (2002). The toxicology of aflatoxins as a basis for public health decisions. *Mutagenesis* 17: 471–481.

8

Climate change, crisis enhancement

30 years ago this month, Dr. James Hansen testified before the U.S. Congress on the need to address climate change—30 years ago this month. He was a top NASA climate scientist . . . Thirty years have passed since then—30 years of added science, 30 years of new science, 30 years of updated reports, and 30 years of mounting evidence of how right Hansen was. Yet, here we still are in Congress still willfully ignoring the unprecedented changes to the climate and the oceans—changes that threaten our planet and its rich array of plant and animal life, changes that put at risk homes, farms, forests, and coasts, changes that affect our very human health and well-being. These are not computer model projections of the distant future but changes we are seeing right before our very eyes now.

Senator Sheldon Whitehouse
(June 5, 2018)

8.1 Consensus on climate change

The extensive body scientific evidence indicating that we are living in a period of significant climate change is the product of the work of thousands of scientists from around the world who have been trained in numerous different physical and social scientific disciples, including climatology, astrophysics, biology, biogeochemistry, archeology, physics, oceanography, meteorology, geology, atmospheric chemistry, glaciology, anthropology, sociology, and political science. Physical scientists have published thousands of rigorously researched and peer-reviewed studies documenting increasing surface, atmospheric, and oceanic temperatures; melting land ice and glaciers; diminishing snow cover; shrinking sea ice and loss of winter lake ice; rising ocean levels; ocean acidification; increasing atmospheric water vapor; and the history of Earth's changing climate. Biological and environmental scientists have assessed the impacts of climate change on biodiversity; species migration; food chain disruptions; ecosystem simplification and fragmentation; and extinction. Social scientists have contributed studies on people's perceptions and experiences of climate change effects; the diversity of attitudes toward and knowledge of climate change; the economic impacts of global warming; the politics and conflicts produced in response to climate change; and mitigative and adaptive responses to climate change. Of all the scientists who study climate-related issues, approximately 97 percent agree that the world is heating up and human activity is the main driver of this change and the related environmental alterations just listed (Oreskes 2004; Cook *et al.* 2016). Other individuals who claim to be climate scientists or climate experts but rarely have the academic credentials to support their assertions do not share in this consensus, but

Ecosystem Crises Interactions: Human Health and the Changing Environment, First Edition. Merrill Singer.
© 2021 John Wiley & Sons, Inc. Published 2021 by John Wiley & Sons, Inc.

the credibility of their opinions is in question because often they are funded to discredit climate science by elite polluters. Broad agreement among scientists across different fields of study is noteworthy. The word "consilience" (from the Latin "jumping together") means coming to agreement about a topic by researchers involved in different areas of study. Climate scientists have achieved consilience on the reality of climate change. People's confidence should be raised in the trustworthiness of scientific conclusions when there is a convergence of evidence and consistent message emerging across disciplines. As Shermer (2005) emphasizes, "proof is derived through a convergence of evidence from numerous lines of inquiry—multiple, independent inductions, all of which point to an unmistakable conclusion."

Within science today, the primary debates are not about whether climate change is occurring, humans are playing a central role in this process, or the health and social consequences of continued planet warming will be grave. Knowledge within science, however, is not binary. Understandings are never fully settled or unsettled. Rather, science knows things with varying degrees of confidence. Lesser degrees do not mean something is unknown, but identify the need for continued research. Indeed, over time, as a result of the ongoing work being done in climate science, there has been significant upgrading in the confidence with which we understand the broad issues of climate change. Continued debate involves issues such as the future timing of adverse climate change effects, local and regional expressions, the full scale of the risks climate change unleashes, the relative roles of all the forcing agents (e.g., CO_2, methane, black carbon) involved at different scales, and the value of various mitigative strategies. When these issues are better known there will still be more questions in need of answering, because the more you know about nature, the more you realize you do not know. The critical question is: When do you know enough to act? On this point there also is a high level of agreement among scientists: We already do (indeed, we have done for years), and the time to act vigorously to both mitigate future climate change and adapt to the effects of existing climate change is now. Scientists have been saying this with increasing urgency. Unfortunately, their voices are met by deaf ears among the people empowered to enact the policy changes that are vitally required. As a result, there is a need for significantly enhanced pressure from below through mass social initiatives for change.

8.2 Driving climate change

What do we know about the forces driving climate change? The level of CO_2 in the atmosphere rose from around 317 parts per million (ppm) in 1958, when Charles David Keeling of Hawaii's Mauna Loa Observatory began atmospheric levels of the most abundant greenhouse gas, to 400 ppm on May 9, 2013. This meant that for every million air particles, 400 of them were CO_2 molecules (i.e., 0.04%). This was the first time that the 400 ppm threshold had been passed in human history. The last time it reached this level was 2.6–5.3 million years ago during the Pliocene, when there were no humans and Earth was a very different place than it is today. Sea level was 16–131 feet higher and the North and South poles were 18 °F hotter. There was a vast spreading of grasslands and savannas. But the 400 ppm record too has been shattered. In April 2018, monthly CO_2 levels exceeded 410 ppm, and in May 2018, a new record of 411 ppm was recorded (Monroe 2018). A year later, this record also was beaten when monthly CO_2 levels reached 415.7 ppm, highlighting yet again the meteoric rise of CO_2 in our atmosphere. In the second decade of the 21st century, more CO_2 emissions have been released than in any other decade in human history (Queally 2019).

Further, scientists have found that the rate of CO_2 increase has accelerated, from an average of 1.6 ppm per year during the 1980s and 1.5 ppm per year during the 1990s to 2.2 ppm per year during the early 2000s (YaleEnvironment[360] 2018). CO_2 concentrations rise and fall somewhat based on the season, with lower levels in late August and early September, when vegetation growth in the Northern Hemisphere is at its zenith, and higher levels in May. From 2016 to 2017, the global CO_2 average increased by 2.3 ppm—the sixth consecutive year-over-year increase greater than 2 ppm. Prior to 2012, back-to-back increases of 2 ppm or greater had occurred only twice. In 2018, CO_2 emissions are set to have hit an all-time high (Fig. 8.1). Researchers at the University of East Anglia and the Global Carbon Project predict a rise of more than 2 percent based on growth in coal use for the second year in a row and sustained growth in oil and gas use. Burning of fossil fuels contributed an additional 37.1 billion tons of CO_2 to atmospheric levels in 2018 (Chestney 2018).

In explaining the force pushing up the proportion of the atmosphere that consists of CO_2, Ralph Keeling, who maintains the longest continuous record of atmospheric carbon dioxide on Earth (since 1958), states: "We keep burning fossil fuels. Carbon dioxide keeps building up in the air . . . It's essentially as simple as that" (quoted in Monroe 2018). While this does cut to the heart of the matter, climate change is hardly a simple issue. For one thing, while very significant, the burning of fossil fuels is not its only driver. The conversion of forest to agricultural land also pushes it, as does the melting of tundra. All of these alterations shift the balance between the amount of the sun's energy radiating down and warming the planet (known as "radiative forcing") and the amount of energy Earth radiates back out into space. Also at play are aerosols—small particles of sulfates, nitrates, ammonium, organic carbon, black carbon, sea salt, mineral dust, trace metals, and water—which are released in to the air during the burning of fossil fuels and through some natural processes. Some aerosols, such as black carbon, drive global warming, while others scatter and absorb solar radiation, causing a degree of cooling during their short atmospheric lifetimes (far shorter than the atmospheric lifetime of CO_2). Moreover, climate change is not just one thing: it is many things, which often interact with one another. Camillo Mora *et al.* (2018) have identified over 450 distinct ways climate change impacts human communities and argue that these threats are likely to overlap, five or more at a time. They include heatwaves, intense precipitation, drought, floods, fires, sea-level rise, and changes in ocean chemistry (see later). Impacts

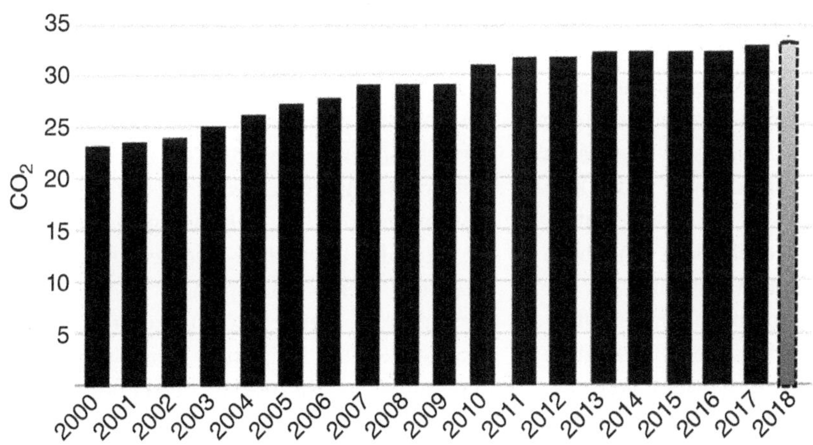

Fig. 8.1 Global energy-related CO_2 emissions.

on human health include malnutrition, mental health problems, infectious epidemics, physical injuries, toxic chemical exposure, stress, and respiratory problems.

In the fall of 2018, the U.N. Environment Program (2018) released its 9th edition of the U.N. Environment Emissions Gap Report. This document, prepared by an international team of climate scientists, provides an assessment of the state-of-the-art of scientific studies on current and estimated future greenhouse gas emissions and compares them with the emission levels that would allow achievement of the goal of the 2015 Paris Agreement to keep average global temperature rise less than 1.8 °F above pre-industrial levels. This difference between current and desired emission levels is known as the "emissions gap." The report indicates that global CO_2 emissions from energy and industry increased in 2017 after 3 years, 2014–2016, of level emission rates. Total greenhouse gas emissions, including from land-use change, reached a record high of 53.5 gigatons in 2017, an increase of 0.7 gigatons compared with 2016. Thus, the emissions gap has widened at a time when negative impacts of global warming from past greenhouse gas release are occurring around the world.

8.3 How serious is climate change?

An international team of ocean researchers and climate scientists from the United States, China, France, and Germany reported during 2018 that their research suggested the oceans of the world had absorbed much more heat than most of the previous estimates of other researchers (Resplandy *et al.* 2018). It has been known for some time that the oceans have the ability to store vast amounts of heat—a thousand times that of the atmosphere. This feature stems from three factors: 1) it takes much more heat to warm water than it does to warm an equal volume of air; 2) the oceans are vast and generally deep (2.5 miles on average), providing a tremendous reservoir for heat absorption; and 3) the oceans are dynamic, being churned by currents, internal waves, eddies, storms, and other circulation mechanisms that mix heat, carbon, oxygen, and other elements exchanged with the atmosphere. It is estimated that since 1955 over 90 percent of the excess heat trapped by the greenhouse gas blanket circling Earth has been deposited in the oceans. Much of it is stored within the top 250 feet due to proximity to the atmosphere and the higher degree of water mixing within this top layer. As the oceans continue to absorb heat, however, it penetrates deeper.

The issue of heat absorption in deeper waters of the ocean is highly pertinent to a public debate that followed the revelation of a slowing of global warming based on land-based measures for the period 1998–2013 despite continued increase in anthropogenic greenhouse gas emissions. Skeptics and deniers of global warming were quick to jump on these findings to declare either that warming had stopped or that the whole idea of global warming was a hoax. Climate scientists who study long-term trends responded that the record shows that there has not been a steady year-by-year increase in average global temperature. Rather, as Richard Muller (2013), Professor of Physics at the University of California, Berkeley, observes: "The planet warmed by 0.6 degrees over the prior of 50 years, but occasional, unexplained temperature fluctuations of as much as 0.3 degrees countered the rise at times and resulted in apparent pauses." In other words, planetary temperatures do not narrowly track carbon dioxide levels in the atmosphere. Other factors intervene and cause dips and pauses, but by stepping back to see the forest for the trees, we can clearly recognize the trend: the planet is warming (Easterling & Wehner 2009). Failure to pay attention to the big picture can give undue importance to short-term fluctuations and their causes. One of these causes may be deep ocean absorption of excess heat. Deep ocean heating did not shows signs of a slow down during the period of concern. This has led some scientists to suggest that the so-called "pause" was driven by heat uptake in the deeper areas of the oceans as a consequence of ocean-layer mixing pushed by El Niño (Hausfather 2013). This is the name given to a warming of surface ocean waters in the eastern tropical Pacific Ocean off the coast of South America that happens every 3–8 years and can affect global temperatures for a year or more.

There is another issue concerning the "pause" that stems from the changing ways ocean temperature assessments have been conducted over the last 165 years. These methodological changes have given rise to systematic errors in calculations of sea surface temperature (SST) (Smith & Reynolds 2005). Until 1940, SST records were based on the temperatures of water samples hauled up in either canvas or wooden buckets lowered from ships. Subsequently, there was a period when measurements were also taken through intake valves installed in ship engine rooms. Researchers came to the realization that these imperfect approaches led to systematic biases in ocean temperature data, which were further complicated by differences between measurements taken by U.S. and U.K. ships that had been entered into the International Comprehensive Ocean Atmosphere Data Set (Worley *et al.* 2005). Since the 1970s, SSTs have increasingly been determined using both moored and drifting buoys, and it is now recognized that ship-based assessments tend to produce warmer temperature readings than buoy-based ones (Emery *et al.* 2001). Additionally, researchers have identified differences between commonly used composite SST reconstructions from the National Oceanic and Atmospheric Administration's (NOAA) Extended Reconstruction Sea Surface Temperature, the Hadley Centre SST, and the Japanese Meteorological Agency's Centennial Observation-Based Estimates of SSTs data sets. These three data sets assimilate information from multiple instrument platforms, including lowered buckets and engine intake valves on ships, buoys, and satellites, to establish long-term records. Efforts have been made to estimate the extent of identified biases and to adjust the SST data accordingly (Kennedy *et al.* 2011a, 2011b). In 2015, Thomas Karl and fellow scientists affiliated with the NOAA published a paper reporting that in order to correct for measurement bias in the engine intake ports of ships, they had somewhat revised the data NOAA had been citing for SSTs for the 1900s and 2000s (Karl *et al.* 2015). These modifications led to the conclusion that global SSTs during the 2000s were actually higher than in previous decades and that there likely had been no pause in global warming after all. Climate change deniers, including Republican Congressman Lamar Smith, who initiated a congressional investigation of NOAA scientists, responded with outrage. They claimed that the adjustment had been politically motivated to support the alleged "hoax of climate change." While deniers still cling to this unsupported assertion, subsequent independent scientific assessment affirms the need to make NOAA's changes to correct bias in the SST records (Hausfather *et al.* 2017). As Cowtan & Hausfather (2017) comment, "the scientific question of how fast the Earth has been warming over the past two decades can be answered by replication from the scientific community, not by a political investigation. And the best evidence we have says that NOAA got it right."

In its 2013 assessment of the existing research on ocean warming, the Intergovernmental Panel on Climate Change (Rhein *et al.* 2013, p. 257) reported that "It is virtually certain that the upper ocean (above 700 m) has warmed from 1971 to 2010." The studies reviewed by IPCC analysts drew somewhat different conclusions, but together they indicate for the time period of concern that rates of ocean heating increase ranged between 74 and 137 TW (1 TW = 10^{12} watts of power). Moreover, the report emphasizes that most of the heating occurring on Earth during this period was in the oceans: "Warming of the ocean accounts for about 93% of the increase in the Earth's energy inventory between 1971 and 2010" (IPCC 2013, p. 257).

The studies reviewed by the IPCC were based on measurements taken by thermometers affixed to battery-powered bullet-shaped Argo floating devices. These machines, of which there are now about 4000 spread across the oceans of the world, take subsurface temperature and salinity measurements every 10 days, and then rise to the surface to transmit their data to the Earth-orbiting Jason-3 satellite. Their name is derived from ancient Greek mythology. The *Argo* was the ship that Jason sailed in his legendary journey to capture the Golden Fleece, a symbol of royal authority. Gaining robust and authoritative knowledge about ocean temperatures, as well as other ocean features, is the goal of the observational and modeling Argo program. As Ocean Scientists for Informed Policy (OSIP 2018) at the Scripps Institution of Oceanography explains, "[t]he complex interactions between continued emissions of greenhouse gases, consequent energy imbalance, and changes in the storage and transport properties of heat in the ocean will largely determine the speed and magnitude of longterm anthropogenic climate change impacts."

The new research by Resplandy *et al.* (2018), based on a different methodology, suggests that ocean warming has actually occurred at the high end of the previous estimates. The new method involves high-precision measurement of oxygen and carbon dioxide in the air. Levels of these gases increase as the ocean warms and releases them into the atmosphere. Using this approach, it has been determined that between 1991 and 2016 the oceans warmed about 60 percent faster than the average rate of warming estimated by the studies reviewed by the IPCC. This means that the oceans took up more than 13 zettajoules of heat energy per year. A joule is the standard unit of energy measurement, and a zettajoule is 1 joule followed by 21 zeroes (and 1 watt = 1 joule per second). The implication is that the harsh impacts projected by the IPCC, including many climate-linked ecocrises interactions, which will occur if Earth's average temperature rises by 2.7 °F above pre-industrial levels, will arrive far sooner than 2035, the IPCC's original estimate. According to Resplandy, "Imagine if the ocean was only 30 feet deep . . . Our data show that it would have warmed by 6.5 degrees Celsius [11.7 °F] every decade since 1991. In comparison, the estimate of the last IPCC assessment report would correspond to a warming of only 4 degrees Celsius [7.2 °F] every decade" (quoted in Kelly & Monroe 2018).

Resplandy *et al.*'s (2018) findings are supported by a review of recent literature on ocean heat content (OHC), which is what oceanographers and climatologists call the amount of heat stored in the ocean. This review (Cheng *et al.* 2019), which covers the work of multiple research teams and numerous lines of evidence, suggests that the oceans are heating at a rate that is about 40 percent faster than was previously estimated by the IPCC. Various studies suggest that since the 1950s, the ocean has been absorbing at least 10 times as much energy each year as humans annually consume worldwide.

In the fall of 2018, the IPCC released a new report, based on thousands of new studies and reviewed by over 1000 climate scientists, which projects that a dangerous global temperature rise of 2.7 °F above pre-industrial levels will occur 5 years earlier (i.e., in 2030) than it had previously anticipated (IPCC 2018). According to Debra Roberts, a biologist and local government official in South Africa, who co-chaired the IPCC working group on impacts: "It's a line in the sand and what it says to our species is that this is the moment and we must act now . . . This is the largest clarion bell from the science community and I hope it mobilises people and dents the mood of complacency" about the mounting risks of climate change (quoted in Watts 2018).

The fall of 2018 was also marked by the release of the congressionally mandated Fourth National Climate Assessment on impacts, risks, and adaptation (U.S. Global

Change Research Program 2018). A product of the Global Change Research Act of 1990, the 1000-plus-page report was authored by various U.S. federal agencies based on existing scientific research. It affirms in an unambiguously strong tone that there is a high degree of certainty that climate change poses a severe threat to the health and economic status of the people of the United States, as well as to the country's aging and deteriorating infrastructure and natural resources. Scheduled for release in December 2018, the Trump administration instead issued the report on Black Friday, the day after Thanksgiving, which has become one of the busiest shopping and slowest news days of the year. After failing to minimize the attention it received, President Trump announced that he completely rejected the report's central finding that global warming is causing ongoing and lasting economic damage, stating to reporters: "I don't believe it" (quoted in Kavelle 2018). Katherine Hayhoe, one of the lead authors, responded: "The science behind climate change has been understood since the 1850s. We cannot afford to have politically motivated people spin the issue any longer" (quoted in Miller 2019).

At the level of American communities, the report stresses that while adverse effects are already being felt, "[m]ore frequent and intense extreme weather and climate-related events, as well as changes in average climate conditions, are expected to continue to damage infrastructure, ecosystems, and social systems that provide essential benefits" (U.S. Global Change Research Program 2018). At special risk are lower-income and other marginalized communities, because they have fewer resources with which to prepare for and cope with extreme weather and climate-related events (Walker *et al.* 2010). Indigenous communities, for example, will increasingly face disruptions of their livelihoods and economies, threats to culturally significant species and resources, and challenges to community stability and cohesiveness (Lynn *et al.* 2013). Some already are confronting decisions about relocation, as discussed in Chapter 7 (Singer 2018). Also at unique risk are coastal communities and coastal property values.

As this suggests, there are significant economic costs to climate change. Regional economies and industries that are heavily dependent on natural resources and favorable weather conditions, such as agriculture, tourism, recreation, and fisheries, are especially vulnerable to the mounting impacts. Higher temperatures, the report stresses, will reduce the efficiency of power generation while increasing energy demands, leading to rising electricity costs (Clarke *et al.* 2018). Barring significant reductions in greenhouse gas emissions, there will be substantial net damage to the U.S. economy throughout the 21st century. Numerous ecosystem benefits, such as clean air and water, natural protection from coastal flooding, wood and fiber, outdoor recreation, and crop pollination, will continue to be degraded (Nelson *et al.* 2013). Also increasingly endangered is the nation's capacity to improve or even sustain current crop productivity, livestock health, and the economic vitality of the rural communities that contain 15 percent of the U.S. population. Over the next several decades, yields from major U.S. crops are expected to decline because of climate-related surges in temperatures, changes in water availability, soil erosion, and disease and pest infestations (Zampieri *et al.* 2017).

Overall, the Fourth National Climate Assessment makes it abundantly clear that even for a comparatively wealthy, large, resource-abundant, and generally geographically well-positioned country like the United States, climate change is deadly serious. Further, and just as importantly, it affirms that many of the adverse effects of climate predicted by scientists are becoming an everyday reality. For example, the previous National Climate Assessment, issued in 2014, projected that as sea levels rose, U.S. coastal cities would suffer more frequent flooding, and that is precisely what has happened in cities like Miami, FL and Charleston, SC. In November 2018,

Charleston experienced its sixth highest tide on record, exceeding levels seen during the 1000-year storm-driven flooding event it endured in October 2015. The 2018 event reached parts of the city that usually escape high-tide flooding, making downtown roads impassable to vehicular traffic. Water levels approached heights usually only seen during hurricane or tropical storm conditions and took longer than usual to recede. Dry-weather flooding due to tides alone (rather than precipitation) now occurs at least 40 times per year in Charleston, a fourfold increase compared to the 1960s (Gyee 2018). In 2015, the previous record number of days with tidal flooding was broken when the city flooded 38 times, and this new record was broken again in 2016 when it flooded 50 times during high tide (Peterson 2017). Extreme weather events like those suffered by Charleston, and others described later, are among the most tangible indications that the effects of climate change are already taking a significant toll.

Notably, just days after losing the 2020 presidential election, Trump fired Michael Kuperberg, head of the National Climate Assessment program (Blest 2020).

8.4 Drought and heatwaves

In 2017, the Energy and Climate Intelligence Unit (ECIU), a nonprofit organization that supports science-based debate on energy and climate change issues in the United Kingdom, published a review of the scientific literature on the role of climate change in amplifying extreme weather events, such as heatwaves, droughts, and floods for the two-year period 2016–2017 (ECIU 2017). The study identified 59 climate science papers published during those years that examined possible links between climate change and specific extreme weather events. Of these, 41 (69 percent) found a positive association. While only five of the individual extreme weather events analyzed in these studies reported costs in terms of lives lost or economic damage, those that did suggested that climate change caused about 4000 deaths (in the short run) and about $8 billion in economic damage. Were data available for all of the extreme weather events analyzed in the 59 studies, the costs of climate change would be revealed as being far higher.

At the end of 2018, the ECIU repeated this analysis with papers published since its prior report. The additional year added 43 relevant papers into the climate science literature, of which 32 (74 percent) reported a positive link between climate change and the likelihood or intensity of an extreme weather event. Four of the new publications reported that climate change decreased the chances of a particular extreme event, all of these involving snow or cold temperatures. Among the 2018 studies were new reports on heatwaves in France, China, Thailand, and the United States and drought in the United States and parts of Africa (ECIU 2018).

What is a drought? In everyday usage, it is a period of unusually dry weather that causes problems in terms of insufficient water for human consumption, agriculture, or livestock rearing (Fig. 8.2). Because dry conditions develop for various reasons and under varying conditions, there is no universally acceptable definition of "drought"; rather, there are multiple ones. In fact, in the mid-1980s, two researchers, Wilhite & Glantz (1985), identified more than 150 published definitions. In an effort to bring some order to discussing droughts, they grouped these under four labels: meteorological, hydrological, agricultural, and socioeconomic. In the first three types, drought is conceived as a physical phenomenon, while in the last category it is a problem of human need versus availability. Meteorological approaches define drought on the basis of the extent of dryness and the duration of the dry period. Hydrological approaches focus on the effects of dry periods on surface or subsurface water levels. Agricultural approaches link meteorological drought factors to their agricultural impacts. Finally, socioeconomic approaches focus on

Fig. 8.2 Drought. *Source:* Lukas/Adobe Stock Photos.

the nature of the socioeconomic effects of extended dry periods (e.g., famine, migration, reduced energy production from hydroelectric power plants, social conflict). As Wilhite & Glantz (1985, p. 119) conclude, drought has multiple adverse consequences "with pervasive societal ramifications."

While increasing concentrations of atmosphere greenhouse gases are widely expected to influence global climate, establishing the fine-scale impact on the creation of drought conditions is difficult because of the complexity of the processes involved in the hydraulic cycle both regionally and globally (Hagemann *et al.* 2013; Mukherjee *et al.* 2018). To address this challenge, hydraulic researchers combine an ensemble of global hydrological models and a range of global climate models to synthesize alternative possible impacts on water availability. Computer modeling research of this sort by Prudhomme *et al.* (2014), focused on identifying regions where the impact of climate change on drought shows a strong likelihood of change between the ends of the 20th and 21st centuries, suggests that Southern Europe, the Middle East, the Southeast United States, Chile, and South West Australia are possible hot spots for increases of more than 20 percent in the frequency of drought.

Scientific advances over the years have allowed movement from an initial focus on the development and use of numerical models of Earth's climate and the assemblage of extensive collections of observational data to the ability to both identify and attribute specific impacts and events to anthropogenic climate change. As a result, analysts can now establish the role of climate change in various meteorological and hydrological occurrences around the world, including drought. Research in California by Diffenbaugh *et al.* (2015) adds to the accumulating evidence that anthropogenic climatic change is influencing the frequency, magnitude, and duration of drought in the state. These researchers show that the increasing co-occurrence of dry years with warm years more than doubled the risk of drought despite limited evidence of a downward trend in precipitation. The occurrence of drought years was greater from 2005 to 2015 than in the preceding century. Fossil fuel emissions have increased the probability that years with reduced rain and snow are also warm years, suggesting that anthropogenic warming is increasing the probability of the co-occurring warm–dry conditions that are driving drought.

As noted, one of the potential but not well understood socioeconomic impacts of drought is conflict, an occurrence some researchers already see occurring in the Middle East hot spot. In the years 2007–2010, Syria suffered its worst drought on record. The observed drying and warming that occurred during these years is consistent with climate model studies of the water cycle effects of increases in greenhouse gases. The drought caused widespread crop failure and a mass migration of devastated farming families to the peripheries of urban centers, resulting in a 50 percent increase in the number of urbanized Syrians between 2002 and 2010 (Kelly *et al.* 2015). By 2009, an estimated 800 000 Syrians had lost their livelihoods due to the drought, and within 2 years 1 million were suffering food insecurity and 2–3 million had been driven into extreme poverty (IRIN News 2009; Worth 2010). The rapidly growing urban peripheries were made up of illegal settlements that were overcrowded, had weak infrastructure (e.g., electricity, potable water, hygienic systems), and suffered high levels of unemployment and crime. These settlements "were neglected by the Assad government and became the heart of . . . developing unrest" (Kelly *et al.* 2015).

Exacerbating the impact of the drought were government policies intended to expand agricultural production without a consideration for sustainability under changing climatic conditions. Syria was forced to begin importing large quantities of wheat, coincident with an unprecedented rise in food prices. During 2007–2008, consumer prices for wheat, rice, and livestock feed more than doubled. Livestock herds were wiped out. Moreover, in the northeast provinces, there was a dramatic increase in nutrition-related diseases among children. The government responded in neoliberal fashion by cutting the fuel and food subsidies on which many Syrians had become dependent.

In short, as various analyses have shown (Gleick 2014; Werrell *et al.* 2015), the devastating civil war that began in Syria in March 2011, which has cost the lives of half a million people and forced more than 7 million to flee the country, was sparked by climate-related factors in conjunction with government mismanagement and rampant social inequality. More broadly, the intense social unrest and sustained street protests that swept across the Middle East and North Africa during 2010–2012, popularly known as the Arab Spring, which led to the fall of a number of governments, though driven by both regional and national political and economic factors, were likely strongly influenced by climate change and a consequent spike in global food prices (Johnson & Mazo 2013). Increasing numbers of researchers have thus come to recognize that a significant social effect of climate change is conflict (Smith & Vivekananda 2007; Buhaug *et al.* 2008; Rubin 2010). As with other factors and processes with which it interactions, climate change is an effect enhancer that magnifies already existing intergroup tensions leading to social conflict.

One aspect of these tensions is seen in the "Dry Corridor" of Central America—an ecological region that runs from southern Mexico through Guatemala, El Salvador, Honduras, Nicaragua, Costa Rica, and Panama—where climate change is forcing small farmers and other rural residents to leave their homes and migrate northward. This region has endured severe droughts and floods that threaten the physical and financial well-being of the people who live there. Pushing climate effects in the region is the El Niño Southern Oscillation, which makes it particularly vulnerable to ever more disrupted and extreme precipitation. The last 6 months of 2018 witnessed a drought in the Dry Corridor that affected the lives of over 77 000 corn farmers in El Salvador and more than 65 000 families in Honduras. Both the

U.N. Food and Agriculture Organization (FAO) and the U.N. World Food Program (WFP) expressed concern that "total or partial loss of crops means that subsistence farmers and their families will not have enough food to eat or sell in coming months" (quoted in Climate Reality Project 2019). This followed a previous drought that began in 2014 and lasted until the end of 2017, causing widespread loss of crops and the deaths of thousands of cattle, changes in food supply that put 2.5 million people at risk of food insecurity. Also pushing Central Americans northward to seek sanctuary in the United States were drug-related violence, engrained structures of inequality in their home countries, political corruption, and dwindling opportunities to escape abject poverty at home. According to the World Bank (2019), El Salvador had the highest murder rate in the world in 2016, at 83 homicides per 100 000 inhabitants (compared to 5.3 per 100 000 in the United States). Honduras was in second place, while Guatemala was in tenth. More broadly, there was the historic role of the United States in propping up oppressive right-wing military regimes, supporting the activities of death squads, advancing corporate plunder, and imposing neoliberal restructuring that has sapped the resources of Central America. As Borger (2018) states, "The families in the migrant caravans trudging towards the U.S. border are trying to escape a hell that the U.S. has helped to create." This adverse "assistance" has come in three forms: economic, political, and climatic, all of which push Central Americans out of their homes and toward the U.S. border.

Globally, there has been an increasing frequency and intensity of heatwaves since the 1950s, associated with climate change. Heat waves—periods of excessively hot weather relative to what is normally expected, often caused by trapped air—pose a significant threat to human health and well-being. While there is no international agreement about the definition of a heatwave, the World Meteorological Organization (WMO 2016, p. 10) defines one as "a marked unusual hot weather . . . over a region persisting at least two consecutive days during the hot period of the year based on local climatological conditions, with thermal conditions recorded above given thresholds."

Prolonged or intense exposure to very hot temperatures without intervening cooling periods can cause a number of heat-related health conditions, such as heat exhaustion (heavy sweating and rapid pulse), heat cramps (intense pain in the muscles), and heat stroke (dramatic elevation of body temperature, commonly accompanied by dehydration). When exposed to high temperature, the human body has evolved mechanisms to cool itself that include signaling from the hypothalamus to dilate blood vessels in order to push blood to the skin and release the excess heat into the environment. This produces a reduction of blood to the brain, muscles, and other organs that can interfere with bodily function and mental capacity. It is estimated that over 1 billion people are at heat risk because of a lack of air conditioning and refrigeration to keep them cool and preserve their food and medicines as global warming advances. The World Health Organization (WHO) reports that heat stress linked to climate change is likely to cause 38 000 extra deaths globally each year between 2030 and 2050. In May 2018, for example, more than 60 people died in Karachi, Pakistan when the temperature rose above 104 °F (Doyle 2019). Most of the dead were working-class people from poorer neighborhoods.

To establish just how significant waves of extreme heat can be, Mora *et al.* (2017a) reviewed the medical literature for instances of deadly heatwaves published between 1980 and 2014. They identified over 78 cases of excess human mortality associated with heat (i.e., greater than would naturally have occurred without the heatwave) from 164 cities in 36 countries. Using the climatic conditions that prevailed during these deadly heat events, they established a global threshold beyond which daily mean surface air temperature and relative humidity become lethal. Research suggests that currently about 30 percent of the human population of the world is exposed to conditions that exceed this deadly threshold for at least 20 days per year. Based on projected planet heating, by the year 2100 Mora *et al.* (2017a) maintain this proportion will increase to about 48 percent even with drastic reductions in greenhouse gas emissions, or about 74 percent under a scenario of continued growth in emission levels.

A number of powerful heatwaves during the 21st century have resulted in high losses of life. These include the 2003 European heatwave, which took 70 000 lives, several North American heatwaves in 2006, which were responsible for hundreds of deaths in California and Canada, the 2010 Russian heatwave, which killed 10 000 people, and the 2015 heatwave in Karachi, India, which claimed about 1000 lives. The high death tolls from these and multiple other heatwaves "are staggering demonstrations that extreme climatic conditions are already exceeding human thermoregulatory capacity" (Mora *et al.* 2017b, p. 1). A special risk applies to people suffering from cardiovascular, respiratory, and cerebrovascular diseases, low-income ethnic minorities, women, seniors over 65 years of age, the isolated, and infants and young children (Basu 2009). In our changing world, access to cooling is no longer an issue of comfort, it is one of social equity.

The 2003 heatwave in Europe has garnered special attention because it bears a close resemblance to what many climate computer models project for summer seasons in the latter part of the 21st century. Thus, it has been informally dubbed the "shape of things to come." Research by Pfister *et al.* (1999) suggests that 2003 was probably the warmest summer in Europe for 460 years. The heatwave did not occur in a vacuum. During that summer, Europe was under the influence of a strong high-pressure system positioned over the English Channel, which produced an extended drought and higher-than-normal heat levels throughout much of the year. Many rivers ran dry by the end of the summer, while crops wilted prior to harvest, creating dire conditions in terms of food supply and livelihood issues for many communities. These are the very conditions—soil moisture depletion, lack of heat-moving rainfall, periods of intense heat, reduced river discharge, crop failure, and resulting heat-related illness and death—that are projected to occur with greater frequency in the future (Beniston 2004).

Weather conditions (i.e., heat and humidity) that prevent the body from dissipating heat are dangerous because they cause unsafe physiological responses or break down pathways in the human body. To identify these pathways, Mora *et al.* (2017b) systematically searched online medical databases. They found reports of five different physiological mechanisms that are triggered by heat exposure, namely ischemia (inadequate blood supply to an organ or body part), heat cytotoxicity (heat-related cell death), inflammatory response (cell damage, cell leakage, and tissue swelling), disseminated intravascular coagulation (overactivation of the proteins—called thrombins—that cause blood clotting), and rhabdomyolysis (death of muscle fibers and release of their contents into the bloodstream, triggering kidney failure). Further, they identified seven vital organs that can be damaged by overheating: the brain, heart, intestines, kidneys, liver, lungs, and pancreas. Of 35 possible combinations (i.e., five physiological mechanisms times the seven vital organs at risk), they found there is medical evidence for 27 different pathways through which physiological mechanisms triggered by excess heat can lead to organ failure and possible death. Importantly, all of the physiological responses listed "are interrelated such that dysfunction in [one] organ has negative effects on others, triggering a vicious cycle of multiorgan deterioration, which often leads to death, permanent disabilities, or lengthy recovery" (Mora *et al.* 2017b, p. 2).

Exemplary of climate-related heatwave trends are the changes occurring in the United Kingdom, a country one usually does not associate with tropical weather. British weather is known for being erratic, but rarely is it extreme. In the past, the average temperature during summer ranged from 48 to 64 °F. Weather in the United Kingdom is shaped by three key factors. First, being geographically positioned at the edge of the Atlantic Ocean, it sits at the end of a phenomenon known as a storm track, a pathway traveled by storms as they are driven by the prevailing winds. The storms form as warm air from the equator encounters cold air from the Arctic. Second, there is the Gulf Stream, an ocean current originating in the Gulf of Mexico; in particular, a part of the current called the North Atlantic Drift provides the country with with a climate milder than would be expected given its northern latitude. Finally, being made up of 136 inhabited islands, the United Kingdom is surrounded by water, which means there is a lot of moisture in the air. Together, these factor make British weather famously unpredictable and a constant topic of everyday conversation. As British anthropologist Kate Fox (2010, p. 33) observes, "we are accustomed to the variability of our weather, and we expect it to change quite frequently." She believes there is some truth in the old saw that the British are weather-obsessed. According to her survey research, 94 percent of British respondents reported they had had a conversation about the weather during the 6 hours prior to being interviewed, while 38 percent said they had had such a conversation in the previous 60 minutes (Fox 2010). At any moment in time, Fox estimates, "at least a third of the population is either talking about the weather, has already done so or is about to do so" (quoted in Geddes 2015).

One aspect of the body of sentiments about the weather in the United Kingdom, namely a nostalgia for the weather of the past, is noted by Trevor Harley, Emeritus Professor of Psychology at the University of Dundee in Scotland and operator of the website The British Weather (Geddes 2015). While his records on U.K. weather patterns over the years contradict people's yearning for a "remembered" better weather of the past, the changes being introduced to the country by climate change may well provide a reality base for British weather nostalgia. According to the Met Office, the United Kingdom's national weather service, heatwaves are lasting twice as long as they did 50 years ago. The Met Office Hadley Centre Climate Programme, which has tracked weather trends since the 1960s, shows that British cold spells are not as chilly as in the past, while summer days are hotter (Met Office 2018a). The hottest day of each year over the last decade was on average 1.4 °F warmer than the hottest days of the years between 1961 and 1990. The top 10 warmest years on record have all occurred since 2002. The highest-ever single day temperature (101.7 °F) occurred in Cambridge in July 2019, while the months of June and July 2018 were the hottest on record for a more extended period. Additionally, the Met Office (2018b) points out that "[h]uman-induced climate change . . . made the 2018 record-breaking U.K. summer temperatures about 30 times more likely than [they] would be naturally." Tropical nights—nights when minimum temperatures do not fall below 68 °F—are still uncommon in the United Kingdom, and are mostly confined to the southern sectors of the country, but they have been more frequent since 1995 and are projected by the Met Office to become even more so in the future. Between 2008 and 2017, a cluster of tropical nights occurred in the South East, the Midlands, and South Wales. Heavy rainfall is also on the increase, with extremely wet days up 17 percent during the period 2008–2017 compared to 1961–1990.

8.5 Melting land ice and tundra

Land ice at both the North and South Poles is melting at an unprecedented rate (Fig. 8.3). The Arctic Report Card 2018 (Osborne *et al.* 2018), issued by the NOAA, affirms that

Fig. 8.3 Melting glacier. *Source:* skeeze/Pixabay.

surface air temperatures in the Arctic have continued to warm at twice the rate of the rest of the planet, with those for the period 2014–2018 exceeding all previous records since 1900. On land, atmospheric warming continues to drive broad, long-term trends in declining terrestrial snow cover, melting of the Greenland Ice Sheet and lake ice, increasing summertime Arctic river discharge, and expansion and greening of Arctic tundra vegetation.

Signs of this process include the calving of massive icebergs such as the one that broke off Antarctica's Pine Island Glacier in October 2018. Five times the size of Manhattan, the enormous chunk of ice severed from the glacier only a month after a crack was first spotted by Sentinel-1, a satellite operated by the European Space Agency. Pine Island Glacier is calving at a heightened rate, having spawning four icebergs since 2013—a notable jump from its previous average rate of one every 6 years (Geggel 2018).

The scientific ability to quantify glacial loss and track the contribution of land ice to sea-level rise has improved in recent decades as a result of various satellite missions and evolving technologies. The key findings—as well as the inconsistencies—in this trove of data were summarized in 2013 in the IPCC's Fifth Assessment Report, which concluded, for example, that the average rate of ice loss from the Antarctic ice sheet likely increased from about 30 gigatons a year over the period 1992–2001 to about 147 gigatons a year over the period 2002–2011, mostly from the northern Antarctic Peninsula and the Amundsen Sea sector of West Antarctic (Stocker *et al.* 2013).

What would happen if all the ice on Earth—5 million cubic miles—melted? What would the world look like? National Geographic (2013) attempted to answer these questions by using the available information to develop a series of cryosphere-less maps. First, the sea would rise by about 215 feet, assuming that all of the resulting water drained into the ocean rather than lakes and other inland water catchments. Second, in North America, most of the Atlantic coast, Florida, and the Gulf Coast would be submerged. Cities like Miami, Tampa, New Orleans, Houston, and New York would be underwater ruins. The agriculture-rich Central Valley of California would be a giant bay and San Diego would be gone. Third, Europe would lose London, Venice, Barcelona, Copenhagen, Amsterdam, Stockholm, Lisbon, and Brussels. Fourth, in Asia, major urban centers like Tokyo, Kolkata (Calcutta), Bangkok, Shanghai, and Beijing would be swamped. Land that is now inhabited by 600 million people in China would flood, as would all of Bangladesh, which currently is home to 160 million people. Fifth, in South America, large tracks of Argentina, including Buenos Aires, as well as most of Uruguay, Paraguay, and parts of northern Brazil would sink below the ocean surface. Sixth, Africa would lose less of its land than other continents, but Egypt, Alexandria, and Cairo would be flooded by the enhanced Mediterranean. Seventh, Australia would gain an inland sea and lose much of the coastal area where four out of five Australians now live. Finally, low-lying island nations such as the Maldives, the Marshall Islands, Tuvalu, and Nauru would no longer exist.

Another, more detailed glimpse of the changes wrought by ice-melt can be seen by focusing specifically on a currently frozen landscape. In 2015, the U.S. National Aeronautics and Space Administration (NASA) launched a program called the Arctic Boreal Vulnerability Experiment (ABoVE) in Alaska and western Canada to study the implications of the vulnerability and resilience of Arctic and boreal ecosystems to environmental change. ABoVE-funded studies link field-based, process-level data collection with geospatial data from airborne and satellite sensors. There is scientific recognition that in northern Arctic and boreal regions, climate change is occurring faster than anywhere else on the planet, "resulting in reduced volume and area of sea ice in the Arctic Ocean during summer, warming and thawing of permafrost, increases in the frequency and severity of climate-driven disturbances, and widespread changes to surface water extent, soil moisture, and vegetation structure and function" (ABoVE 2018). In 2018, an ABoVE study found that carbon in Alaska's North Slope tundra was spending about 13 percent less time locked in frozen soil than had been the case 40 years ago. This means that the carbon cycle is accelerating; it is beginning to resemble the pace of a North American boreal forest, rather than that of a largely treeless tundra ecosystem (Jeong *et al.* 2018). An implication of this study is that rather than being an area that stores CO_2, the region is increasingly—and at a faster pace than previously projected—releasing atmospheric greenhouse gases, driving rather than limiting global warming.

Pico Humboldt, Venezuela's second-highest mountain, reaches to over 16 000 feet above sea level. This massive Andean peak is named for the celebrated German explorer, naturalist, and promoter of science Alexander von Humboldt. Born to an elite family of minor Prussian aristocracy with ties to the court of King Frederick the Great, as a young man Humboldt acquired a passion for travel and a purpose as a scientific explorer. At the turn of the 19th century, he made a number of prolonged exploratory trips to the tropical areas of the Americas, and later wrote up his observations in a series of books. In the United States, during the 1850s, Humboldt's "popularity approached cult status" (Walls 2009a). His writings earned him the sobriquet "the second Columbus," as the scientific discoverer of the Americas (Walls 2009b, p. 12). Most famous is the multi-volume *Cosmos*, in which Humbodlt articulates his belief that the collective study of nature creates empowering bonds within the human community that allow us to overcome the social conflicts that threaten to reduce us to warring tribal factions. The idea for this book first began to take shape in Humboldt's mind as he studied nature on the slopes of the Andes.

Atop the Venezuelan mountain that bears his name is a glacier that is also named for Humboldt. In 1910, five alpine glaciers spanning an area of 4 square miles could be found in the Andean mountains of northwestern Venezuela. In 2018, less than 1 percent of that glaciated area has survived, all of it confined to the shrinking Humboldt glacier. As Mandel (2018) notes, "[t]hanks to climate change, Venezuela has found itself a frontrunner in a somber race, with others such as China, to see which country will lose its glaciers first." In a decade or two, it will be glacier-free.

Also on the dismal list of countries rapidly losing their glaciers is Tanzania. The glaciers of Tanzania, and Africa generally, lack the attention provided their cousins in the polar regions. African glaciers, which are limited to East Africa, date to the ongoing Pleistocene glaciation, the last of the five major glacial periods in Earth's history, popularly called "ice ages." Beyond these comparatively cooler periods, Earth appears to have been ice-free, even at the highest latitudes. All of Tanzania's glaciers are located on the summit of Mount Kilimanjaro. In the middle of the 20th century, there were 16 named glaciers on the mountain, as well as three ice fields or interconnected glaciers. By the 1990s, four glaciers, including Drygalski Glacier and Great Penck Glacier, had disappeared, and all of the remaining ones were in retreat. Furtwangler Glacier, for example, located near the summit of Kilimanjaro, is but a trifling vestige of an ice cap that once crowned the mountain. Even this remnant is expected to disappear before long as a result of climate change (L. Thomspon *et al.* 2002). Research by Lonnie Thompson and coworkers (2009) shows that the areal extent of ice cover at the summit decreased about 1 percent a year from 1912 to 1953, increasing to 2.5 percent a year from 1989 to 2007. Thompson, whose glacier ice core research is considered foundational to the contemporary science of global warming, was a focus of the fêted book *Thin Ice* by Mark Bowen (2005), an accessible introduction to the key figures in the history of the science of climate and their work.

Finally, as noted, unhappily competing in the glacial disappearance race is China, a country with a significant glacier-viewing tourist industry. Baishui No. 1 Glacier on Yulong Snow Mountain in southern China attracts several million visitors each year, but it is one of the world's fastest melting glaciers, having lost 820 feet of ice since 1982. A research team has tracked Baishui's retreat at about 90 feet per year over the past decade (Yue *et al.* 2018). Other glaciers in China evince signs of retreat as well in terms of their area, volume, length, and mass. To assess the status of glaciers in the country, glaciologists compiled the first Chinese Glacier Inventory (CGI) from 1978 to 2002 (Y. Shi *et al.* 2009). Every glacier in China was measured across 34 parameters using aerial photographs and topographical maps. In response to recognized rapid shrinkage of the cryosphere globally, a second CGI, using Landsat TM/ETM+ remote-sensing satellite data, was completed in 2014 (Guo *et al.* 2014). Comparisons show that the glaciers in China have been retreating for a number of decades (Che *et al.* 2017)—with several exceptions, such as glaciers in the Karakoram Mountains that span the borders among Pakistan, India, and China. The average mass loss is larger than for glaciers globally. Some glaciers, such as those in the Hengduan Mountains in western Sichuan province and northwestern Yunnan province, have shrunk at a particularly high rate.

Signifying the global loss of glaciers, in August 2019 researchers from Rice University in Houston, TX installed a plaque in Borgarfjörður, Iceland memorializing the demise of Okjökull, the first Icelandic glacier lost to climate change. In preparation for the installation, Rice University anthropologist Cymene Howe, who studies the social life of ice in the Arctic, said of the plaque: "By marking Ok's [short for Okjökull] passing, we hope to draw attention to what is being lost as Earth's glaciers expire" (quoted in Kim & Gerretsen 2019). Iceland, which has 400 glaciers, is losing 11 billion tons of ice mass every year, and is expected to lose more glaciers as the planet warms. The plaque, titled "A Letter to the Future," reads in part: "This plaque is to acknowledge that we know what is happening and what needs to be done. Only you know if we did it."

Unlike alpine glaciers, ice sheets (also called continental glaciers), such as those that cover most of Antarctica and Greenland (as discussed in Chapter 6), are not limited to mountainous areas. They form as broad domes where snow that falls in winter but does not melt entirely over the summer piles up over thousands of years and spreads out under its own weight in all directions. As they spread, ice sheets are capable of blanketing all land features, from valleys to mountains, deeply under enormous quantities of ice. The ice sheets of Antarctic and Greenland contain about 99 percent of the freshwater ice on Earth. Together, they cover an area of over 7 million square miles. If they both melted away, sea level would rise by 200 feet. Based on combined satellite observations and modelling of surface mass balance, Shepherd *et al.* (2018) show that the Antarctic ice sheet alone lost between 1300 and 4100 billion tons of ice between 1992 and 2017. Ice loss during this period, and earlier, was concentrated in areas closest to warm, salty, subsurface, circumpolar deep water, a finding that is consistent with the enhancement of circumpolar countertrade winds that blow deep water toward Antarctica, causing the melting of floating ice shelves, the destabilization of glaciers, and sea-level rise (Rignot *et al.* 2019).

Since 2007, Michael Bevis of Ohio State University has been using GPS networks to study ice loss in both Greenland and Antarctica. His team found that the Greenland ice sheet lost more than 400 billion tons of ice during that period, nearly four times the rate recorded in 2003 (Bevis *et al.* 2019). Bevis' study focuses on Greenland's lesser-observed southwest region. Moreover, based on stratigraphic analysis of central west Greenland ice cores, Trusel *et al.* (2018) found that the magnitude of recent melting is exceptional over at least the last 350 years. While increases in Greenland ice sheet melting followed the onset of the Industrial Revolution in the mid-1800s, the scale of melting has only recently gone beyond the range of natural variability.

Global glacier and ice sheet retreat have multiple impacts, including changing ocean currents, contributing to sea-level rise, releasing pollutants (which landed on the ice and were buried) into freshwater sources, changing the jet stream, changing ocean salinity, speeding climate change (because less solar light is reflected), reducing the biodiversity of high-altitude environments, flooding mountain settlements, removing the summer melt water needed for drinking, livestock herding, and agriculture, promoting rural to urban migration, and heightening social conflict.

Another environment in which ice routinely forms as temperatures dips is lakes. Lake ice formation during winter has long been of concern to human societies because of its utility in transportation, fishing, recreation, and various cultural practices (e.g., sport, art, literature, film) (Dodds 2018). Concern of a new kind has arrived with global warming, as loss of freshwater ice is among the earliest observed impacts of Earth's warming climate. Observations conducted at over 500 lakes in the Northern Hemisphere by Sharma *et al.* (2019) provide clear evidence of mounting lake vulnerability to ice-free winters. Thousands of lakes that previously used to regularly freeze during winter are now remaining ice-free year-round or seeing a growing number of winters without lake ice formation. Sharma *et al.*'s (2019) findings suggest that an extensive loss of lake ice will occur within the next generation, with significant impacts on surrounding human communities.

8.6 Coastal flooding

Large populations around the globe live in the coastal areas where they are exposed to potential flooding (Fig. 8.4). It is estimated by the United Nations (2017) that over 600 million people, or about around 10 per cent of the world's population, live in coastal areas that are less than 32 feet above sea level. Further, about 2.5 billion people, making up about 40 per cent of the world's population, live within 60 miles of a sea coast. Almost two-thirds of

Fig. 8.4 Flooding. *Source:* Pixabay.

the world's cities with populations of over 5 million, including 8 of the 10 largest urban centers (e.g., Tokyo, Mumbai, Sáo Paulo, New York), are located in areas at risk of sea-level rise. In wealthier nations, there is some degree of structural protection that has been constructed for exposed populations. London, for example, a city built on soggy flood plain, is protected by the Thames River Barrier. Built over a 10-year period at the cost of over $2 billion (current equivalent), and made up of 10 steel gates reaching 1700 feet across the river, the barrier shields the city from flooding by high tides and storm surges. It had to be closed 179 times between 1984 and 2017 to prevent London from flooding. In middle-income and poorer nations, flood defenses are less advanced and exposed coast-dwelling populations face flooding leading to injury and loss of life, social and infrastructural disruption, and economic loss. Dhaka, for example, the capital of Bangladesh and one of the largest and most densely populated cities in South Asia, considered among the most climate-vulnerable megacities in the world, suffered major flooding 10 times between 1954 and 2009. During the intense floods of 1998, parts of the city were inundated for over 2 months (GFDRR 2018). The flooding that year, which overwhelmed a far broader area than just Dhaka, affected the lives of tens of millions of people and caused over 1000 deaths. An estimated 500 000 homes and 10 000 miles of roads were severely damaged, as were thousands of schools and bridges (Shah 1999). Agricultural areas could not be worked, impacting food availability, and both smaller and larger manufacturing facilities were shut down, threatening people's livelihoods. The short-term social impact of the floods was a steep rise in unemployment, a dramatic drop in wages, and an increase in food prices and the cost of other essentials. As is usual in such events, the poor were hit the hardest.

Conditions are likely to grow far worse over time, because climate change and poverty is a deadly combination. According to BRAC (2016), a nongovernment sustainable development and education organization founded in Bangladesh:

> Climate change is a concern now affecting every one of us and Bangladesh is

one of the most vulnerable to its effects. Two-thirds of Bangladesh is less than five metres above sea level, and floods are increasingly destroying homes, croplands and damaging infrastructure. Approximately 10 000 hectares of land is lost every year due to riverbank erosion. Agriculture land is shrinking by 1% annually while the population is growing by 1.2%. This is creating a rise in demand for food, while increasingly unpredictable weather conditions pose a growing challenge to farmers trying to meet those demands.

Even in wealthy counties with resources to fund coastal protective infrastructure, flooding is having growing impacts. David Titley, a meteorologist and Director of Pennsylvania State University's Center for Solutions to Weather and Climate Risk, as well as a former Rear Admiral, assessed the risk of flooding at Naval Station Norfolk in Virginia, the largest naval complex in the world, in blunt terms: "These guys are in a whole heap ton of trouble . . . I think Norfolk is, in the long term, fighting for its existence, its very existence . . . And this is the part of climate change that I don't think most Americans have really come to grips with—that virtually every coastal city is in a fight for its existence. They just don't know it yet" (quoted in Kusnetz 2017). The Shipyard at Norfolk provides repair and upgrade services for every category of ship in the U.S. Navy, especially nuclear-powered submarines and aircraft carriers and their support vessels. But rising seas and extreme weather driven by climate change threaten the base. Since 2008, Norfolk has suffered nine major floods, which have damaged facilities and equipment worth millions of dollars, and high-tide flooding is contributing to extensive delays in ship repairs. The sea level in Norfolk has risen 1.5 feet in the past century, in part because of shifting ocean currents and a slowly sinking coastline. Flooding is so frequent that fish have become a common victim of roadkill deaths. A Federal Emergency Management Agency (FEMA) computer-simulated Category 4 hurricane with 140 mph winds making a direct hit on the area indicated storm surge flooding that would likely submerge the entire naval facility (Kusnetz 2018). Such a storm in a time of global warming is not a distant and unlikely possibility. In 2016, Hurricane Matthew, a Category 5 storm when it devastated Haiti, moved offshore up the Atlantic coast, bringing heavy rain and flooding to the area of the Norfolk base, including 2 feet of water in one building at the facility. Two years later, the Navy evacuated 30 ships from the Station to avoid potential damage from Hurricane Florence. According to Titley, "It's not a question of if, it's a question of when . . . And when it hits, how vulnerable are we going to be, and are we going to be standing there saying, 'oh, we woulda, coulda, shoulda?'" (quoted in Kusnetz 2018). While climate change is denied by some, because the effects are already being felt by the Navy (as well as other branches of the military), and not just in Virginia but at facilities around the country, it has become increasing difficult to ignore. However, this has not stopped the Trump Administration attempting at least to stop use of the term. Referring to a Pentagon document on environmental threats to U.S. military bases and installations, Mooney & Ryan (2018) report:

> The earlier version of the document, dated December 2016, contains numerous references to "climate change" that were omitted or altered to "extreme weather" or simply "climate" in the final report, which was submitted to Congress in January 2018. While the phrase "climate change" appears 23 separate times in the draft report, the final version used it just once.

In October 2019, an improved global elevation dataset produced by Climate Central (Kulp & Strauss 2019) revealed that many coastal elevations are significantly lower than scientists thought. The new digital elevation model, called CoastalDEM, shows that by 2050, sea-level rise could push chronic floods higher than land that is currently home to 340 million people. The threat is especially great in coastal Asia, affecting land occupied by more than 10 percent of the populations of Bangladesh, Vietnam, and many Small Island Developing States.

8.7 The polar vortex

The term "polar vortex" denotes the persistent upper-level low-pressure areas of circulating air extending over about 600 miles near both the North and South Poles. Beneath these vortices are large masses of cold, dense air. They change in strength from year to year. When the Arctic is weak, there is a flow of Arctic air toward the equator that can usher in outbreaks of severe cold weather in the Northern Hemisphere (Song & Robinson 2004). Such an outbreak occurred in North America during what came to be colloquially referred to as the Deep Freeze of 2019. Plummeting temperatures claimed over 20 lives, and more than 650 low-temperature records were broken in the Midwest. The unincorporated community of Cotton, MN, 36 miles north of the city of Duluth, experienced freezing—without wind chill—of −56 °F, making it for a time the coldest place in the United States (Fritz 2019).

The intense cold prompted President Trump, a denier of climate change, to tweet on January 28, 2019:

> In the beautiful Midwest, windchill temperatures are reaching minus 60 degrees, the coldest ever recorded. In coming days, expected to get even colder. People can't last outside even for minutes. What the hell is going on with Global Waming [sic]? Please come back fast, we need you! (quoted in Pierre-Louis 2019)

As Trump's climate-educated critics were quick to point out, very cold temperatures in the Midwest are quite expected in the midst of global warming, reflecting the important difference between local weather at one point in time and long-term, planet-wide trends in climate. On the day of Trump's cynical tweet, for example, in the middle of an intense heatwave, temperatures reached 94 °F in Melbourne, Australia (AccuWeather 2019).

In fact, aspects of climate change may have contributed to the weakening of the polar vortex that brought on the Midwest's bone-chilling cold spell of early 2019. Many climate scientists believe that reduced land snow cover and sea ice resulting in the reflection of less sunlight causes an upsurge in evaporation of moisture and transpiration in plants. Increasing moisture in the atmosphere modifies the pressure and temperature gradient of the polar vortex, causing it to weaken or even collapse. The same pattern steered the punishing winds and rain of Hurricane Sandy into the northern Mid-Atlantic states in 2012 (Greene *et al.* 2013). Moreover, while North America endured torturous cold waves in 1936, 1970, 1977, 1983, 1989, and 1996, there has been a decline in winter storm intensity since the mid-1990s. At the beginning of the Great Arctic Outbreak of 1899, for example, the temperature hit 9 °F in Portland, OR and 33 °F in Los Angeles, CA. On February 13 of that year, temperatures dropped to an all-time low of −2 °F in Tallahassee, FL; 2 days earlier, they had reached −61 °F in Fort Logan, MT. During the Outbreak, Cape May, NJ recorded 34 inches of snow, as did Washington, D.C. Many snowfall records set during February 1899 have never been broken. Ice was even seen flowing down the Mississippi River, past New Orleans and into the Gulf of Mexico. Snow weighed down the fronds of palm trees in parts of Florida and lifelong Tampa residents witnessed the first measurable snow they had ever seen in their city. About 100 died due to the frigid weather (Kocin *et al.* 1988). Given the severity of these Arctic blasts, it is of note that the frequency of intense cold waves nearing those of the Great Arctic Outbreak reached record-low levels in the 2000s (Peterson, *et al.* 2013). Despite the fact that that extreme cold periods of snow and ice will continue to occur, overall the planet continues to warm.

8.8 Hurricanes, cyclones, typhoons, and tropical storms

In the terminology of meteorology, the regionally applied terms "hurricane," "cyclone," and "typhoon" are all called "tropical cyclones" (National Ocean Service 2018). This is the term used to label rotating, organized systems of clouds and thunderstorms that originate over tropical or subtropical waters and have closed, low-level circulation. Tropical cyclones come in different levels of intensity based on the maximum sustained winds speeds (mph) they achieve. The weakest types, with wind speeds of under 39 mph, are called tropical depressions. Those with speeds between 39 and 73 mph are called tropical storms. When winds surpasses 74 mph, a storm is classified, depending on location, as a hurricane (North Atlantic, central North Pacific, and eastern North Pacific), a typhoon (Northwest Pacific), or a cyclone (South Pacific and Indian Ocean). A Western Hemisphere tropical cyclone with maximum sustained winds of at least 111 mph is further classified in one of the major storm groupings on the Saffir-Simpson Hurricane Wind Scale: Category 3 (111–129 mph), 4 (130–156 mph), or 5 (157 mph or higher). This scale was developed in the early 1970s by Herbert Saffir, a structural engineer, and Robert Simpson, a meteorologist and former director of the U.S. National Hurricane Center. In other regions, other scales are used (e.g., the Joint Typhoon Warning Center classifies typhoons with 150 mph or greater winds as super typhoons). Where a storm ranks on a scale is not a useful gauge of how much damage it will cause, however, as this is heavily determined by where it has landfall (e.g., a major urban area vs. a lightly populated one) and where it travels before dissipating (Fig. 8.5).

Globally, between 1971 and 2015 tropical cyclones claimed the lives of about 470 000 people, or roughly 10 000 lives per year

Fig. 8.5 Hurricane. *Source:* WikiImages/Pixabay.

(CRED 2016). Some storms have been particularly impactful. In 2017, Hurricane Maria struck Puerto Rico and killed 2975 people (in excess of the number of expected deaths for the period September 2017 through February 2018), about 500 times the initial estimate of 64 (Santos-Burgoa *et al.* 2018). During and after the storm, the risk of dying was highest among the urban poor. Maria was among the top 10 deadliest storms recorded in the hurricane databases (dating to 1870), managed by the National Hurricane Center (2018).

In assessing the impact of climate change on tropical cyclones, it is critical to differentiate frequency from intensity. While climate scientists are uncertain whether climate change will increase the number of storms per storm season, in that tropical cyclones derive their energy from the heat of ocean temperatures and their surge at landfall from the ocean level, it is believed to be intensifying storm impacts. In other words, while the overall number of storms may fluctuate year to year, the frequency of Category 4 and 5 storms is increasing due to climate change. According to the Center for Climate and Energy Solutions (C2ES 2018), an environmental policy think tank, based on computer modeling it is expected that there will be a 45–87 percent increase in the frequency of Category 4 and 5 hurricanes in the U.S. Atlantic Basin in coming decades. Further, there will be a 2–11 percent increase in their average maximum wind speed. Rainfall during future storms is projected to increase by about 20 percent.

These projections are supported by the work of Elsner *et al.* (2008), who used log-linear regression to model lifetime maximum wind speeds based on satellite-derived data for 171 tropical cyclones around the world for the period 1981–2006. With the exception of the South Pacific Ocean, all the common tropical cyclone areas show increases in the lifetime maximum wind speeds of the strongest local storms. The increases are greatest for tropical cyclones over the North Atlantic and northern Indian oceans, but only over the North Atlantic is there a significant increase in average tropical cyclone intensity. Rising North Atlantic SSTs associated with climate change help "explain the recent upswing in frequency and intensity of Atlantic tropical cyclones" (Elsner *et al.* 2008, p. 92).

Even if Earth's climate was not warming, annual economic damages from hurricanes could double to $56 billion in the next century. The reason, according to analyses by Mendelsohn *et al.* (2012), is that global population is projected to rise to 9 billion by 2100, resulting in significant infrastructural, manufacturing, residential, and other developments along hurricane-prone coastlines. Add global planetary warming and a resulting increase in the frequency of high-intensity storms to this equation and the annual hurricane damage could quadruple to $109 billion by 2100. Further, these researchers have found that the distribution of damage is not likely to be evenly distributed across the planet. Rather, their model suggests that climate change-related hurricane damage will be concentrated in North America. Other work by the same team (Emanuel 2016) suggests that a rapid intensification just before a storm makes landfall is likely to become increasingly frequent and severe as global warming advances.

8.9 Infectious diseases

Humans evolved in a world already long populated by pathogens, microscopic entities that are capable of causing infection. This means that pathogens have been causing human disease and death, and hence having an impact on the evolution of our species, for all of human (and pre-human) history. The distribution of pathogens that are found among both human and nonhuman primates suggests longstanding hominid–pathogen relationships that predate the divergence of the hominid lineage. As humans have evolved, so have their infectious diseases. Some diseases that once were rare have become common, while others have disappeared only to be replaced by new infectious agents. It is estimated that over 15 million people die of infectious diseases every year. The immediate cause of these deaths is one (or more) of the roughly 1400 infectious agents, a diverse set of microscopic entities that includes viruses, bacteria, fungi, prions, protozoa, and helminths (van Doom 2014). While the world and our own bodies are filled with an extraordinary of number microorganisms, human pathogens account for less than 1 percent of the total number of microbial species on Earth.

It is standard epidemiological understanding that three elements are necessary in order for most infectious diseases to occur: an agent (a pathogen), a host (the infected organism), and a transmission environment (a set of conditions that bring the host and agent together and facilitate infection, also known as a transmission pathway). This trilogy is referred to as the epidemiological triangle. A fourth element required by some pathogens is a vector or intermediate host allowing the pathogen to reach the host and/or complete its life cycle. A central element of the transmission environment is climate. Climate and other environmental conditions impact hosts (including their selection of behaviors that facilitate or hinder infection), pathogens, and vectors. Mosquitos, which are vectors for a number of infectious diseases, for example, can only live and reproduce within certain temperature ranges (varying by species) and require access to water for egg laying and larval development.

Changes in climate or weather conditions have profound impacts on infectious diseases. As Epstein (2001, p. 747) indicates, "[c]limate constrains the range of infectious diseases, while weather affects the timing and intensity of outbreaks." The appearance, spread, and health impact of new infectious diseases, as well as the rejuvenated spread of known diseases, are significantly influenced by changes in the environment (McMichael & Martens 2002). Eisenberg *et al.* (2007) have proposed a framework for tracking the relationship between environmental change and the transmission of infectious disease. Their scheme unites three features of the environment–infectious disease relationship: 1) environmental change increasingly through human history has become the product of a complex web of ecological and social factors that help to shape the landscape of infectious disease; 2) the transmission dynamics of pathogenic agents mediate the effects of environmental changes on infectious disease impact; and 3) the infectious disease burden carried by a population—measured by the incidence and severity of infection—is a product of the interplay between environmental change, including changes wrought by human activity, and the transmission cycle or route of transmission of a pathogen. Transmission routes vary and include: direct human-to-human transmission (e.g., HIV infection; COVID-19); vector-borne transmission (e.g., malaria); environmental contact transmission (e.g., food-borne infectious diseases); and zoonotic transmission involving nonhuman-to-human disease movement (e.g., Lyme disease). Climate change can influence all of these transmission routes by: forcing human movement, resulting in new human-to-human contacts (e.g., in refugee camps); allowing vectors to inhabit new environments (e.g., global spread of disease-bearing ticks); complicating temperature control during summer food shipping (e.g., imported seafood); and changing animal locations, resulting in new human–animal contacts (e.g., bat-borne diseases).

Specific examples of climate impacts on infectious disease transmission routes are beginning to accrue in the scientific record (R. Thompson *et al.* 2002; Pascual *et al.* 2006; Singer & Bulled 2016). For example, though not well known in other areas of the country, the small to medium-sized hispid cotton rat (*Sigmodon hispidus*) resides primarily in the Southern United States as well as in parts of Central and South America. Its name, hispid, means "covered in short, stiff hair." Populations of this rodents are also found in Nebraska and New Mexico. It has wide-ranging food habits (e.g., grain, leaves, wood, insects, birds and birds' eggs, crayfish, and carrion), allowing it to colonize different types of ecosystems. Over the last 50 or more years, hispid cotton rats have been extending their range northward and to higher elevations in response to warmer climate conditions. In 2008, members of the species were found for the first time in central Virginia (Francl & Meikle 2009). Notably, they are hosts of both the Black Creek Canal hantavirus, a known cause of hantavirus pulmonary syndrome (HPS), which is a severe, sometimes fatal, human respiratory disease, as well as Muleshoe hantavirus, which also has been linked to HPS (Rollin *et al.* 1995). As Klempa (2009, p. 518) concludes:

> It can reasonably be concluded that climate change should influence hantaviruses through impacts on the hantavirus reservoir host populations. We can anticipate changes in the size and frequency of hantavirus outbreaks, the spectrum of hantavirus species and geographical distribution (mediated by changes in population densities), and species composition and geographical distribution of their reservoir hosts.

This expectation applies as well to other infectious diseases that will put humans increasingly at risk as a result of climate change. One such disease is dengue, a mosquito-borne viral infection. The dengue viruses, of which there are four related serotypes, is transmitted by female mosquitoes primarily of the species *Aedes aegypti*. Sequential infection by two or more serotypes increases the risk of developing severe dengue (also known as dengue hemorrhagic fever), a leading cause of serious illness and death among children in a number of Asian and Latin American countries. The incidence of dengue has grown dramatically around the world in recent decades. One estimate is that there are 390 million new dengue infections per year, of which about 96 million manifest clinically at some level of disease severity (Bhatt *et al.* 2013). Infection can produce severe flu-like symptoms, including a high fever, headache, pain behind the eyes, muscle and joint pains, nausea, vomiting, swollen glands, and rash. Other research on the prevalence of dengue estimates that 3.9 billion people living in 128 countries are at risk of infection (Brady *et al.* 2012).

The geographical limits of the mosquitos responsible for dengue fever transmission are strongly determined by climate. Global warming is a factor in determining how many dengue cases will occur in a year, increasing both the length of the dengue transmission season and the number of areas around the globe where the climate is suitable for dengue. As Hales *et al.* (2002, p. 833) conclude: "climate change will contribute to a substantial increase in the number of people and proportion of global population at risk of dengue fever."

Another sign of the effects on potential vectors is the discovery in the United States—initially in New Jersey, but now in eight states—of the Asian longhorned tick (*Haemaphysalis longicornis*), a species with a native range in East Asia, but which has spread to Australia and a number of Pacific islands (Rochlin 2018). This species, which thrives in temperate, humid climates with forest cover and sufficient precipitation, is a transmitter of severe fever and thrombocytopenia syndrome (SFTS), a newly emerging infectious disease in East Africa caused by the SFTS virus. Symptoms and laboratory abnormalities are fever, low blood platelet count, lowered white blood cell count, fatigue, gastrointestinal problems, and elevated serum enzyme levels. Multiorgan failure occurs in very severe cases. In a study of the first 69 cases in South Korea, Yun *et al.* (2014) reported a case–fatality rate of 47 percent. Infection was associated with warm weather, as cases occurred in the relatively warm southern provinces during the warm season. In its introduced range in Australia and New Zealand, the Asian longhorned tick has been found to transmit the potentially fatal cattle disease theileriosis, caused by the pathogen *Theileria orientalis*.

Though specimens collected in the United States have not been found to carry known pathogens thus far, the United States is home to a relative of SFTS known as the Heartland virus, creating the concern that these ticks might eventually begin carrying and spreading this or other tick-borne pathogens in North America.

8.10 Food loss to heat and insect pests

The FAO (2018) reports that the number of people around the world facing chronic food deprivation increased to nearly 821 million in 2017, up from around 804 million in 2016 (Fig. 8.6). The proportion of people who are inadequately nourished is believed to be about 10.9 percent. Areas where undernourishment is rising fastest are Africa and South America. Meanwhile, though the extent of the food problem in Asian countries has fallen, the pace of this fall has slowed significantly. In 2017, over 7 percent of children globally under the age of 5 years (i.e., 50.5 million children) suffered from wasting (having low weight for height, an indicator of acute malnutrition), putting them at heightened risk for food-related mortality. Analysis by Black *et al.* (2013) concluded that 875 000 deaths (or 12.6 percent of all deaths) among children under age 5 were associated with wasting, more than 500 000 of which were related to severe wasting. Having unreliable access to food (food insecurity) can contribute to child wasting, stunting (low height for age), and micronutrient deficiencies. Inadequate diet hinders fetal, infant, and child growth and development. Insufficient food intake contributes to maternal undernutrition and to higher risk of low birthweight, both of which are risks for child stunting. Moreover, food-insecure children are more likely to face adverse health outcomes and developmental risks (Pereira *et al.* 2017). Widespread food insecurity and undernourishment have multiple causes (e.g., social conflict), but climate change is an increasingly important one due to its impact on food supply.

The relationship between climate change and global food production involves several factors, including the effects of atmospheric

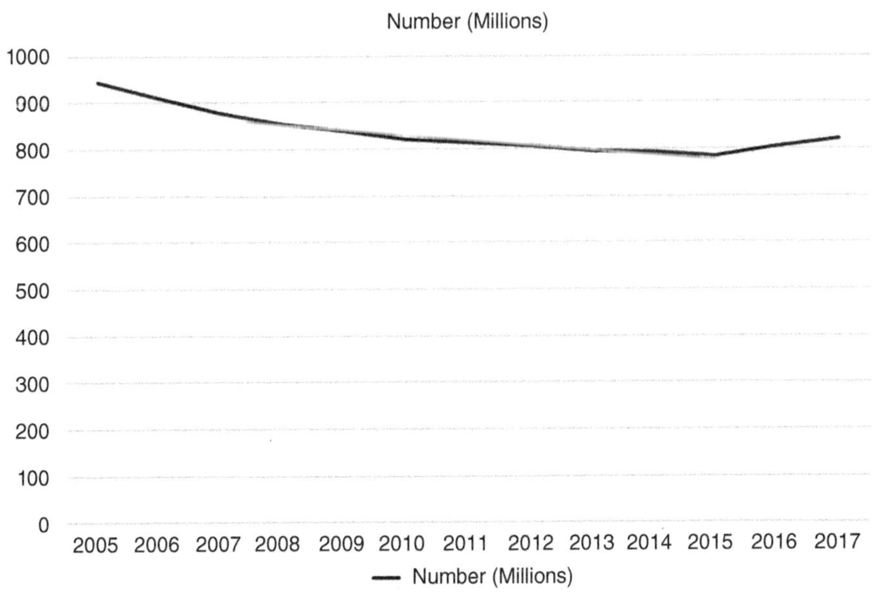

Fig. 8.6 The rising number of undernourished people in the world. *Source:* Modified from the Food and Agricultural Organization of the United Nations (FAO 2018).

CO_2 on desirable and undesirable plant growth, the extension of growing seasons, the warming of cooler regions, the warming of warmer regions, changes in the physiology and demography of insect pest populations, and crop destruction by extreme storms. Each of these will be addressed in turn.

Research using long-term satellite green leaf area index records and 10 different global ecosystem computer models (Zhu *et al.* 2016) reveals that between a quarter and half of Earth's vegetated lands have undergone a significant increase in plant growth since the early 1980s, about 70 percent of which is thought to be due to the rising levels of CO_2 in the atmosphere. This has occurred because green plants use photosynthesis, fueled by energy from the sun, to take in CO_2 (through tiny holes, called stomata, found especially in leaves) and chemically combine it with water and nutrients drawn from the ground in order to produce the food (sugars) they need to grow, as well as oxygen as a byproduct (which is vital to air-breathing organisms like us). Increasing CO_2 concentrations spur photosynthesis and plant development (Long *et al.* 2004). Thus, growers involved in the commercial production of plants in controllable environments like greenhouses often use CO_2 generators and the release of purchased pure liquid CO_2 to enhance plant exposure during photosynthesis, increasing their yield and profits (Mortensen 1987). Even many aquarium hobbyists release CO_2 in their tanks to improve the growth of decorative water plants.

In testing CO_2 enrichment, researchers have learned several key lessons that are relevant to understanding the impact of rising atmospheric CO_2 levels in a time of global warming. First, the effect is highly species-dependent and sensitive to environmental conditions. A critical factor differentiating plant species involves rubisco, a primary enzyme in photosynthesis. About 85 percent, including key global food crops such as rice, wheat, potatoes, barley, soybeans, peanuts, and oats, use only the rubisco pathway in photosynthesis. These are called C_3 plants, because they produce the stable 3-carbon compound 3-phosphoglyceric acid. Other food crops such as corn, sugarcane, sorghum, and millet use an additional pathway and produce a 4-carbon compound. They are thus known as C_4 plants (Sage 2004). This seemingly small difference is important because the photosynthetic efficiency of C_3 plants is less than that of C_4 plants due to photorespiration—"accidently" taking in some oxygen and giving off some CO_2, a reversal of the general pattern of photosynthesis. Photorespiration wastes plant energy and resources (Starkey 1988). As Earth gets hotter, C_3 plants are more prone to such errors. Additionally, in high temperatures, water evaporates at a faster rate, forcing plants to limit water loss from their leaves. This botanical action has the unfortunate byproduct of limiting the amount of CO_2 that is taken in. Under these conditions, C_3 plants waste even more of their energy by taking in more oxygen instead of CO_2. As the temperature rises, the problem worsens considerably. This significantly changes the seemingly rosy picture of plant growth for many critical food crops (Schlenker & Roberts 2009; Reich *et al.* 2018). As an old adage indicates, "Too much of a good thing can be a bad thing."

Second, leaf injuries are sometimes observed in connection with CO_2 enrichment. This includes chlorosis (reduction or loss of the normal green coloration, indicating lack of the chlorophyll needed for photosynthesis), leaf death, and leaf curling. Bush beans (*Phaseolus vulgaris* L.), for example, grown in atmospheres enriched with CO_2, have been found by Ehret & Jolliffe (2011) to suffer both reductions in photosynthetic capacity and loss of primary leaves. Detrimental effects of heightened CO_2 concentrations have also been reported in various other crop plants, including cucumbers (Wittwer 1967), soybeans (Chang 1975), and tomatoes (Madsen 1974), among others (Van Berke 1984). Further, C_3 plants tend to thicken their leaves by as much as one third. While the exact reason for this response is not yet clear, its effect may be to exacerbate climate change, because thicker leaves are less efficient in sequestering atmospheric carbon (Kovenock & Swan 2018).

Third, in addition to changes in food production levels, research shows that when some crops are grown at high levels of atmospheric CO_2, they lose nutritional value. Myers *et al.* (2014), for example, grew crops, including wheat and corn, under two conditions: normal (current) CO_2 levels and elevated CO_2 levels (representing the concentration it is estimated will pertain in the atmosphere in 50 years). They found that the crops grown under elevated levels had lower protein, zinc, and iron contents. This could worsen the public health problem of nutrient deficiencies. In the future, 200 million more people worldwide could develop a zinc deficiency and the 1 billion people who already have one could see it worsen due to these nutritional changes. Research shows that there would be similar effects for iron and protein deficiencies as well. Nor is the loss of nutrient value under conditions of rising levels of CO_2 limited to wheat and corn. Rice, a primary food source for more than 2 billion people, has been found in experimental research to suffer declines in protein, iron, and zinc, and in the availability of vitamins B1, B2, B5, and B9 (though some increase in vitamin E did occur), in response to free-air CO_2 enrichment (Zhu *et al.* 2018).

Finally, while CO_2 is sometimes portrayed as a harmless gas, research suggests it can be injurious to humans. The adverse effects vary based on concentration, duration of exposure, and population vulnerability (e.g., those with pre-existing breathing disorders, children, and the elderly). Symptoms commonly associated with prolonged or concentrated exposure include dizziness, headache, confusion, labored breathing, dimmed vision, vomiting, cardiac arrhythmias, and impaired consciousness (Langford 2005). In experimental research, Satish *et al.* (2012) found that healthy study participants given tests that required decision-making suffered moderate to large deficits in their performance ability depending on the level of CO_2 to which they were exposed. Findings on decreases in functional ability have been supported by other research (Vehviläinen *et al.* 2016).

Beyond CO_2, there is the issue of the climate change effects on the plant and animal toxin, ground-level ozone (O_3). Ground-level ozone, which causes billions of dollars in lost plant productivity annually, is derived primarily from chemical reactions between the fossil fuel emissions of two chemical compounds, nitrogen oxides and volatile organic compounds, in sunlight. It reduces plant productivity by entering leaves through the stomata and causing oxidative stress, which decreases photosynthesis, plant growth, and biomass accumulation (Ainsworth *et al.* 2012). Surface ozone formation increases with rising temperature, especially above 90 °F. Open-air experiments indicate that ozone concentrations at the levels found in currently polluted regions decrease yields by 8–25 percent in rice, soybean, and wheat (Morgan *et al.* 2006; G. Shi *et al.* 2009; Zhu *et al.* 2011). Globally, levels of ozone pollution in 2000 were estimated to have suppressed corn, wheat, and soybean yields by 6–9 percent (Avnery *et al.* 2011a). Research by Avnery *et al.* (2011b) comparing the estimated impact of ozone on global agriculture in the year 2000 with the projected impact in 2030 concluded that there would be further losses of 5.4–26.0 percent for wheat, 15–19 percent for soybean, and 4.4–8.7 percent for corn. Their results, as well as those of other researchers (Fuhrer 2009), affirm that surface ozone poses a growing threat to global food security associated with climate change.

The warming of cooler regions has been cited as a benefit to global food production. This is a theme found in what has been called the "winners and losers" paradigm in climate change discourse (Baer & Singer 2018). In this understanding, while some areas of the world will endure growing hardships as the planet warms, others will benefit. Agricultural researchers in Denmark, for example, have noted potential benefits of global warming. Jørgen Olesen, who studies climate change and agriculture at Aarhus University, claims: "A warmer climate will mean that we can grow new types of crop, such as sweetcorn, and also achieve better yields. This way Danish farmers may benefit from climate change" (quoted in Brix 2014). John Roy Porter, who studies agro-ecosystems at the University of Copenhagen, agrees: "Temperature increase, up to two degrees or perhaps a little more, would boost agricultural production in Denmark. This would be positive for sweetcorn, wheat and grass. Maybe one day it will also be possible to grow soya beans" (quoted in Brix 2014). The problem with the "winners and losers" paradigm is that it focuses on one factor (e.g., local temperature) while ignoring the multidimensional global turmoil that climate change will bring. Denmark, for example, imports protein food sources from countries that may be hard hit by the adverse effects of climate change. Whatever benefits might accrue to Danish farmers will thus be more than offset by the losses of Danish consumers. Moreover, climatologists project that Denmark will experience more frequent heavy rain and cloudbursts in the future, as well as heatwaves—events that present threats to Danish agriculture (Madsen *et al.* 2009).

One of the severest threats to food supplies is the warming of warmer regions. The effects of climate change are already being felt in the tropics. As the planet warms, there will be increasing environmental stress imposed on food crop production due to erratic shifts in the timing and amount of rainfall and, in particular, high-temperature spells (Rosenzweig *et al.* 2014). In 2011, the Consultative Group on International Agricultural Research Program on Climate Change (Ericksen *et al.* 2011) commissioned a rapid assessment across the global tropics using a range of indicators to identify "hot spot" locations where food insecurity was already a concern and where climate change impacts were projected to become increasingly severe by 2050. The study projected declines in the length of the growing season by 5 percent or more across a broad swath of the tropics, including heavily planted parts of Mexico, Brazil, southern and West Africa, the Indo-Gangetic Plains, and South East Asia. Declines to fewer than 120 days were projected for Mexico, northeast Brazil, southern and West Africa, and India; this is the critical threshold required for certain crops to mature, below which agriculture begins to become too risky to pursue as a major livelihood strategy. Abandonment of agriculture can already be seen in parts of West Africa, East Africa, and the Indo-Gangetic Plains. Further, the study found that high temperature stress will become widespread in southern and East Africa, northern and southern India, South East Asia, northern Latin America, and Central America. These results are supported by Battisti & Naylor (2009), whose research using observational data and global climate computer models found a greater than 90 percent probability that by the end of the 21st century, growing season temperatures in the tropics and subtropics will surpass the highest seasonal temperatures recorded during the period from 1900 to 2006.

In Africa, where 70 percent of the population directly relies on the land for its livelihood, the amount of arable land is decreasing at a time when the number of people who need food is growing (Brown 2017). In the central African countries of the Democratic Republic of the Congo, Burundi, Chad, and Rwanda, undernourishment increased from 33.5 percent of the population in 1990–1992 to 41 percent in 2014–2016. The proportion that was undernourished was largest in the Central African Republic (almost 48 percent) and second largest in Namibia (42 percent) (Balasubramanian 2018). It is projected that in coming years, "climate change will lead to changes in yield and area growth, higher food prices and therefore lower affordability of food, reduced calorie availability, and growing childhood malnutrition in Sub-Saharan Africa" (Ringler *et al.* 2010, p. v). It is projected that in 2025, Sudan's food consumption need will be 5685 million tons but in-country production will be only 544 million tons, while in Nigeria food consumption need will be 5302 million tons but production will be only 83 million tons (Balasubramanian 2018). The extent of future climate change impacts on agriculture in sub-Saharan Africa depends on several factors, including crop types, the scale of farming operations, whether a farm is operated for commercial production or for subsistence, and the availability of natural resources such as ground water. However, for various parts of Africa and other tropical areas around the world, the risks of a declining food supply, food insecurity, and the grave indicators of malnutrition are mounting.

Currently, crop pests are responsible for 25–40 percent of all crop loss worldwide (Flood 2010), and as climate change continues they are expanding their distribution and damaging impact (Fig. 8.7). Home ranges are expanding as a result of warmer and wetter conditions. Moreover, while insect populations may be dropping in other ecosystems, pests inhabit human-managed ecologies that concentrate on host plants maintained through intensive monocropping (Björkman & Niemela 2015).

Through normal evolutionary processes, plants tend to develop a notable array of defenses against their common insect predators, including various morphological, biochemical, and molecular mechanisms that counter or otherwise respond to attack. Morphological changes include the premature abscission of plant parts that are under attack from insects. Biochemical responses involve several strategies, such as the production of: 1) toxic secondary metabolites, such as nicotine and furanocoumarins, that are capable of killing insect pests; 2) defensive proteins, such as protease inhibitors and polyphenol oxidases, that slow down insect growth; and 3) volatile organic compounds that attract the natural enemies (e.g., birds, carnivorous insects) of insect herbivores. Plants and their common herbivorous insect pests tend to coevolve offensive and defensive strategies over time (Ehrlich & Raven 1964). Molecular defenses, involving internal signaling pathways (e.g., wound signals), can act to decrease harm by triggering compensatory growth following herbivore attack (Baldwin & Preston 1998; Kessler & Baldwin 2002). Novel diffusing pests, however, with behaviors and other traits that are unfamiliar, can better avoid the natural defenses plants have evolved to ward off or contain insect threats. From an agricultural standpoint, novel herbaceous pests pose an added threat capable of increasing the already considerable crop damage wrought by familiar ones.

Fig. 8.7 Grasshopper. *Source:* yanivmatza/Pixabay.

In sum, the multiple net effects of climate change, many of which are interactive in producing crop destruction (e.g., extreme storms and pest diffusion), will mean a decrease in food yield and an increase in food insecurity and malnutrition around the world, but especially in those environmental hot spots of greatest vulnerability. Often, climate-change hot spots are places where local contribution to greenhouse gas remission may be comparatively low but the consequences in terms of lives will be particularly high. Some hot spots, increasingly, will no longer be able to meet basic requirements for human survival. In such places, people will face famine, an extreme state of food scarcity accompanied by high rates of malnutrition and starvation.

Crop loss is not only a risk to the food supply, but is also a threat to the livelihoods of billions of people. As emphasized by the Climate Institute (Watts 2016, p. 8), "[a]griculture is not simply crops and commodities, but the farmers, farmworkers, and those who rely on them." This extends beyond food crops to other cultigens also jeopardized by climate change.

This broader pattern can be seen among coffee farmers in Colombia. In April of 2017, Luis Gilberto Murillo, Colombia's Minister of Environment and Sustainable Development, reported that climate change had put nearly 12 million Colombians at risk from disasters like flooding and landslides. According to Murillo, "Colombia is very vulnerable to phenomena of extreme climate variability and climate change" (quoted in Acosta 2017). Colombia's coffee industry is built primarily on modest-sized (mostly between 5 and 12 acres) family-owned farms, with only about 5 percent of Colombian coffee producers having plantations larger than 12 acres. These small-farm coffee growers take great pride in their product and the contributions they make to the coffee culture of their country. However, they are increasingly facing climate-induced extreme weather and invasive pests, endangering their livelihoods (Eise & White 2018).

Coffee production and export are central to the Colombian economy. Colombia is one of the world's largest coffee-producing nations, with over 500 000 producers selling over $2.5 billion dollars of high-altitude Arabica coffee beans around the world each year (Fig. 8.8). These harvested beans pass through many hands on their journey to consumers. More than 4 billion cups of coffee are drunk in the world every day, making coffee (which contains the stimulant caffeine) the most popular drug in the world (Publications International 2015). Only Brazil and Vietnam export more than Colombia. According to records amassed by the National Federation of Coffee Growers of Colombia (2013), a trade association of independent coffee growers and small farmers, of the nearly 12 million acres of land used for agriculture in Colombia, about 20 percent is devoted to coffee production. This is 1.5 times the area planted in corn, twice that planted in rice, and over four times that planted in both sugar cane and oil palm, the other leading crops grown in the country. Coffee growing is the source of about 800 000 jobs in Colombia's rural sector.

Fig. 8.8 Coffee beans. *Source:* Couleur/2288 images/Pixabay.

The trees that grow the red beans that are roasted to make coffee are indigenous to Ethiopia. The word "coffee" is derived from Kaffa, the name of a highland region of Ethiopia where coffee trees may have first been domesticated. Catholic Church records indicate that Jesuit priests brought coffee beans with them to Colombia in the early 18th century. The first commercial production and export in the country began in 1835. From that time, coffee production has spread along almost 2000 miles of inter-Andean valleys, from the extreme south of the country to the far north, across several different coffee-growing regions (National Federation of Coffee Growers of Colombia 2013).

One such region in the western central part of the country is Risaralda, a mountainous area known for the high quality of the coffee it produces. It lies in the *Triángulo del Café* (the Coffee Triangle), the district that grows the largest share of Colombia's coffee. To assess the experiences of small rural farmers living there, researchers from Purdue University (Eise & White 2018) interviewed 45 local residents about their understanding of climate change. In responding to questions such as

"What is climate change?" and "How, if at all, has climate change affected you as a farmer?", the study participants demonstrated a clear awareness of the climate-related fluctuations the region was experiencing. Over 90 percent described changes in average temperature, 74 percent reported that droughts had gotten longer and more intense, and 61 percent said there was an increase in mountainside erosion and landslides associated with more rain. These environmental changes unfavorably impacted their coffee crops, including via changes in the flowering and fruiting cycles of the coffee trees (reported by 91 percent of participants), an increase in crop pests (71 percent), and an increase in crop disease (59 percent). These impacts caused the farmers to feel uncertain about what previously had been routine farming decisions. Because the planting and harvesting seasons are no longer regular or predictable, for example, many farmers said they could not count on traditional seasonal indicators to guide them in deciding when to plant, harvest, or tend to their crops. Organizing labor to pick the coffee beans also has become a problem, because the trees often do not flower due to unstable seasonal conditions. The researchers found that the farmers they interviewed "saw climate change as nothing less than an existential threat" to their livelihoods (Eise & White 2018). One farmer lamented: "Our ability to counteract the effects of climate change is minimal . . . It is a threat capable of greatly incapacitating us. So we must be very attentive to the little we can do to mitigate" (quoted in Eise & White 2018).

Research in Brazil (Assad *et al.* 2004) and Mexico (Gay *et al.* 2006) indicates that even small increases in average temperature due to climate change can have calamitous consequences for coffee production, in some cases reducing the area presently suitable for growing by over 90 percent. Climate-induced stress can weaken plants, making them more vulnerable to opportunistic herbivores. Effects on insect pests include changes in the number of generations they can produce per year, their degree of winter mortality, and their distribution and range (Ayres & Lombardero 2000; Karban & Strauss 2004; Gomi *et al.* 2007).

The most important coffee pest throughout the world is a small dark beetle known as the coffee berry borer (*Hypothenemus hampei*), or in Colombia as *barrenador del café*. In their research on this pest, Jaramillo *et al.* (2009) assessed its thermal tolerance in order to make inferences regarding the effects of climate change on coffee production in the tropics. The coffee berry borer is one of the few herbivores that feeds on the coffee endosperm, the part of the seed that serves as a food store for the developing plant embryo. It is able to do this because of its ability to detoxify caffeine, a coffee biochemical defense that is successful in warding off many other pests. A native of the coffee areas in East Africa, it was introduced into Colombia in 1988. Given rising temperatures there, Jaramillo *et al.* (2009) determined that the potential number of generations per year for the insect is 3.4 in Colombia, compared with 2.03–3.13 in East Africa. An increase of this magnitude could profoundly influence the pest's population dynamics. Areas with higher seasonal temperature close to the optimum development of the beetle could experience high pest pressure. In that there are a limited number of coffee berries at any point in time, this could lead to increased borer dispersal, as more females compete for egg-laying sites in coffee plants. Jaramillo *et al.* (2009) conclude that "[s]uch a scenario could have devastating effects in coffee growing areas like Colombia where, because of well distributed precipitation leading to multiple flowering of the plants, there is yearlong supply of coffee berries." Jaramillo (2005)' estimated that for the coffee-growing areas in Colombia to avoid the coffee berry borer following an increase in temperature of 1.8 °F, plantations would have to move about 550 feet up mountainsides in order to maintain their current level of productivity and quality. Research in Uganda and Indonesia shows that the beetle has already expanded its altitudinal range and is attacking coffee plants at sites as high as 1864 miles above sea level (Jaramillo *et al.* 2009).

Because of the El Niño and La Niña disruptive weather patterns between 2008 and 2013, Colombia's coffee production fell by approximately 33 percent as rains, cloudy days, and very hot periods all increased. It was during this time that the disease called coffee leaf rust began to spread in Colombia. While the country subsequently increased production, it did not meet its national production goal of 14.7 million bags of coffee, a shortfall the National Federation of Coffee Growers of Colombia attributes to excessive rains and cloudiness. The low prices coffee growers receive for their product on the global market—a constant theme in the efforts of the Federation—means that they operate on a slim profit margin. The precariousness of life as a coffee grower increased when the global coffee chain changed dramatically as a result of the collapse in 1989 of international pricing agreements between coffee-producing and coffee-consuming nations. This shift involved movement "[f]rom a fairly balanced contest between producing and consuming countries within the politics of international coffee agreements" (Ponte 2002) to dominance by consuming countries and multinational corporations of the most heavily traded commodity in the world after oil in terms of monetary value. As the planet temperature ticks upward, the potential for devastating crop failure that could destroy the livelihood of thousands of family coffee growers is increasing in a socio-economic context of stark income inequality. The prices charged at the consumer end for a 12-ounce cup of distinctively flavored and prepared coffee are determined by the roasters and retailers, not by the farmers.

In August 2019, a special report entitled "Climate Change and Land" was released by the IPCC that reaffirms the critical risks posed by climate change to food security. Prepared by more than 100 experts from 52 countries, the report indicates that half a billion people live in places that are turning into desert, with cultivatable soil being lost between 10 and 100 times faster than it is forming. The growing risk is that food crises will occur on several continents at the same time. The report stresses that climate change is affecting all four pillars of food security: availability (crop yield and production), access (prices and ability to afford food), utilization (nutrition and cooking), and stability (disruptions to availability). According to Priyadarshi Shukla, Co-Chair of IPCC Working Group III: "Food security will be increasingly affected by future climate change through yield declines—especially in the tropics—increased prices, reduced nutrient quality, and supply chain disruptions . . . We will see different effects in different countries, but there will be more drastic impacts on low-income countries in Africa, Asia, Latin America and the Caribbean" (quoted in IPCC 2019). Adds Pete Smith, Professor of Plant and Soil Science at the University of Aberdeen and a lead author on the report: "People's lives will be affected by a massive pressure for migration . . . People don't stay and die where they are. People migrate" (quoted in Flavelle 2019).

The impacts of climate change include the ensnarement of small farmers in debt bondage—the most prevalent form of modern slavery. This has been found to be the case in Cambodia, where the sustainability of agriculture-based livelihoods is being undermined by climate change. Weather conditions there are increasingly capricious. When the rains do not fall, rural families struggle to survive. As one small farmer near the capital city of Phnom Penh lamented to a reporter, "I've never experienced this before . . . There's no rain. There's no flood. There's no rice in the dry season. There's no rice in the rainy season. I don't know what to do. I might try to find some vegetables to pick here and there, but if I can't get any, I might beg for money" (quoted in Bopha 2015). Facing such conditions, many farmers are forced to take out loans, but if the rains still do not arrive, they have little choice but to move to the city. One pattern that has emerged is that brick manufactures in Phnom Penh agree to repay farmers' debts and offer them a consolidated loan, but the farmers and their families are compelled to enter into debt bondage with them until the loan is repaid. Work at brick kilns is so poorly paid that the loans are never repaid, and those in debt are not allowed to leave. Bonded laborers who run away are often tracked down by the kiln owners with assistance from state police. In a report entitled "Blood Bricks: Untold Stories of Modern Slavery and Climate Change from Cambodia," based on a mixed-method study, Brickell *et al.* (2018, p. 53) found that as the small farming way of life "becomes unsustainable, debt may become debt bondage and thus modern slavery: a path of exploitation exacerbated by climate change but rooted in structures of poverty and inequality."

Other surprising consequences of climate change are manifest at the biological level, such as in male/female sex ratios. This has been demonstrated, for example, among reindeer (*Rangifer tarandus platyrhynchus*) living on the Svalbard archipelago of Norway. The annual variation in the proportion of males to total sexed adults among Svalbard reindeer is affected by both winter climate variability and animal density, with increased winter precipitation leading to a strong decrease in the male adult population growth rate and a strong increase in male adult mortality compared to females (Peeters 2014). In some species, sex ratios are shaped by temperature during development. In sea turtles, the temperature of the sand in which eggs are laid and embryos develop determines whether members of the clutch are born male or female. High sand temperature leads to more female hatchlings, causing some biologists to ponder whether a rise in global temperatures will result in a large proportion of baby sea turtles being born female (Schwanz & Janzen 2008). In the American alligator, the reverse pattern occurs: higher sand temperatures lead to more male hatchlings. In both cases, global warming could tip the scales in such a way that there is an inadequate balance of the sexes to maintain reproduction rates. Evidence suggests that human sex ratios may not be immune to the effects of climate change. Research by Misao Fukuda and colleagues (2014) found that more boys are born in places where temperatures are rising and fewer are born in places undergoing environmental changes such as drought and wildfire. One causal factor in this pattern may be that male fetuses are especially vulnerable to external stress factors.

Early on, this chapter raised the question: How serious is climate change? The answer, as addressed in the rest of the chapter, is that it is deadly serious. The threat of climate change is expressed in the multiplicity of ways that the planet is changing and how each of the domains of alteration, from drought to coastal flooding, interacts adversely with other ecocrises. Given all that is going wrong for the well-being of Earthlings, it is easy to feel overwhelmed, powerless, and depressed. However, if you were driving a car and realized you could not avoid careening into a wall, that realization would not convince you to take your foot off of the brake. People survive crashes, and the slower your car is going, the better your changes of coming through alive. As the WHO (2004) affirms, the likelihood of death is 20 times greater for the occupants of a car in a crash with an impact speed of 50 mph than for those in a crash at 20 mph. The consequences of anthropogenic climate change are at this point unavoidable; indeed, they are already occurring. The question is how much they can be limited by the emergence of aggressive countermeasures like a significant reduction in greenhouse gas emissions. Central to this are the actions of powerful groups and individuals who seem committed to a pathway of business as deadly usual, as addressed in the next chapter.

References

ABoVE. (2018). Intro to the Arctic-Boreal Vulnerability Experiment—ABoVE. NASA. https://above.nasa.gov/about.html?#about/.

AccuWeather. (2019). Australian weather. https://www.accuweather.com/en/australia/melbourne/a/january-weather/26216?monyr=1/1/2019&view=table.

Acosta, L. (2017). Millions of Colombians at risk to climate change: minister. *Reuters*. https://www.reuters.com/article/us-colombia-environment-idUSKBN17R2FT.

Ainsworth, E., Yendrek, C., Sitch, S., Collins, W., Emberson, L. (2012). The effects of tropospheric ozone on net primary productivity and implications for climate change. *Annual Review of Plant Biology* 63: 637–661.

Assad, E., Pinto, H., Zullo, J. & Helminsk, A. (2004). Climatic changes impact in agroclimatic zoning of coffee in Brazil. *Pesquisa Agropecuária Brasileira* 39: 1057–1064.

Avnery, S., Mauzerall, D., Liu, J. & Horowitz, L. (2011a). Global crop yield reductions due to surface ozone exposure: 1. Year 2000 crop production losses and economic damage. *Atmospheric Environment* 45(13): 2284–2296.

Avnery, S., Mauzerall, D., Liu, J. & Horowitz, L. (2011b). Global crop yield reductions due to surface ozone exposure: 2. Year 2030 potential crop production losses and economic damage under two scenarios of O3 pollution. *Atmospheric Environment* 45(13): 2297–2309.

Ayres, M. & Lombardero, M. (2000) Assessing the consequences of climate change for forest herbivores and pathogens. *Science of the Total Environment* 262: 263–286.

Baer, H. & Singer, M. (2018). *The Anthropology of Climate Change: An Integrated Critical Perspective*. Abingdon, Oxford: Routledge, Earthscan.

Balasubramanian, M. (2018). Climate change, famine, and low-income communities challenge Sustainable Development Goals. *Lancet Planetary Health* 2(10): E421–E422.

Baldwin, I. & Preston, C. (1998). The eco-physiological complexity of plant responses to insect herbivores. *Planta* 208: 137–145.

Basu, R. (2009). High ambient temperature and mortality: a review of epidemiologic studies from 2001 to 2008. *Environmental Health* 8: 40.

Battisti, D. & Naylor, R. (2009). Historical warnings of future food insecurity with unprecedented seasonal heat. *Science* 323: 240–244.

Beniston, M. (2004). The 2003 heat wave in Europe: a shape of things to come? An analysis based on Swiss climatological data and model simulations. *Geophysical Research Letters* 31(2).

Bevis, M., Harig, C., Khan, S., Brown, A., Simons, F., Willis, M. *et al.* (2019). Accelerating changes in ice mass within Greenland, and the ice sheet's sensitivity to atmospheric forcing. *Proceedings of the National Academy of Sciences of the U.S.A.* 116(6): 1934–1939.

Bhatt, S., Gething, P., Brady, O., Messina, J., Farlow, A., Moyes, C. *et al.* (2013). The global distribution and burden of dengue. *Nature* 496(7446): 504–507.

Björkman, C. & Niemela, P. (2015). *Climate Change and Insect Pests*. Boston, MA: CABI.

Black, R., Victora, C., Walker, S., Bhutta, Z., Christian P, de Onis, M. *et al.* (2013). Maternal and child undernutrition and overweight in low-income and middle-income countries. *Lancet* 382(9890): 427–451.

Blest, P. (2020). Trump just fired one of the country's top climate scientists. *Vice*. https://www.vice.com/en/article/akdgak/trump-just-fired-one-of-the-countrys-top-climate-scientists.

Bopha, P. (2015). Underprepared Cambodia vulnerable to climate change. *VOA*. https://www.voanews.com/a/changing-climate-has-major-impacts-for-under-prepared-cambodia/3075404.html.

Borger, J. (2018). Fleeing a hell the U.S. helped create: why Central Americans journey north. *Guardian*. https://www.theguardian.com/us-news/2018/dec/19/central-america-migrants-us-foreign-policy.

Bowen, M. (2005). *Thin Ice: Unlocking the Secrets of Climate in the World's Highest Mountains*. New York: Henry Holt.

BRAC. (2016). BRAC Annual Report 2016. http://www.brac.net/publications/annual-report/2016/#ecosystem.

Brady, O., Gething, P., Bhatt, S., Messina, J., Brownstein, J., Hoen, A. *et al.* (2012), Refining the global spatial limits of dengue virus transmission by evidence-based consensus. *PLoS Neglected Tropical Disease* 6(8): e1760.

Brickell, K., Parsons, L., Natarajan, N. & Chann, S. (2018). *Blood Bricks: Untold Stories of Modern Slavery and Climate Change from Cambodia*. London: Royal Holloway.

Brix, L. (2014). Global warming may benefit some farmers. ScienceNordic. http://sciencenordic.com/global-warming-may-benefit-some-farmers.

Brown, E. (2017). Climate change is aggravating food insecurity in Africa. Environmental and Energy Study Institute. https://www.eesi.org/articles/view/climate-change-is-aggravating-food-insecurity-in-africa.

Buhaug, H., Gleditsch, N. & Theisen, O. (2008). *Implications of Climate Change for Armed Conflict*. Washington, D.C.: World Bank.

C2ES. (2018). Hurricanes and climate change. Center for Climate and Energy Solutions. https://www.c2es.org/content/hurricanes-and-climate-change/.

Chang, C. (1975). Carbon dioxide and senescence in cotton plants. *Plant Physiology* 55: 515–519.

Che, Y., Zhang, M., Li, Z., Li, H. Wang, S., Sun, M. *et al.* (2017). Glacier mass-balance and length variation observed in China during the periods 1959–2015 and 1930–2014. *Quaternary International* 454: 68–84.

Cheng, L., Abraham, J., Hausfather, Z. & Trenberth, K. (2019). How fast are the oceans warming? *Science* 363(6423): 128–129.

Chestney, N. (2018). Global carbon emissions set to rise further this year: study. *Reuters*. https://www.reuters.com/article/us-climate-change-accord-emissions-idUSKBN1O42D0.

Clarke, L., Eom, J., Marten, E., Horowitz, R., Kyle, P. & Link, R. (2018). Effects of long-term climate change on global building energy expenditures. *Energy Economics* 72: 667–677.

Climate Reality Project. (2019). How the climate crisis is driving Central American migration. https://www.climaterealityproject.org/blog/how-climate-crisis-driving-central-american-migration.

Cook, J., Oreskes, N., Doran, P., Anderegg, W., Verheggen, B., Maibach, E. *et al.* (2016). Consensus on consensus: a synthesis of consensus estimates on human-caused global warming. *Environmental Research Letters* 11(4): 048002.

Cowtan, K. & Hausfather, Z. (2017). Political investigation is not the way to scientific truth. *Scientific American*. https://blogs.scientificamerican.com/guest-blog/political-investigation-is-not-the-way-to-scientific-truth/.

CRED. (2016). EM-DAT: The International Disaster Database. Centre for Research on the Epidemiology of Disasters. https://www.emdat.be/.

Diffenbaugh, N., Swain, D. & Touma, D. (2015). Anthropogenic warming has increased drought risk in California. *Proceedings of the Academy of Sciences of the U.S.A.* 112(13).

Dodds, K. (2018). *Ice: Nature and Culture*. Chicago, IL: University of Chicago Press.

Doyle, A. (2019). Over a billion people struggle to stay cool as Earth warms. Thomson Reuters Foundation. http://news.trust.org/item/20180716040058-2j73u/.

Easterling, D. & Wehner, M. (2009). Is the climate warming or cooling? *Geophysical Research Letters* 36: L08706.

ECIU. (2017). Even heavy weather: tracking the fingerprints of climate change, two years after the Paris Agreement. Energy and Climate Intelligence Unit.

ECIU. (2018). Even heavier weather: tracking the fingerprints of climate change, three years after the Paris Agreement. Energy and Climate Intelligence Unit.

Ehret, D. & Jolliffe, P. (2011). Leaf injury to bean plants in carbon dioxide enriched atmospheres. *Canadian Journal of Botany* 63(11): 2015–2020.

Ehrlich, P. & Raven, P. (1964). Butterflies and plants: a study in coevolution. *Evolution* 18: 586–608.

Eise, J. & White, N. (2018). Coffee farmers struggle to adapt to Colombia's changing climate. *The Conversation*. https://theconversation.com/coffee-farmers-struggle-to-adapt-to-colombias-changing-climate-97916.

Eisenberg, J., Desai, M., Levy, K., Bates, S., Liang, S., Naumoff, K. & Scott, J. (2007). Environmental determinants of infectious disease: a framework for tracking causal links and guiding public health research. *Environmental Health Perspectives* 115(8): 1216–1223.

Elsner, J., Kossin, J. & Jagger, T. (2008). The increasing intensity of the strongest tropical cyclones. *Nature* 455(4): 92–95.

Emanuel, K. (2016). Will global warming make hurricane forecasting more difficult? *Bulletin of the American Meteorological Society* 98(3): 495–501.

Emery, W., Baldwin, D., Schlüssel, P. & Reynolds, R. (2001). Accuracy of in situ sea surface temperatures used to calibrate infrared satellite measurements. *Journal of Geophysical Research* 106(C2): 2387–2405.

Epstein, P. (2001). Climate change and emerging infectious diseases. *Microbes and Infection* 3: 747–754.

Ericksen, P., Thornton, P., Notenbaert, A., Cramer, L., Jones, P. & Herrero, M. (2011). Mapping hotspots of climate change and food insecurity in the global tropics. CCAF Report No. 5. Copenhagen: CGIAR Research Program on Climate Change, Agriculture and Food Security.

FAO. (2018). *The State of Food Security and Nutrition in the World: Building Climate Resilience for Food Security and Nutrition.* Rome: Food and Agriculture Organization of the United Nations.

Flavelle, C. (2019). *Climate change threatens the world's food supply*, United Nations warns. *New York Times*. https://www.nytimes.com/2019/08/08/climate/climate-change-food-supply.html.

Flood, J. (2010). The importance of plant health to food security. *Food Security* 2: 215–231.

Fox, K. (2010). *Watching the English: The Hidden Rules of English Behaviour*. London: Nicholas Brealey.

Francl, K. & Meikle, D. (2009). A range extension of the Hispid cotton rat, *Sigmodon hispidus*, in Virginia. *Banisteria* 33: 54–55.

Fritz, A. (2019). Minus-56 degrees and more: brutal cold-air outbreak is smashing records in the Midwest. *Washington Post*. https://www.washingtonpost.com/weather/2019/01/31/brutal-cold-outbreak-is-breaking-records-midwest-thursday-morning/?utm_term=.311273859f3d.

Fuhrer, J. (2009). Ozone risk for crops and pastures in present and future climates. *Naturwissenschaften* 96: 173–194.

Fukuda, M., Fukuda, K., Shimizu, T., Nobunaga, M., Mamsen, L. & Andersen, C. (2014). Climate change is associated with male:female ratios of fetal deaths and newborn infants in Japan. *Environment and Epidemiology* 102(5): 1364–1370.

Gay, C., Estrada, F., Conde, C., Eakin, H. & Villers, L. (2006). Potential impacts of climate change on agriculture: a case of study of coffee production in Veracruz, Mexico. *Climatic Change* 79: 259–288.

Geddes, L. (2015). Why do Brits talk about the weather so much? *BBC*. http://www.bbc.com/future/story/20151214-why-do-brits-talk-about-the-weather-so-much.

Geggel, L. (2018). Antarctica's Pine Island Glacier just lost enough ice to cover Manhattan 5 times over. *LiveScience*. https://www.livescience.com/63974-pine-island-iceberg-calves-2018.html.

GFDRR. (2018). Flood risk management in Dhaka: a case for eco-engineering approaches and institutional reform. Global Facility for Disaster Reduction and Recovery. https://reliefweb.int/report/bangladesh/flood-risk-management-dhaka-case-eco-engineering-approaches-and-institutional.

Gleick, P. (2014). Water, drought, climate change, and conflict in Syria. *Weather, Climate, and Society* 6(3): 331–340.

Gomi, T., Nagasaka, M,. Fukuda, T. & Higahara, H. (2007). Shifting of the life cycle and life history traits of the fall webworm in relation to climate change. *Entomologia Experimentalis et Applicata* 125: 179–184.

Greene, C., Francis, J. & Monger, B. (2013). Superstorm Sandy: a series of unfortunate events? *Oceanography* 26(1): 8–9.

Guo, W., Xu, J., Liu, S., Shangguan, D. Wu, L. & Yao, X. (2014). The second glacier inventory dataset of China. *Cold and Arid Regions Science Data Center at Lanzhou* 20.

Gyee, G. (2018). High tide that inundated roads around Charleston was 6th highest on record. *Post and Courier*. https://www.postandcourier.com/news/high-tide-that-inundated-roads-around-charleston-was-th-highest/article_ef63d8d0-efec-11e8-bd34-cb7f131264c2.html.

Hagemann S., Chen, C., Clark, D., Folwell, S., Gosling, S., Haddeland, I. *et al.* (2013). Climate change impact on available water resources obtained using multiple global climate and hydrology models. *Earth System Dynamics* 3(2): 1321–1345.

Hales, S., de Wet, N., Maindonald, J. & Woodward, A. (2002). Potential effect of population and climate changes on global distribution of dengue fever: an empirical model. *Lancet* 360: 830–834.

Hausfather, Z. (2013). Global warming pause? *Berkeley Earth*. http://berkeleyearth.org/global-warming-pause/.

Hausfather, Z., Cowtan, K., Clarke, D., Jacobs, P., Richardson, M. & Rohde, R. (2017). Assessing recent warming using instrumentally homogeneous sea surface temperature records. *Science Advances* 3(1): e1601207.

IPCC. (2013). Climate Change 2013: The Physical Science Basis. Intergovernmental Panel on Climate Change. https://www.ipcc.ch/report/ar5/wg1/.

IPCC (2018). Climate Change: The IPCC Scientific Assessment. *Final Report of Working Group* 1. https://www.ipcc.ch/site/assets/uploads/2018/03/ipcc_far_wg_I_full_report.pdf.

IPCC. (2019). Land is a critical resource, IPCC report says. Intergovernmental Panel on Climate Change. https://www.ipcc.ch/2019/08/08/land-is-a-critical-resource_srccl/.

IRIN News. (2009). Drought driving farmers to the cities. *The New Humanitarian*. https://www.thenewhumanitarian.org/feature/2009/09/02/drought-driving-farmers-cities.

Jaramillo, A. (2005). Andean climate and coffee in Colombia. *Colombian National Federation of Coffee Growers. Chinchiná: Centro Nacional de Investigaciones de Café—Cenicafé.*

Jaramillo, J., Chabi-Olaye, A., Kamonjo, C., Jaramillo, A., Vega, F., Poehling, H.-M. *et al.* (2009). Thermal tolerance of the coffee berry borer *Hypothenemus hampei*: predictions of climate change impact on a tropical insect pest. *PLoS One* 4(8): e6487.

Jeong, S.-U., Bloom, A., Schimel, D., Sweeney, C. & Parazoo, N. (2018). Accelerating rates of Arctic carbon cycling revealed by long-term atmospheric CO2 measurements. *Scientific Advances* 4(7): eaao1167.

Johnson, S. & Mazo, J. (2013). Global warming and the Arab Spring. In: C. Werrell & F. Fermia (Eds.), *The Arab Spring and Climate Change*, pp. 14–21. Washington, D.C.: The Center for Climate and Security.

Karban, R. & Strauss, S. (2004). Physiological tolerance, climate change and a northward rage shift in the spittlebug *Philaenus spuramis*. *Ecological Entomology* 29: 251–254.

Karl, T., Arguez, A., Huang, B., Lawrimore, J., McMahon, J., Menne, M. *et al.* (2015). Possible artifacts of data biases in the recent global surface warming hiatus. *Science* 348(6242): 1469–1472.

Kavelle, M. (2018). *Unable to bury climate report,* Trump & deniers launch assault on the science. *Inside Climate News*. https://insideclimatenews.org/news/30112018/fact-check-trump-climate-science-denial-national-assessment-sanders-global-warming.

Kelly, C., Mohtadi, S., Cane, M., Seager, R. & Kushnir, Y. (2015). Climate change in the Fertile Crescent and implications of the recent Syrian drought. *Proceedings of the National*

Academy of Sciences of the U.S.A. 112(11): 3241–3246.

Kelly, M. & Monroe, R. (2018). *Earth's oceans have absorbed 60 percent more heat per year than previously thought.* Princeton University. https://www.princeton.edu/news/2018/11/01/earths-oceans-have-absorbed-60-percent-more-heat-year-previously-thought.

Kennedy, J., Rayner, N., Smith, R., Parker, D. & Saunby, M. (2011a). Reassessing biases and other uncertainties in sea surface temperature observations measured in situ since 1850: 1. Measurement and sampling uncertainties. *Atmospheres* 116(D14).

Kennedy, J., Rayner, N., Smith, R., Parker, D. & Saunby, M. (2011b). Reassessing biases and other uncertainties in sea surface temperature observations measured in situ since 1850: 2. Biases and homogenization. *Atmospheres* 116(D14).

Kessler, A. & Baldwin, I. (2002). Plant responses to insect herbivory. *Annual Review of Plant Biology* 53: 299–328.

Kim, A. & Gerretsen, I. (2019). Scientists memorialize the first glacier lost to climate change in Iceland. *CNN.* https://www.cnn.com/2019/07/22/world/iceland-glacier-climage-change-trnd/index.html.

Klempa, S. (2009). Hantaviruses and climate change. *Clinical Microbiology and Infection* 15(6): 518–523.

Kocin, P., Weiss, A. & Wagner, J. (1988). The great arctic outbreak and east coast blizzard of February 1899. *Weather and Forecasting* 3: 305–318.

Kovenock, M. & Swann. A. (2018). Leaf trait acclimation amplifies simulated climate warming in response to elevated carbon dioxide. *Global Biogeochemical Cycles* 32(2).

Kulp, S. & Strauss, B. (2019). New elevation data triple estimates of global vulnerability to sea-level rise and coastal flooding. *Nature Communications* 10: 4844.

Kusnetz, N. (2017). Rising seas are flooding Norfolk Naval Base, and there's no plan to fix it. *Inside Climate News.* https://insideclimatenews.org/news/10252017/military-norfolk-naval-base-flooding-climate-change-sea-level-global-warming-virginia.

Kusnetz, N. (2018). Rising seas threaten Norfolk Naval Shipyard, raising fears of "catastrophic damage." *NBC News.* https://www.nbcnews.com/news/us-news/rising-seas-threaten-norfolk-naval-shipyard-raising-fears-catastrophic-damage-n937396.

Langford, N. (2005). Carbon dioxide poisoning. *Toxicological Reviews* 24(4): 229–235.

Long, S., Ainsworth, E., Rogers, A. & Ort, D. (2004). Rising atmospheric carbon dioxide: plants FACE the future. *Annual Review of Plant Biology* 55: 591–628.

Lynn, K., Daigle, J., Hoffman, F., Lake, N., Michelle, D. & Ranco, C. (2013). The impacts of climate change on tribal traditional foods. *Climatic Change* 120(3): 545–556.

Madsen, E. (1974). Effect of CO2-concentration on growth and fruit productionof tomato plants. *Acta Agriculturae Scandenavia* 24: 242–246.

Madsen, H., Arnbjerg-Nielsen, K. & Mikkelsen, P. (2009). Update of regional intensity–duration–frequency curves in Denmark: tendency towards increased storm intensities. *Atmospheric Research* 92(3): 343–349.

Mandel, K. (2018). Venezuela's last glacier is about to disappear. *National Geographic.* https://www.nationalgeographic.com/environment/2018/11/venezuela-humboldt-glacier-melting-disappearing-climate-change/.

McMichael, A. & Martens, P. (2002). *Environmental Change, Climate and Health: Issues and Research Methods.* Cambridge: Cambridge University Press.

Mendelsohn, R., Emanuel, K., Chonabayashi, S. & Laura Bakkensen, L. (2012). The impact of climate change on global tropical cyclone damage. *Nature Climate Change* 2: 205–209.

Met Office. (2018a). U.K. climate projections 2018. https://www.metoffice.gov.uk/research/collaboration/ukcp/about.

Met Office. (2018b). Chance of summer heatwaves now thirty times more likely. https://www.metoffice.gov.uk/about-us/

press-office/news/weather-and-climate/2018/2018-uk-summer-heatwave.

Miller, B. (2019). Trump tweets climate change skeptic in latest denial of science. https://www.cnn.com/2019/03/12/us/trump-climate-change-tweet-patrick-moore/index.html.

Monroe, R. (2018). Carbon dioxide in the atmosphere hits record high monthly average. Scripps Institution of Oceanography. https://scripps.ucsd.edu/programs/keelingcurve/2018/05/02/carbon-dioxide-in-the-atmosphere-hits-record-high-monthly-average/.

Mooney, C. & Ryan, M. (2018). Pentagon revised Obama-era report to remove risks from climate change. *Washington Post*. https://www.washingtonpost.com/news/energy-environment/wp/2018/05/10/pentagon-revised-obama-era-report-to-remove-risks-from-climate-change/?utm_term=.a5ba8dd03e10.

Mora, C., Dousset, B., Caldwell, I., Powell, F., Geronimo, R. & Bielecki, C. (2017a). Global risk of deadly heat. *Nature Climate Change* 7: 501–506.

Mora, C., Counsell, C., Coral, R., Bielecki, C. & Louis, L. (2017b). Twenty-seven ways a heat wave can kill you: deadly heat in the era of climate change. *Circulation: Cardiovascular Quality and Outcomes* 10: e004233.

Mora, C., Spirandelli, D., Franklin, E., Lynham, J., Kantar, M., Miles, W. *et al.* (2018). Broad threat to humanity from cumulative climate hazards intensified by greenhouse gas emissions. *Nature Climate Change* 8: 1062–1071.

Morgan, P., Mies, T., Bollero, G., Nelson, R. & Long, S. (2006). Season-long elevation of ozone concentration to projected 2050 levels under fully open-air conditions substantially decreases the growth and production of soybean. *New Phytologist* 170: 333–343.

Mortensen, L. (1987). Review: CO2 enrichment in greenhouses. Crop responses. *Scientia Horticuturae* 3: 1–25.

Mukherjee, S., Mishra, A. & Trenberth, K. (2018). Climate change and drought: a perspective on drought indices. *Current Climate Change Reports* 4(2): 145–163.

Muller, R. (2013). A pause, not an end, to warming. Op-Ed. New York Times. https://www.nytimes.com/2013/09/26/opinion/a-pause-not-an-end-to-warming.html.

Myers, S., Zanobetti, A., Kloog, I., Huybers, P., Leakey, A., Bloom, A. *et al.* (2014). Increasing CO2 threatens human nutrition. *Nature* 510: 139–142.

National Federation of Coffee Growers of Colombia. (2013). Coffee is a driving force of the economy and a guarantee of social stability and peace. *Colombian Coffee Insider*.

National Geographic. (2013). What the world would look like if all the ice melted. https://www.nationalgeographic.com/magazine/2013/09/rising-seas-ice-melt-new-shoreline-maps/.

National Hurricane Center. (2018). NHC Data Archive. https://www.nhc.noaa.gov/data/.

National Ocean Service. (2018). What is the difference between a hurricane and a typhoon? https://oceanservice.noaa.gov/facts/cyclone.html.

Nelson, E., Kareiva, P., Ruckelshaus, M., Arkema, K., Geller, G. & Girvetz, E. (2013). Climate change's impact on key ecosystem services and the human well-being they support in the U.S. *Frontiers in Ecology and the Environment* 11(9): 483–893.

OSIP. (2018). Ocean warming. Ocean Scientists for Informed Policy at the Scripps Institution of Oceanography.

Oreskes, N. (2004). The scientific consensus on climate change. *Science* 306(5702): 1686.

Osborne, E., Richter-Menge, J. & Jeffries, M. (2018). Arctice Report Card. National Oceanic and Atmospheric Administration. https://www.arctic.noaa.gov/Report-Card.

Pascual, M., Ahumada, J., Chaves, L., Rodó, X. & Bouma, M. (2006). Malaria resurgence in the East African highlands: temperature trends revisited. *Proceedings of the National Academy of Sciences of the U.S.A.* 103(15): 5829–5834.

Peeters, B. (2014). *Effects of Climate and Animal Density on Adult Sex Ratio Variation in a Large*

Herbivore. Trondheim: Norwegian University of Science and Technology.

Pereira, A., Handa, S. & Holmqvist, G. (2017). Prevalence and Correlates of Food Insecurity Among Children Across the Globe. *Innocenti Working Paper 2017-09*. Florence: UNICEF Office of Research.

Peterson, B. (2017). Downtown Charleston broke record for flooded days last year; rain, floods continue to swamp Lowcountry. *Post and Courier*. https://www.postandcourier.com/news/downtown-charleston-broke-record-for-flooded-days-last-year-rain/article_266c0424-4b9d-11e7-959e-53ca41732680.html.

Peterson, T., Heim, R., Hirsch, R., Kaiser, D., Brooks, H., Diffwnbaugh, N. *et al.* (2013). Monitoring and understanding changes in heat waves, cold waves, floods and droughts in the United States: state of knowledge. *Bulletin of the American Meteorological Society* 94: 821–834.

Pfister, C., Brázdil, R., Glaser, R. Barriendos, M., Camuffo, D., Deutsch, *et al.* (1999). Documentary evidence on climate in sixteenth-century Europe. *Climate Change* 43: 55–110.

Pierre-Louis, K. (2019). Why is the cold weather so extreme if the earth is warming? *New York Times*. https://www.nytimes.com/interactive/2019/climate/winter-cold-weather.html.

Ponte, S. (2002). The “Latte Revolution”? Regulation, markets and consumption in the global coffee chain. *World Development* 30(7): 1099–1122.

Prudhomme, C., Giuntolia, I., Robinson, E., Douglas B. Clark, D., Arnell, N. *et al.* (2014). Hydrological droughts in the 21st century, hotspots and uncertainties from a global multimodel ensemble experiment. *Proceedings of the Academy of Sciences of the U.S.A.* 111(9): 3262–3267.

Publications International. (2015). *The Book of Incredible Information*. Morton Grove, IL: Publications International.

Queally, J. (2019). “We don’t know a planet like this”: CO2 levels hit 415 ppm for first time in 3 million+ years. *Common Dreams*. https://www.commondreams.org/news/2019/05/13/we-dont-know-planet-co2-levels-hit-415-ppm-first-time-3-million-years.

Reich, P., Hobbie, S., Lee, T. &, Pastore, M. (2018). Unexpected reversal of C3 versus C4 grass response to elevated CO2 during a 20-year field experiment. *Science* 360(63860): 317–320.

Resplandy, L., Keeling, R., Eddebbar, Y., Brooks, M., Wang, R., Bopp, L. *et al.* (2018). Quantification of ocean heat uptake from changes in atmospheric O2 and CO2 composition. *Nature* 563: 105–108.

Rhein, M., Rintoul, S., Aoki, S., Campos, E., Chambers, D., Feely, R., *et al.* (2013). Observations: ocean. In: T. Stocker, D. Qin, G.-K. Plattner, M. Tignor, S. Allen, J. Boschung *et al.* (Eds.). *Climate Change 2013: The Physical Science Basis. Contribution of Working Group I to the Fifth Assessment Report of the Intergovernmental Panel on Climate Change*, pp. 255–316. Cambridge: Cambridge University Press.

Rignot, E., Mouginot, J., Bernd Scheuchl, B., Michiel van den Broeke, M., Melchior, J., van Wessem, M. *et al.* (2019). Four decades of Antarctic ice sheet mass balance from 1979–2017. *Proceedings of the National Academy of Sciences of the U.S.A.* 116(4): 1095–1103.

Ringler, C., Zhu, T., Cai, X., Koo, J. & Wang, D. (2010). Climate change impacts on food security in Sub-Saharan Africa. International Food Policy Research Institute. http://www.ifpri.org/publication/climate-change-impacts-food-security-sub-saharan-africa.

Rochlin, I. (2018). Modeling the Asian longhorned tick (Acari: Ixodidae) suitable habitat in North America. *Journal of Medical Entomology* 56(2): 384–391.

Rollin, P., Ksiazek, T., Elliott, L., Ravkov, E., Martin, M., Morzunov. S. *et al.* (1995). Isolation of Black Creek Canal virus, a new hantavirus from *Sigmodon hispidus* in Florida. *Journal of Medical Virology* 46: 35–39.

Rosenzweig, C., Elliott, J., Deryng, D., Müller, R., Arneth, A., Boote, K. *et al.* (2014). Assessing agricultural risks of climate change in the 21st century in a global gridded crop model intercomparison. *Proceedings of the National Academy of Sciences of the U.S.A.* 111: 3268–3273.

Rubin, O. (2010). Social perspective on the symbiotic relationship between climate change and conflict. *Social Development Issues* 32(2): 29–41.

Sage, R. (2004). The evolution of C4 photosynthesis. *New Phytologist* 161(2): 341–370.

Santos-Burgoa, C., Goldman, A., Andrade, E., Barrett, N., Colon-Ramos, U., Edberg, M. *et al.* (2018). *Ascertainment of the estimated excess mortality from Hurricane Maria in Puerto Rico. The George Washington University*, Global Health Faculty Publications. https://hsrc.himmelfarb.gwu.edu/sphhs_global_facpubs/288.

Satish, U., Mendell, M., Shekhar, K. Hotchi, T., Sullivan, D., Streufert, S. *et al.* (2012). Is CO2 an indoor pollutant? Direct effects of low-to-moderate co2 concentrations on human decision-making performance. *Environmental Health Perspectives* 120(12): 1671–1677.

Schlenker, W. & Roberts, M. (2009). Nonlinear temperature effects indicate severe damages to U.S. crop yields under climate change. *Proceedings of the National Academy of Sciences of the U.S.A.* 106(37): 15 594–15 598.

Schwanz, L. & Janzen, F. (2008). Climate change and temperature-dependent sex determination: can individual plasticity in nesting phenology prevent extreme sex ratios? *Physiology and Biochemical Zoology* 81(6): 826–834.

Shah, S. (1999). *Coping with Natural Disasters: The 1998 Floods in Bangladesh.* Washington, D.C.: World Bank Summer Research Workshop on Poverty and Development.

Sharma, S., Blagrave, K., Magnuson, J., O'Reilly, C., Oliver, S., Batt, R. *et al.* (2019). Widespread loss of lake ice around the Northern Hemisphere in a warming world. *Nature Climate Change* 9: 227–231.

Shepherd, A., Ivins, E., Rignot, E., Smith, B., van den Broeke, M. & Velicogna, I. (2018). Mass balance of the Antarctic Ice Sheet from 1992 to 2017. *Nature* 558: 219–222.

Shermer, M. (2005). The fossil fallacy. *Scientific American.* https://www.scientificamerican.com/article/the-fossil-fallacy/.

Shi, G., Yang, L., Wang, Y., Kobayashi, K., Zhu, J., Tang, H. *et al.* (2009). Impact of elevated ozone concentration on yield of four Chinese rice cultivars under fully open-air field conditions. *Agriculture, Ecosystems & Environment* 131: 178–184.

Shi, Y., Liu, C. & Kang, E. (2009). The glacier inventory of China. *Annals of Glaciology* 50(53): 1–4.

Singer, M. (2018). *Climate Change and Social Inequality: The Health and Social Costs of Global Warming.* Abingdon: Routledge.

Singer, M. & Bulled, N. (2016). Ectoparasitic syndemics: polymicrobial tick-borne disease interactions in a changing anthropogenic landscape. *Medical Anthropology Quarterly* 30(4): 442–461.

Smith, D. & Vivekananda, J. (2007). *A Climate of Conflict.* London: International Alert.

Smith, T. & Reynolds, R. (2005). A global merged land air and sea surface temperature reconstruction based on historical observations (1880–1997). *Journal of Climatology* 18(12): 2021–2036.

Song, Y. & Robinson, W. (2004). Dynamical mechanisms for stratospheric influences on the troposphere. *Journal of the Atmospheric Sciences* 61(14): 1711–1725.

Starkey, T. (1988). Estimating the rate of photorespiration in leaves. *Physiologia Plantarum* 73(1): 147–152.

Stocker, T., Qin, D., Plattner, G.-K., Tignor, M., Allen, S., Boschung, J. *et al. (2013). IPCC, 2013: summary for policymakers. In*: Climate Change 2013: The Physical Science Basis. Intergovernmental Panel on Climate Change.

Thompson, L., Mosley-Thompson, E., Davis, M., Henderson, K., Brecher, H., Zagorodnov, V. *et al.* (2002). Kilimanjaro ice core records: evidence of Holocene climate change in tropical Africa. *Science* 208: 589–593.

Thompson, L., Brecher, H., Mosley-Thompson, E., Hardy, D. & Mark, B. (2009). Glacier loss on Kilimanjaro continues unabated. *Proceedings of the Academy of Sciences of the U.S.A.* 106(47): 19 770–19 775.

Thompson, R., Wellington de Oliveira Lima, J., Maguire, J., Braud, D. & Scholl, D. (2002). Climatic and demographic determinants of American visceral leishmaniasis in northeastern Brazil using remote sensing technology for environmental categorization of rain and region influences on leishmaniasis. *American Journal of Tropical Medicine and Hygiene* 67(6): 648–655.

Trusel, L., Das, S., Matthew B., Osman, M., Evans, M., Smith, B., *et al.* (2018). Nonlinear rise in Greenland runoff in response to post-industrial Arctic warming. *Nature* 564: 104–108.

U.N. Environment Program. (2018). *U.N. Environment Emissions Gap Report*. Nairobi: U.N. Environment.

United Nations. (2017). Factsheet: people and oceans. The Ocean Conference. https://www.un.org/sustainabledevelopment/wp-content/uploads/2017/05/Ocean-fact-sheet-package.pdf.

U.S. Global Change Research Program. (2018). Fourth National Climate Assessment. https://nca2018.globalchange.gov/.

Van Berke, N. (1984). Injurious effects of high CO2 concentrationson cucumber, tomato, chrysanthetnum and gerbera. *Acta Horticulturae* 162: 101–112.

van Doom, H. (2014). Emerging infectious diseases. *Medicine* 42(1): 60–63.

Vehviläinen, T., Lindholm, H., Rintamäki, H., Pääkkönen, R., Hirvonen, A., Niemi, O. *et al.* (2016). High indoor CO2 concentrations in an office environment increases the transcutaneous CO2 level and sleepiness during cognitive work. *Journal of Occupational and Environmental Hygiene* 31(1): 19–29.

Walker, R., Keane, C. & Burke, J. (2010). Disparities and access to healthy food in the United States: a review of food deserts literature. *Health & Place* 16: 876–884.

Walls, L. (2009a). Introducing Humbodlt's cosmos. *Minding Nature*. https://www.humansandnature.org/introducing-humboldt-s-cosmos.

Walls, L. (2009b). *The Passage to Cosmos: Alexander von Humboldt and the Shaping of America*. Chicago, IL: University of Chicago Press.

Watts, C. (2016). *A Brewing Storm: The Climate Change Risks to Coffee*. Sydney, NSW: Climate Institute.

Watts, J. (2018). We have 12 years to limit climate change catastrophe, warns U.N. *Guardian*. https://www.theguardian.com/environment/2018/oct/08/global-warming-must-not-exceed-15c-warns-landmark-un-report?CMP=Share_iOSApp_Other.

Werrell, C., Femia, F. & Sternberg, T. (2015). Did we see it coming?: State fragility, climate vulnerability, and the uprisings in Syria and Egypt. *SAIS Review* 35(1): 29–46.

Whitehouse, S. (2018). Time to wake up: the polar distress. Speech before the U.S Senate. https://www.c-span.org/video/?c4738314/114-sen-sheldon-whitehouse-climate-change.

WHO. (2004). Road safety—speed. World Health Organization. https://www.who.int/violence_injury_prevention/publications/road_traffic/world_report/speed_en.pdf.

Wilhite, D. & Glantz, M. (1985). Understanding the drought phenomenon: the role of definitions. *Water International* 10(3): 111–120.

Wittwer, S. (1967). Carbon dioxide and its role in plant growth. *Proceedings of the 17h International Horticultural Congress* 3: 311–322.

WMO. (2016). *Guidelines on the Definition and Monitoring of Extreme Weather and Climate Events*. Geneva: World Meteorological Organization.

World Bank. (2019). Intentional homicides (per 100 000 people). https://data.worldbank.org/indicator/VC.IHR.PSRC.P5?/.

Worley, S., Woodruff, S., Reynolds, R., Lubker, S. & Lott, N. (2005). ICOADS. Release 2.1: data and products. *International Journal of Climatology* 25: 823–842.

Worth, R. (2010). Earth is parched where Syrian farms thrived. *New York Times*. https://www.nytimes.com/2010/10/14/world/middleeast/14syria.html.

YaleEnvironment[360]. (2018). CO2 levels break another record, exceeding 411 parts per million. Yale School of Forrestry & Environmental Studies. https://www.fasterthanexpected.com/2018/06/07/co2-levels-break-another-record-exceeding-411-parts-per-million/.

Yue, L., Shen, H., Yu, W. & Zhang, L. (2018). Monitoring of historical glacier recession in Yulong Mountain by the integration of multisource remote sensing data. *IEEE Journal of Selected Topics in Applied Earth Observations and Remote Sensing* 11(2): 388–400.

Yun, S., Lee, W., Ryou, J., Yang, S., Park, S. & Roh, J. (2014). Severe fever with thrombocytopenia syndrome virus in ticks collected from humans, South Korea, 2013. *Emerging Infectious Diseases* 20: 1350–1353.

Zampieri, M., Ceglar, A., Dentener, F. & Toreti, A. (2017). Wheat yield loss attributable to heat waves, drought and water excess at the global, national and subnational scales. *Environmental Research Letters* 12(6): 064008.

Zhu, C., Kobayashi, K., Loladze, I., Zhu, J., Jiang, Q., Xuet, X., *et al.* (2018). Carbon dioxide (CO2) levels this century will alter the protein, micronutrients, and vitamin content of rice grains with potential health consequences for the poorest rice-dependent countries. *Science Advances* 4(5): eaaq1012.

Zhu, X., Feng, Z., Sun, T., Liu, X., Tang, H., Zhu, J. *et al.* (2011). Effects of elevated ozone concentration on yield of four Chinese cultivars of winter wheat under fully open-air field conditions. *Global Change Biology* 17: 2697–2706.

Zhu, Z, Piao, S., Myneni, R., Huang, M., Zeng, Z., Canadell, J. *et al.* (2016). Greening of the Earth and its drivers. *Nature Climate Change* 6: 791–795.

9

Business as deadly usual: resisting environmental science

In 2018, although some companies have publicly supported the Paris climate agreement to limit harmful warming, none of these companies has set company-wide emissions reduction targets consistent with this goal. Many continue to downplay or misrepresent climate science and the dangers of carbon emissions, and all continue to support trade groups that spread climate disinformation and work to stymie needed climate policies.

Union of Concerned Scientists (2018)

9.1 A consistent pattern of climate change denial

Why are we plunging headlong into a world of anthropogenic ecocrisis chaos? A peek into the socioeconomic realm of climate change denial offers part of the explanation. Under two administrations, those of George W. Bush and Donald Trump, denial in word and deed has shaped U.S. policy. On March 12, 2019, for example, Trump continued his already several years long denial of climate change by supportively tweeting a quote from Patrick Moore, whom he identified as a co-founder of Greenpeace: "The whole climate crisis is not only Fake News, it's Fake Science. There is no climate crisis, there's weather and climate all around the world, and in fact carbon dioxide is the main building block of all life" (quoted in Miller 2019). Moore's statement was made during an episode of Fox News' Fox & Friends TV program. According to Greenpeace (2019), an international environmental organization that not only recognizes the reality of climate change but, as discussed in Chapter 11, actively seeks to fight against it, "Patrick Moore often misrepresents himself in the media as an environmental 'expert' or even an 'environmentalist', while offering anti-environmental opinions on a wide range of issues and taking a distinctly anti-environmental stance. He also exploits long-gone ties with Greenpeace to sell himself as a paid consultant and pro-corporate spokesperson, usually taking positions that Greenpeace opposes." Without question, for 30 years Moore has sold himself well to timber, mining, chemical, nuclear, and aquaculture industries as a quotable source with a Ph.D. who denies the harmful effects of industrial pollution and climate change. But when he was 24, in 1971, Moore, then a Canadian graduate student in ecology, applied for and was offered a berth on the *Phyllis Cormack* (renamed *Greenpeace*), an aging fishing vessel that the founders of Greenpeace sent across the North Pacific to protest the U.S. testing of a 5-megaton nuclear bomb. Subsequently, Moore participated in Greenpeace campaigns against whaling, seal hunting, pesticides, supertankers, uranium mining, and toxic waste dumping. He remained active in Greenpeace until he quit in 1986, claiming "the environmental movement had abandoned science and logic in favor of emotion and

Ecosystem Crises Interactions: Human Health and the Changing Environment, First Edition. Merrill Singer.
© 2021 John Wiley & Sons, Inc. Published 2021 by John Wiley & Sons, Inc.

sensationalism" (quoted in Kasprak 2019). As Bennett (2004) notes, however, "[h]e didn't just retire; he joined the other side. Today, he's a mouthpiece for some of the very interests Greenpeace was founded to counter." That opposition is reflected in Moore's response to Trump's tweet: "Thanks President Trump. I am in D.C. for a meeting of the CO2 Coalition founded by William Happer, who I admire greatly. CO2 is entirely beneficial. The EPA finding that CO2 is pollution must be reversed" (Moore 2019).

Moore has long been associated with a small cadre of academics like Happer who are vocal climate change deniers. It is for this reason that Trump selected Happer, a retired oil industry-funded professor with no formal training in climate science, to chair his Presidential Committee on Climate Security. Ostensibly tasked with determining whether climate change poses a national security threat, establishment of the Committee was seen by environmentalists as an effort by Trump to undermine the scientific and U.S. intelligence and military consensus that climate change must be taken seriously to avert severe consequences (Ward 2019). According to Francesco Femia, chief executive of the Council on Strategic Risks and co-founder of the Center for Climate and Security, Trump's panel was "the equivalent of setting up a committee on nuclear-weapons proliferation and having someone lead it who doesn't think nuclear weapons exist. It's honestly a blunt-force political tool designed to shut the national security community up on climate change" (quoted in Eilperin & Ryan 2019).

Revealingly, in December 2015 Happer was targeted in a sting operation by Greenpeace. Posing as consultants for a Middle Eastern oil and gas company, Greenpeace activists met with Happer and asked him to write a paid report pushing the benefits of rising carbon emissions. Happer readily agreed, and proposed that his fee for this work ($8000) be donated to the CO2 Coalition. When the Greenpeace operatives contacted the CO2 Coalition they were told to talk to the Donors Trust—a conservative "dark money"-funneling organization—so that the source of the donation could remain a secret. Happer admitted that his report would probably not pass peer review with a scientific journal (Goldenberg 2015). John Sauven, the director of Greenpeace U.K., reported that the clandestine operation exposed how prominent climate change deniers like Happer "can be sponsored by foreign fossil fuel companies to write reports that sow doubt about climate change and that this sponsorship will then be kept secret. Down the years, how many scientific reports that sowed public doubt on climate change were actually funded by oil, coal and gas companies?" (quoted in Goldenberg 2015).

The examples of Moore and Happer, as well as President Trump, typify a consistent pattern in the climate change denial industry that is examined in this chapter. On the one hand, this pattern involves fossil fuel and other wealthy pollution manufacturers quietly bankrolling individuals and organizations with claims of climate expertise to produce written reports and oral testimony that seeks to discredit the voluminous body of actual scientific work describing the human causes and human consequences of climate change and other ecocrises. On the other hand, it involves politicians who receive fossil fuel industry funding using the refuted arguments of climate change deniers to push for energy and environmental policies and programs that allow for business-as-usual degradation of the global environment. To achieve this objective, deniers in and out of government have attempted to tarnish science as a reliable source of knowledge about the environment. Fueled by such sentiments, there have been "concerted efforts to rewrite high school science textbooks, restrict science at the Environmental Protection Agency . . . and the National Science Foundation . . . and defund earth research by the National Aeronautics and Space Administration," developments that "provide tangible signs of a partisan 'war on science'" (Hamilton *et al.* 2015).

Though the climate crisis has never been seen as the dire threat it is by any U.S. administration, and certainly has been intentionally overlooked by some, the "Trump administration has taken its war on climate crisis action to a new level" (Wolf 2019). As Wolf (2019) observes, Donald Trump mocked the very idea of climate change (by saying that sustainable wind turbines and not pollution cause cancer), trolled climate activists (by peddling ocean-polluting plastic straws as a campaign fundraiser), blocked efforts to limit climate pollution (by threatening energy producers that try to raise their fuel economy standards in conformity with the requirements of the laws of the state of California and by having the Environmental Protection Agency [EPA] assert that California's tougher emission rules violate federal law), and whitewashed evidence of climate change (by removing mention of the term from government reports wherever possible). Regarding the last point, the Environmental Data and Governance Initiative (EDGI) released a series of reports itemizing changes to government websites between 2016 and 2018 that reflect the Trump administration's clear-cut anti-science agenda (Sabin Center for Climate Change Law 2019a). In September 2019, Democrats in the U.S. Senate released a report outlining dozens of instances in which the Trump administration censored or minimized climate science across the federal government, including at the EPA, Federal Emergency Management Agency (FEMA), Department of Agriculture, and Department of Homeland Security (Evich 2019). Exemplary of this pattern, a news release written by geological researchers about their findings was sanitized by the Trump administration by removing all references to the calamitous effects of climate change due to sea-level rise by the end of the century. The researchers found that three to seven times more people and businesses than previously believed would be exposed to severe flooding. The final government news release, however, focused on the methodology used rather than the major findings (Waldman 2019). According to Joel Clement, a former climate staffer for the Department of the Interior, "It's an insult to the science, of course, but it's also an insult to the people who need this information and whose livelihoods and in some cases their lives depend on this . . . What's shocking about it is that this has been taken to a new level, where information that is essential to economic and health and safety—essentially American well-being—is essentially being shelved and being hidden" (quoted in Waldman 2019).

While Trump lost the 2020 presidential election, in part due to his fierce climate denial, as Davenport (2020) observes: "the damage done by the greenhouse gases unleashed by President Trump's rollbacks may prove to be one of the most profound legacies of his single term." His policies eliminated or restricted almost 100 rules and regulations on pollution in the air, water, and atmosphere. Exemplary are changes in policies that had restricted the release of industrial soot, a pollutant that can have lasting health effects, especially in minority communities, where it tends to be concentrated. Greenhouse pollution enhanced by Trump policies accumulates in the atmosphere, where it will remain doing harm for decades, regardless of any policy changes made by the Bidden administration.

9.2 A time of questioning environmental science

Brembs *et al.* (2013) affirm a basic premise, but one that is not often unpacked and explored: "Scientific research is largely a public endeavor, requiring public trust." But is science trustworthy? In this era of attacks on science and scientists marching in the street to defend their integrity (Fig. 9.1), this seems a question that merits addressing. In his book, *The Republican*

Fig. 9.1 March for Science. *Source:* kantonia/Pixabay.

War on Science, written during the George W. Bush administration, journalist Chris Mooney (2005) argues that since the 1990s an alliance of conservative Republican office holders, private industry, and fundamentalist Christian organizations has routinely disparaged, undermined, and misrepresented scientific information on topics relevant to public policy. Research findings on a wide range of issues have been targeted, including acid rain, climate change, the efficacy of condoms in preventing the spread of sexually transmitted diseases, the health impacts of excess consumption of sugar and fat, the status of endangered species, the efficacy of abstinence-only sex education programs, and the potential therapeutic value of stem cell research. In an editorial in *Science*—the journal of the American Association for the Advancement of Science—editor Donald Kennedy (2005) remarks that members of Congress "suddenly know how to evaluate individual grants, even defunding those that deal with touchy subjects. One banned grant dealt with the psychology of romance—apparently too hot to handle these days. A number of current legislators have also become amateur neurobiologists, developing an unexpected command of difficult topics like 'persistent vegetative state'. The Senate's chief surgeon, Dr. Bill Frist (R-TN) established a record for definitive long-range TV diagnosis on that subject." His point is that research that does not square with conservative ideology is being attacked as "bad science" by politicians who want to treat science as opinion. As the Union of Concerned Scientists (Carter *et al.* 2019)—which documented 80 attacks on science during the first 2 years of the Trump administration—points out, leading the attack on science are "individuals with limited scientific credentials and significant conflicts of interest." By comparison with prior administrations, this level of hostility toward inconvenient science is unprecedented (Berman & Carter 2018). The goal of this effort, now decades old, is to raise doubt in the public's mind about the believability of science. And among some sectors of the population, it has worked, though other factors (e.g., the Internet) are also at play.

A 2016 Pew Research Center survey of adults in the United States found that many held skeptical views of climate scientists, with only 39 percent reporting "strong" trust in the information they convey (Funk 2017). An equal share of respondents said that they had "some" trust in information from climate scientists, while about 22 percent were more doubtful, saying they did not trust information from climate scientists at all or not very much. Only 28 percent indicated that they had strong trust in scientists' understanding of the causes of climate change, while 32 percent said that climate scientists did not have a good understanding of climate change causation. Other research has found several factors are associated with the nature of attitudes toward climate scientists among adults. Politically more conservative and more religious individuals are less trusting of climate scientists and the research they produce (McCright *et al.* 2013; Blank & Shaw 2015; Gauchat *et al.* 2017; Stanley *et al.* 2017). While a number of studies show that more politically conservative individuals tend to defend the industrial capitalist system from the claims of some environmentalists and environmental scientists that capitalism causes serious ecological and public health problems (Conway & Oreskes 2012), conservatives have greater trust than liberals in production scientists (McCright *et al.* 2013). Kahan *et al.* (2017), moreover, demonstrate that an individual's level of science curiosity increases awareness of the risks associated with global warming independent of their political ideology. Additionally, research by Motta (2018) shows that having had science interest in young adulthood is associated with about a 7 percent increase in trust toward climate scientists in adulthood.

Of historic note, the politicization of scientific assertions is nothing new. Albert Einstein's theory of general relativity, for example, initially provoked significant political backlash. Einstein wrote at the time (quoted in van Dongen 2010, p. 78):

> This world is a strange madhouse. Currently, every coachman and every waiter is debating whether relativity theory is correct. Belief in this matter depends on political party affiliation.

He penned these reflections in 1920, 3 weeks after an emotional rally in Berlin's Philharmonic Hall at which he was denounced as a fraud and scientific barbarian. In 1922, Einstein decided to cancel his plenary lecture at the centennial meeting of the Society of German Natural Scientists and Physicians because he feared that he would become a target of violence, a view shared by local authorities. While several factors contributed to the rise of an antirelativist movement, "the conservative convictions of many antirelativists dovetailed . . . comfortably with their opposition to Einstein" (van Dongen 2010, p. 80).

Returning to the original question: Is science, or more specifically in light of this discussion, are the climate and environmental sciences trustworthy? Without question, individual scientists and research teams make mistakes. A recent error with notable consequences occurred when the Recombinetics company used DNA editing in an attempt to produce hornless dairy cattle. Researchers on the project took skin cells from Holstein dairy bulls and used plasmids—a type of circular mini-chromosome present in bacteria—to convey instructions to them. While the plasmids were meant to persist only temporarily as messengers, they in fact integrated into the genome of the calves born of the recombinant process. Though it is not clear if the bacterial DNA posed a direct health risk to the dairy cattle or to someone drinking their milk, there was considerable concern that the antibiotic resistance gene could be taken up by any of the billions of bacteria present in the cows' bodies. This was deemed a sufficient risk to lead to a culling of all animals produced by the process and a suspension of the research program. Recombinetics, which had rigorously fought regulation by the government, was forced to admit that ultimately it was government scientists who caught the costly error (Regalado 2019).

Some major scientific blunders have occurred among climate scientists, and they serve as a reminder about the need for a high level of caution and methodological rigor in the scientific endeavor. For example, 2 weeks after a prominent study on the pace of ocean warming was published in the premier scientific journal *Nature* (Resplandy *et al.* 2018), the authors submitted corrections to the publication. The error, discovered by a researcher not affiliated with the study (Lewis 2018), occurred in the way the authors dealt with uncertainty in their measurements. As one of the contributors, Ralph Keeling, explained: "Unfortunately, we made mistakes here . . . I think the main lesson is that you work as fast as you can to fix mistakes when you find them . . . I accept responsibility for what happened" (quoted in Mooney & Dennis 2018). While the pre-publication peer-review process maintained by scientific journals is intended to catch such errors, in this case it was left to post-publication review—which also is a normal part of the development of science. After a reexamination of their data, the authors and the journal readily agreed there was a mistake and publicly acknowledged the problem (Hamilton 1992; Keeling 2018). Based on their recalculations, the authors concluded that the ocean still appeared to be warmer than the estimate used by the Intergovernmental Panel on Climate Change (IPCC), but found that the increase in heat had a greater range of probability than they had initially thought—between 10 and 70 percent, as found in other research (Cheng *et al.* 2019). The lesson for almost all scientists is the need for sustained scrutiny. The peer-review process does not end with publication; rather, publication exposes a paper to far wider review. Openness to reexamination by other researchers and to the question-raising outcomes of subsequent research is, in fact, a strength of science and a reason for trusting the scientific process. This example suggests the importance of differentiating an error made on a specific climate study from the rejection of all studies that indicate the planet is warming.

In the climate change denial blogosphere and related domains, however, incidents like this are commonly pointed to as evidence that, if not all of climate science, then at least any part of it that is based on computer modeling of future conditions is faulty. Often, denier criticisms go beyond a critique of climate science methodologies to outright accusations of hoax, fraud, and intentional falsification of findings. For example, according to the Editorial Board of Investor's Business Daily (2018), a politically conservative business newspaper and website committed to keeping "climate charlatans from bankrupting our nation with spurious demands":

> The National Oceanic and Atmospheric Administration may have a boring name, but it has a very important job: It measures U.S. temperatures. Unfortunately, it seems to be a captive of the global warming religion. Its data are fraudulent.

While individual scientists may make mistakes, or even, under pressure to publish, engage in the fabrication of data, it is incredibly rare for the scientific consensus as a whole to be wrong. Scientists get rewarded in money and reputation for finding fault with problematic statements about the nature of reality made by other scientists. That the large majority (approximately 97 percent) of them agree on an issue—for there to be a scientific consensus, as is the case with consequential anthropogenic climate change (Anderegg *et al.* 2010; Cook *et al.* 2013)—is an important reason for trusting in its reality (Tsipursky 2018).

There is no denying that some degree of fraud occurs in scientific research (Smith 2005; Fang *et al.* 2012). Such behavior is defined as "reporting data for which no records of experiment or population are present; manipulating research materials, equipment, or procedures to arrive at the desirable result; adding, changing, or omitting results, which positively or negatively relate to the hypothesis that the research intends to test; and incorporating ideas, statements, procedures of others' work without permission and appropriate credit to the source" (Eliades *et al.* 2005). There are various reasons why fraud may be committed, including the pressures of a "publish or perish" promotional system, employer expectations that scientists will commercialize their research and produce institutional income, the desire for prestige or remuneration, and the pressure of time in fast-paced scientific fields and loaded professional schedules. These constitute underlying counterproductive incentives that can negatively influence the work of scientists. Additionally, there is what is sometimes referred to as the "gray zone," which includes industry efforts to influence the outcomes of scientific research, leading to the writing of partisan publications (Brown 2013; Keltan 2013; George & Buyse 2015).

While it is widely recognized in the scientific community that the primary determinant of scientific conduct is the individual investigator's integrity, a final reason why it is sensible to have cautious trust of scientific findings is that there are institutions and systems in place for monitoring and enhancing scientific veracity. One such institution, the World Conferences on Research Integrity (WCRI 2019), seeks to advance veracity globally by setting standards and agreeing on principles for the organization and conduct of research. In the United States, as in other countries, there are also national structures that perform similar functions, such as the Office of Inspector General of the National Science Foundation (OIG NSF 2019). This body is responsible for preventing and detecting fraud, waste, and abuse. In part, it does this by actively investigating and identifying individuals who attempt to abuse the public trust or dupe government science funding programs. These safeguards are not foolproof, but they provide additional support for accepting the genuineness of climate science that is rarely acknowledged by its detractors. No such internal institutions and procedures are in place for diminishing the fabrication of claims by climate change deniers.

9.3 Skirting accountability: polluters, innocence, and the victim slot

If carbon emission is creating a global health and social crisis, who is responsible? Based on the words of representatives of high-level emitter countries as they address either aggregate or per capita levels of carbon release: everyone else, anyone else, but not them. The United States deflects blame by pointing at China's high emissions volume. China deflects blame by citing its low per capita emission rate compared to Western nations. The global South accuses the more fully industrialized global North, which, in turn, points at the rising carbon release of rapidly developing nations like India and Brazil. Everyone portrays themselves as innocent, and, if a way can be concocted to do so, as victims, not culprits. It would appear that we are all whistling past the graveyard in mindless blamelessness.

Guilt and innocence can be seen as cultural slots into which self and other can be pigeonholed in the assignment of blame and virtue. The notion of the cultural slot has gained popularity in the field of anthropology in recent years. In "Anthropology and the Savage Slot: The Poetics and Politics of Otherness," Michel-Rolph Trouillot (2003) defines his key term, "savage slot," as the enduring category of thought and inquiry used in the West to construct indigenous peoples; a cultural mold that canalizes scholarly study and textual expression. Existence of the savage slot has long shaped how indigenous peoples have been perceived by dominant sectors of society. In describing how scientists, health officials, and politicians explained the causes of the deadly 1992–93 cholera outbreak in the Orinoco Delta of eastern Venezuela, for example, Charles Briggs and Clara Mantini-Briggs (2004) note that authorities racialized the disease to deflect blame away from the dominant society and on to the "indigenous culture" of those most impacted by the epidemic, people of the Warao, an Amerindian ethnic group. The disease raged in this population, it was claimed, not because they were subject to structural violence, poverty, and discrimination by the dominant society but because of their own unhealthy cultural practices. Based on her research in Indonesia, a somewhat different formulation is offered by Tania Li (2000), who uses "slot" to refer to a resilient political tool for marking and separating what the International Labor Organization calls "indigenous and tribal peoples" from other populations living nearby. Hughes (2013) combines these meanings in his use of the term "victim slot" in discussing how institutions and corporations—including oil companies—evade accountability for carbon emissions in the Caribbean petrostate of Trinidad and Tobago. He observes:

> Under climate change, emitters of carbon dioxide—even high emitters—have deliberately occupied or accidentally fallen into this compartment. Like the "savage slot" or the "tribal slot," the victim slot artificially clarifies an inherently murky moral situation. It whitewashes—as innocent—societies, firms, and industrial sectors otherwise clearly complicit with carbon emissions and climate change. To the extent that the slot persuades us, it allows good people to do bad things to the biosphere. (Hughes 2013)

It is little known that the modern hydrocarbon industry had its origins in Trinidad in 1859 where one of the first commercial wells (the other being in Pennsylvania) was drilled by Walter Darwen, founder of the Paria Oil Co. By 2010, Trinidad and Tobago's per capita carbon emissions ranked fourth among nation-states, though, in aggregate, Trinidad and Tobago emitted only 0.1 percent of the global total (IEA 2010). On this basis, national leaders have sought to portray the country as a small island victim of the rising seas, heat waves, and burning forests of the climate change caused by the world's large industrial powers. The oil industry there has laid claim to its own victim slot, because rising seas threaten its offshore rigs. A representative of Petrotrin, the national oil company at the time, told Hughes (2013, p. 578): "Trinidad is a small island developing state so we are vulnerable . . . We have wells that could fall into the sea." More broadly, the international oil industry, through industry lobbying organizations like the Independent Petroleum Association of America, has tried to paint itself as an innocent victim of a conspiracy by city and county governments, environmental groups, and environmentally focused scholars to use U.S. state laws to link fossil fuels and climate change (Hiltzik 2019). As these examples suggest, claiming the victim slot is sometimes a strategy used by those who are most culpable in the social struggle over climate change.

9.4 Fighting for the "right" to pollute

In 2012, the state of California launched a climate pollution cap-and-trade program as part of an initiative to lower its greenhouse gas emissions. Like similar programs in the European Union, the Republic of Korea, and Guangdong province in China, the state government decided to freely allocate a limited number of permits to discharge specific quantities of a specific pollutant per time period. Under the rules, polluters must have enough permits to cover the amount of their emissions. Polluters that need to increase their emissions can buy permits from other companies that are willing to sell their allotment or acquire them at permit auctions held every 3 months by the California Air Resources Board. As of 2018, permits were selling at around $14 per ton of CO_2 emitted. Each year, the number of new permits that are issued declines, creating an incentive for companies to reduce their emissions. When the program was first being introduced, the California Chamber of Commerce and the Pacific Legal Foundation sued the state to stop the auctioning of carbon allowances and use of the revenue to reduce climate pollution. As the Environmental Defense Fund (Morehouse 2016) characterized these ultimately unsuccessful lawsuits, they appeared to be "arguing that they have a right to pollute for free." In practice, this orientation, called "externalization," has long guided the ways companies have dealt with the toxic byproducts of their manufacturing, mining, transportation, and other waste-producing activities. Corporate externalization consists of three components: 1) avoiding significant business expenses (e.g., the cost of installing smokestack scrubbers that limit air pollution by removing harmful particles, gases, or chemical byproducts from industrial exhaust streams); 2) shifting environmental cleanup costs to tax-funded government bodies; and 3) downplaying or outright denying responsibility for the adverse health and other consequences of their pollution practices. Efforts to restrict corporate polluting are vigorously opposed despite the harmful consequences because these:

> are not the concern of the companies or the investors. [They] are the concern of the government. The government will be the one that will have to deal with the floods, the hurricanes, the droughts, the fires, the mudslides, the erosion, the rising seas, the displacement, the battle for resources, the heat-induced deaths, the starvation. (Nolan 2018)

These efforts by corporations and their political allies in California to block regulations on pollution and carbon emissions are not an isolated event. Political scientist C.J. Polychroniou (2018) reports that "[r]ecord amounts of money [have been] spent trying to derail the efforts of activists and labor unions fighting to save the environment." In Washington State, the oil industry invested over $30 million to oppose a ballot initiative that would have placed an initial (but subsequently rising) carbon fee of $15 per metric ton on carbon pollution, with the profits used to provide heating and electrical assistance to low-income families and income, re-training, pension, and health benefits to dislocated fossil fuel workers. Companies including ExxonMobil, BP, Chevron, and Shell banded together through the Western States Petroleum Association to inundate state residents with ads saying the pollution-control measure would lead to higher fuel costs for consumers. The initiative failed, and environmental activists learned that "fossil fuel companies still wield a great deal of power" (Anzilotti 2018). Efforts like this must be resisted over and over by environmentalists and common citizens because of the privileged "right to pollute" attitude of the corporate giants.

Expressing a usually unstated right to pollute by engaging in externalization is not the peculiar behavior of a small set of shady corporate entities, but rather is common among the most elite and wealthiest of companies. By any standard, 3M is elite and wealthy. Moreover, it is a multinational corporate giant with global reach (with operations in over 65 countries) and considerable political clout. It produces a very wide array of products, many of them well known to consumers, including tapes and glues, abrasives, laminates, passive fire-protection materials, personal protective equipment, dental and orthodontic items, electronic materials and circuits, medical products, dish-scrubbing sponges, car-care products, picture-hanging implements, healthcare software, and optical films. Headquartered in Maplewood, Minnesota, a suburb of St. Paul, 3M took in $31.7 billion in total sales in 2017, a year during which it ranked number 97th on the Fortune 500 list of the largest U.S. corporations based on total revenue.

3M's name comes from its original title and focus. Founded as the Minnesota Mining and Manufacturing Company in 1902, the goal of the new enterprise was to mine for corundum, a crystalline form of aluminum oxide used for making sandpaper and grinding wheels. But the company's mine only produced anorthosite, which does not make a useful abrasive. Nonetheless, the business survived, in no small part because of the demand for abrasives during World War I, and even expanded. In 1921, workers at the company invented a sandpaper with silicon carbide, which was used first in automotive paint refinishing, and later in 1925 in the development of masking tape, and 3M's fortunes began to soar.

Like other manufacturers, as its facilities and productivity have grown, 3M has had to deal with the byproducts, many of them toxic chemicals, of its industrial processes. How it has responded to this challenge is reflected in its handling of a contamination crisis involving the drinking water of tens of millions of Americans. The contamination involves two of 3M's nonstick and water-repellent chemicals, which have been found in the water supplies of over 1500 U.S. communities. Known as perfluorooctanoic acid and perfluorooctanesulfonic acid, these anthropogenic perfluorinated chemicals (PFCs) have been used for decades in household products such as Teflon and Scotchgard, as well as in outdoor clothing, food packaging (e.g., fast food wrappers and pizza boxes), and firefighting foam. Both take years to break down in the environment, earning them the nickname "forever chemicals." Both are known to accumulate in human bodies, where they have been linked to kidney and testicular cancer, hormonal problems, immune disorders, miscarriages, and other ailments. It is estimated that they can be found in the bodies of more than 99 percent of Americans today (Grandjean & Clapp 2015; Walker & Rundquist 2017).

Early in 2018, 3M settled a $5 billion-dollar lawsuit that had been filed against it by the state of Minnesota. The suit involved the company's responsibility for enormous amounts of waste containing perfluorinated chemicals that it had dumped in the state over a 40-year period. It charged that 3M "acted with a deliberate disregard for the high risk of injury to the citizens and wildlife of Minnesota when it dumped PFC-containing wastes into the Minnesota environment" (State of Minnesota 2017). The state maintained that 3M discarded these wastes into unlined pits and trenches, despite awareness since the early 1960s that the chemicals would pollute groundwater and since the 1970s that PFCs were widely present in the blood of the general U.S. population. In order to continue reaping billions of dollars in annual revenue from its sale of PFCs, the suit further alleged, 3M misled scientists who were trying to determine the source of PFCs in people's blood. The company went to great lengths to distort the broader scientific community's understanding of the serious health effects posed by PFCs, including funding friendly research while simultaneously paying to ensure that less favorable studies were suppressed. Further, for decades, 3M failed to report legally required information regarding the adverse health effects of PFCs to the EPA. In settling the suit, 3M was able to avoid any admission of liability and, despite the multimillion-dollar settlement, its senior vice president for research and development, John Banovetz, publicly denied there was a PFC-related public health problem (Lerner 2018). In typical pernicious corporate speak intended to spin a negative situation as a positive one, Banovetz lauded 3M's "long history of environmental stewardship" (Lerner 2018).

While a major lawsuit for Minnesota, the settlement covered only one of over 35 PFC-related cases filed against 3M across the country alleging contamination of drinking water. Also, the company is facing strong interest and public demand for tighter state regulations. It has responded with "an aggressive campaign to stave off potentially billions of dollars in damages [including] putting its lobbying muscle to work in Washington and courting the support of state attorneys general as it faces potentially massive financial liability for toxic pollution" (Snider 2018a). Central to 3M's defensive campaign to fight off government action is a new industry advocacy group called the Responsible Science Policy Coalition (RSPC), which embraces the argument that PFC chemicals do not pose a health threat. The group first came to light in July 2018 at a meeting of the Council of Western Attorneys General where it expressed a commitment to helping states with their environmental problems. In its presentation, the RSPC argued that there were many problems with the research showing a causal link between PFCs and health problems. It also submitted a comment signed by James Votaw, a lawyer for Keller and Heckman, an anti-regulatory industry service law firm, raising hackneyed industry complaints about inconvenient science, such as the need for more "high-quality" studies, better peer review of scientific research, and an increased role for outside "experts." In fact, the RSPC comprises 3M, Johnson Controls (a multinational Ireland-based manufacturing company), and other as yet unnamed corporations (Halpern 2018). In other words, it is a front group that claims autonomy but is anything but independent of corporate interest. As Halpern (2018) indicates, groups like the RSPC "principally exist to create doubt and confusion about the state of the science to avoid regulation of the products their members create." Such entities tend to borrow heavily from the playbook the tobacco industry developed to muddy the science linking cancer to smoking and, more recently, from the campaign by the fossil fuel industry to spread confusion about the causes of climate change, as discussed later. Help also from the Trump administration, which tried, unsuccessfully, to block a federal study that found PFCs pose a health danger at far lower levels than the EPA had previously said were safe. Recognizing that the study would be "a public relations nightmare" for 3M, the White House sought to keep it under wraps (Snider 2018b).

3M's political action committee (3M Company PAC) gave money to various candidates running for state-level offices in 2018, including attorney general candidates in Michigan, Ohio, California, and Alabama (Snider 2018a). All of these states have major PFC contamination problems. In Alabama, where the chemicals made by 3M's manufacturing plant in Decatur have contaminated the Tennessee River, a source of drinking water for over 4.5 million people, including some of the state's poorest communities, the newly reelected Republican attorney general, Steve Marshall, refused to join West Morgan and East Lawrence Water and Sewer Authority's lawsuit against the company. Marshall had received $2000 from 3M Company PAC in support of his primary campaign. 3M's multipronged initiative to avoid bearing the costs of its history of pollution reflects not just a "right to pollute" attitude, but also a willingness to fight by any means necessary to protect that right.

9.5 Deadly business: Big Energy and the denial of climate change

Viewed from the perspective of environmental and social sustainability, it is distressing to realize how far extremely wealthy corporations and individuals will go, within and beyond legal boundaries, to guarantee a maximization of wealth. One framework through which to assess this behavior involves an eco-justice perspective. Within the sphere of criminal justice, there has emerged, despite "at times ferocious opposition from powerful business and state interests," an international green criminological movement "towards viewing environmental harm as a serious social and ecological crime" (White 2016, p. xvi). "Environmental crime" refers to misconduct related to the pollution of the air, water, and land and perpetrated against wildlife. Some such practices are targeted by an array of international protocols and conventions (Forni 2010). Transgressions are harmful to humans, environments, and nonhuman animals regardless of their legality within a particular legal system (as wrongdoing is often perpetrated by the very nation-state systems that make the laws). Environmental harm is a highly contested arena because the behaviors involved are claimed by their perpetrators to be legitimate, safe, and lawful. They achieve this through "a combination of embedding harmful practices into everyday activities (such as animal food production), pervasive propaganda efforts about the value of certain types of environmental . . . exploitation . . . and political manoeuvring around and manipulation of legislation, regulations and rules that allow the destruction and degradation of the environment" (White 2013, p. 1). The legal construction of harm involves several conceptions. The term *malum prohibitum* refers to conduct that is prohibited by law but is not deemed as being a highly serious offense. Many environmental harms fall into this category because the benefits of the practices (e.g., mass trawling for eatable fish) overwhelm recognition of the consequences (i.e., ocean depletion of species). These behaviors are often addressed legally through regulation rather than eradication. The term *malum in se* refers to behavior that is inherently wrong and is seen as serious and in need of legal prohibition. An example is the banning of ozone-depleting substances under the Montreal Protocol.

In the effort to protect and expand wealth, a playbook of duplicitous tactics has evolved among polluters and their political allies, aimed at allowing them to block or eviscerate regulations and escape the enforcement of outright prohibitions. The term "playbook" refers to a set of strategies (plays, as in football) for getting things done (e.g., beating a rival team). The kind of concern here, which has been used in various corporate domains over many years, is intended to protect the privilege of the powerful over the interests of the vast majority. This disinformation playbook was initially developed by the tobacco industry beginning in the 1950s, and has since been adopted and adapted by the pharmaceutical industry (in response to the opioid crisis), the chemical industry (to defend PFAS, chromium, and other chemicals), the National Football League (around the issue of concussion), and fossil fuel manufacturers (Michaels 2019). In each instance, these industries have taken the legal principle (which applies to people, not products), established in 1895 in the Supreme Court case of *Coffin* v. *United States*, that a defendant is innocent until proven guilty, and stood it on its head. While from a caution-driven health perspective a reasonable standard is that products should be considered potentially dangerous until proven safe, these industries have attempted to make their dangerous products appear to be safe while insisting that any harms they might cause be proven beyond a shadow of a doubt before they will take steps to protect public health.

According to the Union of Concerned Scientists (2017), the most often used plays in the industry disinformation playbook are: "the fake" (passing off counterfeit science as legitimate); "the blitz" (harassing scientists whose research exposes risk); "the diversion" (manufacturing uncertainty); "the screen" (buying credibility); and "the fix" (manipulating public policy); to which we might add "the wall" (hiding in plain sight). Michaels (2019, p. 2) notes that:

> The industry defending its deadly product will commission scientific studies with flawed methodologies, guaranteed to generate predetermined results. It will demand the raw data from studies done by independent scientists [and] then re-analyze the data to reach different conclusions—alchemy that is easy for any experienced if unethical data cruncher. It will hire scientists, often way outside their fields of expertise, to make pronouncements that their product is safe, or that there is not enough evidence to take action. It will plant articles, often ghostwritten, in legitimate scientific journals, which selectively promote the results of studies that favor their product, while underreporting results showing harm. These scientists

will publish studies in "vanity journals," essentially of their own creation, but which easily fool the public, the press, and policymakers. The industry will hire opinion leaders and publish op-eds under their names, and create front groups to make it look like there is widespread opposition to controlling the exposure in question.

Specific components of the climate change denial playbook utilized by the fossil fuel industry to erect a "denial regime" (Derber 2010), also discussed in Chapter 8, include:

- Demeaning of climate science and climate scientists.
- Sowing public confusion about the reality of climate change through the dissemination of unscientific information.
- Denying inside company knowledge of climate change impacts.
- Funding professional climate change denier organizations and individuals.
- Financially supporting the political campaigns of climate change denier politicians.
- Promoting and applauding the appointment of climate change deniers to public offices and cutbacks on the regulation of environmental degradation.

Active in implementing the climate change denier regime are three intertwined groups: Big Energy corporations, conservative politicians, and an array of think tanks, policy institutes, and media outlets. Prominent among these is ExxonMobil, the largest direct descendant of John D. Rockefeller's Standard Oil Company and among the largest and most profitable publicly traded energy companies in the world (Poole 2018). This massive multinational corporation was formed in 1998 with the merger of Exxon, the world's biggest energy company and a prominent member of the rogue's gallery of environmental polluters since the colossal 1989 *Exxon Valdez* oil spill in Alaska, and Mobil, the second largest oil and gas company in the United States. Derber (2010) identifies two phases in the history of this climate change denier powerhouse: 1) claiming global warming is a hoax; and 2) admitting global warming is real, denying its urgency. Since his book, *Greed to Green*, was published, a third phase has begun: 3) arguing we're all in it together (i.e., denial of particular responsibility).

9.5.1 Phase I: claiming global warming is a hoax

Exxon entered full force into the denial funding business before its merger with Mobil in the period after 1988 and the formation of the IPCC. Created to provide global policy-makers with regular scientific assessments on climate change, over time the IPCC increasingly was able to affirm global warming, its anthropogenic causes, and its multiple consequences—news seen as threatening by Big Energy megacorporations. Exxon's own scientists had predicted that the day would come when carbon-related global warming would reach the point where world governments would seek to control the release of greenhouse gases into the atmosphere. Establishment of the IPCC was a sign that the moment had arrived, and the rollout of headlines announcing IPCC assessments of global warming and the role of emission releases in this adverse planetary change affirmed this was the case. In response, instead of accepting the findings of climate change science and vigorously investing in the development of sustainable energy sources, Exxon and later ExxonMobil took the opposite path by aggressively developing new sources of fossil fuel and helping to assemble the Global Climate Coalition (GCC), a lobbying program backed by other Big Oil and Big Auto corporations intent on blocking government controls on carbon release (Wieners 2014). Frumhoff *et al.* (2015, p. 162) report that "[f]rom 1989 to 2002, the GCC led an aggressive lobbying and advertising campaign aimed at achieving these goals by sowing doubt about the integrity of the IPCC and the scientific evidence that heat-trapping emissions from burning fossil fuels drive global warming. They worked successfully to prevent the U.S. from signing the Kyoto Protocol after it was negotiated in 1997." Further, in the years between 1988 and 2005, ExxonMobil poured over $16 million into the coffers of climate change denier groups that either were already established or were newly emerging (Frumhoff *et al.* 2015). *Mother Jones* magazine "tallied some 40 ExxonMobil-funded organizations that either . . . sought to undermine mainstream scientific findings on global climate change or . . . maintained affiliations with a small group of 'skeptic' scientists who continue[d] to do so. Beyond think tanks, the count also includes quasi-journalistic outlets like Tech CentralStation.com . . . a FoxNews.com columnist, and even religious and civil rights groups. In total, these organizations received more than $8 million between 2000 and 2003" (Mooney 2004). Additionally, translating its considerable economic power into political power, the company lobbied the administration of President George W. Bush to remove top climate scientists from leadership roles in the IPCC and the U.S. National Climate Assessment and to promote federal policies designed to strengthen the United States' reliance on fossil energy.

In 2007, Kert Davies, a climate activist and the founder and director of the Climate Investigations Center, wrote an Internet post in which he asked: "Why would Exxon pay a known global warming denial scientist to spread doubt about the impact of global warming on polar bears? . . . More to the point, why would Exxon want to keep the public from connecting fossil fuel combustion and greenhouse gas emissions to the bear's demise?" (Davies 2007). The answer had to do with the emergence of this species as a symbol in the public imaginary. During the first decade of the 21st century, polar bears—which had evolved to flourish on Arctic ice—became a poster child of global warming. Exemplary was a well-watched *National Geographic* video of a severely emaciated polar bear with the subtitle, "This is what climate change looks like." For this reason, countering the idea that polar bears were at risk became a mission for climate change deniers. To overturn the imagery by implementing "the fake" play (passing off counterfeit science as legitimate), Markus Dyck, Timothy Ball, Willie Soon, David Legates, and Sallie Baliunas, all vocal climate change deniers, published a commentary in the journal *Ecological Complexity* (Dyck *et al.* 2007) in which they claimed that spring air temperatures around the Hudson Bay basin for the period 1932–2002 show no indications of a warming trend. Consequently, they concluded, "[a]ny role of external forcing by anthropogenic greenhouse gases remains difficult to identify. We argue, therefore, that the extrapolation of polar bear disappearance [due to global warming] is highly premature" (Dyck *et al.* 2007, p. 73). The writing of the paper, which was an opinion piece that was not subject to scientific peer review, was funded by several energy companies and associations, including ExxonMobil Corporation.

The assertions made by Dyck and colleagues were quickly refuted by researchers at the U.S. Geological Survey and the National Snow and Ice Data Center, along with other scientists who maintained that "[l]ong-term trends in the population ecology of polar bears in western Hudson Bay in relation to climate change . . . continues to be consistent with the thesis that climate warming in western Hudson Bay is the major factor causing the sea ice to breakup at progressively earlier dates, resulting in polar bears coming ashore too fast for several months in progressively poorer condition, resulting in negative effects on survival of young, subadult, and older (but not prime) adults and reproduction" (Stirling *et al.* 2008, p. 193). The effort to deny this response was shown to be riddled with errors and omissions—a common refrain among scientists who have examined the speeches and texts produced by climate change deniers (e.g., Hunter *et al.* 2010).

Climate scientist Michael Mann, who has been at the forefront of sounding the alarm on climate change, published a paper in 1998 on global temperature patterns and climate forcing during the last six centuries, which contained what has come to be called the "hockey stick graph," a representation of data suggesting an unprecedented level of global warming in the 20th century (Mann *et al.* 1998). The IPCC prominently featured the hockey stick graph in its 2001 report on global climate change and linked warming to fossil fuel emissions. Publicity about Mann generated by this report triggered a vitriolic response from the climate denial regime (Mann 2013). Science educator Bill Nye (2013, p. xiii), popularly known as the Science Guy, summarized some of what Mann has been subjected to over the years for publishing his now repeatedly validated findings:

> Have you ever found a truck parked outside your home with a satiric cartoon of yourself on the side, put there by people who plain hate you? Have you ever had someone threaten you and your family's lives with a mysterious powder in an envelope sent in the mail? Have you had your livelihood threatened by the largest corporation on the planet? Dr. Mann has endured all of this because some people just can't bring themselves to accept the science in their midst.

This litany of intended punishments and intimidations is an expression of "the blitz" play (i.e., harassing scientists whose research exposes risk). Not mentioned on Nye's list are the thousands of hate-filled and threatening emails Mann received, or his being accused of using Stalinist tactics by a Congressman, Rep. Dana Rohrabacher, a supporter of the assertion that climate change is "liberal claptrap" (quoted in Mooney 2004)—who later was defeated in his run for re-election in 2018 by a candidate who supports prioritizing green energies over fossil fuels.

The blitz against Michael Mann also involved a well-funded campaign to discredit the validity of his work, including attempts to gain access to his private correspondence using Freedom of Information Act (FOI) requests. The effort was initiated in 2010 by Virginia Attorney General Ken Cuccinelli, 5 years after Mann moved from the University of Virginia to Penn State University. Citing Virginia's Fraud Against Taxpayers Act, Cuccinelli attempted to bully the University of Virginia into releasing documents related to Mann's work because he viewed Mann's published scientific papers' support for the existence of global warming as a possible swindle of the Virginia taxpayer. State courts ultimately ruled that Cuccinelli's effort was an abuse of power. An editorial published by the journal *Nature* (Nature Editiorial 2010) called Cuccinelli's tactic "an ideologically motivated inquisition that harasses and intimidates climate scientists." Cuccinelli subsequently ran for the governorship of Virginia but was defeated by Terry McAuliffe, who as governor responded to the recognized dangers of climate change through a number of executive actions, including joining the U.S. Climate Alliance. Cuccinelli was later appointed by President Trump as acting Director of the Citizenship and Immigration Services.

As Cuccinelli's case was wending its way through the courts, the American Tradition Institute (ATI) (subsequently renamed the Energy and Environmental Legal Institute)—a conservative group that describes itself as a watchdog on environmental regulation—sought the same documents from the University of Virginia by filing with the Virginia FOI. As an analysis of ATI's tax forms for 2010 reveals, the group received almost 25 percent of its funding from its sister group, American Tradition Partnership, an entity funded by the fossil fuel industry, as well as a \$15 000 contribution from the Atlas Economic Research Foundation, a Virginia-based think tank that accepted over \$1 million in funding from ExxonMobil beginning in 1998 (DeSmog 2015). Like the Cuccinelli effort, ATI's FOI gambit was blocked by the courts.

Another play utilized by ExxonMobil is "the screen." In football, the goal of a screen is to fool the defense into thinking that a long pass is about to be thrown when what is actually planned is a short pass to a receiver protected by a screen of blockers. In the corporate playbook, the screen involves hiding or diminishing public focus on deceptive practices by establishing credit-building alliances with prestigious academic institutions or professional societies. In the case of ExxonMobil and climate science, one way the screen was implemented was through the corporation's sponsorship for many years of the annual meeting of the American Geophysical Union (AGU), one of the nation's most prominent scientific associations. The AGU describes its mission as advancing and communicating science "and its power to ensure a sustainable future" (AGU 2019). Between 2001 and 2015, ExxonMobil donated over $620 000 to the organization, paying $35 000 in 2015 to sponsor a student breakfast at its fall meeting, the world's largest gathering of Earth and space scientists. As the Union of Concerned Scientists (2017) asserts, "[e]ven though these funds only represent a small percentage of AGU's budget, the sponsorship allows the company to recruit future employees under a veneer of scientific credibility, despite having spent over 40 years deceiving the public about the risks of fossil fuel extraction on our climate." An open letter from over 100 AGU member scientists, including Michael Mann, appealed to AGU's president to end the association's relationship with the fossil fuel company: "AGU has established a long history of scientific excellence with its peer-reviewed publications and conferences, as well as a strong position statement on the urgency of climate action. But by allowing Exxon to appropriate AGU's institutional social license to help legitimize the company's climate misinformation, AGU is undermining its stated values as well as the work of many of its own members" (quoted in Hirji 2016a). A trio of AGU members (Achakulwisut *et al.* 2016) added: "This issue has been an elephant in the room for too long. It's high time AGU protects the integrity of climate science and the message it sends to the public by cutting ties with a climate-denying corporation and by denouncing Exxon for its disinforming ways." After twice voting to continue its relationship with ExxonMobil, despite the objection of many of its members, the AGU's Board of Directors announced in 2016 that the fossil fuel giant had decided not to fund the association's annual conference (Hirji 2016b).

An unusual set of strategies in football, called "hide-a-player" techniques, includes a play known as "the wall," commonly associated with Auburn University head coach Gus Malzahn. The objective is to hide a player in plain sight behind a wall of their teammates. ExxonMobil adopted a version of this play to hide information it knew about climate change—what has come to be known as the "Exxon Knew" story. The starting point of this narrative, dating to the period before the merger with Mobil, was a set of studies by Exxon's scientists that accurately predicted rising temperatures and melting ice in the Arctic. Recognizing that these findings were not good for business, company spokespersons consistently denied or hid them. As the *Los Angeles Times* reported, "The gulf between Exxon's internal and external approach to climate change from the 1980s through the early 2000s was evident in a review of hundreds of internal documents, decades of peer-reviewed published material and dozens of interviews conducted by Columbia University's Energy & Environmental Reporting Project and the Los Angeles Times" (Jerving *et al.* 2015). The company's primary strategy was to downplay both the significance and the certainty of global warming. Its plan, hatched in alliance with the American Petroleum Institute and Chevron, included a multimillion-dollar, multiyear effort to conceal what company officials knew was happening with the intent of promoting doubt about climate change in the public policy arena. Targeted audiences identified for the denier message included the media, policymakers, and science teachers. The plan found expression, for example, in a speech given by Lee Raymond (ClimateFiles 2017a), the head of Exxon in China, at the World Petroleum Congress in 1997. During his talk, Raymond addressed three critical climate-related questions: Is the Earth really warming? Does burning fossil fuels cause global warming? and Do we now have a reasonable scientific basis for predicting future temperatures? The answers he provided were quite different from what he actually knew, based on the findings of Exxon's own scientists. He stressed, for example, that Earth was experiencing a cooling spell rather than an increasing warming trend. Further, he claimed that almost all greenhouse gases were caused by nature, not by human activities. Finally, he asserted that there was no scientific basis for predicting future temperatures. All of these points were reiterated in a pamphlet entitled "Global Climate Change: Everyone's Debate" that ExxonMobil issued in 1998. This document began with the specious statement: "Exxon shares with people around the world the goal of protecting Earth's Environment. We take seriously our responsibility to conduct our operations in an environmentally sound way. For that reason, for many years, we've carefully studied and worked to increase understanding of the issue of global climate change . . . It's one of the knottiest and most contentious scientific subjects" (ClimateFiles 2017b).

During the same period, ExxonMobil researchers and engineers were in fact incorporating their findings on melting and projections about future melting into their planning for drilling in the Arctic. A senior ice researcher at an Exxon Canadian subsidiary, Ken Croasdale, reported at an engineering conference in 1991 that greenhouse gases were rising:

> due to the burning of fossil fuels . . . nobody disputes this fact . . . Certainly any major development with a life span of say 30–40 years will need to assess the impacts of potential global warming . . . This is particularly true of Arctic and offshore projects in Canada, where warming will clearly affect sea ice, icebergs, permafrost and sea levels. (quoted in Jerving *et al.* 2015)

Croasdale noted the hazards of warming in the Arctic, including higher seas and larger waves, which could damage the company's existing and planned coastal and offshore drilling platforms, artificial islands, processing plants, pumping stations, and pipelines. His projections have proved to be accurate, with large losses of sea ice occurring just as his models indicated they would.

A geographer, Stephen Lonergan of McMaster University, whom Exxon hired to examine climate change in the Mackenzie River Delta, also came to an inauspicious conclusion that the company did not make public. Using NASA's Goddard Institute for Space Studies and Canadian Climate Centre climate models, Lonergan determined that the area would get considerably warmer and that this fact should not be ignored as it would sizably increase the cost of maintaining roads, pipelines, and engineering structures. Melting permafrost, he reported to the company, would damage buildings, processing plants, and pipelines. Flooding, he added, could be expected, as could more intense storms (Jerving *et al.* 2015). Lonergan realized that his report came as somewhat of a disappointment to his corporate funder, which chose not to present his findings to the public.

The actual risky state of climate change, well known to ExxonMobil-employed scientists, extended to issues far beyond the Arctic (Rich 2019). This conclusion is supported by a study based on quantitative consensus analysis of anthropogenic global warming in the scientific literature and document-by-document textual content analysis of peer-reviewed and non-peer-reviewed scholarly publications written by company scientists, including internal ExxonMobil documents and paid, editorial-style advertisements (called advertorials) the company published in *The New York Times* during the period 1977–2014 (Supran & Orekes 2017). The authors examined the thematic consistency of company communications concerning the state of climate science and its implications for human communities. Specifically, they compared the positions expressed in company communications on whether: climate change was real; it was caused by human activities; it should be responded to as a serious threat; and it was a solvable problem. Their analysis revealed that the clear and undeniable thrust of both peer-reviewed publications by ExxonMobil scientists and internal, nonpublic documents was that climate change was both occurring and caused by humans. Despite this internal knowledge, when ExxonMobil communicated with the public through its advertorials, it conveyed a message of scientific uncertainty about climate change. While accounting for expressions of reasonable doubt, Supran & Orekes (2017) found that 83 percent of peer-reviewed papers by company scientists and 80 percent of relevant internal company documents acknowledged that climate change is real and human-caused, yet only 12 percent of ExxonMobil's advertorials did so, with 81 percent expressing doubt about climate change, such as the following:

> Let's face it: The science of climate change is too uncertain to mandate a plan of action that could plunge economies into turmoil . . . Scientists cannot predict with certainty if temperatures will increase, by how much and where changes will occur. (quoted in Supran & Orekes 2017)

One of the cited advertorials published in *The New York Times* in 1997 referred to a "high degree of uncertainty," "debate," and a "knowledge gap," as well as the need for further "fact-finding" and "additional knowledge" before U.N. negotiators meeting in Kyoto could make decisions. It stressed the goal "of achieving a consensus view" among scientists before any actions should be taken—2 years after the IPCC had already shown that consensus was extremely high.

With regard to the question of whether ExxonMobil misled nonscientific audiences about climate science, Supran & Orekes (2017, p. 12) find that their analysis "supports the conclusion that it did." Further, they point out that "ExxonMobil's advertorials included several instances of explicit factual misrepresentation" (Supran & Orekes 2017, p. 13). They view their findings as contributing to our understanding of misinformation strategies and the social construction of ignorance. As Supran (quoted in A. Kaufman 2019) notes: "The evidence points only one way, that these companies and trade associations funded misinformation to stifle policymakers."

These conclusions support an investigative probe, begun in 2016, by New York Attorney General Eric Schneiderman's office, which issued multiple subpoenas to ExxonMobil under the state's Martin Act alleging that the company's accounting of climate risk "may be a sham" (Larson 2017). ExxonMobil's effort to block the investigation were stymied when the U.S. Supreme Court declined to take up its case in 2018. Freelance journalist Steve Hanley (2019) addresses what will happen next:

> ExxonMobil will slow walk the release of documents as only a group of $1000 an hour attorneys can do. It will fight, it will feint, it will twist and turn itself into a pretzel to avoid having its innermost secrets made public . . . Every lawyer knows that delay benefits one of the parties to every lawsuit. Every day the company can delay this process is another day it can continue to rape the Earth in pursuit of profits. Over time, politicians and judges change. Maybe it will get a better deal in the future. Maybe it can spend billions trying to elect people more sympathetic to its position. In the meantime, it will continue to pump oil as fast as it can find it

> on the misguided basis that it owes nothing less to its stockholders.

As aptly summarized by Lynn (2019):

> The fossil fuel industry's role in actively preventing effective global climate action has become increasingly clear with the publication of books like Nathaniel Rich's *Losing Earth* and Naomi Oreskes and Erik M. Conway's *Merchants of Doubt*. For about half a century, fossil fuel corporations manufactured doubt, attacked science and bought political loyalty. They effectively immobilized political will to implement any kind of legislation that would threaten their profits.

9.5.2 Phase II: admitting global warming is real, denying its urgency

As damning exposés in the mass media on the role of ExxonMobil in denial funding began to multiply, the company decided to reduce its contributions to the denier machine organizations, claiming to have completely ceased such funding in 2007 (Brulle 2014). In a corporate responsibility report, issued under the leadership of Rex Tillerson, the man Donald Trump subsequently appointed as Secretary of State, ExxonMobil (2007, p. 39) affirmed: "In 2008, we will discontinue contributions to several public policy groups whose position on climate change could divert attention from the important discussion on how the world will secure energy required for economic growth in an environmentally responsible manner." Acknowledging "the risks posed by rising greenhouse gas (GHG) emissions" (ExxonMobil 2010), the company further maintained that it had turned the corner and aligned its positions with the findings of climate science on global warming and carbon release.

Despite public statements about its acceptance of the reality and anthropogenic origins of climate change, in 2015 ExxonMobil was still funding denier organizations like the American Enterprise Institute (AEI) in the amount of \$325 000. The oil giant also continued to fund the American Legislative Exchange Council (ALEC) and the Manhattan Institute for Policy Research, as well as other active players in climate change denial and in other efforts to block legislation designed to protect the environment from corporate despoilment (Readfearn 2016a). Moreover, it began a campaign to undercut the 2015 United Nations Climate Change Conference in Paris by supporting preparation of a quickly debunked report that claimed the Paris accord would cost the U.S. economy almost \$3 trillion over several decades, while eliminating 6.5 million industrial sector jobs by 2040 (Kessler & Lee 2017). The report failed to mention the economic benefits to U.S. businesses that would accrue from building and operating extensive renewable energy projects. In 2017, ExxonMobil continued to make donations to climate science denier groups like the U.S. Chamber of Commerce, the Manhattan Institute, the National Taxpayers Union Foundation, the Washington Legal Foundation, and ALEC (Negin 2018). In fact, "ExxonMobil has continued to do so to this day, even as it places national television advertisements featuring attractive young scientists experimenting with green algae" (Rich 2019, p. 188). The company's 2019 grant-making report reveals that it gave \$690 000 to eight climate science denier groups during the year—only a 10 percent reduction from 2018. Moreover, it continued to fund federal lawmakers who oppose a carbon tax, despite its public support for the idea. Campaign contributions in the 2019–2020 election cycle have gone to 115 of the 150 climate science deniers still holding public office (Negin 2020).

During Phase II, ExxonMobil's commitment to corporate responsibility appears to have more or less remained what it always was: staying responsible to its big shareholders and making enormous profits in part by offloading significant production costs in the form of carbon release and environmental pollution. Applying "the fix" (manipulating public policy), an often-used strategy from the climate change denier playbook, ExxonMobil has donated over $2 million in campaign contributions since 2007 to more than 100 members of Congress who deny climate change, such as Texas Senators Ted Cruz and John Cornyn (Banerjee 2016). Cruz is noteworthy as a politician because the oil and gas industries have been his biggest corporate financial supporters since he first campaigned for the Senate in 2012. In 2018, when Cruz only narrowly was re-elected to his seat in the Senate (by 2.6 percent), his campaign was the largest recipient of oil and gas industry funding in the nation (OpenSecrets 2018). Moreover, at least a quarter of his personal wealth comes from investments in fossil fuel company stock. His 2016 financial disclosure report shows that he was given a $30 000 charter flight from Houston to the Cayman Islands for his family by Michael Bleyzer, president of the private-equity investment firm SigmaBleyzer, who had previously held executive positions with ExxonMobil (Downs 2016). As an elected official, Cruz has strongly supported policies that benefit the oil and gas sector, while opposing ones intended to develop renewable energy (Atkin 2015). For example, he was a vocal opponent of the "Exxon Knew" investigation, co-authoring a letter with four other Republican senators that demanded U.S. Attorney General Loretta Lynch drop any federal investigations into whether oil companies committed fraud in past statements by downplaying what they knew about climate change (Lee *et al.* 2016). At one time or another, Cruz has publicly given voice to all of the key (and thoroughly discredited) climate change denier talking points (Readfearn 2016b), including claiming "[c]limate change is not science. It's religion" (quoted in Page 2015). This assertion is rather ironic in that for adherents like Cruz, for whom no elements of the ideology of climate change denial are subject to rejection in the face of contrary evidence, it is actually denialism that is most like a religion. Hypocritically, despite being a climate change denier and an outspoken critic of federal spending, Cruz supported using taxpayer funds to build a nearly 60-mile stretch of concrete seawalls and other barriers on the Texas Gulf Coast to protect the world's largest concentration of petrochemical facilities from sea rise and other effects of climate change. He referred to the fuel industry-led initiative as "a tremendous step forward" (quoted in Weissert 2018).

In 2009, at the Woodrow Wilson International Center for Scholars, Rex Tillerson, then head of ExxonMobil, gave a speech in which he announced that his company supported a carbon tax to help fight climate change. He stated (quoted in Samuelsohn 2009) that, "[a]s a businessman, it's hard to speak favorably about any new tax, but a carbon tax strikes me as a more direct, transparent and effective approach" than the cap-and-trade system that was supported by then President Barack Obama and Democratic Congressional leaders. This speech is often pointed to as the fork in the road that began ExxonMobil's acceptance of climate change and the need to do something about it. Why favor a carbon tax? Unlike cap-and-trade, which puts an increasingly lowered ceiling on overall carbon release, a carbon tax makes corporations pay a fee, but if—through lobbying—the tax rate can be set low enough, energy companies can keep expanding endlessly and releasing carbon as a byproduct. At a point, with the election of Obama, when political pressures were mounting for a policy response to climate change, the carbon tax looked like the better option for elite polluters.

In fact, over time, Tillerson has walked a crooked line on the issue of climate change, sometimes admitting the planet is warming, sometimes—like at the 2013 company shareholders meeting—incorrectly asserting that science really does not understand how climate conditions are changing (Bradley 2013).

Meanwhile, under his leadership, the company did not change its pattern of contributing heavily to greenhouse gas emission. Instead, "his actions as chief executive . . . continued to ignore the climate threat, per se, and to view any regulatory or legislative effort to curtail carbon emissions as a threat to the corporate profit margin and to be fought against" (Polansky 2017). In the view of Kert Davies, under Tillerson ExxonMobil "deliberately changed their stripes on climate, but it was all P.R." (quoted in Schwartz 2016). Ultimately, the cap-and-trade bill was unsuccessful. While it passed the House in 2009, it failed to reach the Senate floor. There were a number of reasons for its failure, but among them was the intense opposition to it by energy companies, including ExxonMobil. Nor was lobbying during Tillerson's reign limited to averting cap-and-trade. It also involved (unsuccessful) opposition to the Dodd-Frank Wall Street Reform and Consumer Protection Act, which included (in Section 1504) anti-corruption legislation that would require oil, gas, and mining companies to declare the payments they made to foreign governments, including bribes. Why would ExxonMobil and its lobbyists at the American Petroleum Institute spend 8 years vehemently fighting passage of an anti-corruption law? Because, according to Global Witness (2018), it did not want to have to disclose its complicity in the corruption of Liberia's oil sector. Ultimately, ExxonMobil skirted U.S. law by inserting a Canadian company (Canadian Overseas Petroleum Ltd.) between itself and crooked Liberian politicians, thereby acquiring a block of land for oil drilling without having to expose its involvement in corrupt practices that fleeced the Liberian people of the benefits of their natural resources.

9.5.3 Phase III: arguing we're all in it together

In Phase III, ExxonMobil's (2019a) guiding slogan is: "We believe that climate change risks warrant action and it's going to take all of us—business, governments and consumers—to make meaningful progress." Stressing that there is no single solution to the risks of climate change, and that no single entity or organization can handle the challenge on its own, ExxonMobil is proud to say it is committed to doing its "part" (Woods 2018). It is the company's public position that humanity faces a two-sided problem: the issue of climate change and the need for energy development to support "human progress" in improving global living standards. In order to make "meaningful progress," ExxonMobil sees its role as being to produce the energy and other products the world needs in a responsible manner. This responsibility does not, apparently, mean moving as swiftly as possible to sustainable green energy systems, but rather making gradual improvements in the efficiency of fossil fuel-based oil, natural gas, plastic, and other petrochemical production. Ignoring the full story of its history on the issue of climate change, the company asserts it has been addressing these problems for decades: "ExxonMobil scientists have been involved in the forefront of climate research for four decades, understanding and working with the world's leading experts on climate" (ExxonMobil 2019a). Moreover, it has begun investing in alternative energy research, such as producing algae that are engineered to bloom into biofuels, biodiesel made from agricultural waste, and carbonate fuel cells that turn emissions into electricity. According to Vijay Swarup, Vice President of Research and Development, however, any commercial breakthrough on these projects is unlikely until at least 2027. In a somewhat ironic twist, the company emphasizes that its approach to energy alternatives differs from such initiatives at other energy producers because its focus is on science (Hirtenstein 2017).

Sidestepping the U.N. report warning that world governments have about a decade to cut global carbon emissions by half or they will face planetary warming that climbs above 2.7 °F, causing cataclysmically destructive

Fig. 9.2 Oil rig. *Source:* jp26jp/Pixabay.

extreme weather, droughts, and flooding (United Nations 2019), ExxonMobil announced in a press release that it had doubled its production in the Permian Basin area in West Texas in 2018 and had plans to "increase" and "accelerate" its output to 1 million barrels per day by 2024, an increase of almost 80 percent. It also announced that it was actively building infrastructure to support a growth in oil production (ExxonMobil 2019b). It had 48 drilling rigs in operation in the Permian Basin in 2018 and planned to increase the count to approximately 55 by the end of 2019 (Fig. 9.2). Among the various oil companies drilling in the Permian Basin, there are plans underway that will add the equivalent of 1000 coal plants' worth of climate pollution to the atmosphere by 2050.

This addition alone would make achieving the emissions cuts called for by the United Nations nearly impossible (A. Kaufman 2019). But ExxonMobil is also planning a second company growth initiative called Growing the Gulf (ExxonMobil 2019c), begun in 2013 and expected to continue through at least 2022, which includes 11 major chemical, refining, lubricant, and liquefied natural gas developments. ExxonMobil views these ventures as "export machines" that will significantly enhance the company's export potential and profitability. Across the board, in fact, it continues a business-as-usual (i.e., endless growth) approach, with the frequent announcement of the construction of new refineries (e.g., Beaumont, TX), the acquisition of new acreage for oil exploration (e.g., Namibia and Argentina), and the discovery of new oil and natural gas reserves (e.g., Pakistan, Guyana, Cyprus, Mozambique, Papua New Guinea, and Brazil). In 2018, ExxonMobil publicized "an aggressive growth strategy" through 2025, including 25 startup projects worldwide, intended to increase company earnings by more than 100 percent to $31 billion a year (ExxonMobil 2018a).

In the end, there are critically important ways in which we are not all in it together. The health-damaging impacts of ExxonMobil's contribution to climate change and the production of other pollutants are not shared equally. These effects fall disproportionately on poorer nations and poorer populations, which generally contribute the least to fomenting ecocrises and have the fewest resources to protect themselves from environmental threats. About 25 corporations have been responsible for half of all carbon emissions since 1988, and ExxonMobil is among them (Griffin 2017). For its part, ExxonMobil—like most energy companies today—is investing a comparatively small amount of its earnings into future alternative energy development and huge amounts into expanded fossil production. The important story in Phase III is that despite touting its new climate change consciousness, "Big Oil like ExxonMobil is continuing to expand operations . . . thereby exacting an enormous toll on humanity and the environment—from air and water pollution to global warming" (Fortuna 2018).

The pattern of duplicity around the issue of pollution by the fossil fuel industry can be found in National Oceanic and Atmospheric Administration (NOAA) Earth Observation Group satellite data from the Permian Basin. The Environmental Defense Fund analyzed these data for the year 2017 and found that Permian operators burned 104 billion cubic feet of natural gas—4.4 percent of all gas produced—but the fossil industry only reported 55 billion cubic feet to the Texas agency responsible for monitoring this practice (Leyden 2019).

During 2017, the Trump Administration entered into a settlement with ExxonMobil to resolve its violations of the Clean Air Act at its eight chemical plants in Texas and Louisiana. While Trump promoted this case as an example of his commitment to environmental law enforcement, as Bakeyah Nelson, Executive Director of Air Alliance Houston, argued: "[t]his settlement follows the current administration's pattern of leniency toward polluters and highlights its disregard for public health . . . Exxon Mobil is a historic repeat violator of the Clean Air Act and was among the worst emitters during Hurricane Harvey" (quoted in Environmental Integrity Project 2017). In the settlement, ExxonMobil was fined $2.5 million for emitting 6000 tons of illegal carcinogenic volatile organic compound (VOC) pollution annually that was never reported as required by law. The action required to correct the violation—installation of flare gas recovery systems and introduction of operation and maintenance procedures designed to increase flare efficiency—would later be claimed by ExxonMobil (2018b) as part of its contribution to protecting the environment.

9.6 The politics of climate change denial

According to Bob Reiss, an author and journalist who has written 14 novels and three nonfiction books:

> I think the important thing that I'm always reminded of when I talk to global warming scientists is that very few of the ones I speak to ever had any belief in global warming, one way or the other, before they started their work. When I give speeches, people are always attacking these scientists, and I find the scientists to be apolitical creatures who are interested in basic science and who are often as surprised as the rest of the people when they get their results. I don't really find that a lot of them have political agendas, certainly before they start their research, and that is more of a surprise than anything else, because people who are not scientists are always pointing their fingers at them saying they have preconceived ideas or that they are setting out to prove one thing or another. And they're not. They're just doing basic research. (quoted in Gambino 2010)

One of the charges made by climate change deniers is that climate scientists are motivated by money to assert the climate is adversely changing and that human actions are the cause. Rick Santorum, the former Republican senator from Pennsylvania, former presidential candidate, and TV pundit, for example, has stated: "If we didn't have climate change we'd have a lot of scientists looking for work. The reality is that a lot of these scientists are driven by the money that they receive" (quoted in CNN 2018). Similarly, Tom DeLay, a Republican from Texas who was the former House majority leader, asserts with reference to the 2018 U.S. Climate Change Report (discussed in Chapter 8): "The report is nothing more than a rehash of age-old 10- to 20-year assumptions made by scientists that get paid to further the politics of global warming" (CNN Transcripts 2018). In fact, participating scientists work on the Climate Change Report voluntarily and are not compensated. Further, "[r] esearchers working in climate change do not receive atypically large paychecks, nor do they strike it rich from grants . . . Professors at public universities who teach earth sciences and environmental studies generally earn more than their peers in humanities and social sciences, but less than faculty in the economics, business and law departments" (Qui 2018).

From a political standpoint, as Nathaniel Stinnett (2019), founder and executive director of the Environmental Voter Project, a nonpartisan nonprofit that uses data analytics and behavioral science to promote voting among environmentalists, claims that: "The real trouble lies with what I call climate liars: the thousands of politicians who secretly accept climate science, but publicly deny it for political gain. If the climate movement is going to prevail, we must be crystal clear about this distinction." According to Stinnett, most anti-climate politicians only pretend to deny climate change because it helps them win elections with conservative voters. The climate movement's real problem, he maintains, is not a lack of scientific acceptance among many politicians but a lack of political power sufficient to sway elections. Hence, the Environmental Voter Project. While many Republicans once publicly discussed climate change as a problem, everything changed with the Supreme Court's Citizens United decision in 2010, which held that the free speech clause of the First Amendment prohibits the government from restricting independent expenditures for political communications by corporations, including nonprofit corporations, labor unions, and other associations. This landmark opened the floodgates of fossil fuel industry money to conservative politicians and "turned climate politics upside down" (Stinnett 2019). As Rhode Island Democratic Senator Sheldon Whitehouse notes: "with its Citizens United weaponry . . . the fossil fuel industry snuffed out Senate bipartisanship on climate change" (quoted in Stinnett 2019).

9.7 The institutions of the climate change denial machine

The 70-year-old TV program "Meet the Press," now hosted by Chuck Todd, is the most-watched Sunday morning public affairs broadcast in the United States. On November 25, 2018, 2 days after the U.S. federal government released the U.S. Climate Assessment, "Meet the Press" focused on the issue of climate change. Included as a guest was Danielle Pletka, vice-president for foreign and defense policy studies at the Washington, D.C.-based AEI. On the telecast, she stated:

> From the standpoint of those who have doubts about this, and I don't think we can have doubts that there is climate change, whether it's anthropogenic I don't know. I'm not a scientist. I look at this as a citizen and I see it so I understand it. On the other hand, we need to also recognize we had two of the coldest years, biggest drop in global temperatures, that we have had since the 1980s, the biggest in the last 100 years. We don't talk about that because it's not part of the agenda The United States has been dropping in CO2 emissions since we pulled out of Paris [i.e., the Paris Climate Agreement]. There are actually good things that are happening.

> We are not using dirty coal anymore. It's the Europeans who are using dirty coal. There actually is some corporate leadership on this. Yes, we need to deal with these problems. Yes, we need to mitigate the things we see. But, we should not be hysterical. (video available in Sinclair 2018)

An insightful assessment of this statement is presented by Hank Green (2018), a well-known vlogger, author, and climate conscious educator, on his YouTube channel SciShow, which features videos focused on several scientific fields, including climate science. Green's objective with SciShow is to dispel the impression that science is an inherently difficult subject—or, in the case of Danielle Pletka's comments on "Meet the Press," to show through a line-by-line deconstruction of the newest and far more subtle public voice of climate science denial that in the name of profit-making, business-as-usual pollution of the environment is still the name of the game. He begins by informing his viewers about Pletka's employer, AEI. As a very large, ultra-conservative, and anti-government-regulation think tank, the mission of AEI is to use the money paid to it by corporations, wealthy individuals, and other deep-pocket entities to influence public policy through lobbying and to affect public opinion on climate-related issues and other topics of concern to its funders.

AEI, now considered as the dominant conservative think tank (Rubin 2014), has been doing this kind of work since 1938, when it was founded by Lewis Brown, former President of Johns-Manville, at one time the world's largest manufacturer of asbestos and asbestos products. In 1981, under Brown's leadership, the company reorganized in order to try and insulate itself against lawsuits from employees suffering from asbestos-related health problems (detailed in Chapter 6). Facing unprecedented record liability for asbestos injury claims, the company filed for bankruptcy in 1982. At the time, court papers indicate that Brown told the board that the directors were to maintain the utmost secrecy about the real reasons for the reorganization. In testimony given in a federal court (Castleman 1996) concerning a business meeting in the early 1940s between representatives of Unarco, a manufacturing company, and both Lewis Brown and Vandiver Brown of Johns-Manville, Charles Roemer (an Unarco employee) stated, "I'll never forget, I turned to Mr. Brown, one of the Browns made this crack (that Unarco managers were a bunch of fools for notifying employees who had asbestosis), and I said, 'Mr. Brown, do you mean to tell me you would let them work until they dropped dead?' He said, 'Yes. We save a lot of money that way'." Additionally, according to a lawsuit filed by George Rosenberg, a Los Angeles lawyer who represented asbestos victims, Johns-Manville's efforts to limit its liability included shopping for a favorable accounting opinion, suppressing studies on the health hazards of asbestos, issuing falsified study data that questioned the risks of asbestos exposure, maintaining a policy of not telling its employees they were suffering from asbestos-related diseases, and claiming that its former general counsel, Vandiver Brown, was dead when in fact he was alive and living in Scotland (Lewin 1983). These accusations were supported by numerous internal company memos, industry letters, and courtroom depositions (Rosenberg 2019). The company was forced to set up The Manville Personal Injury Settlement Trust in 1988 with an initial fund of $2.5 billion to settle asbestos-related claims against it (Mauney 2019).

While it claims to be nonpartisan, the policy positions of AEI almost always mirror those of the Republican Party. Thus, over the years, the organization has advocated for "lower taxes, fewer protections for consumers and the environment, and cuts to the social safety net" (Van Zandt 2018). In 2018, AEI reported having over 225 staff and an annual budget of more than $59 million (AEI 2018). One of the ways it spends its money is on the careful construction of messages for public presentation, such as through pundit comments on TV shows like "Meet the Press." Green (2018) notes that AEI "works very hard on messaging. They do . . . surveys to see what messages resonate most with people, so this is bound to be the best message they could find to work on the broadest number of people . . . Here we have the message they have calculated works best."

One of the components of Pletka's message on "Meet the Press" was the idea that we need to be cautious about thinking the world keeps getting warmer and warmer; after all, she asserts, in recent years we have had two of the coldest years and biggest drops in global temperatures in 100 years. But that is not true. There were no recent years that were the coolest in 100 years; on the contrary, the coldest years during that time span occurred before World War II. All recent years are the hottest in 100 years. The heatwave of 2018 made it a particularly hot year—the fourth-hottest on record (NOAA 2019). Moreover, climate scientists project this pattern will continue through the 2020s and beyond due to a low probability of intense cold events (Sévellec & Drijfhout 2018). Using a new method called PROCAST to predict global mean surface air temperature and sea surface temperature, Sévellec & Drijfhout (2018) suggest that at the global level the years 2018–2022 will be even hotter than previously expected based on the current rate of global warming.

Pletka's problematic claim was based on an article by libertarian journalist Aaron Brown (2018) published in Real Clear Markets under the headline, "Did You Know the Greatest Two-Year Global Cooling Event Just Took Place?" This article states that from February 2016 to February 2018, global average temperatures dropped 0.56 °C (1.008 °F) and that "[y] ou have to go back to 1982–84 for the next biggest two-year drop" of 0.47 °C (0.846 °F). While technically true, as pointed out by Gavin Schmidt, a climatologist and director of NASA's Goddard Institute for Space Studies, "it's just a statistical fluke that people have grabbed because it sounds good . . . It has no relevance for the long-term trends" (quoted in Kruzel 2018). The drop in temperature in 2016 was expected because at the end of the El Niño that year, which warmed the surface of the Pacific Ocean, things cooled down a bit.

Pletka went on to say that colder years are not talked about in the media because it is not "part of the agenda," implying that the media tends to adhere to an agreed plan to push the idea of climate change. Is there such an agenda? Pletka offered no evidence to back up her assertion, but a review by Media Matters of 127 media segments on the global heatwave of 2018 that aired on ABC, CBS, and NBC found that only one story, on "CBS This Morning," even mentioned the connection between climate change and extreme heat events. This finding follows a long-running pattern. As Media Matters, a not-for-profit progressive research and information center, points out: "during the height of hurricane season in 2017, neither ABC nor NBC aired a single segment on their morning, evening, or Sunday news shows that mentioned the link between climate change and hurricanes" (Macdonald 2018). If there was a media agenda on climate change, it would appear it was to minimize the cause of the broad pattern of changes in weather events occurring on Earth rather than to promote the idea of climate change.

Next, Pletka, with a wave of her hand to match the sleight-of-hand in her words, claimed that the United States had been dropping in CO_2 emissions since it announced it was pulling out of the voluntary Paris Agreement to lower global release of this greenhouse gas. Connecting these two events in one sentence implies there is a causal relationship. But as Green emphasizes, CO_2 emissions were dropping before the 2015 Paris Climate Change Conference—where the agreement was negotiated—was even convened. The fall began in 2008, when the U.S. economy entered into a major recession, production and fuel use declined, and over 6 million people lost their jobs. It has continued because of skyrocketing natural gas production, an increase in sustainable fuel facilities, and a resulting drop in the more expensive use of coal as an energy source.

On the heels of her suspect CO_2 comments, Pletka celebrated the fact that unlike in Europe, the United States' use of "dirty coal" has been dropping. Overlooked in this contention is the fact that on a per capita basis, European countries burn far less coal than the United States. In 2015, for example, coal consumption per person in the United States was 1.63 metric tons, as compared to 1.38 in Germany (Statista 2019). While the United States has and exploits large deposits of low-sulfur coal beneath its soil and there is much discussion of "clean coal" production, the question must be asked: Is low-sulfur coal "clean" coal, and are there commercially viable ways of actually stopping carbon emissions from coal-fired electricity plants? Calling somewhat cleaner types of coal "clean coal" is as misleading as saying that e-cigarettes are healthy because they are not as unhealthy as regular cigarettes. Further, efforts by the Obama administration to limit the amount of CO_2 that could be emitted by coal-fired power plants were strongly opposed by the coal industry and climate change denial groups like the U.S. Chamber of Commerce and the Heritage Foundation, and the subsequent Trump administration quickly moved to replace Obama's Clean Power Plan with weaker greenhouse gas rules for power plants.

Finally, having done her best to muddy the waters on climate change, Pletka ended by emphasizing that people should not get hysterical about it. This implies that strong concern about the threat of climate change is, somehow, a hysterical reaction. It bears mentioning that the word "hysteria" originates from the ancient Greek word for uterus and for a long time it was believed that hysterical symptoms were caused by defects of the womb, making hysteria a mental health problem limited to women. This idea is reflective of the underlying sexist cultural belief that women are excessively emotional and less capable of being reasonable than men. Implied in Pletka's advice is that we need to calm down about climate change. While advocating the need for organized, rational, and science-based aggressive collective action to address the growing adverse effects of climate change rather than a disorganized and frantic response, this book does not argue it is something we should calm down about, and for good reason: many lives depend on mitigating climate change.

9.8 Taking climate change deniers to court

While attempts have been made to use existing laws and the courts to silence climate scientists, generally without much success, the laws and courts have also been used to push deniers to prove their claims. A revealing example of this type of counterstrike was undertaken by Public Employees for Environmental Responsibility (PEER) in 2017, when it submitted an FOI request for scientific evidence that could support statements about carbon dioxide produced by humans not being a major factor in climate change made by then EPA Administrator Scott Pruitt during a March 2017 televised interview. The EPA dragged its feet for a year in responding to the request, leading PEER to sue the agency for failing to meet the legal standard for a timely response. The EPA lawyers argued that the evidentiary basis for Pruitt's statement was unknowable, that the EPA administrator's public statements were not a proper focus of a FOI request, that PEER had failed to reasonably describe the records it sought, and that finding and handing over the relevant documents would be unduly burdensome. The D.C. federal court overseeing the matter rejected all of these claims and issued a summary judgment ordering the EPA to turn over any information it had by a court-ordered deadline. The EPA, however, could not find any scientific evidence supporting Pruitt's claim and instead provided PEER with just two documents. The first was a 12-page paper that Pruitt used to prepare for his television interview, which mentions neither carbon dioxide nor climate change, and the second was Congressional correspondence written by Pruitt in he which discussed his desire to hold an internal EPA debate on climate change (Geuss 2017; Sabin Center for Climate Change Law 2019b). Despite many years of climate change deniers asserting that science was on their side, a high-level government official with a long record of rejecting the anthropogenic causes of global warming could not find any scientific evidence supporting his views or informing his policy decisions.

Perhaps hoping to follow in the footsteps of the PEER strategy, in 2019 the Sierra Club filed a FOI request with the EPA to provide scientific documentation supporting Pruitt replacement Andrew Wheeler's March 2019 publicly televised claim that "most of the threats from climate change are 50 to 75 years out" (quoted in E. Kaufman 2019). This statement contradicts the conclusions drawn by leading climate scientists (e.g., the 4th National Climate Assessment).

As reported in this chapter, the political climate war is being fought on many fronts: in the halls of Congress, in the media, in the courts, and, as discussed in Chapter 11, in the streets.

9.9 Fundamentalist denial

The state of Earth's environment has never been a core issue of concern for Christian fundamentalists, participants in an ongoing and influential movement—especially in American Protestantism—that arose in the latter part of the 19th century as a rejection of theological modernism. Since the 1980s, sparked by televangelist Pat Robertson's unsuccessful run for the presidency, fundamentalists have been a strong conservative voice in American politics. Their attitudes on the issue of climate change have tended to echo the missives of the climate denier blogosphere. Research by Arbuckle & Konisky (2015) found that people who affiliate with an Evangelical Protestant church generally are less likely than those who attend mainline Protestant churches to be worried about climate change. Those who accept the Bible as the literal word of God tend to believe that God intends for Earth to end apocalyptically as described in the Book of Revelation and that it is not their responsibility to intervene to stop this destruction. As Baptist pastor Karyn Carlo (2017) notes, many "white fundamentalist Christians oppose the idea of global warming, dismiss it as a hoax, and insist that it is . . . blasphemous worship of Mother Earth." Sarah Palin, for example, the Republican former governor of Alaska and John McCain's selection for vice-presidential running mate, an outspoken fundamentalist, has asserted that climate scientists are "erroneously delivering [data] to the public to make us think that we can somehow change the weather and how they do that is to grow government and allow the government to have more control over us, our homes, our businesses, our families, our lives, and it's quite unfortunate because these people must be purposely doing this, right? Because they are smart enough to know better" (quoted in Goldenberg 2016). Citing Genesis 1:27–28, similarly, Ann Coulter, a constitutional lawyer turned ultra-conservative pundit and author, asserts that the Bible gives dominion over nature to humanity. Specifically, she claims: "God gave us the earth. We have dominion over the plants, the animals, the trees. God said, 'Earth is yours. Take it. Rape it. It's yours'" (quoted in Mencimer 2001). Medieval historian Lynn White (1967, p. 1205) has gone so far as to suggest that "[e]specially in its Western form, Christianity is the most anthropocentric religion the world has seen . . . By destroying pagan animism, Christianity made it possible to exploit nature in a mood of indifference to the feelings of natural objects."

Beginning during the 1990s, a small group who embrace the slogan "creation care" tried to change the views on the environment of their fellow fundamentalists and even became involved in efforts to pass legislation on reducing global warming. Evangelical environmentalism, as this movement has been called, emphasizes biblical directives about humanity as a steward that is charged with taking care of God's creation. Many adherents of this perspective see their work as a ministry of reconciliation. Members of Care of Creation (2016), a global fundamentalist group, for example, assert that ecocrises are real: "People, animals and plants are dying because we human beings have abused God's world. But this is not the primary reason for our work. This is what makes us different from every other 'environmental organization' you have encountered: We do what we do because we love God and his world."

In response to this incipient environmentalism in fundamentalist Christianity, a group of organized evangelical leaders associated with the Christian Right pushed the creation care dissidents to the margins, resulting in climate denial becoming "more deeply anchored as a part of evangelical identity, undoing whatever achievements Creation Care had won in changing minds" (Bean & Teles 2015, p. 2). Climate change denial has maintained a tight grip on the majority of fundamentalists ever since. Nonetheless, researchers have suggested that there remains an untapped potential for environmental activism in the world of evangelical Christianity (Wilkinson 2012). In the era of Trump, and the noted Trump–fundamentalist bond, this potential has been rendered mute.

It is far from clear if climate reality expressed in the growing daily evidence of the effects of climate change will silence climate change denial. Experience of climate effects and the growing body of scientific findings are however beginning to increase popular concern. Additionally, it is noteworthy that in 2019 the libertarian Cato Institute disbanded its climate denial program, the Center for the Study of Science. Moreover, the size of Heartland Institute's annual climate change conference has been shrinking over the years, and the number of sponsoring organizations fell by more than half between 2008 and 2019. As Christine Macdonald (2019) observes, "It seems possible to conceive of a time, not all that far off in the future, when the effects of climate change will become so evident and frequent that denying it's happening will no longer be an option. But that's hardly a comforting thought."

References

Achakulwisut, P., Scandella, B. & Voss, B. (2016). Why is the largest Earth science conference still sponsored by Exxon? *Guardian*. https://www.theguardian.com/environment/climate-consensus-97-per-cent/2016/jan/06/why-is-the-largest-earth-science-conference-still-sponsored-by-exxon.

AEI. (2018). 2018 Annual Report. American Enterprise Institute. http://www.aei.org/wp-content/uploads/2019/01/AEI-2018-Annual-Report.pdf.

AGU. (2019). *Strategic plan*. American Geophysical Union. https://sites.agu.org/leadership/strategic-plan/.

Anderegg, W., Prall, J., Harold, J. & Schneider, S. (2010). Expert credibility in climate change. *Proceedings of the National Academy of Sciences of the U.S.A.* 107(27): 12107–12110.

Anzilotti. (2018). A groundbreaking carbon tax lost at the ballot–again. *Fast Company*. https://www.fastcompany.com/90264251/why-cant-washington-state-pass-a-carbon-tax.

Arbuckle, M. & Konisky, D. (2015). The role of religion in environmental attitudes. *Social Science Quarterly* 96(5): 1244–1263.

Atkin, E. (2015) Ted Cruz's bank account would soar if we increased oil and gas production. *Think Progress*. https://archive.thinkprogress.org/

ted-cruzs-bank-account-would-soar-if-we-increased-oil-and-gas-production-f3e744ad0e25/.

Banerjee, N. (2016) Rex Tillerson's record on climate change: rhetoric vs. reality. *Inside Climate News*. https://insideclimatenews.org/news/22122016/rex-tillerson-exxon-climate-change-secretary-state-donald-trump.

Bean, L. & Teles, S (2015). Spreading the Gospel of Climate Change: An Evangelical Battleground. *New America*. https://static.newamerica.org/attachments/11649-spreading-the-gospel-of-climate-change/climate_care11.9.4f0142a50aa24a2ba65020f7929f6fd7.pdf.

Bennett, D. (2004). Eco-traitor. *Wired*. https://archive.is/dwgUa#selection-669.0-669.11.

Berman, E. & Carter, J. (2018). Policy analysis: scientific integrity in federal policymaking under past and present administrations. *Journal of Science Policy & Governance* 13(1).

Blank, J. & Shaw, D. (2015). Does partisanship shape attitudes toward science and public policy? The case for ideology and religion. *Annals of the American Academy of Political and Social Science* 658: 18–35

Bradley, R. (2013). Rex Tillerson (Exxon Mobil) on climate change (energy/climate realism trumps alarmism). *Master Resource*. https://www.masterresource.org/exxon-mobil/exxon-mobil-trumps-alarmists/.

Brembs, B., Button, K. & Munafò, M. (2013). Deep impact: unintended consequences of journal rank. *Frontiers in Human Neuroscience* 7: 291.

Briggs, C. & Mantini-Briggs, C. (2004). *Stories in the Time of Cholera: Racial Profiling During a Medical Nightmare*. Berkeley, CA: University of California Press.

Brown, A. (2013). Understanding pharmaceutical research manipulation in the context of accounting manipulation. *Journal of Law, Medicine and Ethics* 41(3): 611–619.

Brown, A. (2018). Did you know the greatest two-year global cooling event just took place? *Real Clear Markets*. https://www.realclearmarkets.com/articles/2018/04/24/did_you_know_the_greatest_two-year_global_cooling_event_just_took_place_103243.html.

Brulle, R. (2014). Institutional delay: foundation funding and the creation of U.S. climate change counter-movement organizations. *Climate Change* 122(4).

Care of Creation. (2016). What creation care is all about. http://www.careofcreation.net/what-creation-care-is-all-about/.

Carlo, K. (2017). Against equality: fundamentalism and climate change denial. American Friends Service Committee. https://www.afsc.org/blogs/acting-in-faith/against-equality-fundamentalism-and-climate-change-denial.

Carter, J., Berman, E., Desikan, A., Johnson. C. & Goldman, G. (2019). *The State of Science in the Trump Era: Damage Done, Lessons Learned, and a Path to Progress*. Cambridge, MA: Union of Concerned Scientists.

Castleman, B. (1996). Testimony of Charles H. Roemer, deposition taken April 25, 1984, Johns-Manville Corp. *et al*. v. the United States of America, U.S. Claims Court Civ. No. 465-83C. In: *Asbestos: Medical and Legal Aspects*, 4 edn. Englewood Cliffs, NJ: Aspen Law and Business.

Cheng, L., Abraham, J., Hausfather, Z. & Trenberth, K. (2019). How fast are the oceans warming? *Science* 363(6423): 128–129.

ClimateFiles. (2017a). 1997 Exxon's Lee Raymond speech at World Petroleum Congress. http://www.climatefiles.com/exxonmobil/october-13-1997-exxon-lee-raymond-speech-at-world-petroleum-congress/.

ClimateFiles. (2017b). 1998 ExxonMobil pamphlet: global climate change everyone's debate. http://www.climatefiles.com/exxonmobil/1998-exxon-pamphlet-global-climate-change-everyones-debate/.

CNN. (2018). Santorum claims climate scientists "driven by money." https://www.cnn.com/videos/politics/2018/11/25/sotu-panel-pt-2.cnn.

CNN Transcripts. (2018). CNN right now. http://transcripts.cnn.com/TRANSCRIPTS/1811/26/crn.01.html.

Conway, E. & Oreskes, N. (2012). Why conservatives turned againstscience. *Chronicle of Higher Education.* https://www.chronicle.com/article/The-Conservative-Turn-Against/135488.

Cook, J., Nuccitelli, D., Green, S., Richardson, M., Winkler, B., Painting, R. *et al.* (2013). Quantifying the consensus on anthropogenic global warming in the scientific literature. *Environmental Research Letters* 8(2): 024024.

Davenport, C. (2020). What will Trump's most profound legacy be? Possibly climate damage. *The New York Times.* https://www.nytimes.com/2020/11/09/climate/trump-legacy-climate-change.html.

Davies, K. (2007). Why does Exxon care about polar bears? Greenpeace. https://www.greenpeace.org/usa/why-does-exxon-care-about-polar-bears/.

Derber, C. (2010). *Greed to Green: Solving Climate Change and Remaking the Economy.* Boulder, CO: Paradigm.

DeSmog. (2015). American Tradition Institute. https://www.desmogblog.com/american-tradition-institute.

Downs, C. (2016). Ted Cruz accepted $30 000 flight to Cayman Islands from former Exxon executive during campaign. *Dallas News.* https://www.dallasnews.com/news/2016-presidential-election/2016/08/16/ted-cruz-accepted-30k-flight-cayman-islands-former-exxon-executive-campaign.

Dyck, M., Soon, W., Baydack, R., Legates, D., Baliunas, S., Ball, T. & Hancock, L. (2007). Polar bears of western Hudson Bay and climate change: are warming spring air temperatures the "ultimate" survival control factor? *Ecological Complexity* 4(3): 73–84.

Eilperin, J. & Ryan, M. (2019). White House prepares to scrutinize intelligence agencies' finding that climate change threatens national security. *Washington Post.* https://www.washingtonpost.com/national/health-science/white-house-readies-panel-to-assess-if-climate-change-poses-a-national-security-threat/2019/02/19/ccc8b29e-3396-11e9-af5b-b51b7ff322e9_story.html.

Eliades, T., Athanasiou, A. & Papadopulos, J. (2005). Ethics and fraud in science: a review of scientific misconduct and applications to craniofacial research., *World Journal of Orthodontics* 6(3): 226–232.

Environmental Integrity Project. (2017). Trump administration's settlement with Exxon Mobil appears to require less in pollution controls than advertised. https://www.environmentalintegrity.org/news/trump-administrations-settlement-with-exxon-mobil-appears-to-require-less-in-pollution-controls-than-advertised/.

Evich, H. (2019). Senate Democrats release list of climate studies buried by Trump administration. *Politico.* https://www.politico.com/story/2019/09/19/climate-studies-hidden-by-trump-administration-1753631.

ExxonMobil. (2007). 2007 Corporate Citizenship Report. https://www.globalhand.org/system/assets/a25f2fc01db618e4965a9618b5cd66bc49dfe471/original/2007_Coporate_Citizenship_Report.pdf?1340478058.

ExxonMobil (2010). 2010 Corporate citizenship report. https://corporate.exxonmobil.com/-/media/global/files/sustainability-report/publication/2010-ccr.pdf.

ExxonMobil. (2018a). ExxonMobil outlines aggressive growth plans to more than double earnings. https://news.exxonmobil.com/press-release/exxonmobil-outlines-aggressive-growth-plans-more-double-earnings.

ExxonMobil. (2018b). Baton Rouge area operations. https://corporate.exxonmobil.com/en/Locations/United-States/Baton-Rouge-area-operations-overview#aboutUs.

ExxonMobil. (2019a). Climate change. https://corporate.exxonmobil.com/energy-and-environment/environmental-protection/climate-change.

ExxonMobil. (2019b). ExxonMobil to increase, accelerate Permian output to 1 million barrels per day by 2024. https://corporate.exxonmobil.com/news/newsroom/news-releases/2019/0305_exxonmobil-to-increase-accelerate-permian-output-to-1-million-barrels-per-day-by-2024.

ExxonMobil. (2019c). Growing the Gulf. https://corporate.exxonmobil.com/energy-and-environment/where-we-work/growing-the-gulf.

Fang, F., Steen, R. & Casadevall, A. (2012). Misconduct accounts for the majority of retracted scientific publications. *Proceeding of the National Academy of Sciences of the U.S.A.* 109: 17028–17033.

Forni, O. (2010). *Mapping Environmental Crimes. Freedom from Fear*. Turin: United Nations Interregional Crime and Justice Research Institute.

Fortuna, C. (2018). Should we cheer? ExxonMobil's renewable energy commitments are in the news. *Clean Technica*. https://cleantechnica.com/2018/12/06/should-we-cheer-exxonmobils-renewable-energy-commitments-are-in-the-news/.

Frumhoff, P., Heede, R. & Oreskes, N. (2015). The climate responsibilities of industrial carbon producers. *Climate Change* 132(2): 157–171.

Funk, (2017). Mixed messages about public trust in science. *Issues in Science and Technology* 34(1).

Gambino, M. (2010). Bob Reiss on "welcome to Barrow, Alaska, ground zero for climate change." *Smithsonian Magazine*. https://www.smithsonianmag.com/science-nature/bob-reiss-on-welcome-to-barrow-alaska-ground-zero-for-climate-change-8033405/#sJbFYDyWCLluiOPI.99.

Gauchat, G., O'Brien, T. & Mirosa, O. (2017). The legitimacy of environmental scientists in the public sphere. *Climatic Change* 143: 297–306.

George, S. & Buyse, M. (2015). Data fraud in clinical trials. *Clinical Investigation* 5(2): 161–173.

Geuss, M. (2017). EPA administrator asked to back up climate claims made on TV with science. *Ars Technica*. https://arstechnica.com/tech-policy/2019/04/like-pruitt-before-him-epa-chief-wheeler-is-asked-to-back-up-climate-claims/.

Global Witness. (2018). Catch me if you can: Exxon complicit in corrupt Liberian oil sector. https://www.globalwitness.org/en/campaigns/oil-gas-and-mining/catch-me-if-you-can-exxon-complicit-corrupt-liberian-oil-sector/

Goldenberg, S. (2015). Greenpeace exposes sceptics hired to cast doubt on climate science. *Guardian*. https://www.theguardian.com/environment/2015/dec/08/greenpeace-exposes-sceptics-cast-doubt-climate-science.

Goldenberg, S. (2016). Climate change denier Sarah Palin: "Bill Nye is as much a scientist as I am." *Guardian*. https://www.theguardian.com/us-news/2016/apr/15/sarah-palin-bill-nye-climate-change-hustle-film.

Grandjean, P. & Clapp, R. (2015). Perfluorinated alkyl substances: emerging insights into health risks. *New Solutions: A Journal of Environmental and Occupational Health Policy* 25(2): 147–163.

Green, H. (2018). Deconstructing a genius climate change argument. Hankschannel. https://www.youtube.com/watch?v=_Es4gG9xVf8.

Greenpeace. (2019). Greenpeace statement on Patrick Moore. https://www.greenpeace.org/usa/news/greenpeace-statement-on-patric/.

Griffin, P. (2017). The carbon majors database. CDP. https://b8f65cb373b1b7b15feb-c70d8ead6ced550b4d987d7c03fcdd1d.ssl.cf3.rackcdn.com/cms/reports/documents/000/002/327/original/Carbon-Majors-Report-2017.pdf?1499691240.

Halpern, M. (2018). What is the responsible science policy coalition? Here are some clues. Union of Concerned Scientists. https://blog.ucsusa.org/michael-halpern/what-is-the-responsible-science-policy-coalition-here-are-some-clues.

Hamilton, D. (1992). In the trenches, doubts about scientific integrity. *Science* 255(5052): 1636.

Hamilton, L., Hartter, J. & Saito, K. (2015). Trust in scientists on climate change and vaccines. *SAGE Open* 5(3).

Hanley, S. (2019). Supreme Court rejects ExxonMobil appeal of Massachusetts court ruling. https://cleantechnica.com/2019/01/08/

supreme-court-rejects-exxonmobil-appeal-of-massachusetts-court-ruling/.
Hiltzik, M. (2019). Big Oil as the victim? That's a gas. *Los Angeles Times*. https://www.pressreader.com/usa/los-angeles-times/20190213/281917364340814.
Hirji, Z. (2016a). Union to cut ties with Exxon over climate denial. *Inside Climate News*. https://insideclimatenews.org/news/22022016/scientists-american-geophysical-union-cut-ties-exxon-climate-change-denial.
Hirji, Z. (2016b). Exxon won't sponsor AGU, a win of sorts for oil giant's opponents. *Inside Climate News*. https://insideclimatenews.org/news/05122016/exxon-agu-conference-climate-change-denial.
Hirtenstein, A. (2017). ExxonMobil is spending $1bn a year researching renewable energy. *Independent*. https://www.independent.co.uk/news/business/news/exxonmobil-renewable-energy-research-oil-company-development-biofuels-algae-electricity-a8035496.html.
Hughes, D. (2013). Climate change and the victim slot: from oil to innocence. *American Anthropologist* 115(4): 570–581.
Hunter, C., Caswell, H., Runge, M., Regerhr, E., Arnstrup, S. & Sterling, I. (2010). Climate change threatens polar bear populations: a stochastic demographic analysis. *Ecology* 91(10): 2882–2897.
IEA. (2010) *CO2 Emissions from Fuel Combustion: Highlights*. Paris: International Energy Agency.
Investor's Business Daily. (2018). The stunning statistical fraud behind the global warming scare. https://www.investors.com/politics/editorials/the-stunning-statistical-fraud-behind-the-global-warming-scare/.
Jerving, S., Jennings, K., Hirsch, M. & Rust, S. (2015). What Exxon knew about the Earth's melting Arctic. *Los Angeles Times*. http://graphics.latimes.com/exxon-arctic/.
Kahan, D., Landrum, A., Carpenter, K., Helft, L. & Jamieson, K. (2017). Science curiosity and political information processing. *Political Psychology* 38(S1): 179–199.
Kasprak, A. (2019). Did Patrick Moore, a doubter of anthropogenic climate change, co-found Greenpeace? *Snopes*. https://www.snopes.com/fact-check/patrick-moore-climate-doubter/.
Kaufman, A. (2019). Exxon's climate denial set to face first public scrutiny as legal woes mount. *Huffington Post*. https://www.huffpost.com/entry/exxon-mobil-climate-change-denial_n_5c901482e4b0d50544fee0f2.
Kaufman, E. (2019). Federal report contradicts EPA chief's claims that climate change impacts are decades away, environmental group says. https://edition.cnn.com/2019/04/22/politics/sierra-club-climate-change-scn/index.html.
Keeling, R. (2018). Resplandy *et al.* correction and response. *Real Climate*. http://www.realclimate.org/index.php/archives/2018/11/resplandy-et-al-correction-and-response/.
Keltan, E. (2013). Is Big Pharma addicted to fraud? *Forbes*. https://www.forbes.com/sites/erikakelton/2013/07/29/is-big-pharma-addicted-to-fraud/#5ce7417c15fe.
Kennedy, D. (2005). Silly season on the Hill. *Science*. https://science.sciencemag.org/content/sci/309/5739/1301.full.pdf.
Kessler, G. & Lee, M. (2017). Fact-checking President Trump's claims on the Paris climate change deal. *Washington Post*. https://www.washingtonpost.com/news/fact-checker/wp/2017/06/01/fact-checking-president-trumps-claims-on-the-paris-climate-change-deal/?utm_term=.9207bd737ee5.
Kruzel, J. (2018). Recent cold spells irrelevant to climate change big picture. *PunditFact*. https://www.politifact.com/punditfact/statements/2018/nov/28/danielle-pletka/recent-cold-spells-irrelevant-climate-change-big-p/.
Larson, E. (2017). Schneiderman says Exxon's climate change proxy costs may be a "sham." *Bloomburg News*. https://www.bloomberg.com/news/articles/2017-06-02/n-y-says-exxon-s-climate-change-proxy-costs-may-be-a-sham.
Lee, M., Cruz, T., Sessions, J., Perdue, D. & Vitter, D. (2016). Letter to Loretta Lynch. https://www.scribd.com/doc/313917986/20160526-ClimateChangeLetter.

Lerner, S. (2018). Lawsuit reveals how paid expert helped 3M "command the science" on dangerous chemical. *The Intercept*. https://theintercept.com/2018/02/23/3m-lawsuit-pfcs-pollution/.

Lewin, T. (1983). New suit questions Manville tactics. *The New York Times*. https://www.nytimes.com/1983/11/14/business/new-suit-questions-manville-tactics.html.

Lewis, N. (2018). A major problem with the Resplandy *et al.* ocean heat uptake paper. *Climate Etc.* https://judithcurry.com/2018/11/06/a-major-problem-with-the-resplandy-et-al-ocean-heat-uptake-paper/.

Leyden, C. (2019). Satellite data confirms Permian gas flaring is double what companies report. Environmental Defense Fund. http://blogs.edf.org/energyexchange/2019/01/24/satellite-data-confirms-permian-gas-flaring-is-double-what-companies-report/.

Li, T. (2000). Articulating indigenous identity in Indonesia: resource politics and the tribal slot. *Comparative Studies in Society and History* 42(1): 149–179.

Lynn, P. (2019). Current investigation of ExxonMobil could spur broader climate action. *TruthOut*. https://truthout.org/articles/current-investigation-of-exxonmobil-could-spur-broader-climate-action/.

Macdonald, C. (2019). I went to a climate change denial conference. It made even less sense than you'd think. *In These Times*. http://inthesetimes.com/article/21980/i-went-to-a-climate-change-denial-conference-heartland-institute-trump.

Macdonald, T. (2018). Major broadcast TV networks mentioned climate change just once during two weeks of heat-wave coverage. *Media Matters*. https://www.mediamatters.org/blog/2018/07/12/Major-broadcast-TV-networks-mentioned-climate-change-just-once-during-two-weeks-of-heat-wa/220651.

Mann, M. (2013). *The Hockey Stick and the Climate Wars: Dispatches from the Front Lines*. New York: Columbia University Press.

Mann, M., Bradley, R. & Hughes, M. (1998). Global-scale temperature patterns and climate forcing over the past six centuries. *Nature* 392; 779–787.

Mauney, M. (2019). Johns Manville. Mesothelioma Center. https://www.asbestos.com/companies/johns-manville/.

McCright, A. M., Dentzman, K., Charters, M. & Dietz, T. (2013). The influence of political ideology on trust in science. *Environmental Research Letters* 8(4).

Mencimer, S. (2001). The wisdom of Ann Coulter. *Washington Monthly*. https://washingtonmonthly.com/2001/11/01/the-wisdom-of-ann-coulter/.

Michaels, D. (2019). The Denial Playbook: How Industries Manipulate Science and Policy from Climate Change to Public Health. *Testimony before the House Natural Resources Committee*, Oversight and Investigations Subcommittee, U.S. Congress. https://naturalresources.house.gov/imo/media/doc/Michaels%20Written%20Testimony%20Denial%20Playbook%20Hearing.pdf.

Miller, B. (2019). Trump tweets climate change skeptic in latest denial of science. *CNN*. https://www.cnn.com/2019/03/12/us/trump-climate-change-tweet-patrick-moore/index.html.

Mooney, C. (2004). Some like it hot. *Mother Jones*. https://www.motherjones.com/environment/2005/05/some-it-hot/.

Mooney, C. (2005). *The Republican War on Science*. New York: Basic Books.

Mooney, C. & Dennis, B. (2018). Scientists acknowledge key errors in study of how fast the oceans are warming. *Washington Post*. https://www.washingtonpost.com/energy-environment/2018/11/14/scientists-acknowledge-key-errors-study-how-fast-oceans-are-warming/?utm_term=.408ad9bb6db3.

Moore, P. (2019). https://twitter.com/ecosensenow/status/1105448962641879040?lang=en.

Morehouse, E. (2016). *Who should pay for pollution?* Environmental Defense Fund. http://blogs.edf.org/climatetalks/2016/04/21/who-should-pay-for-pollution/.

Motta, M. (2018). The enduring effect of scientific interest on trust in climate scientists in the United States. *Nature Climate Change* 8: 485–488.

Nature Editorial. (2010). Science subpoenaed. *Nature* 465: 135–136.

Negin, E. (2018). *Why is ExxonMobil still funding climate science denier groups?* Union of Concerned Scientists. https://blog.ucsusa.org/elliott-negin/exxonmobil-still-funding-climate-science-denier-groups.

Negin, E. (2020). ExxonMobil claims it's shifting on climate, but still funding climate science deniers. *CleanTechnica*. https://cleantechnica.com/2020/10/23/exxonmobil-claims-its-shifting-on-climate-still-funding-climate-science-deniers/.

NOAA. (2019). *2018 fourth warmest year in continued warming trend, according to NASA*, NOAA. National Oceanic and Atmospheric Administration. https://www.nasa.gov/press-release/2018-fourth-warmest-year-in-continued-warming-trend-according-to-nasa-noaa.

Nolan, H. (2018). This is how capitalism kills us. *Splinter*. https://splinternews.com/this-is-how-capitalism-kills-us-1830646401.

Nye, B. (2013). Foreword. In M. Mann, *The Hockey Stick and the Climate Wars: Dispatches from the Front Lines,* pp. xi–xiii. New York: Columbia University Press.

OIG NSF. (2019). What Is the Office of Inspector General? National Science Foundation. https://www.nsf.gov/oig/.

OpenSecrets. (2018). Oil & gas: top recipients. https://www.opensecrets.org/industries/recips.php?ind=E01&recipdetail=A&sortorder=U&mem=Y&cycle=2018.

Page, S. (2015). Ted Cruz: "Climate change is not science. It's religion." *ThinkProgress*. https://thinkprogress.org/ted-cruz-climate-change-is-not-science-its-religion-70987f13959c/.

Polansky, A. (2017). Secretary of State Rex Tillerson, career ExxonMobil executive, "hates your children." *Climate Science & Policy Watch*. http://www.climatesciencewatch.org/2017/01/25/secretary-of-state-rex-tillerson-career-exxonmobil-executive-hates-your-children/.

Polychroniou, C. (2018). Oil companies spent millions to defeat green new deal in Washington State. *Truthout*. https://truthout.org/articles/oil-industry-spent-millions-to-defeat-carbon-tax-in-washington-state/.

Poole, C. (2018). The world's largest oil & gas companies 2018: Royal Dutch Shell surpasses Exxon as top dog. *Forbes*. https://www.forbes.com/sites/clairepoole/2018/06/06/global-2000-oil-gas/#70e820f61d1b.

Qui, L. (2018). The baseless claim that climate scientists are "driven" by money. *The New York Times*. https://www.nytimes.com/2018/11/27/us/politics/climate-report-fact-check.html.

Readfearn, G. (2016a). ExxonMobil: new disclosures show oil giant still funding climate science denial groups. DeSmog. https://www.desmogblog.com/2016/07/08/exxonmobil-new-disclosures-show-oil-giant-still-funding-climate-science-denial-groups.

Readfearn, G. (2016b). Checking Ted Cruz's climate science denial howlers. *Guardian*. https://www.theguardian.com/environment/planet-oz/2016/feb/11/checking-ted-cruzs-climate-science-denial-clangers.

Regalado, A. (2019). Gene-edited cattle have a major screwup in their DNA. *MIT Technical Review*. https://www.technologyreview.com/s/614235/recombinetics-gene-edited-hornless-cattle-major-dna-screwup/.

Resplandy, L., Keeling, E., Eddebbar, Y., Brooks, M., Wang, R., Bopp, L. *et al.* (2018). Quantification of ocean heat uptake from changes in atmospheric O2 and CO2 composition. *Nature* 563: 105–108.

Rich, N. (2019). *Losing Earth: A Recent History*. New York: Farrar, Straus and Giroux.

Rosenberg, J. (2019). Asbestos cover-up. Mesothelioma Center. https://www.asbestos.com/featured-stories/cover-up/.

Rubin, J. (2014). Why the American Enterprise Institute chief is so popular. *Washington Post*. https://www.washingtonpost.com/blogs/

right-turn/wp/2014/04/02/why-the-american-enterprise-institute-chief-is-so-popular/?utm_term=.9fdde9c09fff.

Sabin Center for Climate Change Law. (2019a). Silencing Science Tracker. https://climate.law.columbia.edu/Silencing-Science-Tracker.

Sabin Center for Climate Change Law. (2019b). *Public Employees for Environmental Responsibility* v. *EPA*. http://climatecasechart.com/case/public-employees-environmental-responsibility-v-epa/?cn-reloaded=1.

Samuelsohn, D. (2009). Carbon tax advocates make their case with warnings of cap-and-trade chaos. *E&E News*. https://www.eenews.net/special_reports/climate_change_domestic/stories/73061.

Schwartz, J. (2016). Tillerson led Exxon's shift on climate change; some say "it was all P.R." *The New York Times*. https://www.nytimes.com/2016/12/28/business/energy-environment/rex-tillerson-secretary-of-state-exxon.html.

Sévellec, F. & Drijfhout, S. (2018). A novel probabilistic forecast system predicting anomalously warm 2018–2022 reinforcing the long-term global warming trend. *Nature Communications* 9: 3024.

Sinclair, P. (2018). Meet the Press devotes hour to climate. Proof that shame works? *Climate Crocks*. https://climatecrocks.com/2018/12/30/meet-the-press-devotes-hour-to-climate-proof-that-shame-works/.

Smith, R. (2005). Investigating the previous studies of a fraudulent author. *British Medical Journal* 331: 288–291.

Snider, A. (2018a). Inside a corporate giant's fight to thwart a massive pollution tab. *Politico*. https://www.politico.com/story/2018/11/23/political-chemicals-cleanup-3m-1012445.

Snider, A. (2018b). *White House*, EPA headed off chemical pollution study. *Politico*. https://www.politico.com/story/2018/05/14/emails-white-house-interfered-with-science-study-536950.

Stanley, S., Wilson, M. & Milfont, T. (2017). Exploring short-term longitudinal effects of right-wing authoritarianism and social dominance orientation on environmentalism. *Personality and Individual Differences* 108: 174–177.

State of Minnesota. (2017). *State of Minnesota* vs. *3M*. https://www.pca.state.mn.us/sites/default/files/c-pfc2-11f.pdf.

Statista. (2019). Per capita coal consumption worldwide in 2015, by select country (in metric tons of coal equivalent). https://www.statista.com/statistics/604946/per-capita-coal-consumption-in-selected-countries/.

Stinnett, N. (2019). Opinion: Forget about the climate deniers. It's the climate liars we need to stop. *BuzzFeed News*. https://www.buzzfeednews.com/article/nathanielstinnett/climate-deniers-and-climate-liars.

Stirling, I., Derocher, A., Gough, W. & Rode, K. (2008). Response to Dyck *et al.* 2007 on polar bears and climate change in western Hudson Bay. *Ecological Complexity* 5(3): 193–201.

Supran, G. & Orekes, N. (2017). Assessing ExxonMobil's climate change communications (1977–2014). *Environmental Research Letters* 12(8): 084019.

Trouillot, M.-R. (2003). Anthropology and the savage slot: the poetics and politics of otherness. In: *Global Transformations*, pp. 7–28. New York: Palgrave Macmillan.

Tsipursky, G. (2018). (Dis)trust in science. *Scientific American*. https://blogs.scientificamerican.com/observations/dis-trust-in-science/.

Union of Concerned Scientists (2017). The disinformation playbook. https://www.ucsusa.org/center-science-and-democracy/disinformation-playbook.

Union of Concerned Scientists (2018). The 2018 climate accountability scorecard: insufficient progress from major fossil fuel companies. https://www.ucsusa.org/sites/default/files/attach/2018/10/gw-accountability-scorecard18-report.pdf.

United Nations. (2019). Only 11 years left to prevent irreversible damage from climate change, speakers warn during general

assembly high-level meeting. https://www.un.org/press/en/2019/ga12131.doc.htm.
van Dongen, J. (2010). On Einstein's opponents, and other crackpots. *Studies in History and Philosophy of Modern Physics* 41(1): 78–80.
Van Zandt, D. (2018). American Enterprise Institute. https://mediabiasfactcheck.com/american-enterprise-institute/.
Waldman, S. (2019). Trump officials deleting mentions of "climate change" from U.S. Geological Survey press releases. *Science*. https://www.sciencemag.org/news/2019/07/trump-officials-deleting-mentions-climate-change-us-geological-survey-press-releases.
Walker, B. & Rundquist, S. (2017). Mapping a contamination crises: PFCs pollute tap water for 15 million people, dozens of industrial sites. Environmental Working Group. https://www.ewg.org/research/mapping-contamination-crisis.
Ward, A. (2019). The Pentagon calls climate change a national security threat. Trump isn't listening. *Vox*. https://www.vox.com/2019/1/18/18188153/pentagon-climate-change-military-trump-inhofe.
WCRI. (2019). Mission. World Conferences on Research Integrity. https://wcrif.org/foundation/mission.
Weissert, W. (2018). *Big oil asks government to protect it from climate change. Associated Press.* https://www.apnews.com/4adc5a2a2e6b45df953ebcba6b63d171.
White, L. (1967). The historical roots of our ecologic crisis. *Science* 155(3767): 1203–1207.
White, R. (2013). *Environmental Harm: An Eco-Justice Perspective*. Bristol: Policy Press.
White, R. (2016). *Transnational Environmental Crime*. Abingdon: Routledge.
Wieners, B. (2014). Leaked: the oil lobby's conspiracy to kill off California's climate law. *Bloomberg BusinessWeek*. http://www.businessweek.com/articles/2014-11-25/revealed-the-oil-lobbys-playbook-against-californias-climate-law.
Wilkinson, K. (2012). *Between God and Green: How Evangelicals Are Cultivating a Middle Ground on Climate Change*. New York: Oxford University Press.
Wolf, Z. (2019). The Trump administration has taken its war on climate crisis action to a new level. *CNN*. https://www.cnn.com/2019/09/06/politics/climate-change-california-carmakers-trump/index.html.
Woods, D. (2018). Working together to tackle climate risks. EngergyFactor. https://energyfactor.exxonmobil.com/perspectives/working-together-to-tackle-climate-risks/.

Part 3

Human health risks with changing environment

10

Crossing boundaries and thresholds

Planetary boundaries define . . . the boundaries of the "planetary playing field" for humanity if we want to be sure of avoiding major human-induced environmental change on a global scale.

Johan Rockström *et al.* (2009)

10.1 Are there biophysical boundaries for humanity?

Are their planetary boundaries or environmental thresholds that, if crossed, constitute the safe limits of human presence on the planet? Further, are there critical environmental tipping points that, if triggered, set off a cascade of ecosystem disruptions which become self-driven without further adverse anthropogenic inputs? These are the questions addressed in this chapter. Asking them is motivated by consideration of the possibility that human-induced damage to Earth's environmental systems might produce a planet like Mars or Jupiter that cannot sustain life, or at least human life. Over the years, various researchers have attempted to conceptualize biophysical boundaries for human development (Ciriacy-Wantrup 1952; Daly & Cobb 1989; Vitousek *et al.* 1997). Vitousek and colleagues, for example, identify diverse, adverse human effects on Earth's ecosystems and biogeochemical cycles, though they do not attempt to specify quantitative boundaries for safe human presence on the planet.

10.2 Key biogeochemical and biophysical Earth system processes

In formulating a novel conception of safe boundaries, a team of 28 accomplished Earth system and environmental scientists led by Johan Rockström of the Stockholm Resilience Center initiated a comprehensive scientific literature search that led to the identification of nine "planetary boundaries" that are essential to maintaining a global ecosystem in which humanity can survive and thrive. The processes or dimensions of concern are: 1) climate change; 2) ocean acidification; 3) stratospheric ozone depletion; 4) disruption of natural nitrogen and phosphorus cycles; 5) human pressure on global freshwater availability; 6) conversion of forests, wetlands, and other vegetation types to agricultural lands; 7) biodiversity loss; 8) atmospheric aerosol loading; and 9) toxic chemical pollution. Surpassing one or more of these boundaries puts pressure on Earth and could "trigger abrupt or irreversible environmental changes that would be deleterious or even catastrophic for human well-being" (Rockström *et al.* 2009). They constitute biophysical constraints or limits to growth on capitalist production/consumption as we know it (Meadows *et al.* 2004).

Boundaries exist at different spatial scales: local, regional, global. At the local (e.g., an estuary, an island, a forest, a coral reef) and

Ecosystem Crises Interactions: Human Health and the Changing Environment, First Edition. Merrill Singer.
© 2021 John Wiley & Sons, Inc. Published 2021 by John Wiley & Sons, Inc.

regional (e.g., rainforests of the Indonesian archipelago, the coral reefs of the Caribbean) scales, as seen in many examples cited in previous chapters, when boundaries are crossed through human action, there may be significant degradation or even total destruction. Changes can produce feedbacks that impact larger-scale processes. Loss of local carbon sinks such as forested areas, for example, helps to destabilize the wider climate system, pushing it toward thresholds (e.g., a major loss of land ice) of global significance. An important research question concerns the degree to which small-scale regime shifts can propagate across spatial scales, including global-level transitions. A related question, asked by Barnosky *et al.* (2012), is "whether global-scale state shifts are the cumulative result of many smaller-scale events that originate in local systems or instead require global-level forcings that emerge on the planetary scale and then percolate downwards to cause changes in local systems." Simply put: Can the tail wag the dog, or is it always the other way around?

Currently, knowledge is limited about how regime shifts transcend scales and whether local or regional shifts pushed by human actions can lead to global transitions. However, evidence is mounting that some local changes trigger critical transitions over regions larger than the directly affected location. It has been shown, for example, that as habitat loss progresses, the remaining habitat in a landscape area becomes increasingly fragmented, and the fragments become increasingly isolated (Swift & Hannon 2010). In the Atlantic forest of Brazil, for example, research by Pardini *et al.* (2010) shows that high levels of forest loss can push native organisms through an "extinction filter," resulting in the abrupt, landscape-wide loss of small forest-specialist mammals (e.g., various rodent species—not including bats, which can fly to new areas).

To the degree possible, quantifications are developed and justified for each of the proposed planetary boundaries. For example, the boundary for ocean acidification is based on ocean saturation with aragonite, one of the most common forms of calcium carbonate. The level of saturation state of aragonite is commonly used as an indicator for ocean acidification. To avoid acidification that converts coral reefs into algae-dominated systems and leads to the regional extinction of various marine organisms, as discussed in Chapter 5, Rockström *et al.* (2009) argue oceans must remain at ≥80 percent of the average global pre-industrial aragonite saturation point of 3.44 as this will cause them to be corrosive to the aragonite and high-magnesium calcite shells secreted by many species of marine organisms. Similarly, these researchers maintain that no more than 15 percent of global ice-free land may be converted to cropland (a 3 percent increase above the current level) as further expansion at a global scale could seriously threaten biodiversity and undermine the regulatory capacities of the Earth system by disrupting both the climate system and the hydrological cycle. Regarding the loss of biodiversity, the framework specifies that it should not exceed an annual extinction rate of 10 species per million known species. Most notable, the boundary for atmospheric carbon dioxide levels is set at 350 ppm, a boundary that is consistent with a stabilization of global temperatures at about 0.9 °F above pre-industrial levels. As indicated in Chapter 1, however, this boundary has already been crossed, as atmospheric carbon dioxide has surpassed 400 ppm and is growing at about 3 ppm per year. Carbon dioxide levels currently are higher than at any point in at least the last 800 000 years.

Why these boundaries? Rockström *et al.*'s (2009) selection is based on an assessment of what constitutes unacceptable anthropogenic global environmental change. In other words, it is contingent on human acceptance of risk: how much danger humanity is willing to put itself in as a result of environmental threats. Moreover, the identified boundaries are based on a sustainability timeline, namely one that is short enough to influence social decisions about environmental practice and policy but long enough to be able to achieve sustainability for future generations; in short, a timeframe in which human actions matter, with the realization that once environmental boundaries are crossed and thresholds are reached, human efforts to stop significant adverse environmental change will be too little, too late to make a difference. In other words, "the planetary boundaries approach focuses on the biophysical processes of the Earth System that determine the self-regulating capacity of the planet" with the intent of providing society with enough time to respond in order to avoid crossing an impending threshold (Rockström *et al.* 2009).

The nine planetary boundaries identified by these researchers, each of which is now being significantly modified by human actions, can be grouped into three types. First are the global biogeochemical cycles of nitrogen, phosphorus, carbon, and water, as discussed in Chapter 2. Second are the major physical circulation systems of the climate, the stratosphere (e.g., ozone depletion), and the ocean systems (e.g., ocean acidification). Third are biophysical features of Earth that contribute to the underlying resilience of its self-regulatory capacity, including marine and terrestrial biodiversity (or biosphere integrity) and the planet's land system cover (e.g., forest vs. monocrop agriculture). Additionally, there are two critical features that interact with anthropogenic global change: aerosol loading (e.g., particulate matter, tropospheric ozone, oxides of sulfur and nitrogen) and chemical pollution, including radioactive compounds, heavy metals, and a wide array of organic compounds. Rockström *et al.* (2009) believe that there is enough scientific evidence on hand to make at least a first attempt at quantifying control variables for the first seven of these boundaries. The last two are included as being of critical importance to sustainability even though it is not possible yet to suggest quantitative boundary levels. Further, two of the boundaries, climate change and biosphere integrity, are considered the core elements, based on their fundamental importance for the Earth system (Steffen *et al.* 2015). On their own, each of the nine boundaries has the potential to drive the Earth system into a new state if it is substantially and persistently transgressed. Moreover, two of them—biosphere integrity and biochemical flows—have already been pushed into higher risk levels.

10.3 Exploring planetary boundaries

Some concepts never achieve much influence, while some lie dormant for a period before being rediscovered and gaining in popularity. Specific examples of the former case are hard to identify precisely because they generated little or no interest beyond their originators. By contrast, the latter is so common it has been given a name. Stigler's "law of eponymy" states that no scientific discovery is named after its original discoverer. Rather, the first discoverer commonly is forgotten, while the person who rediscovered or popularized the concept gains recognition. Adhering to this "law," Stigler attributes its discovery to the celebrated sociologist Robert K. Merton.

Still other new concepts, like the Anthropocene, find an instant audience and promote broad discussion and exploration in one or more fields. This is the case with planetary boundaries. Introduction of the planetary boundaries framework has sparked a range of new research efforts and discussions involving various domains. The reasons for this favorable response are multiple and include timing (as marked by the growing recognition across scientific disciples of mounting environmental threats), stature (the recognized caliber of the formulators), breadth (incorporation of multiple global environmental stresses within one integrated framework), quantitative foundation (specification of measureable boundaries), and urgency (the recognition of needed political action to avert the transgression of dangerous thresholds).

10.3.1 Global environmental governance and planetary boundaries

One area where the planetary boundaries framework has had a sizable influence is in environmental governance. As Galaz *et al.* (2012) point out, it "is rapidly diffusing into a range of policy arenas," including the U.N. High-Level Panel on Global Sustainability, the European Environmental Agency (EEA), the U.K. Parliamentary Office for Science and Technology, Oxfam International, and meetings between scientists and Tibetan religious leaders including the Dalai Lama. It has "stimulated a discussion on the need to reform international environmental governance" (Galaz *et al.* 2012). What is meant by "international environmental governance"? A range of institutions are involved, including nation states, nonstate actors such as nongovernmental organizations (NGOs), and international organizations such as the U.N. Environment Program. The objective is to affirm and support the idea that natural resources, from air to water, from oceans to biodiversity, are public goods that must be collectively managed through agreements and treaties for the global public welfare. For example, all peoples of the world benefit from a breathable atmosphere, a stable climate, and expansive biodiversity, and this benefit must be sustained for human life to flourish. Still, as Elliot (2002, p. 58) argues, all is not even-handed in global governance, as "the congested institutional terrain . . . provides more of an appearance than a reality of comprehensive global governance" and what exists "masks global relations of power and powerlessness" among nations and between corporations and populations.

A basic but important question raised by the framework is: Can planetary boundaries be governed? Attempts to govern complex Earth system interactions must consider not only national policies (e.g., the United States' withdrawal from the Paris Agreement on Climate Change) but also the activities of private entities (e.g., illegal lumber cutting, corporate toxic dumping, lobbying to weaken environmentally protective regulations). As Nilsson & Perrson (2012) observe, even in cases where international agreements are put in place to avoid contravening planetary boundaries, there are "varying degrees of implementation deficits in all the existing regimes." Agreements are threatened by the gap between meaningful and illusory cooperation among nation-states and other actors, as well as by the intentional flouting of agreements (Walker *et al.* 2009).

The E.U. region—comprising over 25-member states—offers a useful case for an examination of why a multilevel governance understanding is needed in the planetary boundaries model, rather than a purely international-level approach. This is so for two reasons: 1) because the European Union has substantial power to address a number of Earth system interactions; and 2) because it has a goal of increasing policy coherence. Nilsson & Perrson (2012) undertake their examination by focusing on four of the nine planetary boundaries: climate change, land use, biodiversity, and freshwater.

A beginning point for this initiative is the downscaling of these four boundaries to the regional (E.U.) level. This entails addressing a second question: Is there an E.U. share of the global boundaries? Nilsson & Perrson (2012) conclude that it is not feasible to calculate shares across nations globally because three of the four boundaries of concern (land system change, water use, and biodiversity) are geographically differentiated, in terms of both the sources of their drivers and the locations of their most significant impacts. For example, one out of seven damselfly and dragonfly species in Europe is threatened with extinction in part due to intensified water extraction for drinking and irrigation. One of these is the yellow-tailed demoiselle (*Calopteryx xanthostoma*), an exquisite blue damselfly endemic to Europe (Fig. 10.1). Would the extinction of this organism constitute the loss of a global public good? Is there a share of the burden to protect

Fig. 10.1 Damselfly. *Source:* skeeze/Pixabay.

this species that could or should be shouldered by Argentina or Ghana? In this light, these researchers conclude that rather than relying exclusively on an international approach for many boundary issues, a multilevel governance approach that includes a regional focus is appropriate. As they stress, "Whilst ecosystems may often suffer from a 'spatial misfit' with the relevant governance system(s)—and indeed the entire Earth system suffers from a misfit since comprehensive global governance is missing—critical drivers, on the other hand, may operate at a scale which is more suitable to sub-global governance" (Nilsson & Perrson 2012).

This raises a third question: Relative to the environment, which are the core functions that governance responses fulfil? A starting point in addressing this is the work of Michel Foucault (1991, 2010) on governmentality. By this neologism, which combines the terms "government" and "rationality," he means government activity intended to shape or guide the conduct of the people being governed, the citizenry. This is achieved through the use of both technologies of government, such as government statistics, census data, or the creation of agencies that monitor aspects of the life experience of the population like disease spread or arrests, and channels of knowledge production and discourse, including who is considered a legitimate source of knowledge and how that knowledge is framed for public dissemination. Part of the art of government, Foucault argues, is defining citizenship. To achieve this goal, the state designs mechanisms that make citizens known and visible, including ways of classifying individuals or groups (e.g., by income, race, profession, and personal categories). These mechanisms allow for the identification, classification, ordering, and control of the population and its constituent parts, including defining residents who are denied access to full or even partial citizenship. In examining government authority, Foucault goes beyond considering the exercise of raw power to include the important role of individuals' active consent and willingness to participate in their own governance by adhering to government dictates even if they objectively contradict their own self-interests.

While Foucault's work covers a wide range of issues, one arena in which it is not well developed is humanity's relationship with the environment. This territory, however, has been addressed by various Foucault-influenced scholars (Darier 1999; Luke 1999; Rutherford 1999) under the label of "ecogovernmentality." This field of study is guided by the application of Foucault's concepts of biopower—diverse techniques and technologies that confer power bodies, living and dead—and governmentality to the analysis of the regulation of social interactions with the natural world. Exemplary is the analysis by Arun Agrawal (2005) of the effort by the colonial Indian government at the beginning of the 20th century to bring forests in Kumaon Himalaya under centralized control. Traditionally, the residents of this area grazed their animals, chopped and collected firewood, felled lumber, and harvested fodder in the forest. Then, at the turn of the century, the British-controlled colonial state fully inserted itself into process of forest-making by creating new agendas to control, manage, and exploit those landscapes it deemed economically valuable. As elsewhere in India, the forest department in Kumaon Himalaya carried out surveys, delineated various categories of forests, developed plans for the planting, management, and rotational harvesting of trees, limited grazing by domestic animals, restricted and punished the unauthorized collection of fodder and firewood, and introduced fire protection procedures. Those who depended on the forests for their livelihoods rebelled by starting numerous forest fires over a several-year period, continuing to engage in traditional forest activities, and refusing to cooperate with the authorities in identifying resisters. Unable to stop the unwillingness of individuals to participate in their own governance through the exercise of power, the colonial government appointed a three-member committee to investigate villager protests. Ultimately, the committee recommended that the villagers be permitted to take formal control over most of the reclassified forested areas, but under a general set of framing guidelines developed by the government. In the new regime—which has continued into the present—local communities became state-authorized protectors of the forest as their members internalized new state-influenced ways of thinking that supported new identities (as good environmental citizens) and novel social actions, while participating in the achievement of state objectives.

In this light, one of the core functions that government fulfills is the making of environmental citizens (Hobson 2013). For example, both proponents and opponents of hunting think of themselves as good environmental citizens, the former because they support habitat protection and want to see wildlife and wild lands protected to insure there will be plenty of game and hunting opportunities, the latter because they maintain that hunters support habitat changes (like clear cutting) that contribute to increases in game populations, introduce poisonous lead into the environment, and, unlike natural predators, cull healthy adult individuals rather than the sick, the young, and the old. Despite these sharp differences, both groups embrace environmental responsibility as a core value and part of what it means to be a good citizen.

Nilsson & Perrson (2012) identify three other core government functions relative to the environment: 1) reduce environmental risks and vulnerabilities through data collection and monitoring, performance measurement, organized knowledge exchange, and regulation; 2) trigger and control economic activity through economic development, including government spending on ecosystem recovery or the development of carbon markets; and 3) build capacity for change by exploring a range of development options, such as supporting new technologies like solar and wind use in carbon-free energy generation. Based on their analysis, they conclude, at least tentatively, that planetary boundaries can be governed, but this is by no means an easy task because biophysical interactions are so complex. One approach that may prove useful is the creation of stronger integration, both between potentially contradictory economic and environmental policies and between different environmental policies (e.g., those intended to reduce greenhouse gas emissions and those that govern land and water management). This could mean implementing policies around hydropower development that also enhance freshwater protection. These are the type of approaches that are possible at regional levels of governance, such as within the European Union. As this discussion affirms, it is critical to enhance understanding of ecocrises interaction if we are to successfully avoid the peril of crossing Earth's biophysical boundaries.

10.3.2 Modification of values used to define specific planetary boundary dimensions

Water is essential for life on Earth, without exception. This is not true for oxygen. Some anaerobic organisms can grow by metabolizing hydrogen, methane, or several other compounds. But water is another matter. It is an excellent medium for the transport of nutrients and waste, and, because of its polarity (having a positive charge from hydrogen and a negative charge from oxygen), an efficient solvent of material that enters the body (e.g., when we eat). Additionally, water is abundant on Earth—potentially more available than other substances to do the vital jobs it does. However, while about 70 percent of Earth's surface is covered in water, only 3 percent is freshwater, and most of that is frozen as sea or land ice. The primary source of freshwater (apart from relatively small amounts from desalinated seawater, fossil groundwater, and engineered water transfers) is evaporation off the ocean surface, which then falls back as precipitation. Four-fifths of precipitation, however, falls on to and is reabsorbed by the ocean. This means that the total useable amount of freshwater for humans and other non-ocean biota is a tiny fraction of the total amount of water on the planet (Gleick 2000; Newlands 2017).

One of the boundaries identified by Rockström *et al.* (2009) is the maximum amount of freshwater that can be appropriated by humans for their use, beyond which there is a high probability of (possibly very rapid) water-induced environmental changes, with grave impacts on human societies. In the original presentation of the planetary boundary model, this threshold was provisionally set at just over 1 quadrillion (a one with 15 zeros after it) gallons a year of global consumption of "blue water" withdrawn from rivers, reservoirs, lakes, and aquifers. No specific level was set for the "green water" available in soils. Blue water consumption by agriculture, households, and industries already uses about half the original estimated planetary boundary for freshwater. In order to enhance the accuracy of planetary boundary estimates for freshwater, Gerten *et al.* (2013) maintain that it is necessary to account for the environmental flow requirements of rivers and streams. An environment flow requirement is the quantity, timing, and quality of water movement needed to sustain a freshwater ecosystem. These researchers used five different methods to calculate the environmental flow requirements needed to sustain rivers in at least a fair ecological condition and then aggregated the results to obtain average annual values. They thus obtained a revised planetary boundary for consumptive freshwater use that is notably lower (by 30 percent) than the original estimate. In short, the risk for exceeding the planetary boundary for this vital resource is considerably greater than was originally realized. This may also be the case with other of the boundary lines originally suggested by Rockström *et al.* (2009).

10.3.3 Sustainable development goals and planetary boundaries

According to the World Commission on Environment and Development (1987, p. 8), a body established by the United Nations, "[h]umanity has the ability to make development sustainable—to ensure that it meets the needs of the present without compromising the ability of future generations to meet their own needs." The Commission is headed by (and informally named after) Norwegian physician, politician, and environmental advocate Gro Harlem Brundtland, who popularized the concept of sustainable development. It grew out of the 1972 U.N. Conference on the Human Environment, where the conflicts between environment and national socioeconomic development came into sharp focus. This conference produced the Stockholm Declaration, which listed 26 principles concerning the environment and development. While some of these principles identified the need for pollution control, the protection of wildlife, and safeguarding of the oceans, principle number 11 stressed that environment policy must not hamper development. Subsequently, in 1980, the International Union for Conservation of Nature, the U.N. Environment Program, and the World Wildlife Fund (WWF) issued a report entitled the World Conservation Strategy, which argued that essential ecological processes and habitats must be protected, genetic diversity must be preserved, and any use of species and ecosystems must be done in a sustainable manner. Based on these influences, Brundtland wrote in the forward to the World Commission on Environment and Development (1987, p. xi) report: "the "environment" is where we live; and 'development' is what we all do in attempting to improve our lot within that abode. The two are inseparable." Of note, the report also sagaciously stated:

> The Commission has sought ways in which global development can be put on a sustainable path into the 21st Century. Some 5000 days will elapse between the publication of our report and the first day of the 21st Century. What environmental crises lie in store over those 5000 days? (World Commission on Environment and Development 1987)

While there were, in fact, many environmental crises that occurred during those 5000 or so days—including the *Exxon Valdez* oil tanker catastrophe, the intentional burning of 650 oil wells in Kuwait during the first Gulf War, and the rupture of a mining dam that led to extensive river pollution in Aznalcollar, Spain—most prominent among them was recognition of the global significance of climate change, an issue not addressed in the Commission report.

The report's publication was followed by the U.N. Conference on Environment and Development in Rio de Janeiro in 1992, an event attended by more than 100 heads of state that came to be known among environmentalists as the Earth Summit. Ten years later, at the U.N. Conference on Sustainable Development, or Rio+20, U.N. member states agreed to launch a process designed to develop a set of sustainable development goals (SDGs). Based on these events, as part of Resolution 70/1, the U.N. General Assembly committed in 2015 to achieving 17 sustainable development goals by the year 2030. Referring to the development of these goals, Ban Ki-moon (2015), U.N. Secretary-General from 2007 to 2016, stated: "We don't have plan B because there is no planet B."

The SDGs included 14 socioeconomic objectives, including eliminating poverty in all its forms everywhere, ending hunger and all forms of malnutrition, reducing under-5 mortality to at least as low as 25 per 1000 live births, and providing women and girls with equal access to education, health care, decent work, and representation. Setting these crucial objectives, however, has raised concerns among some environmental researchers that, should they be achieved by 2030, the human environmental footprint (based on resource consumption indicators) will surpass Earth's sustainable carrying capacity (O'Neill *et al.* 2018). As ecological economist Dan O'Neill (2018) soberly comments:

> Imagine a country that met the basic needs of its citizens—one where everyone could expect to live a long, healthy, happy and prosperous life. Now imagine that same country was able to do this while using natural resources at a level that would be sustainable even if every other country in the world did the same. Such a country does not exist. Nowhere in the world even comes close. In fact, if everyone on Earth were to lead a good life within our planet's sustainability limits, the level of resources used to meet basic needs would have to be reduced by a factor of two to six times.

In light of this conclusion, is it time to rethink the achievability of sustainable development?

No, it is not, respond Rockström and colleagues (Randers *et al.* 2018a), who view the dual adoption of the SDG initiative and the Paris Climate Agreement as representing a (near) universal development plan or roadmap for the welfare of humanity and the planet. Randers *et al.* (2018a, p. 4) offer "one . . . possible, smarter pathway to success through five transformative and synergistic actions." Many agree the time has come time for both concerted and momentous action, supported by extraordinary global, regional, and local policy changes, in order to avoid existential peril for humanity's future. Without transformative change to attain the SDGs—that is, maintaining improvements at the current pace—only 10 of the 17 targeted development goals will be met by 2030, and even this achievement would come at the expense of pushing eight of the nine planetary boundaries beyond their safe zones. Clearly, an alternative to the current path is needed.

Randers *et al.* (2018a) propose just such an alternative plan, which they call SDGinPB. They envisage it as containing five transformational actions:

1) Accelerated renewable energy growth that is sufficient to cut CO_2 emissions by half every decade.
2) Accelerated productivity in sustainable food chains.
3) New development models in the poorer countries.
4) Active reduction in socioeconomic inequality.
5) Significant investments in education for all, gender equality, and family planning.

Getting fully on this alternative path would require a meaningful shift in the mindsets and values found especially in wealthier nations.

Development of the components of the SDGinPB approach was based on a complex quantitative global computer simulation model called Earth3 (Goluke *et al.* 2018). This model simulates linked socioeconomic and biophysical development over time toward the year 2050. It is composed of three interacting submodels: 1) the socioeconomic submodel (including variables such as population, gross domestic production by person, equity, total use of fossil fuels, and freshwater withdrawal), which is based on a division of the world into seven regions or country clusters; 2) the biophysical submodel, which calculates anthropogenic effects on global warming, sea-level rise, ocean acidity, deforestation, extent of permafrost and glaciers, and productivity of biologically active land; and 3) the performance submodel, which uses outputs from the socioeconomic and biophysical submodels to calculate the development over time of three indicators: a) the number of the 17 SDGs achieved (by region); b) the safety margin (with respect to the nine planetary boundaries); and c) an average regional well-being index based on consumption levels, availability of public services, equity, environmental quality, and population hope for the future. The sources of data inputs for Earth3 include U.N. population data, gross domestic production from the Penn World Tables, the BP Statistical Review of World Energy, Oak Ridge CO_2 data, ecological footprint data from the Global Footprint Network, World Bank development indicators, and educational statistics (Randers *et al.* 2018b).

The Earth3 model system was used to run three different scenarios: 1) "business-as-usual," based on the possibility that global decision-makers will continue to perceive and respond to emerging problems in the conventional manner, without taking any extraordinary action; 2) "accelerated economic growth" of 1 percent per year in all regions; and 3) "stronger focus on SDGs" and on reducing the pressure on planetary boundaries, but without any fundamental change in the current world order (i.e., capitalist, consumerist, focused on short-term gains) and without any meaningful redistribution of income or wealth. The outcomes of running these scenarios are as follows: 1) in neither the business-as-usual scenario nor the accelerated economic growth model does world society achieve the SDGs within the planetary boundaries by 2030, or even by 2050; and 2) the results for scenario 3 are better than for the other two scenarios, as a stronger focus on SDGs leads to the achievement of 12 of the 17 goals by 2030, though it does not lead to the achievement of all of the goals even by 2050.

While all computer simulations are limited by the quality and quantity of input data, this system suggests that sustainable development is possible, but only if there is a shared and vigorous commitment to SDGinPB goals in global governance. This will be no easy task. The Earth3 simulation suggests that extraordinary action is needed (Singer 2018). Considering what is at stake, it would seem to be worth it.

10.3.4 Downscaling planetary-level to subglobal boundaries

Environmental issues are intrinsically scale-dependent and need to take local/regional circumstances into account if they are to be acted upon internally by national governments. At the same time, all countries are exposed to environmental impacts from other countries that affect their ability to respond effectively. The considerable challenges of incorporating biophysical dimensions into national development policy are highlighted by a first attempt at a comparative national "environmental performance" assessment conducted by the Stockholm Resilience Centre and the Stockholm Environment Institute for 61 countries on just four of the planetary boundaries (climate change, water, land, and nitrogen) (Nykvist *et al.* 2013).

To enable comparisons between countries, data were drawn from two sources: international databases and existing peer-reviewed metastudies of large sets of countries. The findings suggested a noteworthy divide between poorer and wealthier countries. The former performed well, in that in per capita terms they used sustainable amounts of their natural resources relative to environmental boundaries. In contrast, wealthier countries and some emerging economies went beyond several of their national boundaries, though there was a less clear pattern on the issue of biodiversity loss. In general, the performance of wealthier countries was found to be worse when consideration was given to consumption "footprints" rather than focusing only on national CO_2/pollution emissions and resource use. The undeniable pattern that emerged, linking population income levels and the crossing of environmental boundaries, is pertinent to calls for the "right to develop" within the shared planetary environmental space. This unalienable right, which was affirmed by the United Nations in 1986 by almost all member nations (except the United States, though this was reversed in 1993) (Sengupta 2001), has been defined as the prerogative of every human person and all peoples to participate in, contribute to, and enjoy economic, social, cultural, and political development. Given this right, the focus on controlling consumption patterns must be placed in highly developed and wealthy countries and the wealthier emerging economies.

This discussion raises the critical issue of fair shares of environmental planetary space. Environmental space refers to the total amount of energy, nonrenewable resources, agricultural land, and forests on the planet that we can use in the present without causing irreversible environmental damage (i.e., crossing thresholds) or depriving future generations of the resources they will need to live healthy and satisfying lives, as reflected in the declaration of the Brundtland Commission. The concept of fair shares is that all peoples of the world, within the carrying capacity of the planet, have equal right to resource consumption. This "right" is not realized in the current distribution of consumption globally. The approximately 1 billion people—or about 20 percent of the world population—living in the wealthiest nations use about 86 percent of the fossil fuels, metals, wood, minerals, and other resources that are extracted from the planet each year. As a result, 1.2 billion (or 20 percent) of the world population survives on less than \$1/day and another 1.8 billion (30 percent) on less than \$2/day. It is estimated that 800 million people go to bed hungry every night, and that between 30 000 and 60 000 people a day do not survive due to hunger alone (Meijer 1999; World

Centric 2005). These chilling statistics underline why efforts to control consumption patterns must focus on developed and wealthy nations and the wealthier emerging nations.

In this light, some attempts to downscale the planetary boundaries approach to subglobal scales have used a "safe and just space for humanity" (SJS) framework. At the national scale, Cole *et al.* (2014) applied this framework to set the parameters for sustainable development in South Africa. The goal was to create a monitoring and communication tool for the national government to use in planning an integrated approach to environmental and social-development issues, including the eradication of social deprivation. The safe and just orientation, which has gained U.N. interest, weds environmentally safe boundaries for development to the improvement of 11 indicators of social well-being, including health, education, employment, social equity, access to potable water, and access to a healthy diet. From the safe and just perspective, "[e]nsuring all people's lives are built upon a social foundation is essential for sustainable development, but so is staying below the environmental ceiling: crossing over either of these boundaries can trigger both social and ecological crises. Sustainable development can only succeed if poverty eradication and environmental sustainability are pursued together" (Raworth 2012, p. 8).

South Africa was chosen by Cole *et al.* (2014) for three reasons. First, it has large, good-quality environmental and social datasets and established national research institutes. Second, as the largest economy in Africa, it has continent-wide influence and global reach among countries with emerging economies, such as Brazil and India. Finally, it is ecologically quite diverse, has significant levels of poverty, and has a very high level of socioeconomic inequality.

Methodologically, in conducting their analysis, these researchers changed two of the global dimensions identified by Rockström *et al.* (2009). Instead of ocean acidification, they focused on marine harvesting, while instead of aerosol loading, they focused on air pollution. They then adjusted all of the planetary boundary indicators and boundaries to suit South African national circumstances. They found that it was challenging to define subglobal equivalents to planetary boundaries. Moreover, variables such as global CO_2 and stratospheric ozone concentrations could not be disaggregated to the national scale, though other dimensions, like nitrogen and phosphorous cycles, were less difficult to examine locally. Eutrophication of freshwater resources is widespread in South Africa and is already a national concern, with phosphorous levels in major reservoirs being monitored as a national indicator. With regard to biosphere integrity, Cole *et al.* (2014) used threat of extinction (i.e., identified endangered and critically endangered ecosystems) rather than extinction as their indicator. Finally, they adopted no indicator for chemical pollution, because of the lack of detailed and accurate data. To assess social deprivation, they used the South African Index of Multiple Deprivation and the annual South African Development Indicators report (Noble *et al.* 2009; Parliamentary Monitoring Group 2013). Again, they made some changes in the original dimensions: water and sanitation were separated into individual dimensions; housing, household goods, and safety were added; and resilience, social equity, and sex equality were removed based on consultation with national experts.

Based on their analysis, Cole *et al.* (2014) found that South Africa:

- exceeds its climate change boundary by 2 percent;
- exceeds its freshwater use boundary by 34 percent;
- is 2 percent below its land use boundary;
- is 2 percent below its phosphorous loading boundary;
- is 41 percent below its nitrogen boundary;
- exceeds its biodiversity loss boundary by 37 percent;

- exceeds its marine harvesting boundary by 45 percent;.
- is 6 percent below its air pollution boundary;
- suffers the most deprivation in the categories of safety (63.5 percent), income (52.5 percent), and jobs (36.3 percent) and the least in terms of basic services, such as electricity and water access.

These findings indicate significant areas of national transcendence of environmental boundaries and enduring areas of social deprivation. The momentous challenge for South Africa, a country in which poverty is widespread and population growth is rapid, is how to address pressing social issues while reducing unsustainable environmental harm and not further exceeding safe environmental boundaries. Cole *et al.*'s (2014) findings, at least, provide a broad roadmap for sustainable development in the country.

This case study suggests that a unified planetary boundary/SJS framework is of value because having the tools to "stay within ecological boundaries over longer timescales would help to ensure inter-generational sustainable resource use" (Dearing *et al.* 2014).

The same framework has also been used in research in two low-income rural Chinese communities where agricultural intensification has contributed to poverty reduction at the expense of environmental degradation, incorporating instrument monitoring, survey, remote sensing, and sediment core analysis from the deepest zones of lakes (Dearing *et al.* 2014). The first community, in Yunnan Province, is dependent on wet and dry cropping, dairy farming, fishing, aquaculture, and forestry. The second, in Anhui Province, is provisioned through intensive agricultural activities. Environmental data collection was designed to gather proxy records of water quality, soil stability, air quality, sediment quality, and sediment regulation.

Findings for community 1 indicate a cautious status for emitted heavy metals and dangerous threshold changes in lake water quality due to agricultural runoff. Socially, records indicate that while health care needs are met, the provision of food, water, and sanitation falls short by substantial margins. Findings for community 2 show that water quality, sediment quality, and air quality have moved into having dangerous status, while soil stability has cautious status. Levels of sediment-associated phosphorous stemming from intensive farming and deposited lead from local fossil fuel -owered industries are both high by international standards—80–100 percent higher than pre-1960s levels in the area. Social needs are met in terms of food security, health care, and minimum income. Access to piped water and sanitation, however, still lags behind other measures of poverty alleviation.

While underlining the tradeoffs between environmental sustainability and social justice (i.e., poverty alleviation) issues, this research provides accessible analyses that pinpoint the statuses of environmental factors that potentially threaten sustainability. What to do with these findings? The challenges faced by environmental scientists in communicating complex concepts in real-world situations (e.g., to policy-makers, the news media, the public, non-government environment organizations) are substantial, as has been well documented for climate scientists (Moser 2009; Farrell *et al.* 2019). Some social analysts have suggested "insider–outside" or "two-communities" models of science–policy/public communication based on a perceived divide between a diverse cadre of scientific knowledge "producers" housed in universities and research institutes and an audience of knowledge "consumers" dwelling in political institutions, media centers, and the wider community (Newman 2014). In seeking to overcome this divide in communicating their findings to relevant audiences, Dearing *et al.* (2014) use a color-coded scheme to identify "safe," "cautious," and "dangerous" categories of operating space, with an environmental ceiling set between the safe/cautious and dangerous conditions. This simple imagery, they argue, "condenses powerful complexity concepts and time-series analyses into an easily understood qualitative basis of assessment" (Dearing *et al.* 2014).

As a wealthy nation with a service-oriented economy, the case of Switzerland offers a contrastive set of downscaling (from planetary to nation-state) challenges. More than half of the environmental impacts created by the consumption of Swiss residents occur outside of the country, because they rely on other parts of the world for the goods they consume. Further, the proportion of imported consumables is on the rise (Frischknecht *et al.* 2014). In 2010, the Swiss government adopted a Green Economy Action Plan, backed up by subsequent measures that embraced the planetary boundary concept. Also mentioned as an action area by Swiss planners is the reduction of environmental impacts along the full value chain (i.e., a footprint perspective).

Dao *et al.* (2018) selected five of the planetary boundaries—climate change, ocean acidification, land cover conversion, biodiversity loss, and nitrogen and phosphorus losses (the latter of which was split into two different boundaries, creating six total)—to use in assessing the Swiss case. While the original planetary boundary framework focused on when boundaries might be crossed, the indicators selected in the Dao *et al.* (2018) study considered whether current yearly footprints are respecting the boundaries. Of concern to these researchers was the responsibility of Switzerland in crossing a global planetary boundary based on a modified "equal share per capita" approach (Nykvist *et al.* 2013). Modification addressed geographic and other differences across locations (e.g., living in Northern European countries requires heating houses for a longer period than in Southern Europe).

Results of this analysis showed that from a global perspective, three of the six computed indicators—climate change, ocean acidification, and biodiversity loss—suggested a clearly unsafe situation. Similar results were found in the case of Switzerland, though nitrogen losses also crossed into unsafe levels. As the researchers report, "the current footprints strongly exceed the limits, by a factor 4.1 for the World and 22.7 for Switzerland" (Dao *et al.* 2018). These findings affirm the correctness of the measures enacted by the Swiss government to significantly reduce its ecological footprint, though the timing of the reductions per capita—approximately 10 percent by 2020, 18 percent by 2030, and 29 percent by 2050—may fall into the "too little, too late" dilemma in light of global patterns.

Other efforts aimed at downscaling planetary-level to subglobal boundaries have focused on regional ecological systems. To begin assessing the relationship of subglobal to global boundaries in Colombia, Vargas *et al.* (2018) examined identified boundaries at the level of a large river basin linked to the Orinoco river. This regional system, lying between Colombia and Venezuela, was selected because of the rapid ecosystem change going on there. Intrusive human activities, such as the development of oil palm plantations and hydropower generation, have grown over several decades, placing mounting pressure on large expanses of relatively undisturbed tropical forest. The area has also seen increasing withdrawal of water resources for use in irrigation (of oil palms, soy, and rice), cattle herding, fodder planting, and hydropower generation, as well as growing eutrophication of lakes and rivers. The researchers focused on the Colombian side of the river basin, an area of over 135 000 square miles. The basin collects water from three sources: the Andes Mountains, an area of mountains and rivers known as the Guyana shield, and part of the Amazon river system. It is an important reserve of tropical forests, covering more than 30 000 square miles.

To assess the utility of the planetary boundaries framework for adaptive management of natural resources at the regional level in the river basin, Vargas *et al.* (2018) used three Earth system processes: disruption of natural nitrogen and phosphorus cycles, human pressure on freshwater availability, and conversion of forests, wetlands, and other vegetation types to agricultural lands. At the global level, specific planetary boundaries have been developed for each of these changes (e.g., 62 teragrams of nitrogen fixated per year). This is not the case at subglobal spatial levels. The researchers found that "[t]urning planetary boundaries into management actions in the Orinoco river basin is challenged by difficulties in identifying stakeholders [willing to turn research into protective policy], uncertainties in defining the level of the boundaries and associated thresholds, and difficulties in developing monitoring strategies based on uncertain thresholds" (Vargas *et al.* 2018). Their work, however, revealed that "ecosystem accounting," an integrated statistical framework used to organize biophysical data, measure ecosystem services, track changes in ecosystem assets, and link all of this information to economic and other human activity, can provide useful information to help overcome these challenges. Ecosystem accounting is designed to use a spatial approach to delineate ecosystem assets as spatial areas that contain a combination of biotic and abiotic components and other characteristics that function together systemically. These ecosystem assets provide ecosystem services, which are the contributions and benefits of ecosystems to vital economic and other human activity (United Nations 2014). Findings from an assessment of the ecosystem assets of the Orinoco Basin, argue Vargas *et al.* (2018), could be used to inform sustainable policy formation before critical environmental boundary lines are crossed. The challenge is to move from information to action.

Not all scientists who have encountered the planetary boundaries framework have been convinced of its utility or accuracy. Some, like Montoya *et al.* (2018), argue that "notions of planetary boundaries add no insight into our understanding of the threats to biodiversity and ecosystem functioning, have no evidence to support them, are too vague for use by those who manage biodiversity, and promote pernicious policies." From this perspective, planetary boundaries is a deeply flawed idea. Those who embrace the utility of the planetary boundary framework fire back that Montoya and colleagues confuse planetary boundaries and tipping points, arguing not only that they are not the same thing, but that boundaries are intentionally drawn short of environmental tipping points in order to trigger alarms prior to the point of hitting a rapid, out-of-control, downward spiral. Usually, the exact location of the tipping point of a planetary system is not precisely known, so boundaries are set to allow time for action to avert system collapse in response to early warning signs that it may be approaching.

An interesting example raised in this debate is the collapse of the Newfoundland cod fishery discussed in Chapter 5. According to Montoya *et al.* (2018), Johan Rockström (2017) claims in a World Economic Forum presentation that the collapse of the fishery in 1989 represents "a very precise tipping point" of human actions transgressing global planetary boundaries. Hence, they conclude, from Rockström's perspective, "[h]uman actions were apparently within bounds before 1989. The year 1989 was apparently 'the boundary between the Holocene and Anthropocene'—a notion we find particularly specious" (Montoya *et al.* 2018). However, as noted in Chapter 5, during the 1950s–1980s new, highly efficient fishing technologies were introduced that dramatically increased fish hauls and led to the collapse. What Rockström is in fact arguing is that in 1989 the cost of increasing cod hauls was made dramatically clear—not that nothing that happened before that year mattered. In fact, in his presentation, Rockström says that he believes that in the mid-1950s, with the Great Acceleration of human pressure on planetary resources (Mcneill 2014), there was an exponential rise in adverse impacts across all Earth system parameters that matter to human well-being. While these technological changes began with the Industrial Revolution, they and their consequences exploded in the period after World War II. While Montoya *et al.*'s (2018) critique is flawed, it does raise the issue of tipping points, which is addressed in the next section.

10.4 Environmental tipping points

In his *English Proverbs and Proverbial Phrases: A Historical Dictionary*, George Latimer Apperson (1929, p. 351) documents the early history of the commonly used proverbial phase, "It is the last straw that breaks the camel's back." This saying refers to the imagery of loading a large burden on a camel's back until, under the additional weight of a mere piece of straw, a tipping point (a concept introduced in Chapter 4) is reached that causes the severely distressed creature to collapse. Using other imagery, consider an expensive vase near the edge of a table. If you press your finger on the top of the vase in the direction of the table's edge, at a certain point it will be literally tipped over and fall to the floor. The tipping point, when the center of gravity passes the vase's balance point, is the precise moment when the vase is displaced from a state of stable equilibrium on the table into a new equilibrium state shattered on the floor. The new steady state is qualitatively and adversely dissimilar from the original one. Of note, the vase did not tip over on its own—it took the force of being pushed to tip it on to the floor.

In his celebrated book, *The Tipping Point: How Little Things Make a Big Difference*—which triggered a tipping point in the popularity of this term—Malcolm Gladwell (2000) refers to a tipping point as the dramatic moment that comes in the wake of an accumulation of smaller changes when everything shifts all at once. He illustrates his point with a simple question:

> I give you a large piece of paper, and I ask you to fold it over once, and then take that folded paper and fold it over again, and again and then again until you have refolded the original paper 50 times. How tall do you think the final stack is going to be? (Gladwell 2000, p. 11)

His unexpected and incredible answer, reflecting the process of exponential progression that occurs as small changes each lead to their own small changes, is that the height of the stack will be about the distance to the sun. In this process, the tipping point is "the moment of critical mass, the threshold, the boiling point" (Gladwell 2000, p. 12). The term labels the fact that "sometimes little things can make a big difference to the state and/or fate of a system" (Lenton 2013), including a planetary geophysical system or cycle. The challenge with environmental tipping points is to improve predictability and the dectability of early warning signals. Enhancement of the former is advanced by development of the latter.

The notion of the tipping point did not begin in the geosciences but rather in sociology, where it was used to label white flight from urban areas into suburbia once the percentage of people of color reached a certain percentage in a neighborhood. During the 1950s, researchers in major American cities found that the process of ethnic tipping proceeded more rapidly in some neighborhoods than in others (Grodzins 1957). While white residents would tolerate a few African American neighbors, they began to sell their homes and move out when the proportion living in their neighborhood passed a certain critical point, though the exact threshold varied by neighborhood and by city. Sociological tipping points like this are shaped by human perceptions, assumptions, social relationships, biases, and values; environmental tipping points rest on geochemical or other physical interactions. The challenge in the physical sciences is how to establish a convincing tipping point in an environmental system; that is, how to know which metaphoric piece of straw will break the camel's back. But the real goal is to understand the system well enough to never find out if you were right by actually reaching the point of no return.

In academic discussions of global change, the term "tipping point" has been used to describe a variety of phenomena, including the following four dynamic relationships: 1) the development of a positive feedback loop (in which an increase in A amplifies B, which further increases A, and so on), as seen in the Arctic ice melt and albedo relationship discussed in Chapter 7; 2) phase and reversible phase transitions involving the movement of a substance from a solid, liquid, or gas state to another state by way of melting, freezing, evaporation, condensation, sublimination, or deposition, such as the melting away of a glacier (a first-order phase transition); 3) phase transitions with hysteresis effects involving abrupt phase change and the absorption or release of energy in the form of latent heat, as illustrated by the rapid climate changes recorded in Greenland during the last glacial cycle; and 4) bifurcations in which small smooth changes made to the parameter values of a system cause a sudden topological break and a change in the system's behavior—an expected occurrence within the climate system. Lenton *et al.* (2008) offer a formal definition for the research of these changes, introducing the term "tipping element" to refer to Earth subsystems that are at least subcontinental in scale (i.e., at a minimum of 620 miles) and can be switched—under the right circumstances—into a qualitatively different state by seemingly minor changes. The tipping point is the precise moment in time at which the future state of the system is qualitatively altered, such as the boiling point in the phase transition from liquid to gas. How tipping points are expressed once they are activated varies. They can be: 1) rapid, abrupt, or spasmodic; 2) gradual or episodic; or 3) slow and asymptotic. In nature, conditions are never constant. Random occurrences, like a wildfire sweeping through a shortleaf pine (*Pinus echinataforest*) forest ecosystem (Fig. 10.2), can change the face of the landscape. The pine trees, however, produce dormant buds underground and store nutrients in their extensive root systems. Resprouting enables an eventual reappearance of the prior

Fig. 10.2 Forest fire. *Source:* skeeze/Pixabay.

ecosystem regime and avoids a regime shift or critical transition (Scheffer *et al.* 2001, 2009). As Paine *et al.* (1998, p. 535) explain:

> All species have evolved in the presence of disturbance, and thus are in a sense matched to the recurrence pattern of the perturbations. Consequently, disturbances within the typical range, even at the extreme of that range as defined by large, infrequent disturbances . . . usually result in little long-term change to the system's fundamental character.

In other words, most forest fires are not points of no return. This is illustrated by the devastating 1988 fires in Yellowstone National Park, which began as several individual fires grew into one massive inferno. This destroyed 793 880 acres, or 36 percent of the park's total acreage, and is considered to be not only the largest wildfire recorded in the park, but on an order of magnitude greater than any prior fires (Young 2001). Though it burned for several months and is considered one of the most important events to have taken place in the National Park System, it produced no lasting ecological surprises. Stone (1998, p. 1527) notes: "the postfire ecosystems are shaping up to be essentially the same as those that prospered before the flames."

Radical shifts in environments do occur, however. A socioecological example is the "compost bomb." A compost pile, consisting of dense concentrations of decomposing organic matter (from kitchen scraps to lawn clippings), which is subjected to a critical level of external heating of between 300 and 400 °F from the sun, lightning, or another source can trigger a runaway breakdown involving escalating internal release of heat by composting microbial organisms, resulting in a fire that destroys the pile. Compost bomb instability involving a potentially catastrophic explosive release can occur in naturally decomposing turf like peatland environments under conditions of intensified global warming (Wieczorek *et al.* 2011). This happened in Western Russia in 2010. The fires were associated with record-high temperatures during the hottest summer in Russian history. The CO_2 release from large peat fires is massive.

Additionally, paleoclimatologists have identified climate-related points of critical transition to new states during Earth's deep history. These involve icehouse-to-greenhouse or greenhouse-to-icehouse fluctuations. The term "greenhouse Earth" refers to long periods of time when there are no continental glaciers on the planet; "icehouse Earth" labels periods like the present when the planet is covered to some degree by ice sheets. An extreme version of the latter, when the planet's surface is completely frozen over, is referred to as "snowball Earth," a condition postulated to have occurred at least three times in Earth's long history (Lenton 2013). One of the known climate turnovers occurred during the Cenozoic, about 34 million years ago. It involved a transition from greenhouse to icehouse conditions, expressed as a shift from a relatively deglaciated climate state to one in which the Antarctic ice sheet grew to between 40 and 160 percent of its contemporary size (Coxall *et al.* 2005). A prolonged decrease in atmospheric CO_2 is believed to have been the cause. The issue of contemporary concern is a reverse icehouse to greenhouse transition, or at least significant movement toward such a planetary change. This would entail "a trip forward to the past," driven by human behavior on the planet (National Research Council 2011, p. 68).

Other tipping points in Earth's history were likely driven by the evolution of lifeforms and their acquisition of novel forms of energy. The origin of photosynthesis about 2.7 billion years ago, for example, led to what was likely the most profound of all: the Paleoproterozoic transition of Earth's atmosphere to an oxygen-rich state. This may have occurred in a geologically short span of time coincident with a reduction of atmospheric oxygen-consuming gases (Claire *et al.* 2006). One scenario is that once the atmospheric oxygen concentration built up to a level sufficient to enable stratospheric ozone (O_3) to accumulate, the emerging ozone layer shielded the troposphere below it from rays of ultraviolet light. This activated a runaway positive feedback involving a reduction in the consumption of oxygen by methane, which facilitated the accumulation of more ozone, allowing more oxygen into the system, and so on. A second reduction in oxidation that permitted another rise in planetary oxygen occurred near the end of the Proterozoic, about 0.6 billion years ago. It was at this time that the deep oceans were oxygenated and the first animals began to flourish in the seas. This biological transition caused a significant ecological transition from simple ecosystems dominated by cyanobacteria to an early form of modern predation-based plant and animal food webs encompassing multiple trophic levels. As Butterfield (2011, p. 81) emphasizes:

> Animals do not just occupy the modern biosphere, they permeate its structure and define how it works. Their unique combination of organ-grade multicellularity, motility and heterotrophic habit makes them powerful geobiological agents, imposing myriad feedbacks on nutrient cycling, productivity and environment.

Over time, animals have "engineered the biosphere" (Butterfield 2011, p. 81), most profoundly since the rise of humans. Though species can affect environments, environmental change affects species evolution. Research suggests that the environmental adaptive capacity of species is likely subject to evolutionary tipping points (Botero *et al.* 2015) that result in novel biological and behavioral adaptive traits and extinction.

A case of current movement toward a tipping point is the melting of permafrost discussed in Chapters 6 and 7. The loss of permafrost triggers a closed loop of causal connections that are self-amplifying. Phase transition in permafrost releases two greenhouse gases: carbon dioxide and an even more potent (by roughly 30 percent) heat-trapping gas, methane (CH_4). Also released, but less studied, is billions of tons of nitrogen stored in frozen soils. Research shows that permafrost meltdown could cause nitrogen to be released as nitrous oxide (N_2O), a greenhouse gas that is almost 300 times more powerful than CO_2 (Voigt *et al.* 2017). This could occur from barren peatland hot spot surfaces covering almost one-fourth of the Arctic. Whereas the majority of positive feedbacks in nature are not strong enough to create tipping points, the positive climate change feedback of combined CO_2, CH_4, and N_2O release in large quantities would likely tip "runaway." This means that one trip around the positive feedback loop would produce an additional change that was as large as or larger than the initial deviation, and the next trip around would produce an even larger one, with each further trip driving a new level of planetary warming. This pattern would lead in each cycle to additional permafrost melt, ice-sheet melt, sea-ice melt, sea-level rise, coastal and island flooding, and other adverse changes and consequences of great socioeconomic and health concern.

Of primary interest are the forces that drive environments to stage change tipping points. These vary and can be difficult to determine. This point is affirmed by examining the forces involved in swinging savanna/forest biostability areas like the Serengeti–Mara ecosystem of East Africa in a particular ecological direction. This region, which contains the Serengeti National Park, has become well known because it supports the most diverse migration of grazing mammals on Earth. Every year, at roughly the same time (depending on rain and the availability of grazing), a circular migration of almost 2 million wildebeest begins in the southern Serengeti and moves in a clockwise direction through the Serengeti National Park in Tanzania and north toward the Masai Mara reserve in Kenya. In addition to wildebeest, elephants and fire have been suggested as important drivers contributing to changes in the extent of Serengeti tree cover (Dublin *et al.* 1990; Sinclair *et al.* 2007). Additionally, it has been proposed that the eradication of rinderpest, a deadly infectious disease of cattle and other even-toed ungulates like wildebeest, had a major impact that cascaded throughout the ecosystem. The organized vaccination of cattle against rinderpest in East Africa began in the 1940s and continued through the 1960s. The effects of the disease's eradication included an eruption in the size of the wildebeest population, which led to: 1) a reduction in total standing grass biomass, because of intensified grazing; 2) a reduction in the frequency of fire, because of its relationship to fuel biomass availability; and 3) conversely, an increase in the extent of tree cover, because of the reduction in fire frequency (Dobson 1995).

A reversal of the East African pattern of tipping from savanna to woodland is of concern in the tropical areas of South America, where forest could be replaced by savanna or even semi-desert due to climate change and land clearing. Oyama & Nobre (2003) observe that "[i]f sustainable development and conservation policies [are] not able to halt the increasing environmental degradation . . . then land use changes could, per se, tip the climate-vegetation system towards [a] new alternative drier stable equilibrium state, with savannization of parts of Amazonia and desertification of the driest area of Northeast Brazil." Large-scale forest dieback associated with forest–savanna critical transition in the Amazon is likely to have significant social costs for human populations (Franklin & Pindyck 2018).

In his broad consideration of tipping points, Lenton (2013) addresses the issue of intentionally encouraging them. The EcoTipping Points Project (Marten *et al.* 2012), begun by human ecologist Gerald Marten and journalist Steve Brooks, for example, undertakes assessment and promotion of "small actions that tip the balance from decline to restoration" of ecosystems with the intention of "restoring sustainability to our imperiled environment." In other words, it seeks to develop and encourage an environmental activist conception of environmental tipping. One of the actions highlighted by the project is the restoration of badly degraded mangrove forests around the villages of Leam Markham and Thung Dase in the Trang province of southern Thailand. Residents of these subsistence villages traditionally lived as small-scale fishers who made a living by laying their nets in the waters of coastal fisheries and by extracting medicinal plants, thatch, and lumber for housing and materials for making boats and fishing gear from the coastal mangrove forests. This way of life came to be threatened, however, and "the lives and livelihoods of the small-scale fisherfolk communities along the coast of Trang Province . . . changed beyond all recognition" (Prasertcharoensuk *et al.* 2010, p. 27). Fisheries close to the shore were heavily damaged by large trawler vessels that began fishing the coasts of southern Thailand and encroaching on the villagers' fishing grounds. Their mechanized gear damaged coral, scraped the seabed, and removed high quantities of local fish. Large tracts of marine and coastal areas were ravished. At the same time, the mangrove forests were opened up to commercial outfits, which began clearing them to produce charcoal. Subsequently, the land was given over to commercial shrimp farming. Caught in a trap, "where day-to-day survival strategies eliminated or reduced their future options," villagers were forced to take jobs on trawlers or cutting down mangroves for low wages, "and the result was a self-reinforcing downward spiral into increasing poverty, and social and environmental degradation" (Suutari 2006).

In 1985, Yadfon (which means Raindrop), a small development organization, began working with Leam Markham and Thung Dase villagers to rebuild and protect their mangrove forests. Villagers created a 235-acre community-managed forest and sea-grass conservation zone. Mangrove seedlings were planted in degraded areas of the forest, while sea grass, which provides vital nursery areas and feeding grounds for many species of fish and shellfish, was replanted in the lagoon. Within several years, the environment tipped. Community-managed mangrove forests began regenerating and the coastal fishery revived. This helped spark a regeneration of the village's society and economy.

Martens (2005, p. 75) describes a parallel local case based on the work of Angel Alcala, a marine scientist from the Silliman University Marine Laboratory in the Philippines:

> Apo Island in the Philippines provides an example of environmental tipping points in action. The introduction of destructive fishing methods was a "negative tip" that set the regional fishery on a forty-year downward spiral to virtual collapse. Apo Island escaped the downward spiral with a "positive tip"—the creation of a small marine sanctuary -which set in motion a cascade of ecological and social changes that restored declining fish stocks and returned the island's marine ecosystem to health.

The EcoTipping Points Project views such examples of working with environmental tipping points to recover desirable system states as a turn from vicious to virtuous cycles. Lenton *et al.* (2008) refer to them as "Gaia geoengineering," after the ancient Greek god that personified Earth. While very beneficial at the local scale, however, small projects like those undertaken by Yadfon, the Silliman University Marine Laboratory, and numerous other NGOs and institutions in Thailand, the Philippines, and worldwide are "a stark reflection of our collective inability thus far to act to tackle the [ultimate] causes of climate change" and other anthropogenic ecocrises (Lenton *et al.* 2008). What is needed, in fact, Lenton *et al.* (2008) argue, is for us to reach a social tipping that allows a rethinking of huge corporate and societal investment in fossil fuels as our dominant energy source.

Within news media communication, use of the term "tipping point," which has gained increasing prominence as a form of metaphorical framing in media coverage of climate change (Burgers *et al.* 2016), is itself contributing to social tipping by facilitating the restructuring of public perceptions of the subject (Russill & Nyssa 2009). Scientists often use metaphors to communicate complex scientific phenomena to the public. An example is the use of warfare metaphors to discuss infectious disease (e.g., "fighting infection"). Metaphors are useful because they offer a handy mechanism through which our minds can "conceptualize the world around us, especially in the face of complexity" (Khullar 2014). Simply put, the familiar and perceptible helps us make sense of the unfamiliar and intricate. At the same time, scientists routinely engage in analogical reasoning using metaphors to develop hypotheses and interpret results. Metaphors assist scientists in structuring new conceptual domains facilitating the development of theory. Niebert & Gropengießer (2015, p. 2) provide several examples:

> Robert Hooke was the first to denote the cell using the term "cell" when an image of a piece of cork under his microscope reminded him of the small rooms, or cells, occupied by monks in monasteries. Kepler developed his concept of planetary motion by comparison with a clock. Huygens used water waves to theorise that light is wavelike. Arrhenius described the greenhouse effect by referring to his experience with hot pots. In ever new variations, scientists employ experiences from everyday life to understand scientific phenomena.

In fact, as Taylor & Dewsbury (2018) indicate, "[t]he language of science is largely metaphorical." This is not surprising in light of recognition that the very nature of human cognition is metaphorical (Lakoff & Johnson 1980).

The use of the tipping point as metaphor is made clear by the example of the tipping of the vase. There is an observable physical tipping that occurs with the passing of the vase's center of gravity. In environmental and climatic cases, the tipping (passing an angle of balance), per se, is only metaphorical, though the passage to a new—perhaps far more adverse—state is still very much the case. Communication scientists refer to this relationship as cross-domain mapping, and it involves people using their knowledge about a source domain, such as concrete objects like vases on tables, to better understand a target domain, such as the climate or another earth system (or white flight to the suburbs). In communication science terms, the notion of the tipping point became a discourse metaphor that now functions as a key framing device within both climate and environmental sciences (Zinken *et al.* 2008). Its ready adoption in science discourses is likely a consequence of it utility, linguistic availability, and ubiquity during the early 2000s. Yet, some have questioned this development. Anthropologist Mark Nuttall (2012), for one, argues that:

> Through this shift in scientific thinking [and terminology], climate is talked about and represented as having the power to influence our lives in ways we have never before experienced or imagined, suggesting something transformative, disruptive, and decentering. Yet, in doing so, the complexity of the human world is explained in terms of scientific models that suggest a return to climatic determinism that simplifies our understanding of human–environment relations.

To explore this issue, van der Hel *et al.* (2018) developed a study to assess how the meaning and use of the tipping point metaphor in science and the news media developed over the 10-year period from 2005 until 2014. Specifically, they were interested in the use of linguistic expressions of the metaphor, its framing function in climate change discourse, and its role as a communication bridge between the jargons of the news media and climate science. They carried out their investigation using a search engine examination of all publications in major world newspapers to identify articles for textual analysis. For the scientific literature, the search engine Web of Science was used. The authors found distinctive patterns in the use of the notion of tipping points in discussions of climate change in both the news media and scientific literature. In the former, there was an initial adoption by journalists of tipping points in climate change as a metaphorical scientific concept encased in inverted commas and explained as motion in space. Over time, tipping points became conventionalized as a way to talk about impending climate changes as important events, without drawing attention to the metaphorical status of the phrase. In the scientific literature, movement was from the use of the tipping point as a clear rhetorical device to its use, in a significantly expanding literature, as a conceptually charged technical term in research on climate. Highlighting the achievement of an established position for the term, in 2008 Richard Kerr published a report on the annual meeting of the American Geophysical Union entitled "Climate Tipping Points Come In From the Cold" in the journal *Science* (Kerr 2008). As this discussion suggests, the evolution of the use of the tipping point metaphor has contributed to a social tipping that has helped reconfigure public perceptions of climate change as an issue of concern.

What remains to be determined is the role the metaphor played in the record surge in concern about climate change in the United States during the second decade of the 21st century (Leiserowitz *et al.* 2018). A similar surge occurred in the United Kingdom. Research there ties it to mass media coverage of specific events like protests in London led by Extinction Rebellion, a major BBC documentary presented by Sir David Attenborough, and the visit to London by the Swedish school climate protestor Greta Thunberg. While these events do appear to have pushed popular apprehension about climate change, the level of concern has been rising for some time. Recent polls suggest that much of the U.K. public is coming to share Extinction Rebellion's point of view (Barasi 2019). The importance of social activism in raising public concern over climate change is addressed in the final chapter.

References

Agrawal, A. (2005) *Environmentality: Technologies of Government and the Making of Subjects.* Durham, NC: Duke University Press.

Apperson, G. (1929). *English Proverbs and Proverbial Phrases: A Historical Dictionary.* London: J.M. Dent and Sons Limited.

Barasi, L. (2019). Guest post: Polls reveal surge in concern in U.K. about climate change. *Carbon Brief.* https://www.carbonbrief.org/guest-post-rolls-reveal-surge-in-concern-in-uk-about-climate-change.

Barnosky, A., Hadly, E., Bascompte, J., Berlow, E., Brown, J., Fortelius, M. *et al.* (2012). Approaching a state shift in Earth's biosphere. *Nature* 486: 52–58.

Botero, C., Weissing, F., Wright, J. & Rubenstein, D. (2015). Evolutionary tipping points in the capacity to adapt to environmental change. *Proceedings of the National Academy of Science of the U.S.A.* 112(1): 184–189.

Burgers, C., Konijn, E. & Steen, G. (2016). Figurative framing: shaping public discourse through metaphor, hyperbole and irony. *Communication Theory* 26(4): 410–430.

Butterfield, N. (2011). Animals and the invention of the Phanerozoic Earth system. *Trends in Ecological Evolution* 26(2): 81–87.

Ciriacy-Wantrup, S. (1952). *Resource Conservation: Economics and Policies.* Berkeley, CA: University of California Press.

Claire, M., Catling, D. & Zahnle, K. (2006). Biogeochemical modelling of the rise in atmospheric oxygen. *Geobiology* 4(3): 239–269.

Cole, M., Bailey, R. & New, M. (2014). Tracking sustainable development with a national barometer for South Africa using a downscaled "safe and just space" framework. *Proceedings of the National Academy of the U.S.A.* 111(42): E4399–E4408.

Coxall, H., Wilson, P., Pälike, H. & Lear, C. (2005). Rapid stepwise onset of Antarctic glaciation and deeper calcite compensation in the Pacific Ocean. *Nature* 433(7021): 53–57.

Daly, H. & Cobb, J. (1989). *For the Common Good: Redirecting the Economy Toward Community, the Environment and a Sustainable Future.* Boston, MA: Beacon Press.

Dao, H., Peduzzi, P. & Priot, D. (2018). National environmental limits and footprints based on the planetary boundaries framework: the case of Switzerland. *Semantic Scholar* 52: 49–57.

Darier, E. (1999) Foucault and the environment: an introduction. In E. Darier (Ed.), *Discourses of the Environment*, pp. 1–34. Malden, MA: Blackwell.

Dearing, J., Wang, R., Zhang, K., Dyke, J., Haberl, H., Hossain, M. *et al.* (2014). Safe and just operating spaces for regional social-ecological systems. *Global Environmental Change* 28: 22.

Dobson, A. (1995). The ecology and epidemiology of rinderpest virus in Serengeti and Ngorongoro Conservation Area. In: A. Sinclair & P. Arcese (Eds.), *Serengeti II: Dynamics, Management, and Conservation of an Ecosystem*, pp. 485–505. Chicago, IL: University of Chicago Press.

Dublin, H., Sinclair, A. & McGlade, J. (1990). Elephants and fire as causes of multiple stable states in the Serengeti-Mara woodlands. *Journal of Animal Ecology* 59(3): 1147–1164.

Elliot, L. (2002). Global environmental governance. In S. Hughes & R. Wilkinson (Eds.), *Global Governance: Critical Perspectives*, pp. 57–73. London: Routledge.

Farrell, J., McConnell, K. Brulle, R. (2019). Evidence-based strategies to combat scientific misinformation. *Nature Climate Change* 9(7): 191–195.

Foucault, M. (1991). Governmentality. In: G. Burchell, C. Gordon & P. Miller (Eds.), *The Foucault Effect: Studies in Governmentality*, pp. 87–104. Chicago, IL: University of Chicago Press.

Foucault, M. (2010). *The Government of Self and Others: Lectures at the Collège de France 1982–1983* (A. Davidson, Ed.). New York: Palgrave Macmillan.

Franklin, S & Pindyck, R. (2018). Tropical forests, tipping points, and the social cost of deforestation. *Ecological Economics* 153: 161–171.

Frischknecht, R., Nathani, C., Büsser, S., Itten, R., Wyss, F. & Hellmüller, P. (2014). *Development of Switzerland's Worldwide Environmental Impact, Umwelt-Wissen*. Bern: Federal Office for the Environment.

Galaz, V., Biermann, F., Folke, C., Nilsson, M. & Olsson, P. (2012). Global environmental governance and planetary boundaries: an introduction. *Ecological Economies* 81: 1–3.

Gerten, D., Hoff, H., Rockström, J., Jägermeyr, J., Kummu. M. & Pastor, A. (2013). Towards a revised planetary boundary for consumptive freshwater use: role of environmental flow requirements. *Current Opinion in Environmental Sustainability* 5(6): 551–558.

Gladwell, M. (2000). *The Tipping Point: How Little Things Make a Big Difference*. Boston, MA: Little, Brown.

Gleick, P. (2000). *The World's Water 2000–2001: The Biennial Report on Freshwater Resources*. Washington, D.C.: Island Press.

Goluke, U., Randers J., Collste D. & Stoknes, P.-E. (2018). Earth3 model system. https://osf.io/a8mvf/.

Grodzins, M. (1957). Metropolitan segregation. *Scientific American* 197: 33–41.

Hobson, K. (2013). On the making of the environmental citizen. *Journal of Environmental Politics* 22(1): 56–72.

Kerr, R. (2008). Climate tipping points come in from the cold. *Science* 319(5856): 153.

Khullar, D. (2014). The trouble with medicine's metaphors. *The Atlantic*. https://www.theatlantic.com/health/archive/2014/08/the-trouble-with-medicines-metaphors/374982/.

Ki-moon, B. (2015). Secretary-General's remarks to the press at COP22. https://www.un.org/sustainabledevelopment/blog/2016/11/secretary-generals-remarks-to-the-press-at-cop22/.

Lakoff, G. & Johnson, M. (1980). *Metaphors We Live By*. Chicago, IL: University of Chicago Press.

Leiserowitz, A., Maibach, E., Rosenthal, S., Kotcher, J., Ballew, M., Goldberg, M. *et al.* (2018). Climate change in the American mind: December 2018. Yale Program on Climate Change Communication.

Lenton, T. (2013). Environmental tipping points. *Annual Review of Environment and Resources* 38: 1–29.

Lenton, T., Held, H., Kriegler, E., Hall, J., Lucht, W., Rahmstorf, S. *et al.* (2008). Tipping elements in the Earth's climate system. *Proceedings of the National Academy of Sciences of the U.S.A.* 105(6): 1786–1793.

Luke, T. (1999). Environmentality as green governmentality. In E. Darier (Ed.), *Discourses of the Environment*, pp. 121–151. Malden, MA: Blackwell.

Marten, G., Gregory, R., Morrison, B., Suutari, A. & Nuñez, D. (2012). A new hope for positive change and sustainability. The EcoTipping Points Project. http://www.ecotippingpoints.org/.

Martens, G. (2005). Environmental tipping points: a new paradigm for restoring ecological security. *Journal of Policy Studies* 20: 75–87.

Mcneill, J. (2014). *The Great Acceleration: An Environmental History of the Anthropocene Since 1945*. Cambridge, MA: Harvard University Press.

Meadows, D., Randers, J, & Meadows, D. (2004). *Limits to Growth: The 30-Year Update*. White River Junction, VT: Chelsea Green.

Meijer, A. (1999). A fair share in environmental space. Share International. http://www.share-international.org/archives/environmental/en_amfair.htm.

Montoya, J., Donahue, I. & Primm, S. (2018). Planetary boundaries for biodiversity: implausible science, pernicious policies. *Trends in Ecology & Evolution* 33(2): 71–73.

Moser, S. (2009). Communicating climate change: history, challenges, process and future directions. *WIREs Climate Change* 1(1): 31–51.

National Research Council. (2011). *Understanding Earth's Deep Past: Lessons for Our Climate Future*. Washington, D.C.: National Academies Press.

Newlands, N. (2017). *Future Sustainable Ecosystems: Complexity, Risk, and Uncertainty*. Boca Raton, FL: CRC Press.

Newman, J. (2014). Revisiting the "two communities" metaphor of research utilisation. *International Journal of Public Sector Management* 27(7): 614–627.

Niebert, K. & Gropengießer, H. (2015). Understanding starts in the mesocosm: conceptual metaphor as a framework for external representations in science teaching. *International Journal of Scientific Education* 37(5–6): 903–933.

Nilsson, M. & Perrson, A. (2012). Can Earth system interactions be governed? Governance functions for linking climate change mitigation with land use, freshwater and biodiversity protection. *Ecological Economics* 81: 10–20.

Noble, M., Barnes, H., Wright, G., McLennan, D., Avenell, D., Whitworth, A. *et al.* (2009). The South African Index of Multiple Deprivation 2001 at Datazone Level. Department of Social Development.

Nuttall, M. (2012). Tipping points and the human world: living with change and thinking about the future. *Ambio* 41(1): 96–105.

Nykvist, B., Persson, Å., Moberg, F., Persson, L., Cornell, S. & Rockström, J. (2013). National Environmental Performance on Planetary Boundaries: A Study for the Swedish Environmental Protection Agency. Stockholm Environment Institute.

O'Neill, D. (2018). Is it possible for everyone to live a good life within our planet's limits? *The Conversation*. https://theconversation.com/is-it-possible-for-everyone-to-live-a-good-life-within-our-planets-limits-91421.

O'Neill, D., Fanning, A., Lamb, W. & Steinberger, J. (2018). A good life for all within planetary boundaries. *Nature Sustainability* 1(2): 88–95.

Oyama, M. & Nobre, C. (2003). A new climate-vegetation equilibrium state for tropical South America. *Geophysical Research Letters* 30(23): 2199.

Paine, R., Tegner, M. & Johnson, E. (1998). Compounded perturbations yield ecological surprises. *Ecosystems* 1: 535–545.

Pardini, R., de Arruda Bueno, A., Gardner, T., Prado, P. & Metzger, J. (2010). Beyond the fragmentation threshold hypothesis: regime shifts in biodiversity across fragmented landscapes. *PLoS One* 5: e13666.

Parliamentary Monitoring Group. (2013). Department of Performance Monitoring and Evaluation 2013: Strategic & Annual Performance Plans. Department of Performance Monitoring and Evaluation.

Prasertcharoensuk, R., Shott, J., Weston, D. & Ronarongpairee, W. (2010). Time for a sea change: a study of the effectiveness of biodiversity conservation measures and marine protected areas along Southern Thailand's Andaman Sea coastline. International Collective in Support of Fishworkers.

Randers, J., Rockström, J., Stoknes, P., Golüke, U., Collste, D. & Cornell, S. (2018a). Transformation is feasible—how to achieve the sustainable development goals within planetary boundaries. Report to the Club of Rome from Stockholm Resilience Centre and BI Norwegian Business School. https://www.stockholmresilience.org/publications/artiklar/2018-10-17-transformation-is-feasible---how-to-achieve-the-sustainable--development-goals-within-planetary-boundaries.html.

Randers, J., Rockström, J., Stoknes, P.-S., Goluke, U., Collste, D., Cornell, S. *et al.* (2018b). Achieving the 17 Sustainable Development Goals Within 9 Planetary Boundaries. https://

www.cambridge.org/core/journals/global-sustainability/article/achieving-the-17-sustainable-development-goals-within-9-planetary-boundaries/5934F82F471B751168A0B2AE59AD0319.

Raworth, K. (2012). A safe and just space for humanity: can we live within the doughnut. *Oxfam Policy and Practice: Climate Change and Resilience* 8(1): 7–11.

Rockström, J. (2017). Beyond the Anthropocene. World Economic Forum. https://www.youtube.com/watch?v=V9ETiSaxyfk.

Rockström, J., Steffen, W., Noone, K., Persson, A., Chapin, F., Lamet, E. *et al.* (2009). Planetary boundaries: exploring the safe operating space for humanity. *Ecology and Society* 14(2): 32.

Russill, C. & Nyssa, Z. (2009). The tipping point trend in climate change communication. *Global Environmental Change* 19(3): 336–344.

Rutherford, P. (1999) The entry of life into history. In E. Darier (Ed.), *Discourses of the Environment*, pp. 37–62. Malden, MA: Blackwell.

Scheffer, M., Carpenter, S., Foley, J., Folke, C. & Walker, B. (2001). Catastrophic shifts in ecosystems. *Nature* 413: 591–596.

Scheffer, M., Bacompte, J., Brock, W., Brovkin, V., Carpenter, S., Dakos, V. *et al.* (2009). Early warning signals for critical transitions. *Nature* 461: 53–59.

Sengupta, A. (2001). Right to development as a human right. *Economic and Political Weekly* 36(27): 2527–2536.

Sinclair, A., Mduma, S., Hopcraft, J., Fryxell, J., Hilborn, R. & Thirgood, S. (2007). Long-term ecosystem dynamics in the Serengeti: lessons for conservation. *Conservation Biology* 21(3): 580–590.

Singer, M. (2018). *Climate Change and Social Inequality: The Health and Social Costs of Global Warming*. Abingdon: Routledge.

Steffen, W., Richardson, K., Rockström, J., Cornell, S., Fetzeret, I., Bennnett, E. *et al.* (2015). Planetary boundaries: guiding human development on a changing planet. *Science* 347(622): 1259855.

Stone, R. (1998). Yellowstone rising again from ashes of devastating fires. *Science* 280: 1527–1528.

Suutari, A. (2006). Thailand—Trang Province—taking back the mangroves with community management. The EcoTipping Points Project. http://www.ecotippingpoints.org/our-stories/indepth/thailand-mangrove-restoration-community-management.html.

Swift, T. & Hannon, S. (2010). Critical thresholds associated with habitat loss: a review of the concepts, evidence, and applications. *Biological Reviews* 85(1): 35–53.

Taylor, C. & Dewsbury, B. (2018). On the problem and promise of metaphor use in science and science communication. *Journal of Microbiological Education* 19(1): 19.1.46.

United Nations (2014). System of Environmental-Economic Accounting 2012: Experimental Ecosystem Accounting.

van der Hel, S., Hellsten, L. & Steen, G. (2018). Tipping points and climate change: metaphor between science and the media. *Environmental Communication* 12(5): 605–620.

Vargas, L., Willemen, L., & Hein, L. (2018). Linking planetary boundaries and ecosystem accounting, with an illustration for the Colombian Orinoco river basin. *Regional Environmental Change* 18(5): 1521–1534.

Vitousek, P., Mooney, H., Lubchenco, J. & Melillo, J. (1997). Human domination of the Earth's ecosystems. *Science* 277: 494–499.

Voigt, C., Marushchak, M., Lamprecht, R., Jackowicz-Korczyński, M., Lindgren, A., Mastepanov, M. *et al.* (2017). Increased nitrous oxide emissions from Arctic peatlands after permafrost thaw. *Proceedings of the National Academy of the U.S.A.* 114(24): 6238–6243.

Walker, B., Barrett, S., Polasky, S., Galaz, V., Folke, C. & Engström, G. (2009). Looming global-scale failures and missing institutions. *Science* 325(5964): 1345–1346.

Wieczorek, S., Ashwin, P., Luke. C. & Cox P. (2011). Excitability in ramped systems: the compost-bomb instability. *Proceedings of the Royal Society A* 467: 1243–1269.

World Centric (2005). Social & economic injustice. http://www.worldcentric.org/conscious-living/social-and-economic-injustice.

World Commission on Environment and Development. (1987). *Our Common Future.* New York: Oxford University Press.

Young, L. (2001). Wildland fire education and outreach case studies flames of controversy: interpreting the Yellowstone fires of 1988. Wildland Fire Education and Outreach Case Studies. https://web.archive.org/web/20070623130522/http://www.nifc.gov/preved/comm_guide/wildfire/fire_26c.html.

Zinken, J., Hellsten, I. & Nerlich, B. (2008). Discourse metaphors. In R.M. Frank, R. Dirven, T. Ziemke & E. Bernádez (Eds.), *Body, Language, and Mind. Volume 2: Sociocultural Situatedness*, pp. 363–385. Berlin: Mouton de Gruyter.

11

Time for change? Toward sustainability, toward life

Some people say that the climate crisis is something that we all have created, but that is not true, because if everyone is guilty then no one is to blame. And someone is to blame. Some people, some companies, some decision-makers in particular, have known exactly what priceless values they have been sacrificing to continue making unimaginable amounts of money. And I think many of you here today belong to that group of people.

Greta Thunberg (quoted in Kottasova & Mackintosh 2019)

11.1 Why go to school?

Greta Thunberg is a high school climate change activist from Sweden who helped to launch the global school strike to fight climate change. Her motivation for urging fellow students around the world to go on strike, which they have done in the tens of thousands (Fig. 11.1), she explains was the simple question: "Why should I be studying for a future that soon may not exist?" (quoted in Kottasova & Mackintosh 2019). Inspired by the school strikes organized by the survivors of the Stoneman Douglas High School gun massacre in Parkland, Florida on August 20, 2018, Thunberg stopped going to school and instead sat down in protest outside the Swedish Parliament. She held a sign that read "*Skolstrejk för klimatet*" ("School Strike for the Climate") and handed out fliers with known facts about the climate crisis and an explanation of why she was striking. She returned with other students every school day until the 2018 Sweden general election on September 9, and afterward continued her strike on Fridays. Her story was picked up in the media and she began to garner considerable attention locally and then globally, not all of it friendly. The epigraph to this chapter is from her pointed speech at the 2019 World Economic Forum, a gathering of government leaders and corporate heads, many of whom do make unimaginable amounts of money.

At the close of a TedTalk Stockholm, Thunberg (2018) noted that people usually end speeches on climate change by talking about hope. She declined to do so, saying:

> We've had 30 years of pep-talking and selling positive ideas and I am sorry but it doesn't work because if it would have the emissions would have gone down by now. They haven't. And yes, we do need hope, of course we do but the one thing we need more than hope is action. Once we start to act hope is everywhere. So instead of looking for hope look for action.

At the World Economic Forum, she added that humanity is faced with a choice:

> We can create transformational action that will safeguard the future living

Ecosystem Crises Interactions: Human Health and the Changing Environment, First Edition. Merrill Singer.
© 2021 John Wiley & Sons, Inc. Published 2021 by John Wiley & Sons, Inc.

Fig. 11.1 Climate action marcher with Greta Thunberg sign. *Source:* Markus Spiske/UnSplash.

> conditions for humankind. Or we can continue with our business as usual and fail . . . I want you to feel the fear every day and I want you to act. I want you to act as you would in a crisis. I want you to act as if the house was on fire, because it is. (Thunberg 2019)

A theme guiding Thunberg's understanding of environmental and climatic issues is embodied in a phrase used by Dr. Martin Luther King in several of his most famous speeches: "[the] fierce urgency of now." The first time he used this expression publicly was in his celebrated "I Have a Dream" speech, delivered on August 28, 1963 at the Lincoln Memorial on the Washington, D.C. National Mall. The second time was in his memorable "Beyond Vietnam" sermon of April 4, 1967, given at Riverside Church in New York City. In each of these stirring orations, King conveyed a sense of deep impatience in the face of impending catastrophe. In the latter, he stressed that a failure to act would condemn modern society to the fate of extinct societies: "Over the bleached bones and jumbled residues of numerous civilizations are written the pathetic words, 'Too late'" (King 1967).

The urgency in Thunberg's words also echoes a talk given a quarter of a century earlier by then 13-year-old Severn Suzuki (1992) at the 1992 Rio Earth Summit:

> I have no hidden agenda, I am fighting for my future. Losing my future is not like losing an election or a few points on the stock market . . . I am afraid to breathe the air because I don't know what chemicals are in it . . . Now we hear of animals and plants going extinct, vanishing forever . . . If you do not know how to fixing [the environment] stop breaking it.

Suzuki (now Cullis-Suzuki) went on to become an environmental activist, television host, and author, and, perhaps with a sense that with collective action there can be a world that is not in peril, a mother of two sons.

Another strong youth voice for climate action is that of Xiuhtezcatl Roske-Martinez of Boulder, Colorado, who had organized more than 35 climate rallies and protests by the age of 12. Martin (2014) reports that "Boulder City Council members know him well. Roske-Martinez helped stop the use of pesticides in city parks, and was among the fiercest advocates for a fee on plastic bags. His was a key voice in a project to contain coal ash, and to end a 20-year contract with Xcel Energy, allowing the city to pursue renewable energy as its primary resource." Roske-Martinez himself explains: "I work on climate change, fracking, and indigenous issues because I have a personal investment in them, and because I recognize that this is going to effect all people, but especially people of color, indigenous people, and youth" (Martinez 2018). He is one of

21 young plaintiffs in the landmark *Juliana* vs. *United States of America*, a suit filed at U.S. District Court in Oregon asserting that the government has violated young people's rights by allowing life-threatening greenhouse gas emissions. Actions like this signal the rising tide of awareness of and opposition to climate change. Noting a "palpable global energy around the climate change battle" (Brennan 2019), Roske-Martinez recalls a time when there were few press conferences and little press coverage of climate change or climate change activism.

Marking this change, on Friday, September 20, 2019, the climate school strike movement inspired by Greta Thunberg held the world's largest climate protests to date. As reported by the *New York Times*, "young people poured into the streets on every continent on Friday for a day of global climate protests. Organizers estimated the turnout to be around 4 million in thousands of cities and towns worldwide" (Sengupta 2019). A group of scientists even rallied in Antarctica. Thunberg, who marched along with tens of thousands of others in New York City, told participants: "Right now we are the ones who are making a difference. If no one else will take action, then we will . . . We demand a safe future. Is that really too much to ask?" (quoted in Sengupta 2019).

This final chapter is about climate and environmental change action. It draws inspiration from people like Greta Thunberg and Severn Cullis-Suzuki who, as children, reached the conclusion that anthropogenic climate and environment change is not someone else's problem to solve, nor is it a problem solvable by individual changes in one's own ecological footprint—though reducing personal energy and resource use and individual pollution production is a useful first step. They understand that those people and corporations with the largest ecological footprints often directly benefit from activities involved in greenhouse gas production, are disconnected from the reality of the ecological crisis because they are insulated from its effects, and have considerable resources to use in adapting to future effects. Consequently, even though they are the largest contributors to global warming and environmental pollution, they are the least responsive to initiatives for volunteeristic change. However, while individuals may act selfishly—and the very wealthy can do so on a grand scale—they can be constrained if the social systems in which they participate are geared to support democracy, social justice, and environmental sustainability. This is why we need to understand ecocrises and ecocrises interactions not as things produced by the actions of individuals, but instead as products of socioeconomic systems. For example, while some people are trying to reduce their consumption of single-use bags, "[t]his focus on individual action . . . overlooks the fact that corporations using plastic packaging are subsidised through publicly funded municipal waste programs. And lighter plastic packaging equals cheaper global shipping—further encouraging production and consumption of more cheap plastic" (Greenwood-Nguyen & Roth 2019). Though a large percentage of social science research on climate change focuses on individual-level beliefs, attitudes, experiences, and policy preferences (McAdams 2017), and this is a valuable area of research, there is a need for a better understanding of the development of collective action and sustained group involvement in response to climate peril.

Confined to St. Petersburg's Fortress of St. Peter and Paul in 1862 for opposing the czarist autocracy and calling for revolutionary change, Russian socialist Nikolay Chernyshevsky (1886) wrote a novel entitled *What Is to Be Done*? and succeeded in having it smuggled out of the prison to become popular among students, leading to his being sentenced to 7 years of forced labor in Siberia. Though it was a work of fiction, the book influenced people like Vladimir Lenin and others who believed that to bring down the oppressive political system in which all power and wealth were concentrated in the hands of the nobles required organized collective action. For Lenin, this meant the creation of a political vanguard party to provide direction for a mass rebellion. For the splintered environmental and climate movements of today, it means building collaborative connections, affiliations, and networks in order to sustain an ever-intensifying struggle against entities committed to business-as-usual economics and undemocratic politics.

Organized collective action has the potential to go beyond first baby steps toward the achievement of both democratic and sustainable systems for humanity on Earth. As John Meeker (2018), Professor of Environmental Health Sciences at the University of Michigan, reminds us, "[t]here are many historical and recent examples of environmental threats that we identified after it was too late." For example, as discussed in Chapter 9, the full risks of exposure to asbestos and perfluorinated chemicals were not known by the public until considerable damage had already been done to human health. Today, we have extensive knowledge about the enormous health and social risks of living in a world facing multiple ecocrises. What is missing is the kind of action that reflects the reality of being in a global emergency. Nevertheless, responses are forming that indicate what must be done in the 21st century if we are to avoid further catastrophe. These provide pathways for challenging the business-as-usual ideology of endless profit-driven economic growth and for exchanging an economic system that appears to be incompatible with a healthy, sustainable, and just world for one in which the goal of economic life is to "sustain, nourish, and restore human and natural communities" (Speth 2017). These points raise a question that will be addressed in this chapter: What should be the goal of environment and climate activism?

11.2 Social movements

Organized collective action for change is the essence of social movements (Nash 2005). These dynamic entities emerge through the more or less spontaneous coming together of people for more than a brief duration around a common cause. During their existence, they shift from initial spontaneity to organization-building, though participation in a social movement does not require affiliation with an organization and movements retain a semiformal structure at best. Indeed, as Marshall Ganz, a senior lecturer at Harvard Kennedy School and long-time activist, point outs, social movements can be "unpredictable, messy, contentious" (quoted in Mingle 2013, p. 7).

Melucci (1989) argues that social movements have three core dimensions: 1) actors' recognition of commonalities, including shared cultural understanding, perceived threat, and possibly identity; 2) a (perhaps somewhat varyingly defined) target for change that is defended by institutional or other empowered adversaries; and 3) acceptance of and optimistic willingness to engage in oppositional action. Social movements articulate frames for the comprehension of current events while drawing on charged historical events or cultural traditions to provide templates for action (Hess 2007). Operating at the interface of culture and politics (Alvarez *et al.* 1998), they are energized by the belief that as a collectivity they can achieve desirable or prevent undesirable social change, such as stopping or limiting environmental disruption or improving environmental conditions. While social movements may emerge with quite broad or even muddled objectives for change, to be successful history suggests they need to have a clarity of purpose and a set of strategies or mobilizing structures that facilitate the translation of enthusiasm into effective action, advocate for a broad shift in social norms, and enable them to endure and push back effectively against a self-interested opposition (McAdams 2017; Satell 2019). In light of the significant expansion of electronic media into everyday life in a hyper-connected world, another feature may be recognition that social media can usefully augment—but cannot replace—the tireless work of community organizing and an enduring willingness to see to the details, including attending face-to-face meetings (Klein 2011).

While there are many kinds of social movements and a voluminous social science literature analyzing their features, of specific concern here are environmental and climate movements with broad (even global) reach, as well as local-level expressions of collective action—and between them, at the mid-range geopolitical level, regional and national movements. As Rootes (1997, p. 320) stresses, though "consciousness of environmental is a necessary condition of collective action . . . it is by no means a sufficient one." Other factors are needed at each level to turn knowledge of ecological degradation and ecocrisis development into action.

11.2.1 The local level

Neighborhood- and community-based movements for environmental/climate sustainability and environmental justice have in places emerged as an important force in both rural and urban politics in various parts of the world. Others, labeled victims' movements, have "short-term objectives—frequently seeking monetary compensations for existing injuries, sickness, or loss of people's property—and provided limited opportunities to challenge the dominant power structure" (Lee 2007).

In other cases, these efforts have been labeled "green politics" because of their wider objectives, though this appellation has included in some discussions the issue of "green gentrification," which is known to push out members of the working class and people of color while attracting wealthier residents, thereby increasing inequality. As a result of green gentrification, community and social movement leaders in some cities have demanded that social equity rather than profit accumulation become the core component of urban redevelopment planning (Gould & Lewis 2017). Additionally, they have sought to affirm local identity and autonomy. In such cases, environmentalism is never simply about the environment, but also about community, identity, and democracy. By definition, local environmental movements are place-based, focusing on unique local environmental degradations or local expressions of wider environmental damage or threat. These points are explored in the following four cases.

An example of a community-based environmental social movement is described by Newman (2011), based on an ethnographic case study in northeast Paris. Newman studied a local residents' campaign in a predominately low-income community primarily composed of post-colonial immigrant ethnic minorities. The movement is an example of what has been called the "environmentalism of the poor," the occurrence of which challenges conceptualizations of environmentalism as a narrow middle-class concern (Inglehart 1995). Often, the environmentalism of the poor involves struggles for environmental preservation in order to sustain livelihoods and local well-being against corporations and state governments seeking to profit from local resources (Nixon 2011). In the Paris case, by contrast, an organized community initiative transformed an abandoned freight train station into a 10-acre public park. Known as a "ghost station," it was one of many Paris stations that had been shut down and padlocked. The driving force for this green justice movement, led by a grassroots community organization, was the demand for bottom-up urban planning focused on the construction of a green public space that addressed longstanding social demands for a less polluted urban environment. Women, especially those with children, played a very visible role in the public events and protests focused on winning approval for the park. The protesters challenged the intention of local officials to lease the site for a new industrial use by a beverage distributor, complaining that it would increase the number of pollution-releasing diesel trucks driving through the neighborhood. They also objected to the idea of using the site as a garbage storage facility, as suggested by some politicians, fearing the ecological and stigmatizing impact it would have on the community. This grassroots movement, and growing news coverage of pollution in the area, ultimately succeeded in winning approval for the park. While its final form reflected class and ethnic conflicts in post-colonial France and addressed local rather than global threats to environmental health, it exemplifies what can be done when people heed the oft-quoted reminder attributed to anthropologist Margaret Mead: "Never doubt that a small group of thoughtful, committed citizens can change the world; indeed, it's the only thing that ever has"; or, more to the point, as Leandro Herrero (2016) rephrases Mead's quote: "Never doubt that a small group of thoughtful, committed, *organized* citizens can change the world; indeed, it's the only thing that ever has." Herrero stresses that it is necessary to counter the "naïve belief that all you need is a lot of passion, a lot of commitment, a lot of good intentions and . . . change will happen."

A second case, involving the dual struggle for a livable environment and grassroots democracy, is found in the harbor city of Kaohsiung City in south Taiwan (officially the Republic of China), where worsening environmental problems are tied to the country's no-holds-barred approach to industrialization, nicknamed the Taiwan Economic Miracle. As part of this process, Kaohsiung, Taiwan's second largest city, came to be surrounded by highly polluting, water-depleting industrial plants (Arrigo & Pouleston 2006), linked among residents to the city's high cancer rate. The site of the community struggle was Camp Wei-wu, which initially was constructed as an army barracks during the Japanese occupation, and after World War II served as a Taiwanese military training center. In 1979, the 67-acre site was found by the military to no longer be suitable for this purpose. Instead, both the Ministry of National Defense and Ministry of the Interior decided the land should be redeveloped for military family housing. In opposition to this plan, the Wei-wu Park Promotion Association (WPPA) was founded in 1992, seeking to establish a natural community park on the site. Specifically, the group wanted a park that reflected the region and its ecosystems, with plants that were native to Kaohsiung, bred locally, and not imported from elsewhere in the country. As Camp Wei-wu had once been a marsh, activists urged the recreation of a wetland environment that would attract indigenous geese, ducks, and pheasant-tailed jacana. Another species they hoped to raise was the formosan sika deer, which had been exterminated in the wild (but could still be found in captivity). To achieve its goal, the WPPA engaged in a campaign of government lobbying, public relations building, dissemination of information to the public, and participation in public hearings. Its efforts succeeded, in principle, as in 1993 the nation decided that Camp Wei-wu would be reserved for park use. The city, however, did not provide funding for the project, which led to a prolonged fight by the WPPA and other local activist groups against the local administration. The battle was not won until an opposition political party gained control of both the city (1998) and national (2000) governments. This led to allocation of significant government funding for the construction of what came to be called Wei Wu Ying Metropolitan Park, which opened in 2010, as well as for other green community projects that have garnered attention for the city.

A third example of a local—in this case, indigenous—initiative involves efforts to mitigate the negative social and environmental impacts of the open-pit Ok Tedi copper and gold mine in the rainforest-covered Star Mountains of Papua New Guinea's Western Province. As reported by ethnographer Stuart Kirsch (2007a, p. 305), since the mid-1980s "the mine has discharged more than 1 billion metric tons of tailings, the finely ground material that remains after the valuable metal has been extracted, and waste rock into the Ok Tedi and Fly rivers, causing massive environmental degradation downstream." Local anti-mining movements have emerged increasingly around the world in recent decades as mining companies have moved to more remote regions seeking new ore deposits (e.g., Velásquez 2017). When the Ok Tedi mine was launched, the majority shareholder and manager was the Australian corporation Broken Hill Proprietary (BHP), which in 2001 merged with Anglo-Dutch Billiton to form BHP Group. In 2015, an iron tailings dam owned by BHP in Mariana, Minas Gerais, Brazil failed, causing what has been described as the worst environmental catastrophe in Brazil's history (Rifai 2015). An even earlier landslide at the Ok Tedi mine construction site was used to convince the government to temporarily permit the discharge of tailings and other mine wastes directly into the Ok Tedi River. This dangerous means of disposing of toxic waste was allowed to continue when closure of another mine—the Panguna copper mine in Bougainville, Papua New Guinea (due to violent community protests)—cut government revenues. Local protests against the Ok Tedi mine began in the late 1980s as trees along the river and community fields along the floodplain were destroyed by mine tailings. People complained that they could no longer catch fish, hunt, or grow food because of the polluting effects. The International Institute for Environment and Development reported that the area 40 miles downriver of the mine was "almost biologically dead and species diversity over the next 130 km [80 miles] had been dramatically reduced. Fertile river bank subsistence gardens, plantations and forest [were] destroyed" (Banks 2002). People living in the area of the catastrophe told Kirsch (2014) that it now seemed *moraron*, Yonggom for spoiled, rotten, or dead. But for as long as awareness of the extensive environmental damage remained mainly at the local level, the community protest failed to have any effect in stopping the destruction of the Ok Tedi river environment or winning those adversely impacted any relief or compensation. As Kirsch (1993, p. 27) reported, people living in villages near the mine were "in a state of despair and . . . feel frustrated and ignored in their efforts to obtain restitution." The protest's organizers consequently decided to take their message global. Working with several non-governmental organizations (NGOs), representatives traveled to the Netherlands for the International Water Tribunal, to Germany for meetings with government officials, to Brazil for the Earth Summit, and to the United States for meetings with several environmental organizations. Subsequently, on behalf of 30 000 people affected by the environmental damage, they filed a lawsuit against Ok Tedi Mining, Ltd. and BHP. Following a prolonged public relations battle waged by the affected communities and environmental groups, this suit ultimately led to an out-of-court settlement of $28.6 million, as part of which a dredging operation was begun in the river to clean up some of the damage and an effort was made to rehabilitate the site around the mine. Though the mine continues to operate, and some toxins still empty into the river system, Kirsch (2007a, p. 312) observes that:

> The Ok Tedi campaign and related political movements have enabled communities affected by mining projects to participate in negotiations that were previously limited to resource developers and states . . . Transnational action networks also established new paths and routes for the sharing of tactics, strategies, and experiences among indigenous peoples and other communities affected by mining

These networks take advantage of new electronic technologies, from cell phones to satellite images, to track corporate environmental impacts, and of social media to communicate their findings to a wide audience. Through such activities, they are able to "put pressure on the corporation wherever it operates" (Kirsch 2007b).

Despite these advances, the corporate power of BHP and the revenue the mine brought to the cash-poor coffers of a developing country with a weak governing structure enabled a very wealthy and well-connected multinational company to flout Papua New Guinea's environmental laws and its people's need for a healthy environment. This occurred in several fundamental ways, including with the company having a hand in drafting legislation that favored its own financial interests. Like other corporations facing environmental complaints from local communities, BHP adopted various "corporate social technologies" (Rogers 2012) designed to co-opt and counter critiques of its activities while managing its public image. Such strategies included the manipulation of scientific research, concealment of the company's awareness of the health and environmental damages it had caused, adoption of the discourse of the company's opponents in promoting itself as responsible, transparent, and committed to sustainability, formation of partnerships with NGOs desperate for some environmental improvements, granting of limited concessions, and sewing of division among the company's critics. These tactics affirm that "the corporation is one of the most powerful institutions of our time" (Kirsch 2007b, p. 1) and that social movements must continually develop new counter-tactics to resist corporate harm.

One such approach, the rearticulation of social alliances, was adopted by the Chippewa (Ojibwe) people of Northern Wisconsin in the 1990s in response to plans by Exxon to open a mine at the Crandon sulfite deposit, 1 mile upstream from the wild rice beds of the Mole Lake Chippewa Reservation (Gedicks 1993). Opening a mine there, Frances Van Zile, a Chippewa tribal elder recognized, would likely degrade the environment where his community fished, hunted, and gathered wild plant foods. As Gedicks & Grossman (2010) report, "The Crandon/Mole Lake mine would have extracted approximately 55 million tons of sulphite—used as a preservative and antioxidant additive by the food and pharmaceutical industries—during the thirty-year life of the project. Over its lifetime, the mine would generate 44 million tons of wastes–the equivalent of eight Great Pyramids of Egypt." When sulfite ores or sulfuric wastes are exposed to water and air, a chemical reaction can occur that creates sulfuric acid (also known as battery acid). Sulfite mining also produces high levels of poisonous heavy metals, such as mercury, lead, zinc, arsenic, copper, and cadmium. Moreover, dewatering the underground mine to avoid flooding would reduce groundwater tables, while to process the sulfite Exxon would need to use toxic chemicals including cyanide. In order to protect tribal environmental resources and community health while asserting tribal sovereignty, the Chippewa developed a multifaceted strategy that included: 1) building national and international support by networking with indigenous rights groups; 2) challenging Exxon through resolutions introduced at shareholder meetings; 3) developing a state-wide alliance to educate the non-Indian public about the risks of mining; 4) pushing for passage of a mining moratorium law; and 5) creating local alliances with their non-Indian neighbors, including sportfishers who valued long-term conservation of the local fishery. The latter was of historic note because in previous years non-Indian fishers had aggressively—at times, violently—protested indigenous treaty rights allowing Native Americans to use spear-fishing in off-reservation areas (a conflict that produced the ironic chant "Indians go home"). Seeing Exxon as a common enemy, the Chippewa and white fisherman "began to cooperate to protect the fishery against a common outside threat" (Gedicks & Grossman 2010) and assembled a rural-based multiethnic, middle- and working-class environment sustainability movement. This coalition was able to stand up to the smear tactics, divide-and-conquer schemes, and threats of Exxon, corporate-funded resource development front groups, and Wisconsin's anti-Indian treaty rights governor (1987–2001) Tommy Thompson. It also unleashed a flood of critical emails at BHP when it joined Exxon in the mining initiative.

In a bold move, the environmental alliance proposed the purchase of the Crandon mine site (including 5939 acres of land and its underground mineral rights) to form a conservation area devoted to sustainable land-management practices, tribal cultural values, and environmentally sensitive tourism. The goal was to use money from tribal casinos to permanently foreclose the Crandon mine by removing the land from and its sulfite from the hands of mining companies. In the end, this gambit succeeded, as it became clear to the mining industry that it was time to cut its loses. At the end of October 2003, Mole Lake and Potawatomi indigenous leaders announced that they had jointly purchased the 5939-acre Crandon mine property for $16.5 million and divided the site between them. Gedicks & Grossman (2010) reported: "As he tacked up a giant 'SOLD' sign on the company, Potawatomi mine opponent Dennis Shepherd exclaimed, 'We rocked the boat. Now we own the boat'."

These four generally successful examples of bottom-up community efforts targeting environmental goals highlight the value of public visibility in the news media, effective use of electronic and social media, grass-roots organizing, creative thinking about shared concerns, and coalition-building in confronting the inclination of power-wielders to overlook environmental health in the interest of profit-seeking. The work of Jane Kleeb, a leader of Bold Nebraska, which organized against the Keystone XL pipeline and helped make it a symbol of the battle over climate change and fossil fuel, suggests some additional points about how to target local climate efforts (Deaton 2017). First, frame issues in terms of local values. In a conservative state like Nebraska, this meant framing climate change in a language of freedom, liberty, and property. It also meant avoiding common climate change tropes like threatened polar bears in the Arctic or the impact of rising seas on poor people in Bangladesh, and instead emphasizing how the pipeline would affect people's water, land, and livelihood right there in the state. Second, it involved being able to tell a good story. Statistics, charts, and scientific findings are all very useful, but meaningful, evocative stories that can resonate with people motivate them to act. Of particular utility in the arena of climate change—a topic that can leave people feeling overwhelmed—may be "stories that empower us . . . and capture our imaginations [which can] inspire and galvanise us into action" (Gillespie 2013).

Comparative research on the issue of why environmental movements arise or fail to do so in response to a potentially threatening environmental disruption has been undertaken by Wright & Boudet (2012). Specifically, they were concerned with why different people and communities come to different conclusions about which issues merit or require mobilization. Their focus was on energy infrastructure projects proposed by private companies seeking profit, such as onshore and offshore liquefied natural gas drilling, nuclear power construction, hydroelectric energy generation, and the wind farms. Popular opposition campaigns draw participants who live or work near a proposed site. Their target may be the company(s) involved, the owners of the land where the project will be sited, a regulatory agency responsible for oversight, consumers of a product related to the project, or some combination of these. For their study, the researchers randomly drew 20 cases across 12 states from the Cambridge Scientific Abstracts Illumina Digests of Environmental Impact Statements for the years 2004–2007. Data on these cases came from local newspapers, including letters to the editor and editorials, and from field interviews. For each case, the researchers examined their data for: 1) evidence of collective action; 2) the overall volume of oppositional actions in the community; and 3) the use of non-institutional oppositional tactics. They found that half of the communities in their sample mobilized in response to a proposed energy infrastructure project and half did not. Local context, including past experience, proved to be critical, and they concluded that "the context in which individuals are embedded will shape the average response"

(Wright & Boudet 2012, p. 735). They also stressed the importance of examining combinations of causal conditions—what they referred to as "recipes"—for mobilization, rather than single contextual causes. Three factors—1) having other local industries similar to the proposed project; 2) suffering from economic hardship; and 3) having no prior experience with community mobilizing—tended to inhibit the motivation to mobilize. Mobilization was significantly more likely if these factors were absent.

11.2.2 The regional/national level

The world's first political party formed around environmental issues was the United Tasmania Group of Australia (the precursor to the Tasmanian Greens). It formed in opposition to deforestation and, especially, the damming of the Serpentine and Huon rivers by the Hydro Electric Commission of Tasmania for the purposes of hydroelectric power generation and to feed Lake Pedder (Brimsmead 1983). Though it failed to win much popular support in state elections, it did inspire the formation of green parties elsewhere around the world (Wall 2010).

In Europe, environmental movements have had the greatest gains of any of the student social movements of the 1960s and 1970s (Rootes 1999). It is in Europe, especially, that environmental concerns have been institutionalized in the form of issue-focused political party formation, and it is there that the challenges and dilemmas of party politics and the formation of social policies to address environmental crises have been most engaged. This process has played out with the eventual formation of what have come to be called "green parties," a term first used by Die Grünen (The Greens), a German party formed in 1979 (Burchell 2002). The term "green" was coined by Petra Kelly, one of the founders of Die Grünen and an early theorist of green politics. She was elected to the Bundestag, the German federal parliament, representing Bavaria, in 1983. Eighteen years later, the German Greens attracted intense ire from greens worldwide, as well as at home, when four of their representatives in the Bundestag supported the government in a confidence vote over German participation in the war in Afghanistan (Erlanger 2001).

In 1981, the Agalev Working Group, a creation of an austerity-oriented Catholic environmental movement begun by the Jesuit priest Luc Versteylen, won two seats in the Belgium Chamber of Representatives and one in the Senate. In 1992, Agalev pushed for the passage of a tax on bottles, one of the first ecotaxes. An example of Pigouvian taxes—those that are intended to make private parties feel the social burden of their market actions—beverage container deposit laws were subsequently adopted in many European countries and a number of other nations around the world. By 1999, greens in Belgium held several cabinet-level positions in coalition governments.

To enhance communication, coordination, and influence in light of the creation of the European Union, in 2004 greens were the first group to form a political party at the pan-European level, called the European Green Party. With 34 member parties, the European Greens have pushed to increase investment in renewable energy, improve recycling programs, limit unnecessary use of plastics, ban the use of some harmful pesticides, and adopt the Paris Agreement on climate change. They have also called for the passage of a European climate law that included binding national carbon budgets designed to reduce greenhouse gas emissions by 55 percent by 2030. In a related platform, the party supports the phase-out of coal by 2030 and of other fossil fuels as soon as possible. Further, it wants to establish an international environmental court to address serious violations of international environmental law.

At the national level, the Greens have had more electoral success in Germany than in other European nations, with a particularly strong base in the southern German state of

Baden-Württemberg. Elsewhere in Europe, Amsterdam has elected a green mayor, who began her term in 2018, and party support generally has expanded not only in the Netherlands, but in Belgium and Luxembourg as well, leading some journalists to talk about a future "Green Wave" on the continent. In the 2014 elections to the European Parliament, the Greens won 50 seats and garnered 6.6 percent of the popular vote—the second lowest numbers among the seven political parties that won positions, representing a drop of seven seats from 2009 (European Parliament 2014). This did not, however, signal a downward spiral for the European Greens. In 2019, they unexpectedly increased their number of seats from 52 to 67, though their victories were limited to Western European countries. Voting exit polling suggest that these gains were due primarily to an increase in turnout among younger voters, though the Greens' support was not limited to this constituency (Keating 2019). Moreover, in 2014, Alexander Van der Bellen, formerly a professor of economics at the University of Vienna, became the first Green elected by popular vote to be head of state, representing Austria.

It is not accurate to equate the environmental movement in Europe with green party formation, as the latter did not grow automatically or always directly from the former. In fact, the size of the green party in a country may be the inverse of the size of the environmental movement found there (Rootes 1997). Nonetheless, it is instructive to examine the process leading from public environmental consciousness to the effort to institutionalize environmental concerns through the development of formal political parties. As a starting point, it bears mention that the various national green parties of Europe have never been a homogeneous grouping, though they have sought to establish multicountry regional formations. Each developed within the context of somewhat unique national social, cultural, and environmental conditions, and with regard to the green parties of Western and Eastern Europe, within differing periods. One thing green parties have tended to have in common, however, is that in addition to environmental issues, they have leaned toward the support of advancing social justice. What they have not tended to do is present and sustain a fulsome challenge to the inherent structural causes of a pending environmental catastrophe.

The greens have also been active as political parties and environmental activist groups beyond Europe. Down Under, an issue of green focus has been coal and coal seam gas (i.e., natural gas, primarily methane). Australia is the world's fourth largest producer and one of the largest exporters of coal; 34 percent of its greenhouse gas emissions come from burning coal, and even higher amounts come from exported coal (Mudd 2013). Because of pollution issues, the Greens have consistently campaigned for leaving coal in the ground, which is a rallying cry for the broader environmental movement of Australia. In addition to the Greens, this includes a large and fluxing array of groups such as Rising Tide, Quit Coal, Six Degrees, Socialist Alliance, Coal Terminal Action Group, and Lock the Gate Alliance (Baer 2012; Burgmann & Baer 2012). Generally, these groups share the perspective that coal consumption should be replaced by renewable sources of energy, particularly solar, wind, and geothermal, all of which are available and could be significantly expanded in Australia. For example, it is a world leader in residential rooftop solar, with photovoltaic solar power collection systems installed on over 20 percent of all private houses (Roberts *et al.* 2018).

Groups within the Australian anti-coal/coal steam movement have adopted various action strategies. In the week before the 2009 Copenhagen climate talks, Rising Tide Australia, for example, staged a day-long blockade of the rail line that feeds into the coal port at Newcastle in Southeastern Australia, the world's largest coal-handling facility. For several years, it also organized an annual

People's Blockade of the World's Biggest Coal Port at Horseshoe Beach in Newcastle. During these daylong protests, people used kayaks, rafts, and other vessels to fill up the harbor and shut it down. In addition, the organization initiated public forums and community education events and sought to build alliances with trade unions, indigenous peoples, and other climate action groups. Following several years of organizing, the group became less visible after 2014, to be replaced by the Coal Terminal Action Group, which focused on opposing the opening of a fourth coal terminal (called T4) in Newcastle. In 2015, the group rented a large billboard with the slogan "No More Coal Dust" beside a picture of a young child. Between 1984 and 2012, coal exports from Newcastle increased 10-fold annually, from 21 to 210 million tons. Construction of T4 would have increased this to 330 million (Climate and Health Alliance 2017). Ultimately, plans to build the additional terminal were scrapped because of falling global demand for coal.

Overall, the Australian anti-coal/seam gas movement has followed a pattern seen in other social movements, "with activist groups and coalitions forming, dividing, and reassembling and with significant sectors of their target constituencies remaining on the sidelines" (Edelman 2001, p. 285). The anti-coal/seam gas movement has launched a cycle of protests, with new groups emerging or disappearing with each cyclic wave. To a degree, this pattern reflects the fact the groups within the movement tend to adopt a fluid, nonrigid style that expresses many members' discomfort with formal structure and centralization. One consequence of the tendency toward sporadic, short-term event-provoked responses, and of the failure to adopt structures that allow not only sustained but mounting opposition to the fuel industry generally, has been worsening climatic and environmental conditions not only in Australia but globally.

Commenting on the collective response to climate change in the United States, McAdams (2017) observes that "the lack of sustained popular mobilization around the issue of climate change is striking." To explain this lost period of aggressive and broad grassroots response, McAdams examines the quarter of a century period after James Hansen's watershed testimony before the U.S. Senate Committee on Energy and Natural Resources on June 23, 1988, a powerful and alarming presentation that earned him the honorific, "Father of Global Warming." In his written statement submitted to the Committee, Hansen (1988, p. 3) reported: "Global warming has reached a level such that we can ascribe with a high degree of confidence a cause and effect relationship between the greenhouse effect and observed warming." This testimony, which was reported widely in the news media alongside that of other climate scientists, might have provided the spark for immediate collective climate action among environmental advocates. McAdams believes it did not for three reasons: 1) lack of favorable political conditions (e.g., a hostile Congress buoyed by energy industry campaign donations and an aggressive energy industry-financed climate misinformation campaign, as discussed in Chapter 10); 2) the domination of the climate change movement by moderate, institutionalized organizations seeking to work within rather than to challenge existing power structures; and 3) subjective impediments to mobilization such as the varying salience of climate change in people's everyday lives and identities. Together, these factors have had a chilling effect on grassroots climate change mobilization in the United States. Even lethal extreme weather events such as Hurricane Katrina and Hurricane Sandy, only 7 years apart, did not catalyze climate change activism to the degree that Three Mile Island did in the case of the antinuclear movement.

On September 21, 2014, the U.S. climate movement did succeed in mobilizing over 400 000 concerned activists to participate in the People's Climate March in New York City, as well as thousands of others in parallel actions in other U.S. cities. The core demand of the protests was that world leaders meeting at the U.N. Climate Summit at the United Nations headquarters in New York should take bold action to mitigate climate change. It was at this meeting, in his closing remarks, that U.N. Secretary-General Ban Ki-moon (2014), after acknowledging the demonstrators outside, declared that the nations of the world had to significantly cut greenhouse gas emissions because "[t]here can be no Plan B because we don't have a Planet B." Some have pointed to the massive 2014 rally as a turning point in the U.S. climate change movement (Berman 2014). This remains to be seen. Activism around climate change definitely has increased, but the clock continues ticking. As sustainable consumption researcher Marc Hudson (2018) reminds us:

> Marches can, at their best, help to break the so-called "spiral of silence," where people don't speak out because they assume that nobody else cares. At their worst, they act as a safety valve, substituting mobilisation for movement-building.

When large-scale social movements develop, they tend to cluster in time and space, forming what have been described as "waves of protest" in which contemporaneous movements with different causes (e.g., fighting racism, gun control, LGBTQ rights, climate change) learn from, emulate, and support one another (della Porta 2015). Emerging from a common sociocultural milieu, they express shared values and distinctive generational experiences.

A number of large, nationally active climate-focused groups have emerged in the United States, though some are active beyond U.S. boundaries at the global scale as well. The Climate Mobilization Project, for example, an affiliate of the internationally focused Climate Mobilization, views itself as the "emergency wing" of the broader climate movement. Starting with the premise that now is not the time for incremental action, its goal is to help organize a climate change response that is on par with the massive mobilization on the American home front during World War II. During 2015 and 2016, Climate Mobilization led grassroots campaigns in the United States to promote this goal. After the election of Donald Trump, it launched its city-by-city strategy designed to get city (and county) governments to declare climate change an emergency and commit to urgent action to eliminate fossil fuel use. For this purpose, it developed a boiler-plate climate emergency draft resolution that local, regional, and state governments could customize and adopt (Climate Emergency Movement 2019). This development reflects the fact while the federal government under Donald Trump showed no interest in responding to or even recognizing climate change, city government officials, especially those already feeling the costly impacts of a changing climate, have begun implementing renewable energy initiatives, expanding public transport, and requiring improved building energy use efficiency (Adams 2016).

The city council of Hayward, CA, for example, the governing body of a city of about 150 000 residents located in the East Bay section of the San Francisco Bay Area, adopted a Declaration of a Climate Emergency in January 2019. The city had already implemented a Climate Action Plan in 2009, whose purpose was to make Hayward a more environmentally and socially sustainable community by reducing local greenhouse gas emissions. It included goals for emission reductions by 2020, 2040, and 2050 of 20, 62.7, and 82.5 percent below 2005 baseline levels. In 2016, however, the city reported that "[t]otal energy consumption—electricity and natural gas, for all sectors—has changed very little since 2005," though there was progress toward the achievement of some subgoals (City of Hayward 2016). The following year, the California Ocean Protection Council Science Advisory Team released "Rising Seas in California—An Update on Sea-Level Rise Science" (Griggs *et al.* 2017), in which team scientists projected a conservative estimate of between 1 and 3.4 feet of sea-level rise (and a worst-case scenario of 10 feet of rise) in the San Francisco Bay by 2100. It was in the light of this report, as well as the release in 2018 of the U.N. Intergovernmental Panel on Climate Change (IPCC) special report on the impacts of global warming, that the city decided to adopt the Declaration of a Climate Emergency (City of Hayward 2019).

In July 2016, Climate Mobilization's call for World War II-scale mobilization was adopted into the Democratic Party Platform. This position was reaffirmed 2 years later when the Democratic National Committee's (DNC) Resolutions Committee passed a "Resolution Expressing the Sense of the DNC that Climate Change Poses an Urgent and Severe Threat to our National Security" and called for a national social and economic mobilization to address the climate emergency (Climate Mobilization Project 2019). How urgent this "emergency" really is to the DNC, however, was put into question when it also adopted a resolution indicating that the Democratic Party would "continue to welcome the longstanding and generous contributions of workers, including those in energy and related industries, who organize and donate to Democratic candidates individually or through their unions' or employers' political action committees" (DNC 2018). In other words, at the national level the Democratic Party agreed to continue to receive donations from the big energy companies that have fought so hard to undercut climate change action while profiting from the national dependence on greenhouse gas-producing energy sources. In response, climate activists from Sunrise Movement, an organization that has spearheaded a national movement to get political candidates to pledge not to take any fossil fuel corporation donations, protested outside the DNC headquarters in Washington, D.C. (Mark 2018).

Climate Mobilization, along with many other environmental organizations, also endorsed the Green New Deal. This is a set of proposed national policies that would address climate change and economic inequalities, such as expansion of the Clean Air Act, implementation of a ban on crude oil exports, ending of fossil fuel subsidies and fossil fuel leasing, and a phasing-out of all gas-powered vehicles by 2040. Howie Hawkins, who ran for governor as a Green in New York State in 2010, was one of the first office-seekers to promote the idea of a Green New Deal, modeling his plan after the original New Deal, the economic stimulus program promoted by President Franklin D. Roosevelt during the Great Depression, which included a number of environmental components. Jill Stein, Green Party presidential candidate in 2016, proposed it as one of her goals if elected, and it was subsequently incorporated into the Green Party's national platform. Following the "Blue Wave" sweep of Democrats into the House of Representatives in the 2018 elections, the Green New Deal initiative was embraced by a solid block of members of Congress. Supporters co-sponsored a proposed House Select Committee on a Green New Deal during the 116th Congress with the task of developing an economic and environmental mobilization plan that would make the U.S. economy carbon-neutral and the United States a socially and environmentally just nation. The proposal, introduced by Alexandria Ocasio-Cortez, U.S. Representative for New York's 14th congressional district, and Senator Ed Markey of Massachusetts, calls for legislators to spend a year consulting with a wide range of experts, write and release a detailed plan by January 2020, and draft legislation for Congressional vote by 2 months after that.

The Sunrise Movement, founded in Washington, D.C. in 2017, is a grassroots organization of youth and young adults that advocates political action on climate change in the United States, with a special focus on pushing Congress to fully and urgently address its causes and consequences. As Witt (2018) observes, "[j]ust as the March for Our Lives has changed gun-control activism from a movement of grieving parents to one led by students, Sunrise is part of a generational shift in the environmental movement." Sunshine members are drawn from "a new political generation, whose lived experiences and worldviews sharply distinguish them from previous generations of youthful activists" (Milkman 2017) Differences include that: 1) they are so-called "digital natives," people who have been living with the Internet and other new electronic technologies their whole lives; 2) they routinely use network-based communication; 3) they have been hearing about the mounting risks and approaching tipping points of climate change and other ecocrises since they were children; and 4) they have witnessed the prior generation's failure to stop a soaring level of class inequality and the powerful political influence of corporations and the wealthy on environmental policy.

The group came into public view after the 2018 midterm elections when over 200 activists, 51 of whom were arrested, occupied the office of Speaker of the House Nancy Pelosi to demand action on climate change. After this sit-in, the organization focused its efforts on convincing candidates and lawmakers to refuse donations from the fossil fuel industry and gaining a consensus within the Democratic Party in support of a Green New Deal. Its website explains its organizing goal: "We're building an army of young people to stop climate change and create millions of good jobs in the process" (Sunrise Movement 2019). A squadron of that army went to the Russell Senate Office Building on Capitol Hill in Washington, D.C. on February 25, 2019 to demand that Senate Majority Leader Mitch McConnell and other senators support the Green New Deal. The young protesters, including a contingent of high-school students from McConnell's home state of Kentucky, wanted to tell the Senate Majority Leader to look them in the eyes while he took huge campaign donations from the oil and gas industries and sacrificed their futures. When the demonstrators began singing, 42 were arrested by Capitol Police (Kelly & Fox 2019).

The founders of the Sunrise Movement brought with them organizing skills gained during their prior involvement in social activism. Three of them, Sara Blazevic, Varshini Prakash, and Stephen O'Hanlon, gained organizing experience through the Fossil Fuel Divestment Student Network, a national initiative that seeks to pressure universities and other institutions to divest their stock holdings in unsustainable energy industries. This effort has been described as signaling "a sea change, from individualised sustainability efforts to youth-led collective political action, and recognition of climate change as a social justice issue" (Grady-Benson & Sarathy 2014). As a divestment organizer in 2015, Blazevic led a 33-day sit-in by 200 students at Swarthmore College in Philadelphia, PA. Under a banner calling for Swarthmore to divest by 2020, students occupied a hallway in the college's first building, Parish Hall. A petition drive by campus divestment activists collected the signatures of over 60 percent of the student body and over 1000 alumni and faculty members. Interviewed at the time, Stephen O'Hanlon explained, "It's been a multi-month organizing effort that has involved building both popular support on campus among students and faculty and also with alumni" (quoted in Aronoff 2015). During the occupation, the students maintained a rotating base of 15 people on duty every night. Twitter was used to message outside supporters for food and other supplies. Organizers hosted a screening of the 2004 film *Iron Jawed Angels*, about Alice Paul, a member of Swarthmore's class of 1905, a leading suffragist and one of the main organizers for the successful passage of the 19th Amendment that extended voting rights to women. The student initiative at Swarthmore, which began in 2012 with the campus group Swarthmore Mountain Justice, helped to spark a national fossil fuel divestment movement. It was the initial training ground for a number of committed climate and social activists, and its basic approach was replicated in the Sunrise sit-in at Nancy Pelosi's congressional office.

Varshini Prakash joined the effort by participating in the occupation of Whitmore Administration Building at the University of Massachusetts in 2016. Over several days, 34 students were arrested, but, unlike Swarthmore, the Board of Directors of the University of Massachusetts ultimately agreed to divest its endowment from direct holdings in fossil fuels (Connolly 2016). Raised in a town outside of Boston, MA, Prakash's grandparents came from southern India, where they endured a severe flood that hit their city, Chennai, in late 2015. At least 500 died and almost 2 million were displaced by the disaster. Along with Sara Blazevic, Prakash received training in organizing at Momentum, an institute committed to giving grassroots organizers the tools needed to build mass movements for social justice. Momentum promotes a community organizing model of best practices drawn from the long social history of nonviolent civil resistance to oppression. It differentiates between two approaches to social organizing. The first, embraced by groups like labor unions, pushes instrumental demands, like higher wages, to win concrete reforms. The second, emerging from the mass protest tradition, seeks to inspire masses of people to take to the streets around pressing social issues like civil rights. Momentum combines what it sees as the most effective components of these alternative approaches (Momentum 2016). Part of Momentum training is trust-building within and across movements through group retreats, a strategy adopted by Sunrise.

In late 2015 and early 2016, Blazevic and Prakash began meeting with other young climate activists to discuss how to intensify the impact of the climate movement. They were impressed by Bernie Sanders' ability to generate political energy and consciousness among their peers. For Blazevic, a eureka moment occurred in December 2015 when she read Sanders' announcement that "[it]'s time for a political revolution that takes on the fossil fuel billionaires, accelerates our transition to clean energy and finally puts people before the profits of polluters" (quoted in Fields 2015). It was time to name those most responsible for driving the fossil fuel agenda, the polluting elite.

Another co-founder of Sunrise is Evan Weber. While attending Wesleyan University, he met international environmental scholar and activist Michael Dorsey, then a visiting fellow in Wesleyan's College of the Environment. Together, Dorsey, Weber, and Matthew Lichtash, another Wesleyan student, wrote a successful grant for $30 000, funded by the philanthropic department of Monster Beverage Corporation. This allowed them to spend the summer of 2013 developing a plan for addressing climate change within a social justice framework that was commensurate with the scale of the global climate crisis (Bond & Dorsey 2010/2011). The Sierra Club, in which Dorsey was an active member, provided them working space in its Washington, D.C. headquarters (Mattews *et al.* 2018). The group produced a document called "The Plan: How the U.S. Can Help Stabilize the Climate and Create A Clean Energy Future." In 2014, Weber used this plan to incorporate a small nonprofit organization in Washington, D.C. called the U.S. Climate Plan with the mission of empowering youth and young adults to organize around scientifically-grounded climate policies in order to elevate the national dialogue on just and equitable energy solutions. The 35-page document included a number of elements, including a fee on emissions from all energy sources, elimination of fossil fuel subsidies, a moratorium on the issuing of new nuclear subsidies, and development of a national green bank to provide reduced interest loans for the creation of emissions-reducing technologies.

A key concern of Weber's was how to sustain a youth-led movement. Additionally, he grappled with how to avoid the mistakes made by the Occupy Wall Street protests against economic inequality, which brought the slogan "We Are the 99 Percent" into popular discourse but failed to maintain momentum after capturing the public's attention (Banjo 2013). Having participated in the Occupy movement, he felt it suggested a direction the climate movement could follow, but he recognized that better organization would be necessary to avoid a mere flash-in-the-pan impact. While pondering these challenges, Weber began networking with other environmentalists, including activists involved in the university divestment initiative.

At the 2014 climate protest in New York City, he met Dyanna Jaye, an environmental activist from Virginia, where she had directly experienced coastal flooding as described in Chapter 8. After graduating from the University of Virginia in 2015, Jaye went to Paris for the Conference of Parties meeting (the United Nations Climate Change Conference, COP 21), where she helped lead a youth contingent pushing for zero fossil fuel emissions by 2050. Barred by U.N. rules from employing certain types of visual demonstration, several hundred members of the contingent drew zeros in black and white paint around their eyes instead. Unhappy with the meek nature of the accords agreed to in Paris, Jaye made a decision to recommit herself to the climate movement.

A significant initial hurdle for Weber, Jaye, and other co-founders of Sunrise was how to build a new climate group world with firmly established environmental organizations and a growing set of groups. They began meeting to discuss their goals and the obstacles they faced in July 2016, and decided to hold monthly meetings at different locations for 9 months. At their meetings, they developed their strategic repertoire by telling stories of prior activist experiences, sharing books on progressive mass movements and the tactics of earlier social struggles, and listening to the lessons learned by Blazevic and Prakash during their training at Momentum.

The Sunrise group's first public appearance was at a climate protest in Washington, D.C., where the team reached out to those in attendance to build support. This developed into a strategy they named Sunrise Semesters. Interested students and other activists could join the group's organizing efforts for a 2–3-month period. The first semester's "class" included 85 participants who worked in five states (Florida, Pennsylvania, Michigan, Minnesota, and New York) to make climate change an issue in the 2018 national elections. The Sunrise Movement, for example, backed Ocasio-Cortez in her upset primary victory over Rep. Joe Crowley. In the 2018 elections, group energy was invested in getting candidates at all levels to pledge not to take contributions from the oil, gas, or coal industries, and instead to prioritize health care, climate control, and grassroots democracy over profits. A detailed "how to" manual was posted on the group's website, which activists could follow in order to get a candidate to sign the pledge. If a candidate agreed, Sunrise encouraged that a photograph of them signing the pledge be shared on social media with the hashtag #NoFossilFuelMoney.

In relation to older environmental groups, the Sunrise Movement is relatively small. Its leaders report they have about 15 full-time staff. It rents space from the Sierra Club in Washington, though officials in both groups affirm that Sunrise is independent and makes its own decisions.

At the grassroots level, Sunrise is organizing by building local hubs. Being officially recognized as a hub requires a team of three members and the collection of 25 signatures on a petition directed at getting public office holders to take the no-fossil-fuel money pledge. Hubs can request a weekend-long activist training session from the Sunrise Training Network. Through its various initiatives, "Sunrise has grown from a small and quixotic project to a full-fledged advocacy organization that draws thousands of volunteers across the country and tens of thousands of participants to its events" (Arrieta-Kenna 2019). The group has an organized presence in over 250 cities around the country and a national budget of over $4 million, primarily from individual grassroots donations.

While there are multiple other climate change groups active in the United States today, such as 350.org, the U.S. Climate Action Network, and Rising Tide North America, the two examples described here suggest that these groups are seeking to adopt structures that promote an increase in public awareness of the urgency of major ecocrises and that can respond to well-funded opposition from the fuel fossil industry.

11.2.3 The global level

There is a slogan in the climate movement, "Burn Borders, Not Coal," that underlines the fact that while the extent of contribution to planetary warming and exposure to its effects varies across nations, climate change is inherently a global issue (Vinay 2006). Hite & Seitz (2016) define global issues as critical problems that individual nations cannot solve alone. If left unresolved, these problems are of such magnitude that they will adversely affect the lives of billions of people. Global issues highlight our contemporary interdependence cross-nationally. Moreover, they are complex. As Hite & Seitz (2016) point out, climate change (an environmental issue) is connected to an energy issue, the population issue, the socio-economic inequality issue, the technology issue, and the future issue. This complexity adds to the global nature of climate change.

Government efforts to address global issues commonly take the form of international agreements and treaties, like the United Nations Framework Convention on Climate Change, the Kyoto Protocol, and the Paris Agreement. While these have raised hopes about finally coming to grips with climate change, none of them has achieved the goal of really making an impactful difference on global warming. Powerful philanthropic organizations have taken a different tact. In the spring of 2013, the Rockefeller Foundation—a charitable organization started in 1913 by Standard Oil magnate John D. Rockefeller—commenced a global program known as 100 Resilient Cities to help urban areas adapt to the physical, social, and economic challenges of the 21st century. The initiative was designed primarily to address the growing issue of urban population growth and the threat posed by climate change. To launch the massive project, the Rockefeller Foundation funded the hiring of 83 local chief resilience officers and trained them to work within city governments. Over $160 million was allotted to jumpstart a wide variety of projects in cities on six continents. Among other things, planners designed greener schoolyards in Paris that would stay cool during heatwaves and help contain the city's urban heat island effect, restored urban wetlands in the city of Atlanta, GA, made improvements to storm-water infrastructure in Norfolk, VA to help protect communities along the Ohio Creek against future flooding and continued sea-level rise, and harnessed the power of the Yaque River in the Dominican Republic to control flooding in Santiago De Los Caballeros (Rockefeller Foundation 2019). Taken together, the cities that received initiative funds represented over 20 percent of the world's total urban population.

Despite being one of the largest privately funded climate change initiatives in the world and receiving a glowing independent evaluation by the Urban Institute (2018) 5 years in, the Rockefeller Foundation announced that 100 resilient cities would suspend operations at the close of 2019. It offered scant explanation for its decision, though it has been suggested that a new set of priorities was adopted by its new president, Raj Shah. The move "raised serious questions about the wisdom of relying on [corporate] charitable organizations to fund resilience work—particularly those that were chiefly responsible for the problems to begin with. For groups that—like the Rockefeller Foundation—owe their wealth to the production and sale of fossil fuels, climate change initiatives can walk a fine line between charity and public relations" (Williams 2019). This point has been raised before, with regard to the donation in 2018 by ExxonMobil of $300 000 to aid Mozambique's recovery from Cyclone Idai, the worst humanitarian crisis in the country's history. Shailja Patel (2018), an internationally acclaimed Kenyan poet, playwright, and political activist pointed out that the donation represented less than 7 minutes of the company's 2018 profits of $6 billion. In Patel's view, because of its role in causing climate change, the company owes hundreds of billions in climate debt to the countries of Africa.

Community activist responses to global issues and the failure of governments and philanthropic organizations to effectively address them increasingly involve the creation of transnational activist networks. Examination of these phenomena emerged from the social scientific development of process models in the study of national social movements. These models assert that the emergence of social movements and their success in achieving their goals is in part dependent on the presence of preexisting networks and organizations. Discussions of transnational activism extend these conceptions to the international level. Critical to activist globalization has been the emergence of truly global electronic communications and the Internet-based Information Revolution across geographies and social classes. These developments, along with the rise of intergovernmental organizations such as the World Bank, have created new opportunities for transnational alliance formation (Keck & Sikkink 1998) and transnational collective action, which has been defined as coordinated international campaigns on the part of networks of activists against international actors, other states, or international institutions (della Porta & Tarrow 2005). Transnational activist networks and their activities involve processes of diffusion (the spread of social movement ideas, practices, conceptual frames, and targets from one nation to another), domestication (the playing out on domestic territory of social conflicts that extend to other nations), and coordination (the strategic planning of synchronized or supportive social movement actions across nations). Interaction in the planning and assessment of coordinated international activist campaigns has led to the formation of "transnational organizational structures and the framing of transnational identities" (della Porta & Tarrow 2005, p. 10).

A historic sign of the rise in importance of transnational activist networks and transnational collective action occurred in Seattle, WA in late November 1999 at the third World Trade Organization meeting, an event that has been called the Battle of Seattle (Smith 2001). A mass demonstration to protest global corporate neoliberalism was mounted by a coalition of diverse organizations. Thousands of meetings in numerous countries involving over 1200 groups, including NGOs, trade unions, religious organizations, and environmental organizations, led over 50 000 activists to go to the city to protest. Reflecting a commitment to take people to the streets and express power through numbers, participants in the demonstrations "marched to slogans such as 'our world is not for sale'; 'no globalization without participation'; 'we are citizens, not only consumers'; 'WTO = capitalism without conscience'; 'trade: clean, green, and fair'" (della Porta *et al.* 2006, p. 1). Parallel demonstrations were held in more than 100 cities around the world. Subsequent international summits on global trade issues were consistently met by counter-summits and street protests. The day of transnational activist networks of cross-border alliances and multilevel activist political process had arrived.

Transnational collective action with a specific focus on climate change has become increasingly common (Dietz & Garrelts 2016). This initiative generally is driven by alliances of transnational, national, and local NGOs and activist groups. The Global Day of Action rallies, for example, began on December 3, 2005, in order to coincide with and express support for meaningful decisions at the First Meeting of Parties to the United Nations Framework Convention on Climate Change (UNFCCC) Kyoto Protocol in Canada. Simultaneous rallies and marches took place in 35 countries.

A major demonstration was organized for December 8, 2007 to coincide with the UNFCCC COP 13/MOP 3 conference, which convened in Bali, Indonesia. Organized rallies in support of this initiative took place in over 80 countries. Three groups, Campaign against Climate Change (a London-based organization founded in 2001 that has developed a network of local affiliates throughout the United Kingdom), Stop Climate Chaos (an Irish and Scottish coalition launched in 2007), and Greenpeace (a transnational organization founded in 1971 with offices in 50 countries that employs nonviolent confrontation strategies to expose environmental threats), played leading roles in organizing the demonstration.

By the time of the 2015 Paris Climate Summit (known as COP21), a transnational climate movement was fully established and mobilized to press conferees to make real progress on climate mitigation. The growing consensus within the movement that the United Nations was unable to solve the climate crisis was confirmed by the end of the summit. The failures of COP21 had a profound effect (de Moor 2017). Realization that world leaders would continue to drag their feet despite the urgency of climate change helped facilitate an increased sense of common purpose and unity within a slow-to-muster and previously divided movement. As expressed by Rising Tide North America (2010, p. 2): "The sooner we realize that politicians and corporations will not solve the climate crisis, the sooner we can get to the real work of building a strong grassroots people's movement—our only hope for survival." A countervailing recognition with divisive potential was the growing uneasiness in some sectors of the movement at the cordial relations between leading environmental organizations and multinational corporations:

> The approach of the mainstream climate groups is nearly as disastrous as the corporate approach. In fact, the lines have been blurred between the two so much it is often hard to distinguish them. Gap Inc. director Robert Fisher sits on the board of the Natural Resources Defense Council; J.P. Morgan executive John Rosenwald and former managing director of Morgan Stanley Kirsten Feldman are trustees for [the] Environmental Defense Fund. Before leaving his job at the Sierra Club executive director Carl Pope went so far as to publicly advocate for natural gas alongside CEO of Chesapeake Energy Aubrey McClendon. (Rising Tide North America 2010, p. 6)

The movement was forced to confront how to influence the climate change response if it did not focus its efforts around meetings among world leaders and corporate heads.

At the global level, a notable climate organization is 350.org, which has grown through several stages from a student-run, issue-oriented salon called Sunday Night Group at Middlebury College in Vermont under the leadership of author/activist Bill McKibben (Fig. 11.2), then a visiting professor at Middlebury. In 2006, this small group of activist students, their teacher, friends of McKibben, and environmental activists organized the climate March Across Vermont over Labor Day weekend. As McKibben (2016) recalls, "Walking seemed Vermontish. I sent out an email to people I knew, and they forwarded it to people they knew. Meanwhile, the more practical members of our group figured out what we'd eat and where we'd sleep (farm fields, for the most part)." By the time the marchers reached their destination, the city of Burlington, VT, the demonstration had swelled to over 1000 people, probably the largest U.S. climate protest to that point in time. Believing they needed to take the urgency of climate change to the national level, McKibben and his student collaborators organized the national Step It Up global warming awareness campaign, which inspired several thousand local demonstrations across the country with its slogan: "Step It Up Congress! Cut Carbon 80% by 2050" (van der

Fig. 11.2 Bill McKibben.

Zee 2007). Coordinating from an office in Burlington, the group used the Internet to contact supporters. A central organizing tool was McKibben's weekly posts on the Grist, an online environmental magazine with the tagline, "Gloom and doom with a sense of humor."

McKibben, deemed "probably America's most important environmentalist" by the *Boston Globe*, has gained popularity through his writing of a series of books on global warming for the general public with the objective of promoting a global climate movement. *The End of Nature* (McKibben 1989) is regarded as the first book on the subject aimed at a broad public audience. At the time he wrote it at age 27, he admits he was still naïve and "completely failed to understand the depth of the kind of cultural transformation that we were going to have to make if we were ever going to deal with climate change" (quoted in Skirble 2010).

The next step was the formation of 350.org with May Boeve, one of the Middlebury activists, as its executive director. The group's name stems from McKibben's view, based on calculations carried out by James Hansen, that the world will not be safe from global warming unless the concentration of CO_2 in the atmosphere drops back to 350 ppm. Drawing on experience gained in the Step It Up campaign, the group led the International Day of Climate Action in October 2009 during the lead-up to the U.N. Copenhagen climate change conference. To reach potential supporters, 350.org members emailed thousands of social movement organizations around the world using a list it received from Paul Hawken's Wiser Earth Project. This global organizing effort produced over 5000 climate movement events in 182 countries. The outcome of the U.N. conference—a weak, nonbinding agreement among nations to lower emissions—was generally regarded as a failure by environmental activists, but the event generated the idea that a sustained and uncompromising global climate movement was imperative.

In 2012, 350.org organized Climate Impacts Day, which sought to communicate how extreme weather and climate change are connected. Activities took place in over 100 countries. Supporters were provided with an online Event Toolkit containing action ideas, photo tips, and a 10-step plan to help them organize events in their own communities. Participants at each location were asked to take a photograph with 350.org signs, and these were collected and posted online with the intention of putting a human face on the impacts of climate change. One showed four Pakistani women standing in front of a makeshift structure that had become their home after massive floods devastated Pakistan's Indus River basin. A photograph from Lebanon captured more than 1000 students protesting pollution that had reached concentrations high enough for scientists to call it toxic to human health. Vermont supporters sent in a picture of climate activists gathered at the site of a covered bridge that was swept away by Hurricane Irene flooding. Coordination of these events followed a networked approach that relied less on central control and more on technological interconnectedness. Explains Boeve (quoted in Smith 2014), "We're participants in something called the Network Innovation Lab, which is trying to gather lessons learned and effective models. The boundaries between people who are on staff and people who are not on staff are much more porous in a network. We have lots of local groups who have their own funding sources and are technically independent."

The Break Free 2016 campaign is another global protest that 350.org helped to coordinate with local climate activists at 20 coal-mining sites on six continents. The goal was to use nonviolent civil disobedience to try to shut down the mines and other facilities in order to draw attention to the contribution of fossil fuels to global warming. Will Bates, one of the original Middlebury students who helped found 350.org, and subsequently its global campaigns director, reported that climate activists wanted "to ensure there continues to be momentum to keep fossil fuels in the ground and to confront fossil fuel companies . . . We need to be disrupting business as usual if governments will not put in place the policies that will get the job done" (quoted in Carrington 2016). In Washington state, over 50 protesters were arrested for trespassing after large groups camped out on railroad tracks used to transport oil to the Shell and Tesoro refineries. The action managed to shut down the rail line for 2 days. During the same period, hundreds of activists occupied Ffos-y-Fran in south Wales, the United Kingdom's largest open-face coalmine. The Vattenfall Welzow-Sued mine in Germany, Europe's largest such mine, was also swamped by protesters and forced to temporarily shut down, while an estimated 10 000 people marched in opposition to a new coal-fired power plant in the Philippine city of Batangas (Milman 2016).

Historically, large environmental organizations in the United States have been overwhelmingly white, with members and leaders coming from the middle and upper classes. Along the course of its organizational life, 350.org, like many environmental groups, has had to confront its origins as a creation of privileged white students from an elite college. Recalling its work in the multiethnic fight against the Keystone XL pipeline, and the need to gain awareness of privilege and what W.E.B. DuBois (1903) called the "wages of whiteness," Boeve (2018) explains:

> We assembled a team of some of the most skilled civil disobedience organizers . . . And 1253 people were arrested [in Washington, D.C.]. It was the first major protest targeting Obama after his election . . . Of course, there were also problems. Not all of us knew how to listen to requests, to work differently with tribes along the route. The story that captured public attention did not adequately explain that this fight began with Indigenous Peoples protecting their treaty rights and sacred lands in Canada. Funding poured into our organization to expand our efforts because we had a big profile, despite our late entry into the fight.

Ultimately, 350.org was able to build collaborative relationships with Brave Heart Society, Indigenous Environmental Network, Native Organizers Alliance, Dakota Rural Action, and other indigenous groups.

While transnational struggles for change are not new (e.g., women's suffrage), the rise of the global climate and environment movements has greatly expanded the playing field by uniting sectors from every corner of the world, North and South, through electronic networks and coordinated actions. In challenging the slow-moving apparatuses of transnational political bodies and corporate resistance to any perceived threats to relentless production and unfettered profit-making, elements within the wider global climate and environment movements have begun to question the dominant economic structure and to consider alternatives that lead to just and sustainable social worlds. Developments of this sort are considered in the next section.

11.3 Stepping toward change

The painful awareness of the compelling need for change has generated a range of "first steps" toward a better future. Most of these have been identified by The Next System Project, an initiative of The Democracy Collaborative, a research center at the University of Maryland founded in 2000 with the goal of promoting democratic renewal, increased civic participation, and community revitalization. The Next System Project supports intrepid thinking and action to address the systemic challenges faced by the United States, including the challenges of economic inequality, racial injustice, and climate change. It encourages the development of radically different approaches to replace the failed socio-economic models currently in use.

The first issue of concern in any context is the number of manufacturing, wholesale, or retail businesses that are owned and governed democratically by their workers, on a one-member, one-vote basis. The organization of worker cooperatives of this sort varies depending on the size of the entity, with smaller units adopting direct-democracy forms of governance and larger entities using more hierarchical management structures but reserving rights of ultimate decision-making to the full membership. In the assessment of Esteban Kelly, executive director of the U.S. Federation of Worker Cooperatives, "worker cooperatives may be the most coherent alternative to capitalism as we know it because they put capital at the service of labor rather than the other way around" (quoted in Juda 2019). An example of a successful workers' cooperative is the solar energy company Namasté Solar in Colorado. Founded in 2005, the initially private company transitioned to a workers' cooperative model 5 years later. In 2018, Namasté employed 170 people and had installed over 32 000 solar panels with a combined capacity of more than 75 megawatts of power. A management team oversees everyday operations, but a full-day "Big Picture Meeting" of all members to make decisions, review financial statements, discuss high-level matters, and facilitate team-building is held four times a year (Namasté Solar 2019). While collectively owned units of production and distribution are empowering for worker/owners and may address problems of worker alienation, with some notable exceptions (e.g., Ocean Spray Cranberries, an agriculture cooperative owned by more than 700 cranberry farmers) they have tended to remain local arrangements that currently present no threat to the privatized capitalist world system.

Resident-owned communities (ROCs) have become a leading source of affordable housing in the United States. An example can be found in a low-income neighborhood in Mead, WA. In 2016, the 149 residents of a trailer park known as the Mead Royale Mobile Home Park, 20 minutes from Spokane, decided to take action to fight problems of drug abuse, street crime, and housing unit abandonment—problems that were neglected while the community was controlled by private management

(Green 2018a; Capital Impact 2019). Reflecting popular sentiment, the residents sarcastically nicknamed the site "Meth Royale." It is estimated that 18 million people who earn under $20 000/year live in mobile homes in the United States. Though they are a source of affordable housing for many, in cases where residents rent space from a property owner that charges rent, establishes the rules, and oversees conditions, they face displacement if the owner increases the rent, sells the property, or fails to maintain facilities. At Mead, residents experienced a lack of safety, security, and a sense of belonging. Fed up, they rallied together to form a housing cooperative and purchase the property their homes sat on. As an ROC, each member household has equal ownership in the property and equal say in how the community is run. They have control of the rules, rent, repairs, and community improvement projects. Once they gained ownership, they adopted the name Takes a Village Homeowners Cooperative, Inc. and cleaned up the property to address years of neglect, upgraded the sewer system, and implemented a cycle of community social events and management meetings. Again, while they provide significant benefits to participants, ROCs like the one in Mead remain local initiatives that have had limited overall impact on for-profit housing industries.

Conservation land trusts or land conservancies are charitable organizations that seek to conserve sensitive natural areas, farmland, ranchland, water sources, cultural resources, or valued landmarks. As defined by the Land Trust Alliance (2001), an organization that provides technical support to a network of over 1000 such trusts in the United States, a land trust is a "nonprofit organization that, as all or part of its mission, actively works to conserve land by undertaking or assisting direct land transactions—primarily the purchase or acceptance of donations of land or conservation easements." Two of the largest are The Nature Conservancy and World Land Trust. In 1954, in its first acquisition, The Nature Conservancy, a member organization, purchased a 60-acre hemlock forest in New York. Sixteen years later, it advanced to the global level with its launch of the International Conservation Program, designed to identify natural area and conservation organizations in Latin America in need of assistance. While it promotes land conservation, the Conservancy partners with corporations like The Dow Chemical Company Foundation that are abhorrent to other environmental groups. Well-known writer and activist Naomi Klein (2013) has sharply criticized it for earning money from an oil well on land it controls in Texas and for its continued involvement with fossil fuel companies.

Various strategies are used by conservation trusts, including land purchases, purchases of conservation easements to block development, and purchases of mining, logging, or drilling rights. Some, like The Conservation Land Trust, work with government park programs. Land trusts played a central role in the 2004 creation of Great Sand Dunes National Park in Colorado, for example, and the expansion of Hawaii Volcanoes National Park in 2003. Outside of the United States, they were active in the creation of El Impenetrable National Park in Argentina and Yendegaia National Park in Chile.

Another form of land trust known as a residential community land trust became active during the Civil Rights Movement in an effort to protect marginalized minority communities and reverse rampant poverty. Today, there are over 200 community land trusts in the United States that advocate for communities burdened by the loss of affordable housing and under consistent threat of displacement (Green 2018b). The work of these nonprofit trusts contrasts with that of conventional affordable housing programs, which offer prospective homeowners a one-time assistance subsidy. With trusts, investments are used to allow households to stay in a community in perpetuity while regenerating monies used to support additional families or tenants, furthering the reach and impact of the often-scarce housing resources. The Sawmill Community Land Trust, for example, was formed in 1996 to purchase and remediate 27 acres on the site of a former particleboard factory near downtown Albuquerque, NM in an effort to preserve housing affordability for working families. Sawmill grew to manage 34 acres of reclaimed industrial land, on which it has constructed 93 inexpensive single-family-ownership homes (open to buyers who earn 80 percent or less of the local Area Median Income) and three affordable rental apartment complexes (open to renters who earn 40–60 percent of the AMI), including specialty facilities for senior living, as well as community gardens, playgrounds, and a plaza (Sawmill Community Land Trust 2019).

Investor-owned utilities, like Pacific Gas and Electric in California, provide needed energy services. Like other corporations, they have as their primary focus enhancement of shareholder returns and maintenance of constant growth. Unlike other companies, however, they do not primarily gain profit from buying or making a product and selling it for a higher amount. Rather, state public utility commissions or, in some cases, the federal energy regulatory commission set the rate utilities can charge their customers, which includes a profitable return above costs (called a return on equity). When expenses go up—even if this is caused by company mistakes—utilities seek rate increases to allow them to pass the cost on to their customers. Incorporated in California in 1905, the San Francisco-based Pacific Gas and Electric provides natural gas and electricity to approximately 16 million people across a 70 000-square-mile service area, making it one of the largest combined natural gas and electric energy companies in the United States (Pacific Gas and Electric 2019). Implicated in or suspected of having caused a number of costly forest fires due to fallen power lines, and accused of failing to use the money it obtains from its customers for infrastructure maintenance and safety, the California Public Utility Commission has considered breaking up the giant utility. Advocates of public ownership of energy utilities support an alternative strategy. They point to cases like Nebraska, which has operated on publically owned power since it outlawed investor-owned utilities in the 1930s. They also cite East Bay Community Energy, an electricity supplier in Alameda County, which reinvests the money it makes back into the community by creating new clean power projects, local green jobs, and local energy efficiency programs, and which promotes climate justice in its service area. The company, managed by a board composed of local representatives, procures electricity from clean, renewable sources such as solar and wind on behalf of Alameda County residents and businesses. The energy flows through existing Pacific Gas and Electric power lines, and the for-profit company is responsible for customer billing, maintenance of power lines, and response to new service requests and emergencies.

Unlike privately-owned utilities, which have been criticized for market manipulation and for using profits to inflate executive salaries, publically-owned ones have gained support because they promote community participation in critical energy decision-making, the use of renewable energy sources, investment in local workforces, lowering energy costs, and the delivery of energy services on an equitable basis. According to WPPI Energy (2019), a regional power company that serves 51 locally owned, not-for-profit electric utilities in Wisconsin, Michigan, and Iowa, "In the end, public power does exactly what its name suggests; it puts power in the hands of the public." Thus far, public utilities are found primarily at the city, county, or small regional level.

11.4 Toward changing the system: addressing ultimate causes

Most social movements seek to modify undesirable aspects of reigning social systems. Their goal is system correction, or what André Gorz (1964) calls reformist reforms. Some, however, known as antisystemic movements, define the dominant socio/political and/or socio/economic system itself as the ultimate cause of social suffering and seek to transition to a new sociopolitical and economic order. These movements seek nonreformist reforms, which are positive changes that enhance the struggle for a new social order. In the contemporary world, the dominant global economic regime is capitalism supported by varying state structures and degrees of investment in public welfare (e.g., Finland, the United States, Russia, Nigeria). Challenging the capitalist world system are the alternative visions of antisystemic movements hoping to circumvent the kinds of dystopian ecocrises and interacting ecocrisis scenarios described in previous chapters by engaging in a socioecological revolution. The objective of these movements is to achieve social parity and justice, truly democratic sociopolitical processes, environmental sustainability, a safe climate for humans and other species, and the preservation of biodiversity and vital ecosystems. Participants in environmental antisystemic movements reject the idea that it is possible to morally persuade the polluting elite to curb their social and environmental offenses (which often is the approach of well-meaning system-correction approaches; e.g., Robinson 2018). Rather, their goal, based on the conclusion that capitalism's relentless search for the extraction of profit will ultimately undermine all efforts at reforming it, is system change. Recognizing that not all embraces of environmentalism are progressive (Biehl & Staudemair 2011), and many are system-correcting rather than system-changing, these antisystemic alternatives look to push past tendencies in most green parties and major environmental groups to assist capitalism in being more ecologically sensitive by addressing some pressing ecological issues, but not their ultimate causes (Williams 2010; Magdoff & Foster 2011; Tokar 2014; Löwy 2015). Further, these approaches seek to draw attention to the inherent interrelationship of society and environment under the united banner of ecology and equity (Eiglad 2015). In the choice between reformed capitalism and transition to a post-capitalist society, many social activists believe that only the latter option offers real hope for ecological sustainability and social justice. Thus, their goal is to implement a viable democratic alternative to capitalism. They believe that over the course of human history there have been various economic systems, and hence other economies are not only possible but inevitable (Gibson-Graham 2006).

Mainstream political theorists and social commentators, however, often dismiss as utopian efforts to rethink the reigning structure of social and economic relations. Literary critic and Director of the Center for Critical Theory at Duke University, Fredric Jameson (1994), is credited with observing that it has become easier to imagine the end of the world than the end of capitalism. This narrow view, labeled by Mark Fisher (2010) as the capitalist realism, pervades conventional cultural production (e.g., science fiction), political-economic activity, and public thought. Fisher argues that the bank and auto industry bailouts promoted by the Obama administration in the heat of the 2008 economic crisis, for example, were a quintessential expression of the capitalist realist view that allowing these corporations to fail was unimaginable. While capitalist realism does not claim that capitalism is a perfect system that solves all problems, it asserts that it is the only workable system that is compatible with human nature (e.g., self-interest) and economic law. Fisher used the term "reflexive impotence" to describe the condition of recognizing the significantly flawed nature of capitalism while believing that it is not possible to transcend it because it is part of the natural order of things. Among alternative theorists, capitalist realism represents a crisis of imagination. Exemplary is Erik Olin Wright, author of *Envisioning Real Utopias* (2010), who views a real utopia as the process of imagining the contours of an alternative social world that embodies emancipatory ideals and then seeking the kinds of social innovations that will move our world toward that destination. Wright (2010, p. 1) hopes "to contribute to rebuilding a sense of possibility for emancipatory social change" and a democratic egalitarian social organization. Various visions exist for a new world order, as discussed in the following sections.

11.5 The solidarity economy

While there are different understandings of what is meant by the concept "solidarity economy" (also known as social solidarity economy), the most far-reaching versions comprise a set of strategies and envision a popular struggle aimed at the abolition of capitalism and the hierarchical and unequal structure of social relations that it produces (Pun *et al.* 2015; Razeto 2017). Solidarity economy has been defined as all production, distribution, and consumption activities that contribute to the democratization of the economy based on citizen commitments at both a local and the global level (Alliance 21 2001). Solidarity building within this movement focuses on the valuing of economic justice, unity with oppressed countries and groups of people (e.g., the poor, people of color, LGBT+ people), and working for economically, socially, and environmentally restorative development (Laville 2010). The concept first found expression in Latin America in a social activity known as an *encuentro*, a meeting which emphasizes dialogue and exchange of ideas among activists, during the mid-1980s. These gatherings blossomed into large collective events in the mid and late 1990s, as a convergence of three social trends (Miller 2007). First was the economic exclusion encountered by growing segments of society in the Global South, caused by deepening national debt and the imposition of the structural adjustment programs of neoliberalism by international lenders like the Work Bank. The resulting cuts in the social safety net and government employment forced many communities to develop autonomous and locally-based strategies of horizontal networking to meet basic needs, such as worker and producer cooperatives, neighborhood and community associations, local savings and credit associations (e.g., West African tontines), collective kitchens, neighborhood self-help associations, fair-trade associations, and unemployed or landless worker mutual-aid societies. While not totally independent of the capitalist economy, these grassroots initiatives are rooted in cooperation and poor and working-class solidarity rather than capitalist profit-seeking for the private benefit of the wealthy minority. Second, in the middle classes, enduring economic uncertainty and projections of downward social mobility led to a range of economic security projects such as consumer cooperatives, community gardens, cooperative childcare and healthcare initiatives, free software initiatives, giveaway shops, local exchange trading systems, housing cooperatives, intentional communities, and ecovillages. These entities share a set of operative values: teamwork, independence from centralized authorities, escape from the dominant economic system and its profit-over-people orientation, and participatory self-management. Third, there has been the emergence of local and regional movements seeking to build global connections among poor, working-class, and middle-class initiatives in the pursuit of a democratic alternative to capitalist globalization.

These efforts came together in the First Latin Meeting of Solidarity Culture and Socioeconomy, held in Porto Alegre, Brazil in 1998, which included participants from Brazil, Mexico, Argentina, Peru, Nicaragua, Bolivia, Colombia, and Spain. A product of this conference was the formation of the Latin American

Solidarity Economy Network. Four years later, the first World Social Forum was assembled, which led to the creation of the Global Network of the Solidarity Socioeconomy. Annual meetings of the World Social Forum have been held since, and regional and local social forums are convened around the world. The first U.S. Social Forum took place in Atlanta in June 2007, though the concept is still not well known in the United States. Cándido Grzybowski (2010), Director of the Brazilian Institute of Socioeconomic Analyses and a World Social Forum co-founder, reports that "[t]he numerous recent crises are expressions of the contradictions and limitations of the form of global capitalism that has been imposed on humanity and threatens the sustainability of life on earth. The assertion that another world is possible is now an absolute necessity."

What would a solidarity economy look like? This question cannot be answered by pointing to a nation that has transformed its economic system based on the principles of the solidarity economy ideal. There is no blueprint for building such an economy, and advocates recognize it as an evolving framework. Based on local expressions of initiatives—some old, some new—that are aligned with solidary economy values, in theory this economic system would entail a process of transition from private ownership of the means of production and wage labor to worker self-management of production and distribution (RIPESS 2019). In Brazil, for example, solidarity economy units commonly cooperate as networks by collectively marketing independently produced goods. In the Brazilian apparel industry, a program that adheres to this approach is called Justa Trama, which means "Between us, we will be just, well-entwined, and clear." Justa Trama was founded in the community of Cantagalo in Rio de Janeiro in 2005 by Nelsa Nespolo, a former factory worker and community organizer. Nespolo recalls, "I worked for three factories in the apparel and food industries. But after a while, I realized I didn't want to be an employee anymore. I wanted to control my schedule, my life" (quoted in Changemakers 1999). The initiative links every actor in the clothing production chain—from farm workers who grow pesticide-free organic cotton, to weavers who make fabric, to seamstresses who sew garments, cords, bags and purses, jewelry, and dolls—into a single collective that governs the full process (Veronese & Lacerda 2014). Representatives from every link in the chain sit on Justa Trama's executive board and participate in making democratic decisions about management and fair income distribution. As contrasted with fair-trade associations, which ask consumers to pay a premium for ethically produced products, the Justa Trama model increases workers' wages but also confers management responsibility and ownership on the workers themselves. Justa Trama also links its production chain to consumers, marketing its brand Justa Trama Fibra Ecólogica at regional stores and trade fairs and online. The long-range objective of the solidarity economy approach is to build multiple participatory projects like Justa Trama and link them together in a collaborative web that will dominate sectors of the economy.

Across nations, such a web extends globally through the Intercontinental Network for the Promotion of Social Solidarity Economy (RIPESS), which builds bridges of communication and collaboration among likeminded projects in Latin America and the Caribbean, North America, Europe, Africa, Asia, and Oceania, and between them and various international organizations (e.g., the United Nations Development Program, the International Labor Organization, and the Organization for Economic Co-operation and Development). Moreover, it seeks to promote a true commons of shared and equal (as opposed to privatized) access to "a wealth of valuable assets that [should] belong to everyone. These range from clean air to wildlife preserves; from the [international] judicial system to the Internet. Some are bestowed to us by nature;

others are the product of cooperative human creativity" (Walljasper 2011).

11.6 Stateless democracy

In 2016, the phrase "Google Murray Bookchin" began showing up on websites like 8chan, Reddit, and Facebook. It proposes that the reader search the Internet for information on the ecologically oriented political philosopher Murray Bookchin. The meme ultimately became a protest slogan chalked on the ground at fossil fuel blockades. Another frequently encountered meme is the inspirational phrase (which often appears with a picture of Bookchin), "The assumption that what currently exists must necessarily exist is the acid that corrodes all visionary thinking"—a declaration Bookchin made in his essay, "The Meaning of Confederalism" (1990). His argument is that it is not only faulty to believe that society has become too complex to be governed through face-to-face or participatory democracy, but such a widely held idea functions to quash the belief that there can be a democratic alternative to the nation-state. There are, Bookchin (1990) maintains, a number of very problematical assumptions in the "too complex for horizontal rather than hierarchical governance" assertion, "most disturbingly, [that it rests] on an unconscious acceptance of the economic status quo." The status quo is anything but democratic in a world of enforced gross inequalities of wealth, power, access to resources, and social status. Moreover, it relies on the belief that it is possible to effectively address social and environmental ills merely by tinkering with the operation of dominant bureaucratic institutions. Rather, Bookchin (1992) argues, "[c]apitalism can no more be 'persuaded' to limit growth than a human being can be 'persuaded' to stop breathing. Attempts to 'green' capitalism, to make it 'ecological', are doomed by the very nature of the system as a system of endless growth."

While Bookchin's ideas, which have had influence on both sectors of the modern environmental movement and revolutionary political struggles (e.g., those waged by the Kurdistan Workers' Party), might at first sound similar to the ideals of solidarity, they have different theoretical roots. Bookchin, the child of immigrant communist parents, joined the American Communist movement through the Young Pioneers at the age of 9, during the Great Depression. But he grew disillusioned with Stalin, and gravitated to the movement that followed his rival Leon Trotsky. As a member of the Trotskyist-oriented Socialist Workers Party from 1939 to 1947, he worked and organized workers in several factories. After World War II, he rejected political Marxism but retained his belief in the need to transcend capitalism. His political commitments came to align with those of anti-state anarchism, and when he died in July 2006, the London *Times* called him "the most important anarchist thinker in North America for more than a quarter of a century" (quoted in Biehl 2007). In his view, "[a]n anarchist society should be a decentralized society, not only to establish a lasting basis for the harmonization of man and nature, but also to add new dimensions to the harmonization of man and man" (Bookchin 1971, p. 79). A number of years before his death, however, he no longer identified himself as an anarchist. He spent the final 5 years of his life "theorizing, inspiring, and trying to organize a revolution that would be not only socialist but (unlike Marxian socialism) antihierarchical, democratic, and ecological" (Biehl 2007).

Bookchin's environmental concerns began to emerge a few years after the close of World War II. Beginning in 1952, before the publication of Rachel Carson's *Silent Spring* (1962), he took up the problem of chemicals in food, having recognized that the use of pesticides, herbicides, and other chemicals in agriculture was having toxic effects on human health (Herber 1962). To overcome the need for their use, he argued for local production of food near where it was to be consumed. He also criticized corporate monocrop agriculture and called for crop diversity and crop rotation—all ideas that were well ahead of their time. Further, he denounced large-scale, centralized agriculture, which reduced farmers to exploited laborers. Unifying his anti-capitalist and environmental values, he condemned the absorption of food production into the capitalist economy, because this made maximizing profit rather than human and environmental health the top priority. Additionally, he lamented the alienation of humanity from nature and advocated what he called "social ecology" (Bookchin 1982), the perspective that human domination and destruction of nature follow directly from domination within human societies. Finally, though he misanalyzed the timeframe involved, Bookchin recognized the dangers of greenhouse gas accumulation long before most. Writing in 1964, he observed: "This growing blanket of carbon dioxide, by intercepting heat radiated from the earth into outer space, will lead to rising atmospheric temperatures, to a more violent circulation of air, to more destructive storm patterns, and eventually to a melting of the polar ice caps (possibly in two or three centuries), rising sea levels, and the inundation of vast land areas. Far removed as such a deluge may be, the changing proportion of carbon dioxide to other atmospheric gases is a warning of the impact man is having on the balance of nature" (Herber 1964).

The road to a democratic, stateless society, according to Bookchin, is a libertarian confederation among human communities based on grassroots democratic decision-making: "Libertarian Municipalism constitutes the politics of social ecology, a revolutionary effort in which freedom is given institutional form in public assemblies that become decision-making bodies" (Bookchin 2015, p. 96). However, decentralism and self-sustainability must involve a much politically and geographically broader program of social organization than localism: "Today, what often leads to serious misunderstandings among decentralists," he insisted, "is their failure in all too many cases to see the need for confederation" (Bookchin 1990). It is necessary to draw a "clear distinction between policymaking and the coordination and execution of adopted policies" (Bookchin 1990). Policy-making, he believed, should be exclusively the right of popular and locally elected community assemblies, based on the practices of participatory democracy. Administration and coordination, by contrast, "are the responsibility of confederal councils, which become the means for interlinking villages, towns, neighborhoods, and cities into confederal networks" (Bookchin 1990). In other words, while in nation-states power flows from the top down, in the confederalist assemblages Bookchin envisioned, it would flow from the bottom up. In its fullest level of development, "the economy itself is confederalized by placing local farms, factories, and other needed enterprises in local municipal hands—that is, when a community, however large or small, begins to manage its own economic resources in an interlinked network with other communities" (Bookchin 1990). Unlike the role of workers as decision-makers in linked collectives that form democratic production chains, in Bookchin's stateless societies, people would express their power as members of communities.

Lest decentralization be seen as impossible in the contemporary world, the model of Switzerland, arguably the most decentralized nation in existence, merits consideration. Switzerland is divided into 26 sovereign states, known as cantons, which contain 2600 municipalities. Each of the cantons has its own unicameral parliament, administrative agencies, and court system. Despite some movement toward greater centralization in recent years, modern Switzerland has been dubbed a "shining example" of decentralization by the economist Oliver Hartwich (2015). The Swiss model, Hartwich says, offers an alternative to the "monoculture of centralist bureaucracies."

11.7 Ecosocialism

The *Oxford English Dictionary* reports that the English-language word "eco-socialism" first appeared in the London *Times* in 1985, as a translation of the German word *ökosozialismus*. However, 5 years earlier, in 1980, the Socialist Environment and Resources Association, an affiliate of the British Labor Party, had published a comic strip-style pamphlet that it titled "Eco-Socialism in a Nutshell" (Writers & Readers Publishing Cooperative 1980). This guide defined ecosocialism simply and directly as the fusion of ecological and socialist considerations. How these are to be fused in practice is the focus of an ongoing conversation among people concerned about the dual problems of social suffering and environmental destruction that haunt our world. As part of that conversation, on May 7, 2011, three members of the journal *Capitalism Nature Socialism* held a seminar entitled Ecosocialism at the Historical Materialism Conference at The New School in New York City. The featured speaker was Joel Kovel, who, during the course of his life (1936–2018), was the author of the books *The Enemy of Nature* (Kovel 2002) and the *Ecosocialist Manifesto* (Kovel & Löwy 2001), an activist with the Green Party of the United States, co-founder of the short-lived Ecosocialist International Network, and editor of *Capitalism Nature Socialism*. In his talk, Kovel emphasized that modern globalism was a form of imperialism driven by the inherent need in a capitalist economy for corporations to grow indefinitely or perish, a process that was in contradiction, he believed, with the finite resources of the planet.

In the last years of the 20th century, Kovel recognized that there was a need for an ecological turn among social activists committed to building a more equitable world. During this time, he became convinced that the future must be ecosocialist because without ecosocialism there would be no future (Kovel 2011). In his view, "[t]he test of a post-capitalist society is whether it can move from the generalized production of commodities to the production of flourishing, integral ecosystems. In doing so, socialism will become ecosocialism" (Kovel 2011). Notably, in the Belem Ecosocialist Declaration, Kovel and co-authors Ian Angus and Michael Löwy (Angus *et al.* 2008) asserted that ecosocialism "criticizes both capitalist 'market ecology' [e.g., green capitalism] and productivist socialism, which ignored the earth's equilibrium and limits. It redefines the path and goal of socialism within an ecological and democratic framework." For these advocates, ecosocialism would transform: 1) the social relations of production, by implementing democratic ownership of the means of production; 2) the energy system, by moving from fossil-fuel to sustainable sources; 3) the transportation system, by drastically reducing the use of private vehicles and replacing them with free and efficient public transportation; 4) patterns of production and consumption, by producing only sustainable and recyclable goods and developing green architecture; 5) the food system, by defending local food sovereignty, eliminating polluting industrial agribusinesses, creating sustainable agro-ecosystems, and working actively to renew soil fertility; and 6) the environment, by creating pollution clean-up programs and radically diminishing environmental and climate pollution.

The conceptualization of ecosocialism has been advanced by various scholars and activists. In light of the struggles and limitations of past socialist initiatives, as seen the Soviet Union and Stalinism, it is recognized that a sustainable ecosocialism cannot be created in one country. Rather, the struggle for a safe climate and a healthy environment, as well as a generally healthy population across social classes and geographies, needs to be part of a larger international struggle for social justice and environmental sustainability (Singer *et al.* 2019). At the global level, Hans Baer (2018), who embraces Wright's notion of real utopia, proposes several systemic-challenging reforms to facilitate a transition from the present existing capitalist world system to "the democratic eco-socialist world system." These include: 1) the creation of new left parties designed to capture the state; 2) emissions taxes at the sites of production; 3) public ownership of the means of production; 4) increasing social equality and achieving sustainable population; 5) workers' democracy; 6) meaningful work and shortening the work week; 7) energy efficiency, renewable energy sources, and appropriate technology; 8) sustainable agriculture and forestry; 9) resisting the culture of consumption and adopting sustainable and meaningful consumption patterns; 10) sustainable settlement patterns and local communities; and 11) sustainable transportation and trade. Baer does not believe that democratic ecosocialism is a guaranteed next state in the evolution of human social systems, nor that it is a fully fleshed out socioeconomic conception. Standing at a historic crossroads, however, and given the dismal future to be found on the capitalist-driven path of global warming, continued environmental degradation, and resulting ecocrises interaction, an alternative democratic ecosocialist path, he argues, merits committed and exigent exploration (Baer 2018).

This exploration must be guided by two interwoven imperatives. First, humans, though living in many different countries and embracing many different cultures, must learn to live in relative harmony with one another to ensure equal access to needed resources and good living standards. Second, we must learn to live in relative harmony with nature. Baer emphasizes the idea of relative harmony because he recognizes that all human societies have conflicts and contradictions and that all species (but especially humans) impact the environment. History, he notes, tells us that social changes can occur very quickly once certain social structural and environmental conditions have reached a tipping point. Capitalism, he believes, is driving us to the tipping point where the creation of a democratic ecosocialist world becomes a possibility. Such a transition will not automatically solve all of humanity's problems, but it will bring in to being the kinds of social arrangements that allow humans and the planet to flourish.

As reflected in this discussion, there are a range of ideas about how to achieve a just and sustainable future and a recognition of the responsibility we all bear in collectively working toward that goal. Quite literally, given the possible alternative future of ever-enhanced suffering and increasing environmental degradation, there is no time like the present.

References

Adams, S. (2016). Cities are acting on climate change. *Will the new administration join?* World Resources Institute. https://www.wri.org/blog/2016/12/cities-are-acting-climate-change-will-new-administration-join.

Alliance 21. (2001). Proposal paper on the solidarity economy. https://web.archive.org/web/20140908111404/http://www.alliance21.org/2003/article.php3?id_article=552.

Alvarez, S., Dagnino, E. & Escobar, A. (1998). *Cultures of Politics, Politics of Cultures: Re-Visioning Latin American Social Movements.* Boulder, CO: Westview Press.

Angus, I., Kovel, J. & Löwy, M. (2008). Belem ecosocialist declaration. *Climate & Capitalism.* https://climateandcapitalism.com/2008/12/16/belem-ecosocialist-declaration-a-call-for-signatures/.

Aronoff, K. (2015). Swarthmore students launch first indefinite occupation for fossil fuel divestment. *Common Dreams.* https://www.commondreams.org/views/2015/03/23/swarthmore-students-launch-first-indefinite-occupation-fossil-fuel-divestment.

Arrieta-Kenna, R. (2019). The Sunrise Movement actually changed the Democratic

Conversation. So what do you do for a sequel? *Politico*. https://www.politico.com/magazine/story/2019/06/16/sunrise-movement-boot-camp-227109.

Arrigo, L. & Pouleston, G. (2006). The environmental movement in Taiwan after 2000: advances and dilemmas. In: D. Fell, H. Kloter & C. Yu (Eds.), *What Has Changed? In Taiwan Before and After the Change in Ruling Parties*. Wiesbaden: Harassowitz Verlag.

Baer, H. (2012). *Global Capitalism and Climate Change: The Need for an Alternative World System*. Lanham, MD: AltaMira Press.

Baer, H. (2018). *Democratic Eco-socialism as a Real Utopia: Transitioning to an Alternative World System*. New York: Berghahn.

Banjo, S. (2013). Remember Occupy Wall Street? Probably not. *Quartz*. https://qz.com/421817/remember-occupy-wall-street-probably-not/.

Banks, G. (2002). *PNG—Baseline Study*. Mining, Minerals and Sustainable Development. IIED and World Business Council for Sustainable Development.

Berman, T (2014). Organizers want The People's Climate March to be a turning point for climate change activism. *CBC Radio*. https://www.cbc.ca/radio/thecurrent/organizers-want-the-people-s-climate-march-to-be-a-turning-point-for-climate-change-activism-sept-19-2014-1.2771467.

Biehl, J. (2007). Bookchin breaks with anarchism. *The Anarchist Library*. https://theanarchistlibrary.org/library/janet-biehl-bookchin-breaks-with-anarchism.

Biehl, J. & Staudemair, P. (2011). *Ecofascism Revisited: Lessons from the German Experience*. Porsgrunn: New Compass Press.

Boeve, M. (2018). Climate change is changing the world—now we too must change. *Bioneers*. https://bioneers.org/may-boeve-climate-change-is-changing-the-world-now-we-too-must-change/.

Bond, P. & Dorsey, M. (2010/2011). Anatomies of environmental knowledge and resistance: diverse climate justice movements and waning eco-neoliberalism. *Journal of Australian Political Economy* 66: 286–316.

Bookchin, M. (1971). *Post Scarcity Anarchism*. Berkeley, CA: Ramparts Press.

Bookchin, M (1982). *The Ecology of Freedom*. Palo Alto, CA: Cheshire Books.

Bookchin, M. (1990). The meaning of confederalism. *The Anarchist Library*. https://theanarchistlibrary.org/library/murray-bookchin-the-meaning-of-confederalism.

Bookchin, M. (1992). *Remaking Society*. Boston, MA: Southend Press.

Bookchin, M. (2015). *The Next Revolution*. London: Verso Press.

Brennan, C (2019). Boulder activist lends voice ahead of Global Climate Strike. *Daily Camera*. https://www.dailycamera.com/2019/09/18/boulder-activist-voice-global-climate-strike/.

Brimsmead, H. (1983). *I Will Not Say the Day Is Done*. Chippendale, NSW: Alternative Publishing Cooperative.

Burchell, J. (2002). *The Evolution of Green Politics: Development and Change Within European Green Parties*. New York: Earthscan.

Burgmann, V. & Baer, H. (2012). *Climate Politics and the Climate Movement in Australia*. Melbourne, VIC: Melbourne University Press.

Capital Impact. (2019). Takes a village: how residents took ownership over their community. https://www.capitalimpact.org/stories/takesa-village-how-residents-took-ownership-over-their-community/.

Carrington, D. (2016). Climate activists threaten to shut down world's major coal sites. https://www.theguardian.com/environment/2016/mar/01/climate-activists-shut-down-worlds-major-coal-sites-reclaim-the-power.

Carson, R. (1962). *Silent Spring*. Boston, MA: Houghton Mifflin.

Changemakers. (1999). Enabling producers to become board members. https://www.changemakers.com/nelsa-nespolo-justa-trama.

Chernyshevsky, N. (1886). *What Is To Be Done?* New York: Thomas Y. Crowell & Co.

City of Hayward. (2016). Community greenhouse gas inventory update: building energy usage and emissions data, 2005—2015. https://www.hayward-ca.gov/sites/default/

files/2016%20CSC%20GHG%20Rpt%202016-07-11.pdf.

City of Hayward. (2019). Declaration of a climate emergency. https://hayward.legistar.com/LegislationDetail.aspx?ID=3834475&GUID=10419A87-7AA3-40BA-B294-10D7FF9A2640&Options=&Search=.

Climate and Health Alliance. (2017). The toll from coal—Newcastle seeks to stop T4. http://www.caha.org.au/the_toll_from_coal_newcastle_seeks_to_stop_t4_v7kbncj0et_hdutysmczwg.

Climate Emergency Movement. (2019). Green New Deal climate emergency resolution. https://www.theclimatemobilization.org/blog/2019/02/07/green-new-deal-is-a-breakthrough/.

Climate Mobilization Project. (2019). DNC reaffirms call for climate emergency mobilization. https://www.theclimatemobilization.org/blog/2018/09/05/2018-9-4-dnc-passes-climate-emergency-mobilization-resolution/.

Connolly, R. (2016). *UMass becomes first major public university to divest from direct fossil fuel holdings*. News & Media Relations, University of Massachusetts. https://www.umass.edu/newsoffice/article/umass-becomes-first-major-public.

Deaton, J. (2017). The problem with the climate movement? Too much science. *Nexus Media*. https://nexusmedianews.com/the-problem-with-the-climate-movement-too-much-science-33e03bc2f4e4.

della Porta, D. (2015). *Social Movements in Times of Austerity*. Cambridge: Polity.

della Porta, D. & Tarrow, S. (2005). *Transnational Protest and Global Activism*. Lanham, MD: Rowman & Littlefield.

della Porta, D., Andretta, M., Mosca, L. & Reiter, H. (2006). *Globalization from Below: Transnational Activists and Protest Networks*. Minneapolis, MN: University of Minnesota Press.

de Moor, J. (2017). The "efficacy dilemma" of transnational climate activism: the case of COP21. *Environmental Politics* 27(6): 1–22.

Dietz, M. & Garrelts, H. (2016). *Routledge Handbook of the Climate Change Movement*. Abingdon: Routledge.

DNC. (2018). Resolution reaffirming the DNC's unwavering commitment to workers and unions in industries that power the American economy, and expressing the Democratic Party's appreciation of their profound contributions to our party, our nation's economy, and our way of life. https://gallery.mailchimp.com/b575b9e5364b5673b6f9df3f1/files/222d3906-d801-4d69-8ed6-5e34991ced18/Exec_Labor_Resolution_8.10.18.pdf.

DuBois, W. (1903). *The Souls of Black Folk*. New York: Penguin.

Edelman, M. (2001). Social movements: changing paradigms and forms of politics. *Annual Review of Anthropology* 30: 285–317.

Eiglad, E. (2015). *Social Ecology and Social Change*. Porsgrunn: New Compass Press.

Erlanger, S. (2001). Pressing greens, German leader wins historic vote on sending troops to Afghanistan. *The New York Times*. https://www.nytimes.com/2001/11/17/world/pressing-greens-german-leader-wins-historic-vote-sending-troops-afghanistan.html.

European Parliament. (2014). Results of the 2014 European elections. http://www.europarl.europa.eu/elections2014-results/en/seats-group-member-2014.html.

Fields, L. (2015). Bernie Sanders has a plan to take on the fossil fuel industry. *Vice*. https://news.vice.com/en_us/article/d39xdz/bernie-sanders-takes-on-fossil-fuel-industry-in-new-climate-change-action-plan.

Fisher, M. (2010). *Capitalist Realism: Is There No Alternative?* Alresford: Zero Books.

Gedicks, A. (1993). *The New Resource Wars: Native and Environment Struggles Against Multinational Corporations*. Boston, MA: South End Press.

Gedicks, A. & Grossman, Z. (2010). Defending a common home: Native/non-Native alliances against mining corporations in Wisconsin. Wisconsin Resources Protection Council. http://www.wrpc.net/articles/defending-a-common-home-nativenonnative-alliances-against-mining-corporations-in-wisconsin/.

Gibson-Graham, J.-K. (2006). *The End of Capitalism (As We Knew It): A Feminist Critique of Political Economy*. Minneapolis, MN: University of Minnesota Press.

Gillespie, E. (2013). Sustainable storytelling is a powerful tool that communicates vision. *Guardian*. https://www.theguardian.com/sustainable-business/blog/sustainable-stories-powerful-tool-communicates-vision.

Gorz, A. (1964). *A Strategy for Labor*. Boston, MA: Beacon Press.

Gould, K. & Lewis, T. (2017). *Green Gentrification: Urban Sustainability and the Struggle for Environmental Justice*. New York: Routledge.

Grady-Benson, J. & Sarathy, B. (2014). Fossil fuel divestment in U.S. higher education: student-led organising for climate justice. *Local Environment* 21(6): 661–681.

Green, J. (2018a). Resident-owned community. The Next System Project. https://thenextsystem.org/learn/stories/video-resident-owned-community.

Green, J. (2018b). CLTs are nonprofit organizations that acquire and steward land in a "trust" for the permanent benefit of low-income communities. The Next System Project. https://thenextsystem.org/learn/stories/community-land-trust.

Greenwood-Nguyen, A. & Roth, R. (2019). Plastic in the oceans is not the fault of the global south. *The Conversation*. https://theconversation.com/plastic-in-the-oceans-is-not-the-fault-of-the-global-south-110247.

Griggs, G., Árvai, J., Cayan, D., DeConto, R., Fox, J., Fricker, H. *et al.* (2017). *Rising Seas in California: An Update on Sea-Level Rise Science*. Oakland, CA: California Ocean Science Trust.

Grzybowski, C. (2010). *Beyond the world of the social forum*. Institute for Agriculture and Trade Policy. https://www.iatp.org/news/beyond-the-world-of-the-social-forum.

Hansen, J. (1988). Statement presented to the Committee on Energy and Natural Resources. https://www.sealevel.info/1988_Hansen_Senate_Testimony.html.

Hartwich, O. (2015). *Learning from Switzerland and Schwyz*. Avenir Suisse. https://www.avenir-suisse.ch/en/learning-switzerland-schwyz/.

Herber, L. [pseudonym of Bookchin, M.]. (1962). *Our Synthetic Environment*. New York: Knopf.

Herber, L. [pseudonym of Bookchin, M.]. (1964). Ecology and revolutionary thought. Comment. http://dwardmac.pitzer.edu/anarchist_archives/bookchin/ecologyandrev.html.

Herrero, L. (2016). The missing word in the famous Margaret Mead quote. *Daily Thoughts*. https://leandroherrero.com/the-missing-word-in-the-famous-margaret-mead-quote/Press.

Hess, D. (2007). Crosscurrents: social movements and the anthropology of science and technology. *American Anthropologist* 109(3): 463–472.

Hite, K. & Seitz, J. (2016). *Global Issues: An Introduction*. Malden, MA: John Wiley & Sons.

Hudson, M. (2018). Is the "Zero Hour" youth climate march a turning point, or more of the same? *The Conversation*. https://theconversation.com/is-the-zero-hour-youth-climate-march-a-turning-point-or-more-of-the-same-100173.

Inglehart, R. (1995). Public support for environmental protection: objective problems and subjective values in 43 societies. *PS: Political Science and Politics* 28(1): 57–72.

Jameson, F. (1994). *Seeds of Time*. New York: Columbia University Press.

Juda, J. (2019). Worker cooperatives are businesses that are owned and governed democratically by their workers, on a one-member, one-vote basis. The Next System Project. https://thenextsystem.org/learn/stories/worker-cooperatives.

Keating, D. (2019). The EU election's "green wave" is set to transform energy policy. *Forbes*. https://www.forbes.com/sites/davekeating/2019/05/27/the-eu-elections-green-wave-is-set-to-transform-energy-policy/#4666e91cbc6f.

Keck, M. & Sikkink, K. (1998). *Activists Beyond Borders: Advocacy Networks in International Politics*. Ithaca, NY: Cornell University Press.

Kelly, C. & Fox, L. (2019). Protesters rally at McConnell's office in support of Green New Deal. *CNN*. https://edition.cnn.com/2019/02/25/politics/sunrise-movement-mcconnell-protests/index.html.

Ki-moon, B. (2014). U.N. Secretary-General Ban Ki-moon's remarks at the Climate Summit press conference in New York. United Nations. https://www.un.org/press/en/2014/sgsm16186.doc.htm.

King, M. (1967). *Beyond Vietnam. The Martin Luther King, Jr. Research and Education Institute*. Stanford, CA: Stanford University.

Kirsch, S. (1993). Social impact of the Ok Tedi mine on the Yonggom villages of the North Fly, 1992. Ok-Fly Social Monitoring Programme No. 5. https://dev.crawford.anu.edu.au/rmap/archive/Ok-Fly_social_monitoring/Ofsmp05-Kirsch1993-the-Yonggom-people.pdf.

Kirsch, S. (2007a). Indigenous movements and the risks of counterglobalization: tracking the campaign against Papua New Guinea's Ok Tedi mine. *American Ethnologist* 34(2): 303–321.

Kirsch, S. (2007b). *Reverse Anthropology: Indigenous Analysis of Social and Environmental Relations in New Guinea*. Stanford, CA: Stanford University Press.

Kirsch, S. (2014). *Mining Capitalism: The Relationship Between Corporations and Their Critics*. Berkeley, CA: University of California Press.

Klein, E. (2011). The four habits of highly successful social movements. *Washington Post*. https://www.washingtonpost.com/blogs/ezra-klein/post/the-four-habits-of-highly-successful-social-movements/2011/08/25/gIQAeifVNL_blog.html?utm_term=.fd142c589e74.

Klein, N. (2013). Time for big green to go fossil free. *The Nation*. https://www.thenation.com/article/time-big-green-go-fossil-free/.

Kottasova, I. & Mackintosh, E. (2019). Teen activist tells Davos elite they're to blame for climate crisis. *CNN*. https://edition.cnn.com/2019/01/25/europe/greta-thunberg-davos-world-economic-forum-intl/index.html.

Kovel, J. (2002). *The Enemy of Nature*. London: Zed Books.

Kovel, J. (2011). The future will be ecosocialist—because without ecosocialism there will be no future. *Ecosocialist Horizons*. http://ecosocialisthorizons.com/2011/11/the-future-is-ecosocialist/.

Kovel, J. & Lowy, M. (2001). Ecosocialist manifesto. http://environment-ecology.com/political-ecology/436-an-ecosocialist-manifesto.html.

Land Trust Alliance. (2001). 2001 National Land Trust Census.

Laville, J.-L. (2010). The solidarity economy: an international movement. *RCCS Annual Review* 2(2).

Lee, A. (2007). Southern green revolution: urban environmental activism in Kaohsiung, Taiwan. *City and Society* 19(1): 114–138.

Löwy, M. (2015). *Ecosocialism: A Radical Alternative to Capitalist Catastrophe*. Chicago, IL: Haymarket Books.

Magdoff, F. & Foster, J. (2011). *What Every Environmentalist Needs to Know About Capitalism*. New York: Monthly Review Press.

Mark, J. (2018). When fossil fuel money talks, the DNC listens. *Sierra*. https://www.sierraclub.org/sierra/when-fossil-fuel-money-talks-dnc-listens.

Martin, C. (2014). Xiuhtezcatl Roske-Martinez, 14, wants to save the world. *Denver Post*. https://www.denverpost.com/2014/05/28/xiuhtezcatl-roske-martinez-14-wants-to-save-the-world/.

Martinez, X. (2018). Xiuhtezcatl Martinez explains why he's fighting climate change. *TeenVogue*. https://www.teenvogue.com/story/xiuhtezcatl-martinez-explains-why-hes-fighting-climate-change.

Mattews, M., Bowlin, N. & Hulac, B. (2018). Inside the Sunrise Movement (it didn't happen by accident). *E&E News*. https://www.eenews.net/stories/1060108439.

McAdams, D. (2017). Social movement theory and the prospects for climate change activism in the United States. *Annual Review of Political Science* 20: 189–208.

McKibben, B. (1989). *The End of Nature*. New York: Anchor Books.

McKibben, B. (2016). Walk of ages: how a Vermont march helped launch a climate movement. *Seven Days*. https://www.sevendaysvt.com/vermont/walk-of-ages-how-a-vermont-march-helped-launch-a-climate-movement/Content?oid=3633748.

Meeker, J. (2018). We are guinea pigs in a worldwide experiment on microplastics. *The Conversation*. https://theconversation.com/we-are-guinea-pigs-in-a-worldwide-experiment-on-microplastics-97514.

Melucci A. (1989). *Nomads of the Present: Social Movements and Individual Needs in Contemporary Society*. Philadelphia, PA: Temple University Press.

Milkman, R. (2017). A new political generation: Millennials and the post-2008 wave of protest. *American Sociological Review* 82(1).

Miller, E. (2007). "Other economies are possible!": building a solidarity economy. Grassroots Economic Organizing. http://www.geo.coop/node/35.

Milman, O. (2016). "Break Free" fossil fuel protests deemed "largest ever" global disobedience. *Guardian*. https://www.theguardian.com/environment/2016/may/16/break-free-protest-fossil-fuel.

Mingle, J. (2013). Fighting for the future: activists and scholars debate the role of social movements in climate change. *Environment@Harvard* 5(1): 1–7.

Momentum (2016). About momentum. https://www.momentumcommunity.org/about-momentum.

Mudd, G. (2013) Australia's mining legacies: the impacts of mining waste will be felt for years to come. *Arena Magazine* 124. https://arena.org.au/australias-mining-legacies-by-gavin-m-mudd/.

Namasté Solar (2019). Namasté Solar's 2018 impact by the numbers. https://www.namastesolar.com/2018-by-the-numbers/.

Nash, J. (2005). *Social Movements: An Anthropological Reader*. Oxford: Blackwell.

Newman, A. (2011). Contested ecologies: environmental activism and urban space in immigrant Paris. *City and Society* 23(2): 192–209.

Nixon, R. (2011). *Slow Violence and the Environmentalism of the Poor*. Cambridge, MA: Harvard.

Pacific Gas and Electric. (2019). Company profile. https://www.pge.com/en_US/about-pge/company-information/profile/profile.page.

Patel, S. (2018). https://twitter.com/shailjapatel/status/1108786133654298625.

Pun, N., Ku, B., Yan, H. & Koo, A. (Eds.). (2015). *Social Economy in China and the World*. Abingdon: Routledge/Taylor and Francis.

Razeto, L. (2017). *Los caminos de la economia solidaria*. Chile-Columbia: Ediciones Univérsitas Nueva Civilizacíón.

Rifai, R. (2015). Toxic sludge reaches Atlantic after Brazil dams burst. *Aljazeera*. https://www.aljazeera.com/news/2015/11/brazil-dam-bursts-toxic-sludge-151122125732943.html.

RIPESS. (2019). What is social solidarity economy. Intercontinental Network for the Promotion of Social Solidarity Economy. http://www.ripess.org/what-is-sse/what-is-social-solidarity-economy/?lang=en.

Rising Tide North America. (2010). The climate movement is dead: long live the climate movement. https://risingtidenorthamerica.org/publications/the-climate-movement-is-dead-long-live-the-climate-movement/.

Roberts, M., Copper, J., Bruce, A., Barton, T., Haghdadi, N. & Hu, R. (2018). *Solar trends report for solar citizens*. Australian PV Institute. http://apvi.org.au/wp-content/uploads/2018/12/Solar-Trends-Report-for-Solar-Citizens-FINAL_11-12-18_2_logos.pdf.

Robinson, M. (2018). *Climate Justice*. London: Bloomsbury.

Rockefeller Foundation. (2019). 100 resilient cities. https://www.rockefellerfoundation.org/100-resilient-cities/.

Rogers, D. (2012). The materiality of the corporation: oil, gas, and corporate social technologies in the remaking of a Russian region. *American Ethnologist* 39(2): 284–296.
Rootes, C. (1997). Environmental movements and green parties in Western and Eastern Europe. In: M. Redclift & C. Woodgate (Eds.), *International Handbook of Sociology*, pp. 319–348. Northampton, MA: Edward Elgar.
Rootes, C. (1999). *Environmental Movements: Local, National and Global*. London: Frank Cass.
Satell, G. (2019). *Cascades: How to Create a Movement that Drives Transformational Change*. New York: McGraw-Hill Education.
Sawmill Community Land Trust. (2019). About us. http://www.sawmillclt.org/about-us/people-we-serve/.
Sengupta, S. (2019). Protesting climate change, young people take to streets in a global strike. *The New York Times*. https://www.nytimes.com/2019/09/20/climate/global-climate-strike.html.
Singer, M., Baer, H., Long, D. & Pavlotski, A. (2019). *Introducing Medical Anthropology: A Discipline in Action*, 3 edn. Landham, MD: Rowman & Littlefield.
Skirble, R. (2010). Environmental writer turns words into action: activist launches worldwide campaign to fight climate change. *VOA*. https://web.archive.org/web/20100516032727/http://www1.voanews.com/english/news/american-life/profiles/Environmental-Writer-Turns-Words-into-Action-91461014.html.
Smith, H. (2014). How 350.org went from "strange kid" to head of the green class. *Grist*. https://grist.org/climate-energy/how-350-org-went-from-strange-kid-to-head-of-the-green-class/.
Smith, J. (2001) Globalizing resistance: the Battle of Seattle and the future of social movements. *Mobilization* 6(1): 1–19.
Speth, G. (2017). The joyful economy. The Next System Project. https://thenextsystem.org/learn/collections/new-systems-possibilities-and-proposals.
Sunrise Movement. (2019). We Are Sunrise. https://www.sunrisemovement.org/.
Suzuki (1992). Severn Cullis-Suzuki at Rio Summit 1992. https://www.youtube.com/watch?v=oJJGuIZVfLM.
Thunberg, G. (2018). School strike for climate—save the world by changing the rules. TEDxStockholm. https://www.youtube.com/watch?time_continue=670&v=EAmmUIEsN9A.
Thunberg, G. (2019). Special address, annual meeting of the World Economic Forum 2019. https://www.youtube.com/watch?v=M7dVF9xylaw.
Tokar, B. (2014). *Toward Climate Justice: Perspectives on the Climate Crisis and Social Change*. Porsgrunn: New Compass Press.
Urban Institute. (2018). Institutionalizing Urban Resilience: A Midterm Monitoring and Evaluation Report of 100 Resilient Cities. https://www.urban.org/research/publication/institutionalizing-urban-resilience.
van der Zee, B. (2007). People realise it is time to take action. *Guardian*. https://www.theguardian.com/environment/2007/nov/08/climatechange.ethicalliving.
Velásquez, T. (2017). Tracing the political life of Kimsacocha: conflicts over water and mining in Ecuador's Southern Andes. *Latin American Perspectives* 45(5): 154–169.
Veronese, M. & Lacerda, L. (2014). The case of the solidarity-based supply chain Justa Trama. La transformation sociale par l'innovation sociale. https://blogecosol.files.wordpress.com/2012/04/marc3adlia_justa_trama.pdf.
Vinay, B. (2006). *Global Issues for Global Citizens: An Introduction to Key Development Challenges*. Washington, D.C.: World Bank.
Wall, D. (2010). *The No-Nonsense Guide to Green Politics*. Oxford: New Internationalist Publications.
Walljasper, J. (2011). What, really, is the commons? *Terrain.org* 27. https://www.terrain.org/articles/27/walljasper.htm.
Williams, C. (2010). *Ecology and Socialism: Solutions to Capitalist Ecological Crisis*. Chicago, IL: Haymarket Books.
Williams, D. (2019). The shameful hypocrisy of climate change philanthropy. *Quartz*. https://

qz.com/1631614/the-shameful-hypocrisy-of-climate-change-philanthropy/.

Witt, E. (2018). The optimistic activists for a Green New Deal: inside the youth-led singing sunrise movement. *The New Yorker*. https://www.newyorker.com/news/news-desk/the-optimistic-activists-for-a-green-new-deal-inside-the-youth-led-singing-sunrise-movement.

WPPI Energy. (2019). About us. https://wppienergy.org/about-us/.

Wright, E. (2010). *Envisioning Real Utopias*. New York: Verso.

Wright, S. & Boudet, H. (2012). To act or not to act: context, capability, and community response to environmental risk. *American Journal of Sociology* 118(3): 728–777.

Writers & Readers Publishing Cooperative. (1980). *Eco-socialism in a Nutshell*. London: The Socialist Environment and Resources Association.

Index

Ecosystem Crises Interactions: Human Health and the Changing Environment, First Edition. Merrill Singer.
© 2021 John Wiley & Sons, Inc. Published 2021 by John Wiley & Sons, Inc.

h

i

j